Analyzing Multivariate Data

Analyzing Multivariate Data

Paul E. Green
Wharton School
University of Pennsylvania
Philadelphia, Pennsylvania

with contributions by
J. Douglas Carroll
Bell Laboratories
Murray Hill, New Jersey

The Dryden Press Hinsdale, Illinois

Copyright © 1978 by The Dryden Press
A division of Holt, Rinehart and Winston, Publishers
All rights reserved
Library of Congress Catalog Card Number: 77–81241
ISBN: 0–03–020786–X
Printed in the United States of America
89   038   987654321

Text and cover design by Harry Voigt

To Patty, Bill, Ken, Carol, Gregory and Steven—
the most unpredictable (and lovable) set of
multivariate observations one can imagine.

The Dryden Press
Series in Marketing

Philip Kotler and Paul Green, Consulting Editors

Preface

Most users of multivariate statistical techniques are not professional statisticians. They are applications-oriented researchers—psychologists, sociologists, marketing researchers, management scientists, and so on—who, from time to time, need the techniques to help them in their work. This text has been written for them and for students of these disciplines. The book's technical level assumes knowledge of basic statistics and the fundamentals of matrix algebra, which is pretty much standard for introductory texts of this type.

If the book has novelty (and I hope it does) it lies more in point of view and presentation style than in subject matter. As implied by the title, emphasis on data analysis and the objectives of people who do data analysis has shaped the character of the whole enterprise. Some of the book's features should be pointed out.

First, the writing style tends to be informal, even chatty at times. Mathematical proofs are generally avoided and a lot of the discussion proceeds by numerical demonstration as opposed to more abstract argument. The text is probably more redundant than most books on this topic, and deliberately so. The frequent recapitulation of key ideas may irritate the professional statistician but should be welcomed by most applications researchers.

Second, the book makes liberal use of packaged computer programs for carrying out the larger-scale data analyses. However, unlike many books on multivariate analysis, these are not the author's programs. Rather, they represent one of the most complete and accessible program packages around—the University of California Biomedical programs.

Third, emphasis on data analysis is aided by the inclusion of a modest-size data bank of 252 respondents' evaluations of a set of TV test commercials. While the data bank is drawn from marketing research, its characteristics are general enough to be relevant to almost any discipline within the behavioral and administrative sciences. The data bank serves as a kind of running example throughout the book.

Fourth, the book as a whole is content-neutral. By means of (artificial) examples, I have tried to present some of the flavor of a variety of applications areas. I hope the psychologist and sociologist can benefit as much from the material as the applied statistician or the business researcher.

Each of the book's chapters tends to follow a similar presentation format. The presentation of each technique is motivated by a series of research questions and a small numerical example that is simple enough to be solved on a desk calculator. The computations are carried out in detail and various geometric arguments are used to illuminate the main ideas. The same concept is then examined somewhat more formally and the assumption structure of the technique is outlined.

The scaled-up computational aspects of the technique are then described in the context of the TV commercial study. This provides a data analysis of sufficient size and complexity to require the application of computer package routines. The appropriate Biomedical programs are described briefly and their application is illustrated. The chapter concludes with a substantive interpretation of the computer results and additional discussion on the use of the technique (and various extensions of it) in applied studies.

Part I of the book consists solely of Chapter 1. The objective of Chapter 1 is to describe what multivariate analysis can be used for and who uses it. In addition, the central concept of *linear composite* is introduced and illustrated in this chapter.

Part II consists of Chapters 2 through 5, dealing with various aspects of single criterion, multiple predictor association. Multiple and partial regression, (univariate) analysis of variance and covariance, and two-group discrimination are the main techniques described here. The material is tied together through emphasis on model comparison tests and descriptive measures of association (as well as significance testing). In addition, the general problems of multicollinearity and determining the importance of predictor variables receive attention.

Part III consists of Chapters 6 and 7, dealing with the multivariate normal distribution and multiple criterion, multiple predictor association. Canonical correlation, multiple discriminant analysis, and multivariate analysis of variance and covariance are the main techniques described here. Each of these tools, as it is presented, is related to its univariate counterpart in Part II.

Part IV, consisting of Chapters 8 and 9, is concerned with the analysis of interdependent data structures. Factor analysis, multidimensional scaling, and cluster analysis are the major techniques that are discussed here. Emphasis is placed on the more commonly applied procedures of principal components analysis and the common factor model.

In addition, the book has three appendices. Appendix A provides a capsule description of the TV commercial study and a recapitulation of the various chapter-by-chapter analyses, as well as ancillary descriptions of the data. Appendix B presents statistical and matrix algebra backup material of a review nature. Appendix C provides a list of the more commonly used statistical tables that are referred to in the text.

The reader should also be alert to a stylistic device used in the book. All footnotes appear at the end of their respective chapters, rather than at the bottom of the page. While this may cause some inconvenience, the advantages are that longer notes can be accommodated without the need to resort to extremely small type size. Moreover, the reader who is less interested in technical backup, mathematical proofs, and other side excursions will not be slowed in his perusal of the book's main themes.

In conclusion, this is a book for the data analyzer. I hope that its pragmatic orientation will serve many of the needs of applications researchers, whatever their content specialty or multivariate problem area may be.

Acknowledgements

This book has undergone a long gestation period—so long, in fact, that I have lost track of the exact number of drafts. What I have remembered, however, are the many excellent people who gave me the benefit of their expertise.

Informal reviews of an early version of the manuscript were provided by Professors Leonard M. Lodish, University of Pennsylvania, David B. Montgomery, Stanford University, and Allan Shocker, University of Pittsburgh. Extensive reviews of the penultimate version were made by Professors Terry C. Gleason, University of Pennsylvania, Vithala R. Rao, Cornell University, and Venkataraman Srinivasan, Stanford University.

The book's production is largely the work of five people. Mrs. Joan Leary labored tirelessly over what seemed to be an endless manuscript to type and retype. She also managed to turn my feeble sketches into professional-caliber art work. Numerous computer runs were carried out by Professor Arun K. Jain, University of New York at Buffalo, and Michael T. Devita, Robinson Associates, Inc. Ishmael Akaah and Lynda Kenny of the University of Pennsylvania served as my general style critics and companion proofreaders. Thanks are also due to the editorial staff of Dryden Press for their constant help throughout the reviewing and production process.

Finally, my indebtedness to J. Douglas Carroll, Bell Laboratories, is both extensive and deep. Dr. Carroll provided painstaking and thorough reviews of all chapters. On more than one occasion he showed me how complicated ideas could be explained with a minimum of fuss. Just as helpfully, he was able to spot lingering obscurities in material that I originally thought was simple and clear.

With all of this help it must be a wonder that the book has any faults at all. Like it or not, however, the author is responsible for any mathematical blunders and murky passages that still persist. I would appreciate the reader's help in ferreting them out.

Paul E. Green

Contents

Analyzing Multivariate Data

PART I

Overview

CHAPTER 1

The Process of Data Analysis

1.1 Introduction

The development and application of methods for analyzing multivariate data have burgeoned over the last several years. Whole classes of new techniques have been invented and the dissemination of both old and new methods has increased markedly.

Several reasons underlie this pronounced growth. Researchers have become increasingly aware of the tantilizing complexity of human behavior, including its multivariate characteristics. Training in mathematical methods for applied social scientists has been on the increase. Large-scale sample surveys dealing with various aspects of behavior have added measurably to data inventories. And that wondrous workhorse, the computer, has freed the scientist from most of the labor of computation.

If multivariate data analysis still retains a touch of the arcane in the behavioral and administrative sciences, its mystery is much less tinged with skepticism than it was a decade ago. Today's task is primarily one of learning how to use the new methodology.

This book is concerned with the application of multivariate methods to substantive problems in data analysis. Our aim is to show what multivariate data analysis does and how it goes about doing it. This chapter sets the stage for the rest of the text by describing the rationale for multivariate techniques and how the analysis of data fits into the broader process of research and decision making.

By way of overview we first examine a small set of multivariate data from the standpoint of what questions might be asked of it. This leads to a discussion of the process of data analysis and the role of multivariate procedures in that process.

The data matrix provides the starting point for classifying multivariate techniques. Accordingly, we discuss the characteristics of the data matrix and the types of scales by which multivariate measurements can be expressed. This, in turn, leads to a description of the preliminary operations of multivariate data summarization that are undertaken prior to application of particular models. Multivariate analysis typically deals with means, variances, and covariances; their efficient computation by vector and matrix operations is illustrated here.

The next principal section of the chapter introduces the topic of linear composites of a set of data-based variables. The concept of linear composite (or linear combination) plays a central role in multivariate analysis. Most multivariate techniques are concerned with computing linear composites that exhibit certain properties, such as best matching some criterion-variable values or maximally separating individuals in terms of their scores on the linear composite.

Since multivariate analysis obviously depends on data—and on computer programs for implementing the techniques—we conclude the chapter with an introductory description of a real data bank that is reproduced in Appendix A. This particular set of data has been drawn from the world of survey research and concerns respondent reactions to various test versions of a set of television commercials. The example could just as easily have been drawn from educational research or from social psychology. Its content is less important than its structure inasmuch as these data provide a kind of running case throughout the book. Since all basic data are reproduced in Appendix A, the reader can replicate many of the analyses described here or can try out some new ones on his own.

1.2 The Nature of Multivariate Data

Assume for the moment that you are a social psychologist interested in the study of various characteristics that are related to judged likableness of persons. Your research assistant has searched through various high school yearbooks and scholastic records and prepared a set of ten single-paragraph descriptions, each representing a high school senior. A member of your introductory psychology class has then been asked to rate each of the paragraphs with respect to his perception of the senior's:

$$X_2: \text{Extra-curricular activity and}$$
$$X_3: \text{Scholastic achievement}$$

on a 9-point scale. One week later your student was then shown the same descriptions and asked to rate them in terms of his personal overall liking of the person described in each paragraph. Overall liking is denoted as X_1.

Table 1.1 shows your student's ratings of the paragraphs on each of the three

Table 1.1 Ratings of Stimulus Paragraphs Describing Ten High School Seniors

Paragraph Description of Stimulus Person	Overall Liking			Degree of Extra-Curricular Activity			Degree of Scholastic Achievement		
	X_1	X_{d1}	X_{s1}	X_2	X_{d2}	X_{s2}	X_3	X_{d3}	X_{s3}
a	3	-1.3	-0.45	4	-0.3	-0.11	2	-2	-0.74
b	7	2.7	0.93	9	4.7	1.68	7	3	1.11
c	2	-2.3	-0.79	3	-1.3	-0.47	1	-3	-1.11
d	1	-3.3	-1.13	1	-3.3	-1.18	2	-2	-0.74
e	6	1.7	0.58	3	-1.3	-0.47	3	-1	-0.37
f	2	-2.3	-0.79	4	-0.3	-0.11	4	0	0
g	8	3.7	1.27	7	2.7	0.97	9	5	1.85
h	3	-1.3	-0.45	3	-1.3	-0.47	2	-2	-0.74
i	9	4.7	1.62	8	3.7	1.33	7	3	1.11
j	2	-2.3	-0.79	1	-3.3	-1.18	3	-1	-0.37
Mean	4.3			4.3			4.0		
Standard Deviation	2.908			2.791			2.708		

variables, while Figure 1.1 shows a scatter plot of X_1 versus X_2, X_1 versus X_3 and X_3 versus X_2.[1] (In all cases we assume that higher numbers indicate more of the property in question.)

The raw ratings data of Table 1.1, and the accompanying scatter plots of Figure 1.1 represent, in microcosm, the type of data amenable to multivariate analysis.[2]

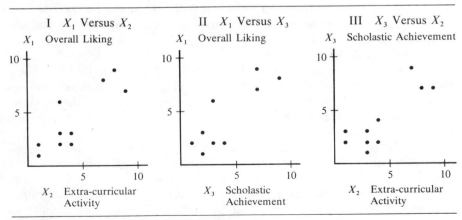

I X_1 Versus X_2
X_1 Overall Liking
X_2 Extra-curricular Activity

II X_1 Versus X_3
X_1 Overall Liking
X_3 Scholastic Achievement

III X_3 Versus X_2
X_3 Scholastic Achievement
X_2 Extra-curricular Activity

Figure 1.1 Two-Variable Scatter Plots of Data From Sample Problem of Table 1.1

We might be interested in any of the following research questions:

1. Are ratings of overall liking related to the rater's judgments about extra-curricular activity and/or scholastic achievement and, if so, how?

2. Can we find a formula for predicting values of overall liking from ratings on extra-curricular activity and scholastic achievement?

3. How accurate are the formula's predictions?

4. Which variable, extra-curricular activity or scholastic achievement, is "more important" in predicting ratings of overall liking?

These questions are illustrative of the sorts of things we might try to find out from the data of this little study. As it stands now, however, about all we can observe from the plots of Figure 1.1 is that:

1. X_1 appears to be positively related to changes in X_2 and X_3, when each is considered separately.

2. X_3 appears to be positively related to changes in X_2.

3. The relationships seem to be approximately linear, although it is evident that there is considerable scatter about these (assumed linear) relationships.

While scatter plots are helpful in giving us a quick overview of some of the data relations, we have no precise information as yet on the various research questions stated above.

1.2.1 Some Definitions

Multivariate analysis is concerned with the study of association among sets of measurements. For purposes of this book we shall define multivariate methods as a *collection of procedures for analyzing association between two or more sets of measurements that have been made on each object in one or more samples of objects.* If only two sets of measurements are involved, the data are typically referred to as bivariate, an important special case of multivariate data. While we shall be interested in this special case, most of our attention will be devoted to cases in which at least three sets of measurements are taken, as in the situation of Table 1.1.

The *objects*—things, persons, events—are the entities on which the measurements are taken and are sometimes called units of association. In this illustration each stimulus paragraph denotes a unit of association on which three measurements have been taken.[3] Units of association are the "carriers" of the measurements; these carriers are not measured in their entirety but only with respect to the variables of interest to the researcher.

The *variables*—sometimes called properties or characteristics—are those aspects of the units of association that are measured. In this illustration they are denoted as X_1, X_2, and X_3 and represent subjective ratings on 9-point scales.

By *measurement* is meant the assignment of one set of entities, usually numbers, to some other set of entities, such as empirically observed behaviors, so that the well-defined relations among the first set can be used to represent the relations assumed to hold across the observed behaviors. Measurement results in various types of *scales,* along which entities are compared.

1.2.2 Observations and Data

As so elegantly expressed by Coombs (1964), there are important differences between observations and data. Each step in the process of going from observation to data requires some underlying theory about the area of interest. First the researcher must confront the problem of what phenomena to observe. In the illustrative problem of Table 1.1 we elected to observe the participant's rating of each stimulus paragraph (according to instructions provided by the researcher) on three verbalized dimensions—liking, extra-curricular activity and scholastic achievement. Alternatively (or in addition) we could have set up an apparatus to record the subject's response time, eye movements, or even galvanic skin response, if desired. One set of choices, then, concerns what is to be observed from the universe of behaviors that could be observed by the researcher.

Figure 1.2 (adapted from Coombs) shows a schema in which the universe of potential observations—already conditioned by the researcher's experimental design or survey instrument—is winnowed down to a subset of observed behaviors. However, even the observed behaviors have to be encoded in terms of relations assumed to hold between stimuli, subjects, or stimulus-subject combinations. This encoding process—usually numerical—produces the data according to some assumed relational (measurement) model. As Coombs stresses, it is the interaction of the researcher (guided by theory) with the environment of potential observations that eventually leads to data.

The last step is to look for associational structure in the data by the fitting of multivariate models. As depicted by the funnel-like sketch in Figure 1.2, each researcher choice has a restrictive effect, as certain aspects of the phenomena being studied are placed in bolder relief at the expense of ignoring other aspects. The search for structure in data, in turn, has two objectives in the context of multivariate analysis:

1. Discovering regularities or patterns in the behavior of two or more variables.

2. Testing alternative models of association between two or more variables, including the determination of whether (and how) two or more groups differ in their multivariate profiles.

In the illustrative problem of Table 1.1 we have already undertaken some exploratory analysis—rudimentary as it may be—by making up scatter plots of the ratings data, shown in Figure 1.1.

As a next step, however, we might wish to see if certain prior-specified multivariate models fit the data and, if so, the parameter values that lead to some best fit, conditioned by functional form and fitting procedure. While this book emphasizes the data-to-structure step of the schema in Figure 1.2, it is relevant to point out that many preliminary decisions on what is to comprise data (i.e., recorded and encoded observations) may be required. Indeed, the scales by

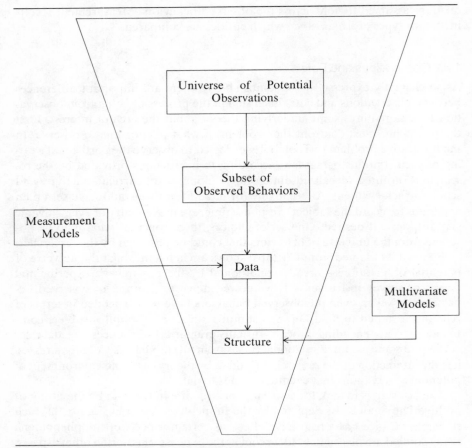

Figure 1.2 The Process of Searching for Structure

which the behaviors are encoded will generally affect the multivariate model that is selected in going from data to structural model.

Finally, the schema of Figure 1.2 is surely an oversimplification. In particular, the relationships among data, multivariate models, and structure are quite fluid in practice. One may try one model, see how it fits, and then try another, and so on. In other cases, one may even recode observations in order to see if some other multivariate model, requiring a different type of data, might account for observed relations. While the validity of statistical inferences may be seriously affected by all of this, the fact remains that the process of data analysis often goes through several cycles of multivariate model selection and fitting in an effort either to discover and codify regularities in the data or to test prior hypotheses about patterns of association.

1.3 Classifying Multivariate Data

The heart of any multivariate analysis consists of the data matrix, or in some cases, matrices.[4] The data matrix is a rectangular array of numerical entries whose informational content is to be summarized and portrayed in some way. For example, in univariate statistics the computation of the mean and standard deviation of a single column of numbers is often done simply because we are unable to comprehend the meaning of the entire column of values. In so doing we often (willingly) forego the full information provided by the column in order to understand some of its basic characteristics, such as central tendency and dispersion. Similarly, in multivariate analysis we often use various summary measures—means, variances, covariances—of the raw data for much the same reason.

In multivariate analysis we are frequently interested in accounting for the variation in one variable or group of variables in terms of *covariation* with other variables. When we analyze associative data we hope to "explain" variation according to one or more of the following points of view:

1. Determining the nature and degree of association between a set of criterion variables and a set of predictor variables, often called "dependent" and "independent" variables, respectively.

2. Finding a function or formula by which we can estimate values of the criterion variable(s) from values of the predictor variable(s). This is usually called the *regression* problem.

3. Assaying the statistical *assurance* in the results of either or both of the above activities, via tests of statistical significance, placing confidence intervals on parameter estimates, or other ways.

In other cases of interest, however, we have no prior basis for distinguishing between criterion and predictor variables. Still, we may be interested in their interdependence as a whole and the possibility of summarizing information provided by this interdependence in terms of other variables, often taken to be linear composites of the original ones (as will be shown later in the chapter).

Since the key notion underlying the classification of multivariate methods is the *data matrix,* a conceptual illustration is shown in Table 1.2. We note that the table consists of a set of objects (the m rows) and a set of measurements on those objects (the n columns). Cell entries represent the value X_{ij} of object i on variable j. The objects are any kind of entity or unit of association with characteristics capable of being measured. The variables serve to define the objects in any specific study. The cell values represent the state of object i with respect to variable j. Cell values may consist of nominal, ordinal, interval, or ratio-scaled measurements, or various combinations of these, as we go across columns.[5]

By *nominal* data we mean data that are described categorically, where the only thing we know about an object is that it falls into one of a set of mutually exclusive and collectively exhaustive classes that have no necessary order vis-a-vis one another. *Ordinal* data are ranked data where all we know is that one object i has more, less, or the same amount of some variable j as some other

Table 1.2 Illustrative Data Matrix

Objects	Variables					
	1	2	3 $\,\ldots\,$ j $\,\ldots\,$ n			
1	X_{11}	X_{12}	X_{13}	$\ldots$ X_{1j}	$\ldots$	X_{1n}
2	X_{21}	X_{22}	X_{23}	$\ldots$ X_{2j}	$\ldots$	X_{2n}
3	X_{31}	X_{32}	X_{33}	$\ldots$ X_{3j}	$\ldots$	X_{3n}
$\vdots$	$\vdots$	$\vdots$	$\vdots$	$\vdots$		$\vdots$
i	X_{i1}	X_{i2}	X_{i3}	$\ldots$ X_{ij}	$\ldots$	X_{in}
$\vdots$	$\vdots$	$\vdots$	$\vdots$	$\vdots$		$\vdots$
m	X_{m1}	X_{m2}	X_{m3}	$\ldots$ X_{mj}	$\ldots$	X_{mn}

object i'. *Interval-scaled* data enable us to say how much more one object has than another of some variable j. Ratio-scaled data also enable us to define an origin (e.g., a case in which object i has a zero amount of variable j), and ratios of scale values are assumed to be meaningful. Each higher scale type subsumes the properties of those below it.[6]

1.3.1 Organizing the Techniques

Three descriptors are often used as bases for multivariate technique organization:

1. Whether one's principal focus is on the objects or on the variables of the data matrix.

2. Whether the data matrix is divided or partitioned into criterion and predictor subsets and, if it is, the number of variables in each.

3. Whether the cell values represent nominal, ordinal, or interval-scale measurements.[7]

This schema results in four major subdivisions of interest:

1. *Single criterion, multiple predictor association,* including multiple regression, analysis of variance and covariance, and two-group discriminant analysis.

2. *Multiple criterion, multiple predictor association,* including canonical correlation, multivariate analysis of variance and covariance, and multiple discriminant analysis.

3. *Analysis of variable interdependence,* including factor analysis and other types of dimension-reducing methods.

4. *Analysis of interobject similarity,* including cluster analysis and other types of object grouping procedures.

The first two categories involve dependence structures where the data matrix *is* partitioned into criterion and predictor subsets; in both cases interest is focused on the variables. The last two categories are concerned with interdependence—either focusing on variables or on objects. Within each of the categories various techniques are differentiated in terms of the type of scale by which the variables are measured.

1.3.2 Variable Types

Traditionally, multivariate methods have emphasized two types of variables:

1. More or less continuous variables, such as interval-scaled (or ratio-scaled) measurements.
2. Binary-valued variables, coded zero or one.[8]

The reader is no doubt already familiar with variables such as length, weight, and height, with scale values that can vary more or less continuously over some range of interest.

Natural dichotomies such as sex (male or female) or marital status (single or married) are also familiar. What is perhaps not as well known is that any (unordered) polytomy, consisting of three or more mutually exclusive and collectively exhaustive categories, can be recoded into dummy variables that are typically coded as zero or one. To illustrate, a person's occupation, classified into five unordered categories, could be coded as:

Category	Dummy Variable			
	1	2	3	4
Professional	1	0	0	0
Clerical	0	1	0	0
Skilled laborer	0	0	1	0
Unskilled laborer	0	0	0	1
Other	0	0	0	0

For example, if a person falls into the Professional category he is coded 1 on dummy variable 1 and 0 on dummies 2 through 4. In general, a k-category polytomy can be represented by $k - 1$ dummy variables, with one category—such as the last category—receiving a value of zero on all $k - 1$ dummies.[9]

Multivariate techniques that are capable of dealing with some or all variables at the *ordinally* scaled level are of more recent vintage. With relatively few exceptions our attention in this book will be focused on either continuous (interval- or ratio-scaled) or zero-one coded variables.

1.3.3 Single Criterion, Multiple Predictor Association

Figure 1.3 shows some of the major ways in which the data matrix can be viewed from the standpoint of technique selection.

In Panel I of the figure we note that the first column of the matrix has been singled out as a criterion variable and the remaining $n - 1$ variables are considered as predictors. For example, the criterion variable could be average

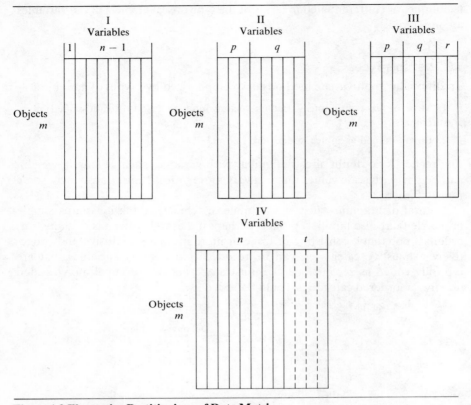

Figure 1.3 Illustrative Partitionings of Data Matrix

weekly consumption of beer by the i-th individual. The $n - 1$ predictors could represent various demographic variables, such as the individual's age, years of education, income, and so on. This is a prototypical problem for the application of *multiple regression* in which one tries to predict values of the criterion variable from a linear composite of the predictors.[10] The predictors, incidentally, can be either continuous variables or zero-one dummies, such as marital status or sex.

In other instances, the single criterion variable could represent the prior categorization of each individual as heavy beer drinker (coded one, arbitrarily) or light beer drinker (coded zero), based on some designated amount of average weekly beer consumption. If so, we could develop a linear composite of the predictors that enables us to classify each individual into either heavy or light beer drinker status. *Two-group discriminant analysis* can be employed for this purpose.

Another possibility can arise in which the criterion variable continues to be average weekly beer consumption but the predictor set consists of a classification of each individual into some occupational group, coded as a set of dummy (sometimes called design) variables. This represents an instance in which the technique of *analysis of variance* could be used. On the other hand, the predictor set could consist of individuals' occupations and their annual incomes (in dollars) as well. If so, *analysis of covariance* could be applied. In this case we would be interested in whether average weekly beer consumption differs across occupations once the effect of income is controlled for statistically.

1.3.4 Multiple Criterion, Multiple Predictor Association

In Panel II of Figure 1.3 the $p = 3$ criterion variables could denote an individual's consumption of beer, wine, and liquor and the remaining variables could denote his demographic characteristics. Then, if we were interested in the linear association between these two batteries of variables we could employ the technique of *canonical correlation.*

Suppose, alternatively, that all individuals had been previously classified into one of four groups: (a) malt beverage drinker only; (b) drinker of spirits (liquor or wine) other than malt beverages; (c) drinker of both spirits and malt beverages; and (d) drinker of neither. We could then develop a linear composite of the demographics that would enable us to assign each individual to one of the four groups in some "best" way. This is an illustration of *multiple discriminant analysis.* (Note that four mutually exclusive groups are classifiable in terms of $p = 3$ criterion dummies.)

Alternatively, we could continue to let the criterion variables denote individual consumption levels of beer, wine, and liquor but now assume that the predictors represent dummies based on an occupational classification. If so, *multivariate analysis of variance* is the appropriate procedure. If income is again included as a covariate, we have an instance of *multivariate analysis of covariance.*

Panel III of Figure 1.3 shows a data structure involving association among three batteries of variables. If so, *generalized canonical correlation* could be employed on mixtures of continuous and zero-one coded (dummy) variables. In this case we would be interested in what all three (or more) batteries exhibit in common and, in addition perhaps, in the strength of association between all pairs of batteries.

1.3.5 Dimension-Reducing Methods

Panel IV of Figure 1.3 shows a set of t appended columns (dotted lines), each of which is assumed to be a linear composite of the original n variables. Cases can arise in which we wish to replace the n original variables by t new variables, where t is much less than n. Suppose we want to portray the association across the m individuals in terms of fewer variables than the original n variables. If so, we might employ *factor analysis* or some other dimension reduction method to represent the original set of n correlated variables by t $(t < n)$ "latent" variables in such a way as to retain as much of the original information as possible. The composites themselves are usually chosen to obey still other conditions, such as being mutually uncorrelated.

Thus, if the original n variables are various demographics characterizing a set of beer drinkers, we might be able to find a set of "more basic" dimensions—social class, stage in life cycle, etc.—where functions of these basic dimensions account for the observable demographic variables.

1.3.6 Interobject Similarity

So far we have confined our attention to the columns of the matrices in Figure 1.3. Suppose now that the n columns represent consumption of various kinds of alcoholic beverages—beers, ales, red wines, white wines, liquors, after-dinner cordials—over some stated time period. Each individual's consumption profile could be compared with every other individual's and we could develop a measure of inter-individual similarity with respect to patterns of alcoholic beverage drinking.

Having done so, we could then proceed to *cluster* individuals into similar groups on the basis of the overall similarity of their consumption profiles. Note here that information on specific variables is lost in the computation of inter-individual similarity measures. Since our focus of interest is on the objects rather than on the variables, we may be willing to discard information on separate variables in order to grasp the notion of overall inter-object similarity more clearly.

All of these techniques—and others as well—have been employed in the behavioral and administrative sciences. As suggested above, we shall see later on that the tools of multivariate analysis form a unified set, based on a relatively few descriptors for distinguishing specific techniques. Moreover, in virtually all applications of multivariate analysis we are attempting somewhere along the way to best match one set of numbers, as obtained by a linear function of the original variables (or some subset of them), to another set of numbers that may also be a linear function of the original variables (or some subset of them).

1.4 Preparing the Data

The researcher wishing to apply multivariate methods rarely works with the raw data themselves. Rather, various summaries of the data, such as matrices composed of:

1. Raw sums of squares and cross products
2. Mean-corrected sums of squares and cross products
3. Variances and covariances
4. Correlations

are used. Each of these matrices is a type of sums of squares and cross products matrix. They differ from each other in terms of corrections made for differences in means and/or standard deviations across the variables of interest.

1.4.1 Preliminary Matrix Calculations

By way of review, the formulas used to compute various statistical measures, such as the mean, standard deviation, covariance, and product moment correlation, are shown in scalar notation in Table 1.3. Each of these formulas has its counterpart in considering the vector-valued variables that comprise multivariate data.

To start off, suppose we denote a matrix of raw (original) data as **X**. In the illustrative problem of Table 1.1, **X** is of order 10×3 (10 rows by 3 columns). In general, a column vector, of order $m \times 1$, is denoted by:

$$\mathbf{x} = \begin{bmatrix} X_1 \\ X_2 \\ \vdots \\ X_m \end{bmatrix}$$

and its transpose $\mathbf{x}'$, a $1 \times m$ row vector, is denoted by:

$$\mathbf{x}' = (X_1, X_2, \ldots, X_m)$$

(Throughout the book matrices are denoted by bold-faced capitals and vectors by bold-faced lower-case letters.)[11]

In terms of the specific data of Table 1.1, the first column of the original data matrix **X** is the 10×1 vector:

$$\mathbf{x} = \begin{bmatrix} 3 \\ 7 \\ \vdots \\ 2 \end{bmatrix}$$

and the first row of **X** is the 1×3 vector:

$$\mathbf{x}' = (3, 4, 2)$$

The row vector of means of **X**, called the centroid, is denoted by:

$$\bar{\mathbf{x}}' = (4.3, 4.3, 4.0)$$

and is computed by the matrix operations:

$$\boxed{\bar{\mathbf{x}}' = \frac{1}{m} \mathbf{1}' \mathbf{X}}$$

Table 1.3 A Review of Common Statistical Formulas for Computing Sample-Based Measures of Central Tendency, Dispersion, and Covariation*

Mean	Mean-Corrected Variate
$$\bar{X} = \sum_{i=1}^{m} X_i/m$$	$$X_{di} = X_i - \bar{X}$$

Standard Deviation	Standardized Variate
$$s = \left[\frac{\sum_{i=1}^{m} (X_i - \bar{X})^2}{m-1} \right]^{1/2}$$ $$= \left[\frac{\sum_{i=1}^{m} X_{di}^2}{m-1} \right]^{1/2}$$	$$X_{si} = (X_i - \bar{X})/s$$ $$= X_{di}/s$$

Covariance	Product Moment Correlation
$$\mathrm{cov}(X_1 X_2) = s_{12} = r_{12}\, s_1 s_2$$ $$= \frac{\sum_{i=1}^{m} X_{di1} X_{di2}}{m-1}$$	$$r_{12} = \frac{\mathrm{cov}(X_1 X_2)}{s_1\, s_2} = \frac{s_{12}}{s_1 s_2}$$ $$= \frac{\sum_{i=1}^{m} X_{si1} X_{si2}}{m-1}$$

*An original observation is denoted by X_i while m denotes sample size. The sample standard deviation is denoted here as s, an estimator of σ, the universe standard deviation.

where $\mathbf{1}'$ denotes a 1×10 unit row vector and $\mathbf{X}$ is the 10×3 data matrix. From Table 1.1 we have:

$$\bar{\mathbf{x}}' = \frac{1}{10} (1, 1, \ldots, 1) \begin{bmatrix} 3 & 4 & 2 \\ 7 & 9 & 7 \\ \vdots & \vdots & \vdots \\ 2 & 1 & 3 \end{bmatrix} = (4.3, 4.3, 4.0)$$

Having found $\bar{\mathbf{x}}'$, we can then compute $\mathbf{X}_d$, the 10×3 matrix of mean-corrected scores by:

$$\boxed{\mathbf{X}_d = \mathbf{X} - \mathbf{1}\,\bar{\mathbf{x}}'}$$

where $\mathbf{1}\,\mathbf{x}'$ denotes the *matrix product* of a unit column vector and the row vector $\bar{\mathbf{x}}'$. In terms of the sample problem, we have:

$$
\mathbf{X}_d = \overset{\mathbf{X}}{\begin{bmatrix} 3 & 4 & 2 \\ 7 & 9 & 7 \\ \vdots & \vdots & \vdots \\ 2 & 1 & 3 \end{bmatrix}} - \overset{\mathbf{1}}{\begin{bmatrix} 1 \\ 1 \\ \vdots \\ 1 \end{bmatrix}} \overset{\bar{\mathbf{x}}'}{(4.3,\ 4.3,\ 4.0)}
$$

$$
= \begin{bmatrix} 3 & 4 & 2 \\ 7 & 9 & 7 \\ \vdots & \vdots & \vdots \\ 2 & 1 & 3 \end{bmatrix} - \begin{bmatrix} 4.3 & 4.3 & 4.0 \\ 4.3 & 4.3 & 4.0 \\ \vdots & \vdots & \vdots \\ 4.3 & 4.3 & 4.0 \end{bmatrix} = \begin{bmatrix} -1.3 & -0.3 & -2 \\ 2.7 & 4.7 & 3 \\ \vdots & \vdots & \vdots \\ -2.3 & -3.3 & -1 \end{bmatrix}
$$

as shown (in separate columns) in Table 1.1.

As already noted, Table 1.3 shows review computations for the standard deviation s, the covariance, $\text{cov}(X_1, X_2) = s_{12}$, and the product moment correlation r_{12}. In this book we shall always be computing the variance s^2 or the standard deviation s as *estimators* of their universe counterparts, σ^2 or σ, respectively. As such, we shall be dividing by some appropriate degrees of freedom (such as $m - 1$, as shown in Table 1.3).

In the context of matrix algebra the analogous formula for s^2, an estimator of the universe variance σ^2, is:

$$
s^2 = \frac{1}{m-1} \mathbf{x}'_d \mathbf{x}_d
$$

in which we find the scalar product of $\mathbf{x}_d$, a column vector of mean-corrected scores with itself.[12] For example, the variance of the first column $\mathbf{x}_{d1}$ of $\mathbf{X}_d$ in Table 1.1 is:

$$
s^2 = \frac{1}{9}(-1.3, 2.7, \ldots, -2.3) \begin{bmatrix} -1.3 \\ 2.7 \\ \vdots \\ -2.3 \end{bmatrix}
$$

$$
= 8.46
$$

The standard deviation s is simply the (positive) square root of the variance, and its formula (in scalar notation) is shown in Table 1.3. Similarly, we can find the variances of the second and third columns of $\mathbf{X}$ (from Table 1.1) as 7.79 and 7.33, respectively. If we place these variances in a diagonal matrix:

$$
\mathbf{D} = \begin{bmatrix} 8.46 & 0 & 0 \\ 0 & 7.79 & 0 \\ 0 & 0 & 7.33 \end{bmatrix}
$$

we have a summary of the three variances of the sample problem. We can next find the square root of **D** as:

$$\mathbf{D}^{1/2} = \begin{bmatrix} 2.908 & 0 & 0 \\ 0 & 2.791 & 0 \\ 0 & 0 & 2.708 \end{bmatrix}$$

which is a diagonal matrix of standard deviations (as also shown in Table 1.1). Finally, let us define $\mathbf{D}^{-1/2}$ as the inverse of $\mathbf{D}^{1/2}$:[13]

$$\mathbf{D}^{-1/2} = \begin{bmatrix} 1/2.908 & 0 & 0 \\ 0 & 1/2.791 & 0 \\ 0 & 0 & 1/2.708 \end{bmatrix}$$

Then, if we desire to find $\mathbf{X}_s$, the matrix of *standardized* data in Table 1.1, we see that it is given simply by:

$$\boxed{\mathbf{X}_s = \mathbf{X}_d\,\mathbf{D}^{-1/2}}$$

In terms of the sample problem of Table 1.1, we have:

$$\mathbf{X}_s = \overset{\mathbf{X}_d}{\begin{bmatrix} -1.3 & -0.3 & -2 \\ 2.7 & 4.7 & 3 \\ \vdots & \vdots & \vdots \\ -2.3 & -3.3 & -1 \end{bmatrix}} \overset{\mathbf{D}^{-1/2}}{\begin{bmatrix} 1/2.908 & 0 & 0 \\ 0 & 1/2.791 & 0 \\ 0 & 0 & 1/2.708 \end{bmatrix}}$$

$$= \begin{bmatrix} -0.45 & -0.11 & -0.74 \\ 0.93 & 1.68 & 1.11 \\ \vdots & \vdots & \vdots \\ -0.79 & -1.18 & -0.37 \end{bmatrix}$$

as can be checked by examining the appropriate columns of Table 1.1. All of this, of course, is simply equivalent to dividing each column of $\mathbf{X}_d$ by its own standard deviation. At this point, then, we have found $\mathbf{X}_d$, the 10×3 matrix of mean-corrected scores and $\mathbf{X}_s$, the 10×3 matrix of standardized scores.

Just as we found it helpful to make up scatter plots of the raw data (Figure 1.1), we could prepare two-variable plots of either $\mathbf{X}_d$ or $\mathbf{X}_s$. Figure 1.4 shows illustrative plots for transformations of the first two variables (X_1 and X_2) in Table 1.1. We note that the mean-corrected scores (in $\mathbf{X}_d$) involve only a translation of the origin to the centroid. Thus, Panel I of Figure 1.4 represents a parallel displacement of the configuration in Panel I of Figure 1.1.

Standardized scores (provided by $\mathbf{X}_s$) not only introduce a centroid-centered origin but also stretch or compress the configuration (along directions parallel

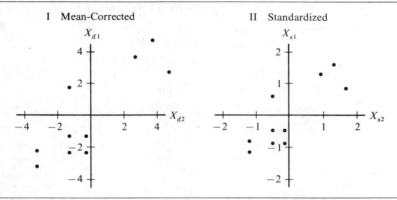

Figure 1.4 Illustrative Scatter Plots of Mean-Corrected and Standardized Data (First Two Variables of Table 1.1)

to the original axis orientation) according to $\mathbf{D}^{-1/2}$, the diagonal matrix whose entries are the reciprocals of the standard deviations of the original variables in $\mathbf{X}$. In Panel II of Figure 1.4 we note a slight change in the configuration since the first axis is squeezed (in the ratio of $1/2.91 : 1/2.79$ or $0.96 : 1$) relative to the second so as to adjust for the fact that the standard deviation of (the variable) X_1 is slightly greater than that of X_2.

1.4.2 The Cross Product Matrices

The various cross product matrices of interest to multivariate analysis are computed quite simply from $\mathbf{X}$, $\mathbf{X}_d$ and $\mathbf{X}_s$. Each cross product matrix is of the same general form and each has its particular role to play in multivariate analysis.

First, the raw sums of squares and cross products matrix is defined as the product of the transpose of the data matrix by the data matrix itself:

$$\boxed{\mathbf{B} = \mathbf{X}'\mathbf{X}}$$

and is shown in Table 1.4 for the illustrative data of Table 1.1. Next, the mean-corrected sums of squares and cross products matrix $\mathbf{S}$ is given by:

$$\boxed{\mathbf{S} = \mathbf{X}'_d\,\mathbf{X}_d}$$

and also appears in Table 1.4. If the entries of $\mathbf{S}$ are each divided by $m-1$, we have the covariance matrix:

$$\boxed{\mathbf{C} = \frac{1}{m-1}\,\mathbf{S}}$$

and this is shown in Table 1.4 as well.

Table 1.4 Cross Product Matrices of Data From Table 1.1

Raw Sums of Squares and Cross Products			Mean-Corrected Sums of Squares and Cross Products		

$$\mathbf{B} = \begin{bmatrix} 261 & 247 & 232 \\ 247 & 255 & 229 \\ 232 & 229 & 226 \end{bmatrix} \qquad \mathbf{S} = \begin{bmatrix} 76.1 & 62.1 & 60.0 \\ 62.1 & 70.1 & 57.0 \\ 60.0 & 57.0 & 66.0 \end{bmatrix}$$

Covariance Matrix			Correlation Matrix		

$$\mathbf{C} = \begin{bmatrix} 8.46 & 6.90 & 6.67 \\ 6.90 & 7.79 & 6.33 \\ 6.67 & 6.33 & 7.33 \end{bmatrix} \qquad \mathbf{R} = \begin{bmatrix} 1.000 & 0.850 & 0.847 \\ 0.850 & 1.000 & 0.838 \\ 0.847 & 0.838 & 1.000 \end{bmatrix}$$

Finally, the correlation matrix $\mathbf{R}$ can be obtained from the matrix of standardized variates as:

$$\mathbf{R} = \frac{1}{m-1} \mathbf{X}_s' \mathbf{X}_s$$

or, equivalently, from the covariance matrix by adjustment for differences in standard deviations:

$$\mathbf{R} = \mathbf{D}^{-1/2} \, \mathbf{C} \, \mathbf{D}^{-1/2}$$

As can be checked from Table 1.3, the latter formula is consistent with the scalar-notation formula for the product moment correlation r_{12}. The numerical values of $\mathbf{R}$ also appear in Table 1.4.[14]

At this point, then, vector and matrix operations have provided a concise set of procedures for finding various statistical quantities of interest that involve a single sample of multivariate data.[15]

1.5 An Overview of Linear Composites

Linear composites assume special status in multivariate methods. Much of what is done in multivariate data analysis involves the computation of a linear composite of one set of (say, predictor) variables that meets certain objectives, such as *best matching* the values of:

1. An original (criterion) variable.

2. Another linear composite which is, itself, computed over a set of other (e.g., criterion) variables.

Multiple regression is an illustration of a technique that is based on the first objective. With additional constraints placed on the relationships among pairs of linear composites, canonical correlation can be taken to be an illustration of the second objective. Accordingly, even at this introductory stage it seems useful to point out some of the main concepts underlying the calculation of linear composites.

As noted earlier, *values* of a linear composite are simply weighted sums of two (or more) variables, expressed (in scalar notation) as:

$$Y = w_1 X_1 + w_2 X_2 + \cdots + w_n X_n$$
$$= \sum_{j=1}^{n} w_j X_j$$

Note that this same idea can be compactly expressed as the scalar product of the vectors $\mathbf{w}$ and $\mathbf{x}$:

$$Y_i = \mathbf{w}' \mathbf{x} = \mathbf{x}' \mathbf{w}$$

The linear composite itself is represented by the *vector* $\mathbf{y}$ with entries Y_i. If we take some matrix of data such as $\mathbf{X}$, the 10×3 matrix of Table 1.1, and some arbitrary vector of weights, such as:

$$\mathbf{w} = \begin{bmatrix} 0.5 \\ 2.0 \\ 0.1 \end{bmatrix}$$

we could find a linear composite by means of the product:

$$\mathbf{y} = \mathbf{X}\mathbf{w}$$

To illustrate this case with the original data of Table 1.1, we have:

$$\mathbf{y} = \begin{bmatrix} 3 & 4 & 2 \\ 7 & 9 & 7 \\ \vdots & \vdots & \vdots \\ 2 & 1 & 3 \end{bmatrix} \begin{bmatrix} 0.5 \\ 2.0 \\ 0.1 \end{bmatrix} = \begin{bmatrix} 9.7 \\ 22.2 \\ \vdots \\ 3.3 \end{bmatrix}$$

for the preceding vector of weights $\mathbf{w}$.

Next, suppose that we have two sets of weights, expressed as columns in the 3×2 matrix:

$$\mathbf{W} = \begin{bmatrix} \mathbf{w}_1 & \mathbf{w}_2 \\ 0.5 & 1 \\ 2.0 & 2 \\ 0.1 & 3 \end{bmatrix}$$

If *separate* linear composites $\mathbf{y}_1$ and $\mathbf{y}_2$ are then calculated, we have the 10×2 matrix:

$$\mathbf{Y} = \begin{bmatrix} \mathbf{X} \\ 3 & 4 & 2 \\ 7 & 9 & 7 \\ \vdots & \vdots & \vdots \\ 2 & 1 & 3 \end{bmatrix} \begin{bmatrix} \mathbf{W} \\ 0.5 & 1 \\ 2.0 & 2 \\ 0.1 & 3 \end{bmatrix} = \begin{bmatrix} \mathbf{y}_1 & \mathbf{y}_2 \\ 9.7 & 17 \\ 22.2 & 46 \\ \vdots & \vdots \\ 3.3 & 13 \end{bmatrix}$$

where each of the columns of $\mathbf{Y}$ denotes a separate linear composite.

Thus, the calculation of linear composites involves rather straightforward operations of matrix multiplication, or linear transformation. Restrictions on the types of linear transformations characterize different classes of linear composites, as will be shown in subsequent chapters.

1.5.1 Centroids and Covariances Under Linear Composites

From univariate statistics we recall the effect that: (a) adding a constant k to a variable X_j or (b) multiplying X_j by a constant k, has on the mean and variance of X_j.

These familiar results can be summarized by saying that if X_j is some set of observations with mean $\overline{X}$ and (estimated universe) variance s^2:

1. Addition of a constant k to each value of X_j results in a new variable $(k + X_j)$ with:
   a. A mean, $\overline{(k + X_j)} = k + \overline{X}$.
   b. A variance, $\text{var}\,(k + X_j) = \text{var}\,(X_j) = s^2$.

2. Multiplication of each value of X_j by a constant k results in a new variable $(k\,X_j)$ with:
   a. A mean, $\overline{(k\,X_j)} = k\,\overline{X}$.
   b. A variance, $\text{var}\,(k\,X_j) = k^2\,s^2$.

These basic results are generalizable to vector-valued random variables $\mathbf{x}$. If $\mathbf{X}$ is an $m \times n$ data matrix with a centroid given by $\overline{\mathbf{x}}'$ and a covariance matrix by $\mathbf{C(X)}$ and $\mathbf{K}$ is an $m \times n$ matrix whose rows are repetitions of a vector of constants $\mathbf{k}'$, then:

1. *Addition* of $\mathbf{K}$ to $\mathbf{X}$ results in a new matrix with:
   a. A centroid, $\overline{(\mathbf{K + X})} = \mathbf{k}' + \overline{\mathbf{x}}'$.
   b. A covariance matrix, $\mathbf{C\,(K + X)} = \mathbf{C(X)}$.

Next, suppose we have just a single scalar k. If so:

2. *Multiplication* of **X** by k results in a new matrix with:
   a. A centroid, $\overline{(k\,\mathbf{X})} = k\,\overline{\mathbf{x}}'$.
   b. A covariance matrix, $\mathbf{C}\,(k\,\mathbf{X}) = k^2\,\mathbf{C}(\mathbf{X})$.

The preceding comments can all be summarized by the table:

	Univariate Case		Multivariate Case	
Operation	Mean	Variance	Centroid	Covariance Matrix
Add k (or $\mathbf{k}'$)	Add k	Do Nothing	Add $\mathbf{k}'$	Do Nothing
Multiply by k	Multiply by k	Multiply by k^2	Multiply by k	Multiply by k^2

1.5.2 Some Numerical Illustrations

To illustrate some of the relations involving linear composites, let us take a particularly simple case involving the 5×2 matrix of scores in Table 1.5 that is

Table 1.5 Illustrative Data (from Table 1.1) for Computing Linear Composites

Original Data

$$\mathbf{X} = \begin{bmatrix} X_1 & X_2 \\ 3 & 4 \\ 7 & 9 \\ 2 & 3 \\ 1 & 1 \\ 6 & 3 \end{bmatrix} ; \quad \mathbf{X} + \mathbf{K} = \begin{bmatrix} X_1 & X_2 \\ 5 & 5 \\ 9 & 10 \\ 4 & 4 \\ 3 & 2 \\ 8 & 4 \end{bmatrix} ; \quad 2\mathbf{X} = \begin{bmatrix} X_1 & X_2 \\ 6 & 8 \\ 14 & 18 \\ 4 & 6 \\ 2 & 2 \\ 12 & 6 \end{bmatrix}$$

$$\overline{X}_j \quad 3.8 \quad 4.0 \qquad \overline{X}_j \quad 5.8 \quad 5.0 \qquad \overline{X}_j \quad 7.6 \quad 8.0$$

Mean-Corrected Data	Covariance Matrix of **X**	Linear Composite

$$\mathbf{X}_d = \begin{bmatrix} X_{d1} & X_{d2} \\ -0.8 & 0 \\ 3.2 & 5 \\ -1.8 & -1 \\ -2.8 & -3 \\ 2.2 & -1 \end{bmatrix}$$
$$\overline{X}_j \quad 0 \quad\quad 0$$

$$\mathbf{C}(\mathbf{X}) = \frac{\mathbf{X}_d'\mathbf{X}_d}{m-1} = \begin{bmatrix} 6.7 & 6 \\ 6 & 9 \end{bmatrix}$$

$X_{d1} + X_{d2}$	$(X_{d1} + X_{d2})^2$
-0.8	0.64
8.2	67.24
-2.8	7.84
-5.8	33.64
1.2	1.44
Sum 0	110.80

$$\mathrm{var}(X_{d1} + X_{d2}) = \frac{110.8}{4} = 27.7$$

Variance of $Y = (X_{d1} + X_{d2})$

$$\mathrm{var}(\mathbf{y}) = (1,1)\begin{bmatrix} 6.7 & 6 \\ 6 & 9 \end{bmatrix}\begin{bmatrix} 1 \\ 1 \end{bmatrix} = 27.7$$

obtained from the first five stimulus paragraphs and variables X_1 and X_2. Table 1.5 shows the matrix of raw data and a (derived) matrix of mean-corrected data X_d. Also shown is the associated covariance matrix $C(X)$.

To demonstrate the preceding assertions, let us proceed to add the matrix K whose rows are each a vector of constants:

$$k' = (2,1)$$

to X in Table 1.5. We find that the new centroid is:

$$\overline{(K + X)} = k' + \bar{x}' = (2, 1) + (3.8, 4.0)$$
$$= (5.8, 5.0)$$

Next, if we multiply each row of X by 2 we see that the new centroid is:

$$\overline{(k\,X)} = k\bar{x}' = 2\bar{x}' = 2(3.8, 4.0)$$
$$= (7.6, 8.0)$$

Similarly, the reader can check the preceding assertions regarding the effects of these simple transformations on $C(X)$, the covariance matrix that appears in Table 1.5. However, what is more interesting is to examine the effect on centroids and covariance matrices of more general linear composites.

To illustrate, suppose we were to take one of the simplest possible functions (of the mean-corrected data in Table 1.5), namely the sum of the X_{d1} and X_{d2} variables:

$$\boxed{Y = X_{d1} + X_{d2}}$$

As X_d takes on its various values we get the 5×1 linear composite:

$$y = X_d\,1 = \begin{bmatrix} -0.8 & 0 \\ 3.2 & 5 \\ \vdots & \vdots \\ 2.2 & -1 \end{bmatrix} \begin{bmatrix} 1 \\ 1 \end{bmatrix} = \begin{bmatrix} -0.8 \\ 8.2 \\ \vdots \\ 1.2 \end{bmatrix}$$

This particular linear composite is simply the sum of two X_d variables and the mean of this linear composite is:

$$\bar{y} = \overline{(X_d\,1)} = \bar{x}'\,1$$
$$= (0 \times 1) + (0 \times 1)$$
$$= 0$$

Note, then, that a *single* linear composite of mean-corrected data also has a mean of 0. Table 1.5 shows this to be the case, since the sum of the five values of $X_{d1} + X_{d2}$ in the table is, indeed, equal to 0.

The variance of the linear composite **y** is given by the expression:

$$\text{var}(\mathbf{y}) = \mathbf{y}'\,\mathbf{y}/(m - 1)$$

Since $\mathbf{y} = \mathbf{X}_d\,\mathbf{1}$ we have:[16]

$$\text{var}(\mathbf{y}) = \mathbf{1}'\,\frac{\mathbf{X}_d'\,\mathbf{X}_d}{m - 1}\,\mathbf{1} = \mathbf{1}'\,C(\mathbf{X})\,\mathbf{1}$$

That is, the variance of this specific linear composite **y** is found by summing all of the entries of the *original* covariance matrix $C(\mathbf{X})$.

This result can be verified from Table 1.5 by noting that the variance of the variable $Y = X_{d1} + X_{d2}$ is 27.7. This is the same value that is obtained from:

$$\mathbf{1}'\,C(\mathbf{X})\,\mathbf{1} = 27.7$$

that is also shown in Table 1.5.

This is a very useful result since it means that we do not have to go through the preliminary calculations of finding **y** in order to know what its mean and variance will be.

Since the present case involves the *simple sum* of X_{d1} and X_{d2}, we have the result from examination of the general entries (see Table 1.3 for a review):

$$\text{var}(\mathbf{y}) = \overset{\mathbf{1}'}{(1, 1)} \overset{C(\mathbf{X})}{\begin{bmatrix} s_{11}^2 & r_{12}s_1 s_2 \\ r_{21}s_2 s_1 & s_{22}^2 \end{bmatrix}} \overset{\mathbf{1}}{\begin{bmatrix} 1 \\ 1 \end{bmatrix}}$$

$$\text{var}(\mathbf{y}) = s_{11}^2 + s_{22}^2 + 2\,\text{cov}(X_1\,X_2)$$
$$= s_{11}^2 + s_{22}^2 + 2\,r_{12}s_1 s_2$$

which agrees with the common formula for the variance of the sum of two random variables that was learned in univariate statistics. Note that the variance of the linear composite **y** depends on the *covariance* of the variables X_1 and X_2 as well as their variances.

Of course, we are not restricted to the special case of a linear composite that entails the simple sum of two variables. Suppose we return to the *two* linear composites, introduced earlier, via the matrix:

$$\mathbf{W} = \begin{bmatrix} 0.5 & 1 \\ 2 & 2 \\ 0.1 & 3 \end{bmatrix}$$

but now retain only the first two rows of this matrix (since only the variables

X_{d1} and X_{d2} are involved at this point); call this 2×2 matrix $\mathbf{W}^*$. Continuing to use the mean-corrected matrix $\mathbf{X}_d$ of Table 1.5, we then have:

$$
\mathbf{Y} =
\overset{\mathbf{X}_d}{\begin{bmatrix} -0.8 & 0 \\ 3.2 & 5 \\ -1.8 & -1 \\ -2.8 & -3 \\ 2.2 & -1 \end{bmatrix}}
\overset{\mathbf{W}^*}{\begin{bmatrix} 0.5 & 1 \\ 2 & 2 \end{bmatrix}}
=
\begin{matrix} \mathbf{y}_1 & \mathbf{y}_2 \\ \begin{bmatrix} -0.4 & -0.8 \\ 11.6 & 13.2 \\ -2.9 & -3.8 \\ -7.4 & -8.8 \\ -0.9 & 0.2 \end{bmatrix} \end{matrix}
$$

First, we check to see that $\mathbf{Y}$ has a centroid of $\mathbf{0}'$, as it should. Then, applying the same rationale as before, we have:

$$
\mathbf{C(Y)} = \mathbf{Y}'\,\mathbf{Y}/(m-1)
$$

$$
= \mathbf{W}^{*\prime}\,\frac{\mathbf{X}_d'\,\mathbf{X}_d}{(m-1)}\,\mathbf{W}^* = \mathbf{W}^{*\prime}\,\mathbf{C(X)}\,\mathbf{W}^*
$$

$$
= \overset{\mathbf{W}^{*\prime}}{\begin{bmatrix} 0.5 & 2 \\ 1 & 2 \end{bmatrix}}
\overset{\mathbf{C(X)}}{\begin{bmatrix} 6.7 & 6 \\ 6 & 9 \end{bmatrix}}
\overset{\mathbf{W}^*}{\begin{bmatrix} 0.5 & 1 \\ 2 & 2 \end{bmatrix}}
$$

$$
\mathbf{C(Y)} = \begin{bmatrix} 49.68 & 57.35 \\ 57.35 & 66.7 \end{bmatrix}
$$

The reader can verify that this same covariance matrix could be obtained from $\mathbf{Y}'\mathbf{Y}/(m-1)$.

The preceding method represents a general procedure for finding the centroid and covariance matrix for any type of linear composite of $\mathbf{X}$, or $\mathbf{X}_d$, as the case may be. Moreover, the same general procedure can be applied to various submatrices of interest, such as pooled within-groups or among-groups covariance matrices.[17]

1.5.3 Concluding Comments on Linear Composites

The unweighted linear composite $\mathbf{y}$ has entries Y_i that involve a simple sum of the original (mean-corrected) variables:

$$
Y_i = X_{di1} + X_{di2} + \cdots + X_{din}
$$

We should note that the variance of an unweighted linear composite (in scalar notation) is:

$$
\boxed{\mathrm{var}(\mathbf{y}) = \sum_{j=1}^{n} s_j^2 + 2\sum_{j=1}^{n-1}\sum_{k=j+1}^{n} r_{jk}s_j s_k}
$$

Furthermore, in the more general case in which values of **y** are given by the weighted sum:

$$Y_i = w_1 X_{di1} + w_2 X_{di2} + \cdots + w_n X_{din}$$

the variance of **y** is given (in scalar notation) by:

$$\text{var}(\mathbf{y}) = \sum_{j=1}^{n} w_j^2 s_j^2 + 2 \sum_{j=1}^{n-1} \sum_{k=j+1}^{n} w_j w_k r_{jk} s_j s_k$$

Both of these formulas will be useful in subsequent chapters. (The first, of course, is a special case of the second with $w_j = 1$ for all j.)

Linear composites will play a central role in subsequent chapters. For example, in the next chapter dealing with multiple regression we shall seek a linear composite of a set of predictors that minimizes the sum of squared differences between values of the linear composite and the original values of the criterion variable. Least squares theory will be used to determine the specific weights that make up this linear composite.

Later on in the book we shall seek two sets of weights, each set to be applied to its respective set of variables so that the resulting pair of linear composites is maximally correlated. This is the problem of canonical correlation.

Still later in the book we shall seek a set of weights (constrained so that their sum of squares equals unity) whose linear composite results in maximum variance when the weights are applied to a given set of variables. This is the problem of principal components analysis, a type of factor analysis (if one takes the broader view of this term).

In short, linear composites represent a key concept in multivariate analysis and it is appropriate to introduce and illustrate them in the book's beginning chapter.

1.6 Introduction to the Data Bank

In various chapters throughout the book we shall have occasion to apply selected multivariate methods to a set of actual data of realistic size. Problems could, of course, be drawn from a variety of substantive areas—psychology, medicine, education, economics, city planning, and so on. The particular problem chosen here is drawn from the field of marketing research, specifically, the pretesting of what are known as semifinished television commercials.

This study was conducted a few years ago for a nationally known marketer of automotive tires. All data are as they were actually obtained in the study. A brief introduction to the study is presented here, while Appendix A amplifies this initial discussion and integrates the results of various analyses conducted throughout the book.

The pretesting of semifinished television commercials is an activity engaged in by many large firms. Although research procedures may differ somewhat, the general idea can be summarized as follows:

1. One or more semifinished commercials, designed to serve as test stimuli, are prepared by the firm's advertising agency. By semifinished is meant that the video portion of the commercial consists of a series of slides that are mechanically projected on a small, portable viewing machine with synchronized audio. The slides depict what is intended to be covered in the finished production and the audio is recorded by a professional announcer. Cost of the preparation runs about one-twentieth of the cost of a finished commercial.

2. Similar mock-ups are made of one or more existing commercials that are being used by the firm and by competitors. The same announcer usually provides accompanying audio and these stimuli serve as controls.

3. A sample of 200–300 respondents, drawn from five to seven cities throughout the U.S., are interviewed in central location facilities, such as shopping centers. Often, mobile interviewing units that can be transferred from center to center are used for this purpose.

4. While interviewing formats differ, a typical interviewing sequence consists of presenting the respondent with a series of pretest questions regarding interest in the product class, brand purchased last, and so on.

5. Respondents are then exposed to each commercial, in turn, with presentation order randomized across respondents. After exposure to each commercial the respondent is asked a series of questions regarding main ideas conveyed, his likes and dislikes about various things that were said in the commercial, his belief in the claims, and degree of interest in purchasing the brand advertised in the commercial.

6. At the conclusion of all exposures the respondent is asked to choose that brand (or no brand) among the set of brands advertised that he would be most interested in purchasing.

7. The interview is concluded with a short questionnaire in which respondent demographic characteristics are obtained.

The above format, while prototypical of this type of survey, is, of course varied, depending upon the special characteristics of the problem and the product class.

1.6.1 The Alpha Co. Test

The sample problem and data bank used in this book have been drawn from one such test conducted by a national marketer of automobile tires that we shall call the Alpha Co. The product class of interest was steel-belted radial tires, a relatively new product at the time the study was undertaken. The Alpha Co. had been selling such tires as replacement equipment for several years but had used media other than national television in which to advertise its brand. The company's agency had prepared a semifinished test commercial which it believed would fare well if introduced on network television.

The three major competitors in this market were (as we shall call them) Beta Co., Gamma Co., and Delta Co. In contrast to the study's sponsor, these firms had already been advertising steel belted radials on national television. Accordingly, Alpha's agency made up a semifinished version of a commercial used in the past by each competitor. Beta's commercial was a highly "active" commercial that showed a car equipped with Beta's brand moving at high speed over various kinds of road junk. The commercial emphasized the toughness of Beta's tire, its puncture resistance, and its long mileage.

Gamma's commercial was a "low-key" type of commercial that did not involve high visual activity. This commercial emphasized the product benefits of good braking action, good cornering, and long mileage.

Delta's commercial employed a "scientific" appeal that listed five product benefits—smooth ride, protection from puncture, protection against skidding, road contact, and cornering. The commercial demonstrated these features in pictorial (animated) form.

Alpha's test commercial was similar to Beta Co.'s in that it emphasized tire toughness in a visually active, race-car type setting. Other benefits mentioned in the Alpha commercial were road traction and long mileage.

1.6.2 Data Collection

Data for 252 male adults, between the ages of 18 and 64, were collected from five different cities throughout the U.S. Central location facilities were used for the interviewing, as described above. Respondents were contacted randomly in various shopping centers and asked to participate in the survey. To qualify for inclusion in the sample each respondent had to own and be responsible for the care and maintenance of at least one car that was at least two years old. After completing the interview, which typically lasted about 30–45 minutes, each respondent received a small gift for his participation.

As in most studies of this type, the sample could hardly be classed as representative of the whole replacement tire buying market. Although contacted randomly, respondents were obtained on a volunteer basis in various shopping centers in and around each test city. Some attempt was made to control for respondent age but, even at that, no claim could be made that the sample was "representative" of the universe of replacement tire buyers.

1.6.3 Questionnaire and Analysis

Appendix A contains the questionnaire used in this study, a brief overview of the analysis, a number of supporting tables and preliminary tabulations, and the basic data. Also shown is a recapitulation of the findings of various multivariate analyses conducted in later chapters. As indicated earlier, we shall use this small data bank to illustrate various multivariate procedures discussed in the book. The data bank is large enough to enable us to try out a number of things, including split-sample replication and cross validation. On the other hand, the data bank is compact enough to make the various analyses reasonably self-contained and straightforward.

In addition to this running case, a number of small examples—all based on artificial data—are used to illustrate the conceptual aspects of various techniques in later chapters. The employment of small, easy-to-follow examples is deliberate, namely, to show application of the techniques without bogging down the reader in complex numerical detail.

1.7 Format of Succeeding Chapters

The remainder of the book is divided into three main parts. Part II, consisting of Chapters 2 through 5, describes various procedures for examining single criterion, multiple predictor association, including multiple regression, analysis of variance and covariance, two-group discrimination, and Automatic Interaction Detection. The organizing framework for this part is the general linear model, introduced in Chapter 2.

Part III, consisting of Chapters 6 and 7, discusses multiple criterion, multiple predictor association, including canonical correlation, multiple discriminant analysis, and multivariate analysis of variance and covariance. The organizing framework here is the generalized canonical correlation model, introduced in Chapter 6.

Part IV, consisting of Chapters 8 and 9, is concerned with the analysis of interdependence—factor analysis, nonmetric procedures for dimension reduction, clustering, and other types of object grouping procedures.

Appendix A, as briefly described earlier, provides data for the running case example, supporting tabulations, and summary comments on the substantive findings obtained from applications of the various multivariate procedures described in preceding sections of the book. Appendix B represents a technical appendix that provides review material on a variety of statistical and mathematical topics. Appendix C provides a small set of tables related to the more common statistical distributions used in applied research.

Review Questions

1. Examine the literature of your field and select three studies in which multivariate analysis was used.
   a. In each case, what was the substantive problem of interest?
   b. What hypotheses were expressed by the author?
   c. How did the selected technique(s) relate to the content side of the problem?
   d. What other techniques are suggested by the nature of the problem?
2. In the context of your own research try to formulate a problem that appears suitable for multivariate analysis.
   a. How would you classify the problem in terms of the system described in section 1.3 of the chapter?
   b. What multivariate procedures are suggested by this classification?

3. A researcher in the field of educational research has collected the following test scores (coded and hypothetical data) from a group of eight students:

Student	Language Aptitude X_1	Analogical Reasoning X_2	Geometric Reasoning X_3	Sex of Student X_4
A	2	3	15	male
B	6	8	9	male
C	5	2	7	female
D	9	4	3	male
E	11	10	2	female
F	12	15	1	female
G	1	4	12	male
H	7	3	4	female

a. Code the categorical variable X_4 into a dummy variable, using the designation of male $\Rightarrow 1$ and female $\Rightarrow 0$.

b. Find the means and standard deviations of each of the four variables, $X_1, \ldots, X_4$.

c. Compute the 8×4 matrices $\mathbf{X}_d$ and $\mathbf{X}_s$.

d. For variables X_1, X_2 and X_3 (only), compute:

i) $\mathbf{B} = \mathbf{X}'\mathbf{X}$ ii) $\mathbf{S} = \mathbf{X}_d'\mathbf{X}_d$

iii) $\mathbf{C} = \dfrac{1}{m-1}\mathbf{S}$ iv) $\mathbf{R} = \dfrac{1}{m-1}\mathbf{X}_s'\mathbf{X}_s$

4. Referring to the original 8×4 data matrix $\mathbf{X}$ in the preceding question (where sex is coded male $\Rightarrow 1$ and female $\Rightarrow 0$), first compute $\mathbf{X}_d$, the matrix of mean-corrected data.

a. Then find the linear composite: $\mathbf{y} = \mathbf{X}_d\mathbf{w}$, where:

$$\mathbf{w} = \begin{bmatrix} 0.2 \\ 0.3 \\ 0.5 \\ 0.1 \end{bmatrix}$$

b. Find the linear composites based on the transformation $\mathbf{Y} = \mathbf{X}\mathbf{W}$ given by the matrix:

$$\mathbf{W} = \begin{bmatrix} 1 & 2 \\ 2 & 3 \\ 1 & 2 \\ 0 & 1 \end{bmatrix}$$

c. Find the mean and variance of $\mathbf{y}$ and the centroid and covariance of $\mathbf{Y}$ in parts a and b, respectively.

5. Referring to the original 8×4 data matrix $\mathbf{X}$ in question 3:
   a. Compute an 8×2 mean-corrected matrix $\mathbf{X}_d$ for only the first two variables X_1 and X_2.
   b. Plot these eight points in two dimensions.
   c. Find and plot (in the same space as above) the linear composite $\mathbf{y} = \mathbf{X}_d \mathbf{w}$ where:

$$\mathbf{w} = \begin{bmatrix} 0.819 \\ 0.574 \end{bmatrix}$$

   d. Multiply $\mathbf{X}_d$ by 2 and compare $\mathbf{X}'_d \mathbf{X}_d$ with $(2\,\mathbf{X}_d)'(2\,\mathbf{X}_d)$.
   e. Find the mean and variance of $\mathbf{y} = \mathbf{X}_d\,\mathbf{w}$ (from part c) and compare it to the mean and variance of $\mathbf{y} = 2\,\mathbf{X}_d\,\mathbf{w}$.

6. Referring to the 8×2 mean-corrected matrix $\mathbf{X}_d$ (consisting of the first two variables X_1 and X_2) in the preceding problem:
   a. Find the linear composites given by: $\mathbf{Y} = \mathbf{X}_d\,\mathbf{W}$ where:

$$\mathbf{W} = \begin{bmatrix} 0.819 & 0.174 \\ 0.574 & 0.985 \end{bmatrix}$$

   b. Find the covariance matrix $\mathbf{C(X)}$ and compare this to the covariance matrix $\mathbf{C(Y)}$.
   c. Find the correlation matrix $\mathbf{R(X)}$ and compare this to the correlation matrix $\mathbf{R(Y)}$.
   d. Find, and compare, $\mathbf{R(X)}$ and $\mathbf{R(Y)}$ from standardized matrices, $\mathbf{X}_s$ and $\mathbf{Y}_s$, that are also computed for the first two variables (X_1 and X_2) only.

Notes

1. All data of Table 1.1 are fictitious. We shall often use small and simplified sets of data for expository purposes.

2. The columns labeled X_d and X_s in Table 1.1 are described later.

3. One could expand on this notion and call the rater-paragraph combination the unit of association. This amplification is particularly relevant if the data of several raters in the study are to be compared.

4. Most of this section is drawn from *Mathematical Tools for Applied Multivariate Analysis* (Green, 1976).

5. The original data matrix of the illustrative problem (Table 1.1) is of order 10×3 with cell values consisting of interval-scaled ratings for the variables X_1, X_2, X_3.

6. The four principal types of scales—nominal, ordinal, interval and ratio—are assumed to form a progression of increasing "strength" of measurement, where the nominal scale is the weakest form and the ratio scale is the strongest. For example, the ratio scale has all of the properties of the first three and, in addition, possesses a unique origin or zero point.

7. The techniques described in this text in general require no stronger form of measure-

ment than an interval scale. Hence, unless noted otherwise, we shall assume from here on that the data are no more strongly scaled than interval-scaled.

8. Subsequently (Shepard, 1962), research interest has centered on "nonmetric" methods in which some or all of the variables are measured in terms of ordinal scales. Later chapters discuss some of these methods.

9. Notice that the k-th dummy variable is not needed; if no 1 appears in any of the $k - 1$ dummies we know that the object belongs to the k-th class.

10. By linear composite is meant a vector with entries that are obtained from a function of the type $Y = w_1 X_1 + w_2 X_2$ where w_1 and w_2 are weights; as such, a linear composite is simply a vector whose entries are weighted sums.

11. As noted, matrix or vector *entries* are denoted by italicized capital letters. While we use this same notation for denoting variables X_1, X_2, etc., the distinction should always be clear from the context.

12. As recalled from vector algebra, if we have two $m \times 1$ vectors:

$$\mathbf{a} = \begin{bmatrix} a_1 \\ a_2 \\ \vdots \\ a_m \end{bmatrix} \text{ and } \mathbf{b} = \begin{bmatrix} b_1 \\ b_2 \\ \vdots \\ b_m \end{bmatrix}$$

their scalar product is defined as:

$$\mathbf{a'b} = a_1 b_1 + a_2 b_2 + \cdots + a_k b_k + \cdots + a_m b_m$$
$$= \sum_{k=1}^{m} a_k b_k$$

The scalar product is sometimes referred to as the dot product or inner product of two vectors.

13. In words, the inverse of a diagonal matrix is simply another diagonal matrix whose diagonal entries are the reciprocals of the diagonal entries of the original matrix.

14. It should be noted that each of the cross product matrices is of the same form, namely the product of the transpose of a matrix by the matrix itself. The product:

$$\mathbf{B} = \mathbf{Z'} \mathbf{Z}$$

is sometimes called the *minor product moment* of $\mathbf{Z}$ while the product:

$$\mathbf{C} = \mathbf{Z} \mathbf{Z'}$$

is sometimes called the *major product moment* of $\mathbf{Z}$ (Horst, 1963).

15. In Chapter 4 we shall describe counterpart matrix procedures for finding among-groups and within-groups cross product matrices when multivariate data for two (or more) groups (samples) are being analyzed.

16. This result follows from the substitution:

$$\frac{1}{m - 1} (\mathbf{X}_d \mathbf{1})' (\mathbf{X}_d \mathbf{1})$$

and the fact that the transpose of the product of two or more matrices equals the product of their transposes in reverse order. The net result, as shown, is a scalar.

17. The same general procedure can also be used where sums of squares and cross products matrices of either raw or mean-corrected data are involved.

PART II

Single Criterion, Multiple Predictor Association

CHAPTER 2

Multiple and Partial Regression

2.1 Introduction

In section 1.5 of the preceding chapter the topic of *linear composite* was introduced as a central concept of multivariate analysis. In this chapter multiple regression is discussed as the prototypic model of single criterion, multiple predictor association. In multiple regression we seek a particular linear composite of the predictor variables with values that best match those of the criterion. We shall take "best match" to mean that the sum of squared deviations between values on the composite and their counterpart values on the criterion is minimized.

The basic ideas of multiple regression are discussed from an intuitive and geometrically oriented point of view, using the small numerical example that was introduced in the preceding chapter. The response surface or point model, in which observations are represented as points and variables as dimensions, is first described. The vector model, a complementary representation in which the units of association (observations) are dimensions and the variables are vectors, is then illustrated.

Next, attention turns to the more formal aspects of the general linear model. The multiple regression model, representing a prototype of the general linear model, is presented in matrix algebra terms. Solution procedures are described and illustrated with data from the sample problem.

We conclude the chapter with a discussion of various computer program approaches to multiple regression, including stepwise regression, and illustrate their application to portions of the data bank of Appendix A. In so doing, we carry out a double cross-validation regression analysis and check for stability in the signs of the regression coefficients.

2.2 Two-Variable Regression

The search for association among variables is a basic activity in all the sciences. Multiple regression (and, its special case, two-variable regression) are tools for the study of relationships. In general, we are interested in four questions:

1. Can we find a linear composite that will compactly express the relationship between a set of predictors and a criterion variable?

2. If we can, how strong is the relationship; that is, how well can we predict values of the criterion variable from the linear composite of the predictors?

3. Is the overall relationship statistically significant?

4. Which predictors are most important in accounting for variation in the criterion variable; in particular, can the original model be reduced to fewer variables that still provide adequate prediction of the criterion?

In many situations found in the behavioral and administrative sciences, linear relationships of the form:

$$Y_i = \beta_0 + \beta_1 X_{i1} + \beta_2 X_{i2} + \cdots + \beta_n X_{in} + \epsilon_i$$

can do a good job in predicting values of the criterion variable, Y. *The above expression is linear in the parameters.* That is, the contribution of β_j to Y, per unit change in X_j, is assumed not to depend on either: (a) the specific level of $X_j (j = 1, 2, \ldots, n)$ or (b) the level of any other predictor $X_k (k = 1, 2, \ldots, n; k \neq j)$. The intercept β_0 represents the "contribution" to Y when all X_j are zero.

The last term on the right, denoted as ϵ_i $(i = 1, 2, \ldots, m)$ is an error term that is included to reflect our usual inability to predict Y perfectly on the basis of a limited number of predictors and the further possibility of error in the measurement or observation of Y.

Since we shall be assuming that the mean of the error term $E(\epsilon_i) = \bar{\epsilon}_i = 0$, the (conditional) means of the Y_i are given by:

$$E(Y_i | X_1, X_2, \ldots, X_n) = \beta_0 + \sum_{j=1}^{n} \beta_j X_{ij}$$

Hence, with $X_1, X_2, \ldots, X_n$ fixed, the response variable Y is assumed to come from a probability distribution with a mean of:

$$E(Y_i) = \beta_0 + \sum_{j=1}^{n} \beta_j X_{ij}$$

For example, with only one predictor variable X_1, the means of Y—one for each value of X_1—would all fall on a straight line with intercept β_0 and slope β_1.

As a case in point, let us continue with the small numerical example introduced in the preceding chapter. For ease of reference, Table 2.1 reproduces the data of Table 1.1. The only change is in the labeling of variables where, following convention, the first variable, overall liking, is now called Y and the variables: degree of extra-curricular activity and degree of scholastic achievement are now denoted X_1 and X_2, respectively. Thus, in the current formulation of the problem Y is the criterion variable and X_1 and X_2 are the two predictors.

We can approach the multiple regression problem most easily by first considering the special case of two-variable or bivariate regression. Subsequent discussions, involving multiple predictors, then become more straightforward.

2.2.1 Assumption Structure of Two-Variable Regression

As we remember from elementary statistics, in the two-variable case the underlying linear regression model is:

$$Y_i = \beta_0 + \beta_1 X_{i1} + \epsilon_i$$

where Y denotes the criterion variable. Typically, Y is estimated by $\hat{Y}$, whose values are obtained from a linear "composite" involving a single predictor X_1. β_0 the intercept, and β_1 the slope, are also parameters to be estimated. The error term ϵ_i is defined as:

$$\epsilon_i = Y_i - \hat{Y}_i$$

for $i = 1, 2, \ldots , m$, the number of observations. The predictor variable X_1 is assumed to be fixed (nonrandom) and known.

Figure 2.1 shows a conceptualization of the two-variable regression model. We see that the conditional means of Y (one for each specified value of X_1) all

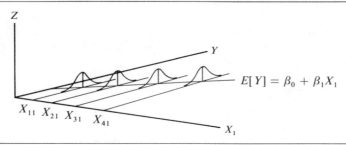

Figure 2.1 Two-Variable Regression Model (The axis Z denotes the conditional density function of Y for each value of X_1)

fall on a straight line given by the linear equations shown in the figure. The ordinate consists of values on a set of conditional probability density functions, one for each value of X_1.

Let us now examine the assumptions regarding the error ϵ. These assumptions are as follows:

1. Since the i-th observation Y_i is assumed to be the sum of a fixed part, $\beta_0 + \beta_1 X_1$ and a random part, ϵ_i, Y_i itself is considered a random variable.

2. The errors ϵ_i are assumed to be independently distributed with parameters:

$$
\begin{aligned}
E(\epsilon_i) &= \bar{\epsilon}_i = 0 \\
\mathrm{var}(\epsilon_i) &= \sigma^2 && \text{for all } i \\
\mathrm{cov}(\epsilon_i, \epsilon_h) &= 0 && \text{for } i \neq h
\end{aligned}
$$

for $i, h = 1, 2, \ldots, m; i \neq h$. That is, the errors are assumed to be uncorrelated with a mean (expectation of ϵ_i) of zero and a constant variance of σ^2.

3. The error term is also assumed to be independent of X_1, the fixed and known predictor variable.[1]

4. Since ϵ_i and ϵ_h are uncorrelated, so are Y_i and Y_h. Furthermore, the parameters of Y_i are:

$$
\begin{aligned}
E(Y_i) &= \beta_0 + \beta_1 X_{i1} \\
\mathrm{var}(Y_i) &= \mathrm{var}(\epsilon_i) = \sigma^2
\end{aligned}
$$

Note that we have not made any *distributional* assumptions about the ϵ_i at this point.[2]

Sample-based estimates of Y, β_0 and β_1, as found from regression procedures, are denoted by:

$$
\hat{Y}_i = b_0 + b_1 X_{i1}
$$

and we shall let e_i denote the sample-based estimate of ϵ_i, the error term. The problem, then, is to solve for b_0 and b_1. The familiar least squares criterion involves minimization of the expression:

$$
\sum_{i=1}^{m} e_i^2 = \sum_{i=1}^{m} (Y_i - \hat{Y}_i)^2 = \sum_{i=1}^{m} [Y_i - (b_0 + b_1 X_{i1})]^2
$$

Since Y_i and X_{i1} are given, the task is to solve for b_0 and b_1 and it is here that one seeks the specific linear composite that makes the $\hat{Y}_i$ best match their counterpart Y_i values (in the sense of minimum sum of squared deviations).

Appendix B shows the derivation of the normal equations that produce the desired minimization. In the two-variable case the estimates b_1 and b_0 are found from the well-known formulas:

$$b_1 = \frac{\sum_{i=1}^{m} (Y_i - \overline{Y})(X_{i1} - \overline{X}_1)}{\sum_{i=1}^{m} (X_{i1} - \overline{X}_1)^2}$$

$$b_0 = \overline{Y} - b_1 \overline{X}_1$$

where $\overline{Y}$ and $\overline{X}_1$ are the means of Y and X_1, respectively.

2.2.2 Computing the Parameters of a Two-Variable Regression

If we turn our attention to the data of Table 2.1 we can illustrate the preceding comments numerically. For example, let us regress Y on X_1. Table 2.2 shows the detailed calculations that lead to the coefficients b_0 and b_1.

Table 2.1 Ratings of Stimulus Paragraphs Describing Ten High School Seniors (from Table 1.1)

Paragraph Description of Stimulus Person	Overall Liking			Degree of Extra-Curricular Activity			Degree of Scholastic Achievement		
	Y	Y_d	Y_s	X_1	X_{d1}	X_{s1}	X_2	X_{d2}	X_{s2}
a	3	-1.3	-0.45	4	-0.3	-0.11	2	-2	-0.74
b	7	2.7	0.93	9	4.7	1.68	7	3	1.11
c	2	-2.3	-0.79	3	-1.3	-0.47	1	-3	-1.11
d	1	-3.3	-1.13	1	-3.3	-1.18	2	-2	-0.74
e	6	1.7	0.58	3	-1.3	-0.47	3	-1	-0.37
f	2	-2.3	-0.79	4	-0.3	-0.11	4	0	0
g	8	3.7	1.27	7	2.7	0.97	9	5	1.85
h	3	-1.3	-0.45	3	-1.3	-0.47	2	-2	-0.74
i	9	4.7	1.62	8	3.7	1.33	7	3	1.11
j	2	-2.3	-0.79	1	-3.3	-1.18	3	-1	-0.37
Mean	4.3			4.3			4.0		
Standard Deviation	2.908			2.791			2.708		

Table 2.2 A Two-Variable Regression of Y on X_1

Case	Y	X_1	Y^2	$X_1{}^2$	YX_1	$\hat{Y}_{(1)}$	Residuals $Y_{r(1)} = Y - \hat{Y}_{(1)}$
a	3	4	9	16	12	4.034	−1.034
b	7	9	49	81	63	8.464	−1.464
c	2	3	4	9	6	3.148	−1.148
d	1	1	1	1	1	1.377	−0.377
e	6	3	36	9	18	3.148	2.852
f	2	4	4	16	8	4.034	−2.034
g	8	7	64	49	56	6.692	1.308
h	3	3	9	9	9	3.148	−0.148
i	9	8	81	64	72	7.578	1.422
j	2	1	4	1	2	1.377	0.623
Sum	43	43	261	255	247	43.000	0.
Mean	4.3	4.3					0.
Standard Deviation	2.908	2.791					1.531

$$b_1 = \frac{\displaystyle\sum_{i=1}^{m}(Y_i - \overline{Y})(X_{i1} - \overline{X}_1)}{\displaystyle\sum_{i=1}^{m}(X_{i1} - \overline{X}_1)^2} = \frac{\displaystyle\sum_{i=1}^{m}Y_iX_{i1} - n\overline{Y}\,\overline{X}_1}{\displaystyle\sum_{i=1}^{m}X_{i1}^2 - n\overline{X}_1^2}$$

$$= \frac{247 - 10(4.3)^2}{255 - 10(4.3)^2} = 0.886$$

$$b_0 = \overline{Y} - b_1\overline{X}_1 = 4.3 - 0.886(4.3)$$

$$= 0.491$$

The resulting regression equation is:

$$\hat{Y} = 0.491 + 0.886X_1$$

where b_1 and b_0 are found from the above formulas.[3] The slope b_1 measures the change in $\hat{Y}$ per unit change in X_1 while the intercept b_0 denotes the value of $\hat{Y}$ when X_1 is zero.

We can substitute the various values of X_1 in the regression equation to find the predicted criterion values $\hat{Y}_i$. These are shown in Table 2.2, along with the residuals $Y_{r(1)} = Y - \hat{Y}_{(1)}$. The residuals sum to zero; this is a property of the least squares procedure.

If we were to correlate the residuals $Y_{r(1)} = Y - \hat{Y}_{(1)}$ with the original predictor X_1, we would find that the correlation is precisely zero. As such, the residuals are often referred to as the *orthogonal complement* of Y with respect to

X_1. We take note of the fact that the standard deviation of the residuals (1.531) is less than that of Y, suggesting that there is some (linear) association of Y with X_1.

2.2.3 Computing the Strength of the Relationship

In virtually any applied study involving regression we compute not only the regression equation's coefficients but some measure of goodness of fit as well. In the two-variable case the measure of fit is r^2, which is called either the coefficient of determination or the squared (simple) correlation. This coefficient varies between 0 and 1 and represents the proportion of total variation of Y_i (as measured about its own mean $\overline{Y}$) that is accounted for by variation in X_1.

If we were to use $\overline{Y}$ to estimate each value of Y_i, then a measure of our inability to predict Y would be given by the sum of squared deviations $\sum_{i=1}^{m} (Y_i - \overline{Y})^2$. On the other hand, if we tried to predict Y_i from a linear regression based on X_1, we could use each $\hat{Y}_i$ to predict its counterpart Y_i. In this case a measure of our inability to predict Y_i is given by $\sum_{i=1}^{m} (Y_i - \hat{Y}_i)^2$. We can define r^2 as the following function of these two quantities:

$$r^2 = 1 - \frac{\displaystyle\sum_{i=1}^{m} (Y_i - \hat{Y}_i)^2}{\displaystyle\sum_{i=1}^{m} (Y_i - \overline{Y})^2}$$

If each $\hat{Y}_i$ predicts its counterpart Y_i perfectly, then $r^2 = 1$ since the numerator of the second term on the right is zero. However, if the use of X_1 in the regression equation does no better than $\overline{Y}$, then the second term on the right is 1 and $r^2 = 0$, indicating no ability to predict Y_i (beyond the use of $\overline{Y}$ itself).[4]

In the case of the regression of Y on X_1, the computed value of r^2 is:

$$r_{y1}^2 = 1 - \frac{\displaystyle\sum_{i=1}^{m} [Y_i - \hat{Y}_{i(1)}]^2}{\displaystyle\sum_{i=1}^{m} (Y_i - \overline{Y})^2}$$

$$= 1 - \frac{21.09}{76.10} = 0.723$$

and we say that 72.3 percent of the variation in Y is accounted for by variation in X_1.

In summary, the regression equation, with estimated parameters b_0 and b_1, is designed to answer the first question of finding a compact functional form for predicting Y_i, while r_{y1}^2 is designed to answer the second question of measuring how good the predictions are.

2.2.4 A Breakdown of Accounted-For Variance

Let us examine the expression r_{y1}^2 somewhat more closely for the case of Y regressed on X_1. We first find from Table 2.2 that:

1. $$\sum_{i=1}^{m} (Y_i - \hat{Y}_{i(1)})^2 = (-1.034)^2 + (-1.464)^2 + \cdots + (0.623)^2$$

$$= 21.09$$

is the sum of squared errors in predicting Y_i from $\hat{Y}_{i(1)}$.

2. $$\sum_{i=1}^{m} (Y_i - \overline{Y})^2 = (3 - 4.3)^2 + (7 - 4.3)^2 + \cdots + (2 - 4.3)^2$$

$$= 76.10$$

is the sum of squared errors in predicting Y_i from $\overline{Y} = 4.3$.

As might next be surmised, there is one additional quantity of interest, namely:

3. $$\sum_{i=1}^{m} (\hat{Y}_{i(1)} - \overline{Y})^2 = (4.034 - 4.3)^2 + (8.464 - 4.3)^2$$

$$+ \cdots + (1.377 - 4.3)^2$$

$$= 55.01$$

which is the sum of squares due to the regression of Y on X_1. Notice that the sum of the terms in 1 and 3 equals that in 2.

One way of arriving at this result is to start with the identity:

$$Y_i - \overline{Y} = (Y_i - \hat{Y}_{i(1)}) + (\hat{Y}_{i(1)} - \overline{Y})$$

in which the total deviation of $Y_i - \overline{Y}$ is the sum of:

1. The deviation of Y_i from its value $\hat{Y}_{i(1)}$ predicted from the regression line.
2. The deviation of $\hat{Y}_{i(1)}$ from $\overline{Y}$.

If we square the expressions on both sides of the equation and sum over i, we get:

$$\sum_{i=1}^{m} (Y_i - \overline{Y})^2 = \sum_{i=1}^{m} (Y_i - \hat{Y}_{i(1)})^2 + \sum_{i=1}^{m} (\hat{Y}_{i(1)} - \overline{Y})^2$$

since the cross product term turns out to be zero. Thus, we have a rather neat decomposition of sums of squares into two additive parts, one part due to errors in prediction and one part due to differences in regression values from the criterion-variable's mean.[5]

Figure 2.2 shows how the deviations are measured in the illustrative case of Y regressed on X_1. Panel I shows the decomposition for a single observation (case g in Table 2.1) while Panels II through IV show the total deviations and the two component parts of the total. Note further that if b_1, the slope of the regression, is zero, nothing is gained over the use of $\overline{Y}$ since, in this case:

$$b_0 = \overline{Y} - 0(\overline{X}_1)$$
$$= \overline{Y}$$

and, hence, each $\hat{Y}_{i(1)}$ equals $\overline{Y}$.

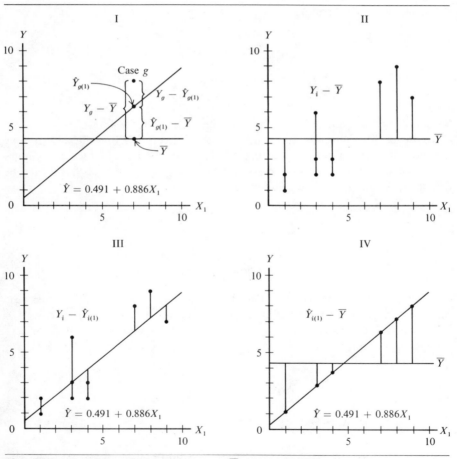

Figure 2.2 A Breakdown of Deviations $Y_i - \overline{Y}$ Into Two Additive Parts

2.2.5 Is r^2 Significant?

There is still another question to be raised about the goodness of fit measure r^2, namely, is it statistically significant? If we recapitulate the results so far, we have:

$$\hat{Y} = 0.491 + 0.886X_1; r_{y1}^2 = 0.723$$

The coefficient $b_1^{(s)} = 0.886$ is an estimate of the slope $\beta_1^{(s)}$, where we now use the superscript (s) to emphasize the fact that a *simple* (two-variable) regression is involved. If we let ρ^2 denote the population coefficient of determination, then $r_{y1}^2 = 0.723$ is an estimate of ρ_{y1}^2.

It is now time to make a distributional assumption about the error term ϵ. The usual assumption is that ϵ is *normally* (and independently) distributed with zero mean and constant variance.

Having made this distributional assumption, we can set up two types of tests that yield equivalent results:

$$H_0: \beta_1^{(s)} = 0 \text{ versus } H_1: \beta_1^{(s)} \neq 0$$

$$H_0: \rho_{y1}^2 = 0 \text{ versus } H_1: \rho_{y1}^2 \neq 0$$

Standard methods are available for carrying out each of these tests.

However, we shall illustrate a different test procedure—one that has greater potential for generalization. This procedure is often called the *models comparison* approach and its results are also equivalent to the two preceding tests. In this approach the hypotheses under test are:

$$H_0: Y = \beta_0 + \epsilon \text{ versus}$$

$$H_1: Y = \beta_0 + \beta_1^{(s)}X_1 + \epsilon$$

The first formulation is usually called the *restricted* or reduced model, since this one excludes $\beta_1^{(s)}X_1$ while the second formulation is called the *full* model.

In the case of the restricted model we already know that β_0 equals $\overline{Y}$. As shown earlier, in this case the sum of squared prediction errors is:

$$\text{SSE}_r = \sum_{i=1}^{m} (Y_i - \overline{Y})^2$$

$$= 76.10$$

Here we use the notation SSE_r to denote the sum of squared errors under the restricted model. However, the full model includes an *additional* parameter estimate $b_1^{(s)}$, with an associated sum of squared prediction errors of:

$$\text{SSE}_f = \sum_{i=1}^{m} (Y_i - \hat{Y}_i)^2$$

$$= 21.09$$

SSE_f, denoting the error sum of squares for the full model, can never be greater than SSE_r. However, if the difference $\text{SSE}_r - \text{SSE}_f$ is "small," then the additional parameter estimate $b_1^{(s)}$ is not doing much good in reducing error in predicting the criterion and we accept the restricted model. If the difference is large, the addition of the extra parameter estimate has materially reduced prediction error. The actual test involves an F statistic[6] and is defined as follows:

$$F = \frac{\text{SSE}_r - \text{SSE}_f}{\text{SSE}_f} \cdot \frac{d_f}{d_r - d_f}$$

where d_f denotes the number of degrees of freedom associated with the full model and d_r denotes the number of degrees of freedom for the restricted model. In the present example, we have:

$$F = \frac{76.10 - 21.09}{21.09} \cdot \frac{8}{1}$$

$$= 20.87$$

In this case $d_f = m - 2 = 8$ to account for the fact that two parameters, β_0 and $\beta_1^{(s)}$, are being estimated.[7]

The degrees of freedom associated with SSE_r are $d_r = m - 1 = 9$ reflect the fact that the constraint equation:

$$\sum_{i=1}^{m} (Y_i - \overline{Y}) = \sum_{i=1}^{m} e_i = 0$$

is to be satisfied. From Table C.4, we note that the 0.05-level F value (under the null hypothesis) associated with $(d_r - d_f) = 1$; $d_f = 8$, is 5.32. Since the computed $F = 20.87$ clearly exceeds the tabular F, we conclude that the full model is to be accepted.

The comparison of models, as described here, would appear to be a rather elaborate procedure compared to tests of the more traditional variety, such as the standard t test or F test.[8] Our motivation for presenting the models-comparison test at an early stage in the exposition stems from its high degree of flexibility in carrying out other kinds of tests in multiple regression (and other types of linear models as well). *We shall emphasize the models-comparison approach throughout the book.*

2.2.6 Additional Two-Variable Regressions

So far we have discussed the regression of Y on X_1 alone. Returning to Table 2.1, we note that Y could also be regressed on X_2. Moreover, we could regress X_1 on X_2 or X_2 on X_1, as well. Table 2.3 shows the resulting regression equations, r^2's and residuals. Figure 2.3 shows the associated scatter plots (including Y versus X_1).

We note from Table 2.3 that Y and X_2 are positively associated, as are X_1 and X_2. The scatter plots of Figure 2.3 also show this positive association; moreover, the plots indicate that linear functions appear to represent the relationships rather well.

Up to this point in the chapter we have not discussed the familiar product moment correlation (see Table 1.3 in Chapter 1) and its relationship to simple

Table 2.3 Additional Two-Variable Regressions Computed from the Data of Table 2.1

Case	Y on X_2 $\hat{Y} = 0.664 + 0.909X_2$ $r^2 = 0.716$ $Y_{r(2)} = Y - \hat{Y}_{(2)}$	X_2 on X_1 $\hat{X}_2 = 0.504 + 0.813X_1$ $r^2 = 0.702$ $X_{r2(1)} = X_2 - \hat{X}_{2(1)}$	X_1 on X_2 $\hat{X}_1 = 0.845 + 0.864X_2$ $r^2 = 0.702$ $X_{r1(2)} = X_1 - \hat{X}_{1(2)}$
a	0.518	−1.756	1.427
b	−0.027	−0.822	2.109
c	0.427	−1.943	1.291
d	−1.482	0.683	−1.573
e	2.609	0.057	−0.436
f	−2.300	0.244	−0.300
g	−0.845	2.805	−1.618
h	0.518	−0.943	0.427
i	1.973	−0.009	1.109
j	−1.391	1.683	−2.436
Mean	0	0	0
Standard Deviation	1.548	1.478	1.523

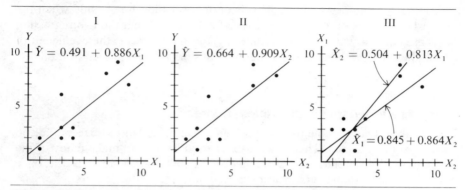

Figure 2.3 Scatter Plots of Two-Variable Regressions from Table 2.1

regression coefficients; it is time to remedy the situation. In Table 1.3 we reviewed one formula for computing the product moment correlation between two variables, such as X_1 and X_2:

$$r_{12} = \frac{\sum\limits_{i=1}^{m} X_{si1} X_{si2}}{m - 1}$$

where $m - 1$ appears in the denominator because the variates X_{si1} and X_{si2} have both been earlier standardized by employing a standard deviation that was also based on a denominator of $m - 1$. This formula simply states that the product moment correlation is an average cross product of two standardized variates.

However, another way to look at the product moment correlation between, say, X_1 and X_2 in Table 2.3 is in terms of standardizing a regression coefficient. In two-variable regression the following relationship holds between the regression slopes and the product moment correlation:

$$r_{12} = b_1^{(s)} \frac{s_1}{s_2} = 0.813 \frac{(2.791)}{2.708}$$

$$= b_2^{(s)} \frac{s_2}{s_1} = 0.864 \frac{(2.708)}{2.791}$$

$$= 0.838$$

where s_1 and s_2 are obtained from Table 2.1 and $b_1^{(s)}$ and $b_2^{(s)}$ are obtained from Table 2.3. Thus, r_{12} is a *symmetric* measure of the association between X_1 and X_2, while $b_1^{(s)}$, the regression of X_2 on X_1, and $b_2^{(s)}$, the regression of X_1 on X_2, are in general not equal to each other.

Panel III of Figure 2.3 shows two separate regression lines. Only when the standard deviations of X_1 and X_2 are equal will we find that the two regression slopes are equal as well.[9] We note that $(r_{12})^2 = 0.702$; this is the coefficient of determination, developed (along different conceptual lines) in section 2.2.3.

2.3 Introductory Aspects of Multiple Regression

The step from two-variable to multiple regression turns out to be less of a leap than might be expected. This is because we shall be able to make many of the same kinds of interpretations as were made in the bivariate regression case, once some adjustments to the variables are carried out.

2.3.1 Partial Regression Coefficients

The simplest case of multiple regression involves the criterion variable Y and two predictors, say X_1 and X_2, in the linear equation:

$$\hat{Y} = b_0 + b_1 X_1 + b_2 X_2$$

where b_0, b_1, and b_2 denote estimates of the parameters β_0, β_1, and β_2, respectively.

In this case b_1 is a *partial* regression coefficient that measures the change in $\hat{Y}$ per unit change in X_1 when all other predictors, namely X_2, are held constant. Similar remarks pertain to b_2, in which X_1 is now held constant.

The theoretical model for multiple regression, including assumptions about the error term ϵ, is fully analogous to the two-variable case. Moreover, solution techniques are also based on the least squares criterion and the resulting normal equations (see Appendix B). While we shall discuss these various facets of multiple regression later in the chapter, let us first concentrate on the meaning of the partial regression coefficients b_1 and b_2.

Tables 2.2 and 2.3 provide the conceptual key. First, assuming that X_1 and X_2 are correlated predictor variables, the partial regression coefficients will *not* equal their simple counterparts, as found from the separate two-variable regressions:

$$b_1^{(s)} = 0.886; \; b_2^{(s)} = 0.909$$

However, let us do two things:

Regress the original values of Y (from Table 2.1) on the residuals $X_{r1(2)} = X_1 - \hat{X}_{1(2)}$ (from Table 2.3).

Regress the original values of Y on the residuals $X_{r2(1)} = X_2 - \hat{X}_{2(1)}$ (from Table 2.3).

For example, to find the first of these coefficients b_1, we can use the standard formula of section 2.2.1; simplifications are possible because the mean of each column of residuals in Table 2.3 is already zero. The formula is:

$$b_1 = \frac{\sum\limits_{i=1}^{m} [Y_i - \overline{Y}][X_{ri1(2)}]}{\sum\limits_{i=1}^{m} [X_{ri1(2)}]^2}$$

Entering the appropriate columns of Tables 2.1 and 2.3 gives us:

$$b_1 = \frac{(3 - 4.3)(1.427) + (7 - 4.3)(2.109) + \cdots + (2 - 4.3)(-2.436)}{(1.427)^2 + (2.109)^2 + \cdots + (-2.436)^2}$$

$$= 0.493$$

By similar computations, we would find that the analogous partial regression coefficient for X_2 is $b_2 = 0.484$.

As it turns out, each of these *is* a partial regression coefficient. That is, a partial regression coefficient is a *simple regression coefficient computed between Y and a set of X_j-residuals after the linear association of all other predictors with X_j has been removed.*

This is what is meant by the familiar phrase "holding the other predictors constant." In the more general case, residuals are found by regressing the particular X_j of interest on *all* of the remaining predictors via a *multiple* regression. While this procedure would, of course, be going about the calculation of partial regression coefficients the hard way, it does serve to show their conceptual linkage to simple regression coefficients.

Having obtained the partial regression coefficients $b_1 = 0.493$ and $b_2 = 0.484$, the intercept term for the multiple regression is found in the usual way as:

$$b_0 = \overline{Y} - b_1\overline{X}_1 - b_2\overline{X}_2$$

$$= 4.3 - 0.493(4.3) - 0.484(4.0)$$

$$= 0.25$$

In summary, partial regression coefficients are equivalent to simple regression coefficients that are found by regressing the original values of Y on a set of predictor-variable residuals which, in effect, hold all other predictors constant.[10] *Only in the case where the predictor variables are uncorrelated to begin with will the partial regression coefficients equal their simple regression counterparts.*

We shall note further parallels between the two-variable and multiple-variable cases when we discuss coefficients of partial and multiple correlation.

Before describing these counterpart measures, however, let us consider some geometric aspects of multiple regression.

2.3.2 The Response Surface Model

Figure 2.3 showed a set of scatter diagrams in which simple (two-variable) regression analysis was used to find a line of best fit in a two-dimensional space. Analogously, in multiple regression involving n predictor variables we can imagine fitting an n-dimensional hyperplane embedded in $n + 1$ dimensions. The $n + 1$st dimension consists of the criterion variable. Each estimated conditional mean of Y_i—denoted by $\hat{Y}_i$—is a linear function of the $X_1, X_2, \ldots, X_n$ values and, hence, lies on the hyperplane. Lack of fit is portrayed by variation in $Y_i - \hat{Y}_i$ in one dimension higher than the dimensionality of the hyperplane.

Figure 2.4 shows the case dealing with X_1 and X_2 in the sample problem. The intersection of the regression plane (response surface) with the Y axis provides the estimate b_0, the intercept term. Next, if we imagine a plane passing perpendicularly through the X_1 axis, we, in effect, hold X_1 constant; hence, b_2 represents the estimated contribution of a unit change in X_2 to a change in Y. Similar remarks pertain to the interpretation of b_1. As we already know, b_1 and b_2 are partial regression coefficients.

The response plane itself is oriented so as to minimize the sum of squared deviations between each Y_i and its counterpart value $\hat{Y}_i$ on the fitted response plane, where these deviations are taken along directions parallel to the Y axis. Similarly, we can find the sum of squared deviations about the mean of the Y_i's by imagining a plane perpendicular to the Y axis passing through the value $\overline{Y}$. Total variation in Y is thus partitioned into two parts. These are found by:

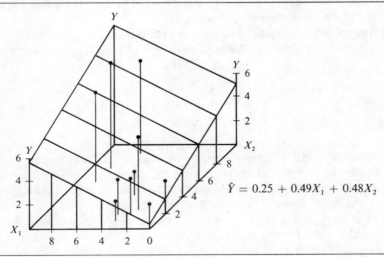

$$\hat{Y} = 0.25 + 0.49X_1 + 0.48X_2$$

Figure 2.4 Scatter Plot and Fitted Regression Plane

1. Subtracting unaccounted-for variation, involving squared deviations $(Y_i - \hat{Y}_i)^2$ about the fitted regression plane, from

2. Total variation, involving squared deviations $(Y_i - \overline{Y})^2$ from the plane imagined to be passing through $\overline{Y}$.[11]

The quantity $\sum_{i=1}^{m} (Y_i - \hat{Y}_i)^2$ represents the unaccounted-for sum of squares and the quantity $\sum_{i=1}^{m} (Y_i - \overline{Y})^2 - \sum_{i=1}^{m} (Y_i - \hat{Y}_i)^2$ represents accounted-for sum of squares. If no variation is accounted for, then we note that using $\overline{Y}$ is just as good in predicting Y_i as introducing the variables X_1 and X_2.

What happens if other functional forms are fitted? Figure 2.5 shows the case where the original linear term X_1 is replaced by a quadratic term in which the first predictor is now assumed to be X_1^2. Figure 2.6 shows the case in which we include a cross product term, X_1X_2, in addition to the two original linear terms. Notice that in these cases the character of the surface embedded in three dimensions changes to reflect the nature of the function being fitted. Notice also that b_1 and b_2 change from their numerical values in Figure 2.4 since a different model is being fitted and the predictors are correlated.

Note, however, that in all three cases (Figures 2.4 through 2.6), the regression model is *linear in the parameters*. Once we enter X_1^2 or X_1X_2 as a predictor variable, the procedure does *not* differ from what would be followed in the case of fitting the plane in Figure 2.4. Rather, it is the character of the response surface that changes. In all three illustrations, the spatial objects (the points in the space) are the observations and the dimensions of the space are represented by the variables—criterion or predictors as the case may be.

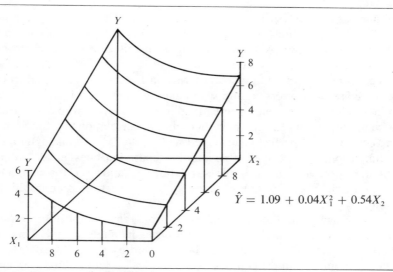

$$\hat{Y} = 1.09 + 0.04X_1^2 + 0.54X_2$$

Figure 2.5 Fitted Regression Surface Involving a Quadratic Term

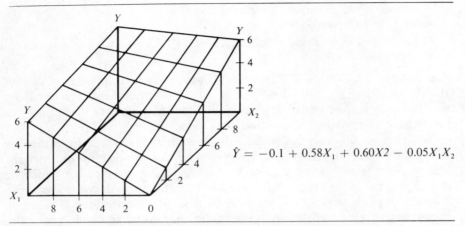

$$\hat{Y} = -0.1 + 0.58X_1 + 0.60X2 - 0.05X_1X_2$$

Figure 2.6 Fitted Regression Surface Involving an Interaction Term

2.3.3 The Vector Model

The foregoing representation of the ten observations in variable space is often called a *point* representation. In this interpretation we imagine that the ten observations are represented by points in three dimensions, where each variable Y, X_1, or X_2 denotes a dimension. Alternatively, we can imagine that each of the ten observations represents a dimension and each of the variables constitutes a *vector* in this ten-dimensional space. However, the three vectors will not span the whole ten-dimensional space but, rather, will lie in (at most) a three-dimensional space.[12]

We may also remember from matrix algebra that if the vectors are assumed to be of unit length, the (product moment) correlation between each pair of vectors is given by the cosine of their angle. In this case we have three two-variable correlations: r_{y1}, r_{y2}, and r_{12}.

This concept is pictured, in general terms, in Figure 2.7. As noted from the figure there are two vectors, x_1 and x_2 emanating from the origin. Each is a 10-component vector of unit length embedded in the observation space. The cosine of the angle separating x_1 and x_2 is their simple correlation r_{12}. Since the criterion vector y is not perfectly correlated with x_1 and x_2, it must extend into a third dimension. (The cosines of its angular separation between x_1 and x_2 are measured, respectively, by its simple correlations, r_{y1} and r_{y2}.) However, one can project y onto the plane formed by x_1 and x_2. The projection of y onto this plane is denoted by $\hat{y}$.

From this viewpoint, the idea behind multiple regression is to find the particular vector in the x_1, x_2 plane that minimizes its angle with y. This vector will be the projection $\hat{y}$ of y onto the plane formed by x_1, x_2. Since any vector in the x_1, x_2 plane is a linear composite of x_1 and x_2, it follows that we want the $\hat{y} = b_1x_1 + b_2x_2$ that minimizes the angle (or maximizes the cosine of the angle, or correlation) with **y**. The cosine of this angle θ (see Figure 2.7) is R, the multiple correlation. The problem, then, is to find a set of b_j's that define a linear composite of the vectors $\mathbf{x}_1$ and $\mathbf{x}_2$ maximizing the cosine R. One achieves the same result by finding the b_1 and b_2 that minimize the squared distance from the terminus of **y** to its projection $\hat{\mathbf{y}}$ into the X_1, X_2 plane and, hence, we have, all over again, the job of finding the b_j that minimize:

$$\sum_{i=1}^{m} (Y_i - \hat{Y}_i)^2 = \sum_{i=1}^{m} (Y_i - b_1 X_{i1} - b_2 X_{i2})^2$$

which leads to the least squares equations described in Appendix B. Since all variables are measured in deviation-from-mean form, the intercept b_0 is zero in this representation.

In general, the $\mathbf{x}_1$, $\mathbf{x}_2$ axes will be oblique, as noted in Figure 2.7. The linear composite of $\mathbf{x}_1$ and $\mathbf{x}_2$ that results in the prediction vector $\hat{\mathbf{y}}$ involves combining oblique axes via b_1 and b_2 in the vector model.

In summary, in the vector model the multiple correlation coefficient R is the *cosine of the angle θ made by* **y** *and* $\hat{\mathbf{y}}$. The b_j's are analogous to partial regression coefficients and represent coordinates of $\hat{\mathbf{y}}$ in the oblique space of the predictor variables. If more than two predictors are involved, the same geometric reasoning applies, although in this case the predictors are represented by higher-dimensional hyperplanes.

2.3.4 Coefficients of Multiple Determination and Correlation

In section 2.2.3 we discussed the interpretation of r^2, the simple coefficient of determination or squared correlation between two variables, such as Y and X_1.

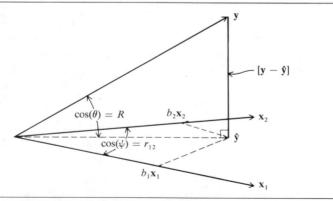

Figure 2.7 Geometric Relationships in Vector Space

As recalled, r^2 varies between 0 and 1 and represents the proportion of total variation in Y, as measured about $\overline{Y}$, that is accounted for by variation in X_1.

Analogously, we can define R^2, the *coefficient of multiple determination* between Y and a set of predictor variables, $X_1, X_2, \ldots, X_n$. R^2 also varies between 0 and 1 and measures the proportion of variation in Y that is accounted for by the best linear composite of $X_1, X_2, \ldots, X_n$ that leads to the set of predicted values $\hat{Y}$.

One way of defining R^2 is in terms of accounted for variation in Y. We can let the variable $\hat{Y}$ denote predicted values of Y from the *multiple* regression. Then we have:

$$R^2_{y\hat{y}} = 1 - \frac{\sum\limits_{i=1}^{m}(Y_i - \hat{Y}_i)^2}{\sum\limits_{i=1}^{m}(Y_i - \overline{Y})^2}$$

just as was the case for the simple r^2, described in section 2.2.3. The square root $R_{y\hat{y}}$ is called the *coefficient of multiple correlation*.

However, we can also calculate R (and R^2) from procedures already introduced in the context of simple product moment correlation as shown in Table 1.3 of Chapter 1. First, we recall the multiple regression equation (that was plotted as a plane in Figure 2.4) of the sample problem:

$$\hat{Y} = 0.25 + 0.493X_1 + 0.484X_2$$

R, the coefficient of multiple correlation, is the *simple correlation* between Y and $\hat{Y}$. From Table 1.3 we recall the product moment correlation formula:

$$R_{y\hat{y}} = \frac{\text{cov}(Y, \hat{Y})}{s_y \cdot s_{\hat{y}}}$$

Table 2.4 shows the set of $\hat{Y}_i$ and the residuals $Y_i - \hat{Y}_i$ for the sample problem. $R_{y\hat{y}}$ and $R^2_{y\hat{y}}$ are then computed quite straightforwardly by either of the two approaches above to get:

$$R_{y\hat{y}} = 0.885; \quad R^2_{y\hat{y}} = 0.783$$

2.3.5 Partial and Part Correlation Measures

In multiple regression, we often wish to find the simple correlation between a criterion variable Y that has been purged of its linear association with all predictors except X_j and that same predictor X_j, similarly purged of its linear association with all other predictors. The *partial* correlation measure fills this need.

Table 2.4 Calculation of the Coefficients of Multiple Determination and Correlation

Case	Y	$\hat{Y}$	$Y - \hat{Y}$	Case	Y	$\hat{Y}$	$Y - \hat{Y}$
a	3	3.185	−0.185	f	2	4.152	−2.152
b	7	8.066	−1.066	g	8	8.048	−0.048
c	2	2.209	−0.209	h	3	2.692	0.308
d	1	1.707	−0.707	i	9	7.574	1.426
e	6	3.176	2.824	j	2	2.191	−0.191

Coefficients of Multiple Determination and Correlation

$$R^2_{y\hat{y}} = 1 - \frac{\sum\limits_{i=1}^{m} (Y_i - \hat{Y}_i)^2}{\sum\limits_{i=1}^{m} (Y_i - \overline{Y})^2} = 1 - \frac{16.49}{76.10} = 0.783; \ R = 0.885$$

$$R_{y\hat{y}} = \frac{\text{cov}(Y, \hat{Y})}{s_y \cdot s_{\hat{y}}} = \frac{6.623}{2.908(2.574)} = 0.885$$

The partial correlation can be defined as the *simple correlation* between a set of Y-residuals and a set of X_j-residuals, where *each is an orthogonal complement with respect to the remaining predictors.*

A less well-known measure of association is the coefficient of *part* correlation. In this case we find the *simple correlation* between a Y that has *not* been purged of linear association with the remaining predictors and the residuals of some predictor X_j that has been regressed on all other predictors.

If we let $k, l \ . \ . \ .$ denote the partialed out predictors, then we can use the notation:

$r_{yj \cdot k, l, \ldots}$: coefficient of partial correlation

$r_{yj(k, l, \ldots)}$: coefficient of part correlation

Tables 2.2 and 2.3 provide the appropriate sets of residuals for finding:

$$r_{y1 \cdot 2}; \ r_{y2 \cdot 1}; \ r_{y1(2)}; \ r_{y2(1)}$$

via the usual product moment formulas of Table 1.3. However, a general computing procedure, which relates the partial and part correlation coefficients to the partial regression coefficient will serve our purposes even better.[13] This is described in the next section of the chapter.

2.3.6 Recapitulation

At this point we have informally described the following measures of association:

1. The coefficient of multiple correlation, R, which is the simple correlation between Y and the best linear composite $\hat{Y}$ of the predictors.

2. The coefficient of multiple determination, R^2, which can be interpreted as measuring the variance in Y accounted for by the best linear composite of the predictors.

3. The partial regression coefficients, b_j, which are actually simple regression coefficients computed between a set of original values of Y and a set of residuals involving X_j (following the removal of its linear association with all other predictors).

4. The partial correlation coefficients, $r_{yj \cdot k, \ell, \ldots}$ which are actually simple correlations between two sets of residuals with the linear association of all other predictors partialed out of *both* Y and X_j.

5. The part correlation coefficients, $r_{yj(k, \ell, \ldots)}$, which are simple correlations between the *original* Y variable and an X_j in which its linear association with all other predictors has been partialed out.

The last three measures—partial regression coefficients, partial correlation coefficients, and part correlation coefficients—exhibit interesting interrelationships which help to show their basic unity.

The starting point for showing the connections among the coefficients is the partial regression coefficient b_j. As we know from preceding discussion, b_j measures the change in Y per unit change in X_j, when each is expressed in terms of its respective *original units*. (Correlation measures, on the other hand, are dimensionless.) First, if we want to transform b_j into a standardized form $b_j{}^*$, multiplication by a simple ratio of standard deviations does the trick:

$$b_j{}^* = b_j \frac{s_j}{s_y}$$

where s_j is the original standard deviation of predictor X_j and s_y is the original standard deviation of the criterion variable Y.[14]

The standardized partial regression coefficient, $b_j{}^*$, measures the change in X_j when both variables have been originally standardized to mean zero and unit standard deviation. If the standard deviation of X_j is very small relative to that of X_k, this difference is adjusted in conversion of the b_j's to the beta weights, $b_j{}^*$'s. In so doing the $b_j{}^*$'s can be readily compared across predictors.

Panel I of Table 2.5 illustrates the computation of $b_1{}^*$ in the context of the sample problem. As recalled from Figure 2.4, the partial regression coefficient of X_1 is 0.493. When this is multiplied by the ratio of standard deviations s_1/s_y (from Table 2.1), we have $b_1{}^* = 0.473$.

Next, let us consider how the partial correlation $r_{yj \cdot k, \ell, \ldots}$ could be expressed in the same general context. The transformation is as follows:

$$r_{yj \cdot k, \ell, \ldots} = b_j \frac{s_j \sqrt{1 - R_{jj}^2}}{s_y \sqrt{1 - R_{yy}^2}}$$

In this case the adjustment ratio consists of two standard deviations, each being the standard deviation of the set of *residuals* found after regressing Y

Table 2.5 Illustrative Calculations of Beta, Partial, and Part Correlation Coefficients

I. Beta Weight	II. Partial Correlation
$b_1{}^* = b_1 \dfrac{s_1}{s_y}$	$r_{y1 \cdot 2} = b_1 \dfrac{s_1 \sqrt{1 - r_{12}^2}}{s_y \sqrt{1 - r_{y2}^2}}$
$= 0.493 \left\lvert \dfrac{2.791}{2.908} \right\rvert$	$= 0.493 \dfrac{(2.791) \sqrt{1 - 0.702}}{(2.908) \sqrt{1 - 0.716}}$
$= 0.473$	$= 0.485$

III. Part Correlation Involving
Residuals of X_1 Only

$$r_{y1(2)} = \frac{b_1 s_1 \sqrt{1 - r_{12}^2}}{s_y}$$

$$= \frac{0.493(2.791) \sqrt{1 - 0.702}}{2.908}$$

$$= 0.259$$

and X_j separately on the remaining $n - 1$ predictor variables. In each case the quantities:

$$\sqrt{1 - R_{jj}^2}; \ \sqrt{1 - R_{y\hat{y}}^2}$$

denote the factor that converts s_j or s_y, respectively, to its standard deviation of residuals.

Table 2.5 shows an illustration of the procedure, based on computing the standard deviation of the columns of residuals, $Y_{r(2)}$ and $X_{r1(2)}$ in Table 2.3.[15] As intuition would suggest, higher R^2's lead to smaller scale factors by which s_j (or s_y) are multiplied; in the limiting case, an R^2 of 1 leads to a standard deviation of residuals of 0, suggesting that X_j is redundant with the $n - 1$ remaining predictors.

The coefficient of part correlation $r_{yj(k,\ell,\ldots)}$ can be computed by a similar type of transformation of the partial regression coefficient:

$$\boxed{\ r_{yj(k,\ell,\ldots)} = \frac{b_j s_j \sqrt{1 - R_{jj}^2}}{s_y}\ }$$

Notice, in this case, that the standard deviation of Y is not altered since, as described earlier, the part correlation represents the simple correlation between one variable (the criterion Y) expressed in original terms and a predictor X_j, expressed as a set of residuals found by regressing X_j on the remaining $n - 1$ predictors.[16]

In brief, beta weights and partial and part correlation coefficients are simply different types of *rescalings of the partial regression coefficient*. And this, in turn, is a simple regression between Y and a set of X_j-residuals obtained after regressing X_j on all of the remaining predictors.

2.4 A Matrix Formulation of Multiple Regression

In the preceding sections we briefly described simple (two-variable) linear regression and multiple regression. Both are illustrations of the general linear model. Not surprisingly, then, we shall be covering some of the same ground covered earlier but this time in the broader context of the general linear model.

2.4.1 The General Linear Model

Since matrix notation is used in various sections of this chapter, at the outset we shall formulate the general linear model in matrix algebra terms. We can let **y** denote an $m \times 1$ (column) vector of criterion-variable values with general entry Y_i ($i = 1, 2, \ldots, m$). The predictor values are represented by an $m \times (n + 1)$ matrix **X** denoting values X_{ij} of observation i on each of the j predictors ($j = 1, 2, \ldots, n$). (The first column of **X** consists of unities and is used for finding β_0, the intercept term.)

We let **β** denote the $(n + 1) \times 1$ vector of regression parameters whose entries are the intercept, β_0 and the partial regression coefficients, β_j ($j = 1, 2, \ldots, n$). The $m \times 1$ vector **ε** denotes the error term. Given these preliminaries, the general linear model can be compactly expressed as:

$$\mathbf{y} = \mathbf{X}\boldsymbol{\beta} + \boldsymbol{\epsilon}$$

This equation, in turn, can be written more explicitly as:

$$
\underset{\mathbf{y}}{\begin{bmatrix} Y_1 \\ Y_2 \\ \vdots \\ Y_m \end{bmatrix}}
=
\underset{\mathbf{X}}{\begin{bmatrix} X_{10} & X_{11} & \cdots & X_{1n} \\ X_{20} & X_{21} & \cdots & X_{2n} \\ \vdots & \vdots & & \vdots \\ X_{m0} & X_{m1} & \cdots & X_{mn} \end{bmatrix}}
\underset{\boldsymbol{\beta}}{\begin{bmatrix} \beta_0 \\ \beta_1 \\ \vdots \\ \beta_n \end{bmatrix}}
+
\underset{\boldsymbol{\epsilon}}{\begin{bmatrix} \epsilon_1 \\ \epsilon_2 \\ \vdots \\ \epsilon_m \end{bmatrix}}
$$

Note that the criterion vector **y** is of order $m \times 1$. The first (by convention) column in **X**, the $m \times (1 + n)$ matrix of predictors, consists of a vector of unities. That is:

$$
\begin{bmatrix} X_{10} \\ X_{20} \\ \vdots \\ X_{m0} \end{bmatrix}
=
\begin{bmatrix} 1 \\ 1 \\ \vdots \\ 1 \end{bmatrix}
$$

which, as will be shown later, is used as a computational device to find the intercept term β_0. The $(n + 1) \times 1$ vector β denotes the vector of parameters, and the $m \times 1$ vector ϵ denotes the error term.

In section 2.2 we described the matrix X as consisting of a set of fixed (nonrandom) constants. We also talked about the Y_i's as being a set of criterion-variable values, conditioned by a particular realization of the X_j's for the ith observation. It is now time to make these comments more precise by discussing two major variants of the general linear model.

The *fixed* version of the general linear model interprets the X matrix as a set of known constants. The error term ϵ has m elements consisting of measurement error and other variables that could affect y but are not included in X. This is the "classical" version of the general linear model that we have considered up to this point.

Now, however, suppose we relax the assumption that the X matrix consists of fixed values. If so, an equally acceptable variant of the general linear model is the *random* model. In this version, the Y's *and* the X_j's are all random variables. However, once a particular realization of the X_j's occurs, for some case i, that realization is held constant and Y_i is observed *conditionally* on this fixed realization.

The critical assumption of the random version of the general linear model is that the multivariate distribution of the X_j's of each observation i *does not involve the parameters β or ϵ*. This same assumption is, of course, implied by the fixed version since the X_{ij}'s are not random to begin with. In particular, the vector of parameters β is assumed to be nonrandom in both versions of the general linear model.

In brief, the fixed and random versions are identical except for the possibility of allowing the X_{ij}'s to be realizations of a set of random variables (in the case of the random version). However, by requiring y to be conditioned by a particular realization of X, the estimators for both versions turn out to be exactly the same. This can be summarized by saying that in both models:

1. The error term ϵ does not depend on X.

2. The vector of parameters β is nonrandom and X does not depend on β.

3. The relationship is linear in the parameters as, of course, is implied by the matrix equation $y = X\beta + \epsilon$.[17]

In addition to these general assumptions, both versions place additional restrictions on the expectation (mean) and variance of ϵ.

$$E(\epsilon) = \mathbf{0}_{m \times 1} \qquad\qquad E(y|X) = X\beta$$
$$\text{or equivalently,}$$
$$\text{cov}(\epsilon) = \epsilon\epsilon' = \sigma^2 I_{m \times m} \qquad \text{cov}(y|X) = \sigma^2 I_{m \times m}$$

A particularly interesting aspect of both versions of the general linear model is that no distributional assumptions have been made, as yet, about the error term ϵ. An important theorem, known as the *Gauss-Markoff theorem,* does not depend on the specific statistical distribution of ϵ. The Gauss-Markoff theorem refers to the *least squares estimators* b_j of the regression parameters β_j. According to this theorem, the vector of least squares estimators $\mathbf{b}$ (of the parameters $\boldsymbol{\beta}$):

1. Is unbiased, in the sense that for each of its entries $E(b_j) = \beta_j$; that is, the expected value of b_j is the parameter β_j that it estimates.

2. Exhibits the least variance of any linear unbiased estimator of β_j that can be obtained from the sample data.

Thus, regardless of the distributional form of the error vector ϵ, the estimator $\mathbf{b}$ provided by least squares procedures is the best (smallest variance) linear unbiased estimator of $\boldsymbol{\beta}$.

As before, we shall often assume that ϵ is normally distributed in order to make various significance tests. However, this assumption is not needed for the Gauss-Markoff theorem to hold.

As long as we assume that $\boldsymbol{\beta}$ is a nonrandom vector of parameters to be estimated by $\mathbf{b}$, we have considerable flexibility in choosing the matrix of predictor variables $\mathbf{X}$. In particular, $\mathbf{X}$ *can consist in whole or in part of dummy variables.* In Chapter 1 we described the basic ideas underlying the coding of k mutually exclusive and collectively exhaustive classes into $k - 1$ dummy variables. (This notion is taken up later in the chapter when we analyze some of the T.V. commercial data. It also appears in Chapter 3's discussion of analysis of variance and covariance.)

Historically, the fixed regression model has dominated the scene and that is why this version was introduced in section 2.2. We shall continue to emphasize the fixed model although it is appropriate to point out that in some applied fields such as econometrics (Johnston, 1972), the random model is receiving quite a bit of attention. Fortunately, the estimators of $\boldsymbol{\beta}$ and $\sigma^2(\epsilon)$ are the same in both versions of the general linear model.

2.4.2 Parameter Estimation

Having formulated the general linear model in the preceding section, we now consider the estimation of $\boldsymbol{\beta}$, the vector consisting of the intercept and the n partial regression coefficients. Continuing to use the same general notation as employed in scalar form in section 2.2, we have:

$$
\begin{matrix} \mathbf{y} \\ \begin{bmatrix} Y_1 \\ Y_2 \\ \vdots \\ Y_m \end{bmatrix} \end{matrix} = \begin{matrix} \mathbf{X} \\ \begin{bmatrix} 1 & X_{11} & \cdots & X_{1n} \\ 1 & X_{21} & \cdots & X_{2n} \\ \vdots & \vdots & & \vdots \\ 1 & X_{m1} & \cdots & X_{mn} \end{bmatrix} \end{matrix} \begin{matrix} \mathbf{b} \\ \begin{bmatrix} b_0 \\ b_1 \\ \vdots \\ b_n \end{bmatrix} \end{matrix} + \begin{matrix} \mathbf{e} \\ \begin{bmatrix} e_1 \\ e_2 \\ \vdots \\ e_m \end{bmatrix} \end{matrix}
$$

In terms of the fixed version of the general linear model, we imagine some specified set of X_{ij}'s. For each row of $\mathbf{X}$ a random observation Y_i is drawn that is conditioned by the particular X_j's in that row. We then wish to estimate β under the assumptions of the general linear model.

The estimating equation (to be found from least squares) is:

$$
\mathbf{\hat{y}} =
\begin{bmatrix} \hat{Y}_1 \\ \hat{Y}_2 \\ \vdots \\ \hat{Y}_m \end{bmatrix}
=
\overset{\mathbf{X}}{\begin{bmatrix} 1 & X_{11} & \cdots & X_{1n} \\ 1 & X_{21} & \cdots & X_{2n} \\ \vdots & \vdots & & \vdots \\ 1 & X_{m1} & \cdots & X_{mn} \end{bmatrix}}
\overset{\mathbf{b}}{\begin{bmatrix} b_0 \\ b_1 \\ \vdots \\ b_n \end{bmatrix}}
$$

The vector of fitted values $\mathbf{\hat{y}}$ can be subtracted from the vector of original values $\mathbf{y}$ to get the error vector $\mathbf{e}$:

$$
\mathbf{e} =
\begin{bmatrix} e_1 \\ e_2 \\ \vdots \\ e_m \end{bmatrix}
=
\overset{\mathbf{y}}{\begin{bmatrix} Y_1 \\ Y_2 \\ \vdots \\ Y_m \end{bmatrix}}
-
\overset{\mathbf{\hat{y}}}{\begin{bmatrix} \hat{Y}_1 \\ \hat{Y}_2 \\ \vdots \\ \hat{Y}_m \end{bmatrix}}
$$

The problem now is to find the vector $\mathbf{b}$ that minimizes the sum of squared deviations:

$$
\sum_{i=1}^{m} e_i^2 = \sum_{i=1}^{m} (Y_i - \hat{Y}_i)^2
$$

But this expression, in turn, can be written in matrix algebra format as the scalar product:

$$
\mathbf{e'e} = (\mathbf{y} - \mathbf{Xb})'(\mathbf{y} - \mathbf{Xb})
$$

The vector $\mathbf{b}$ that minimizes the scalar product $\mathbf{e'e}$ is derived in Appendix B and turns out to be:

$$
\mathbf{b} = (\mathbf{X'X})^{-1}\,\mathbf{X'y}
$$

Thus, to find **b** all we need to do is find the inverse of **X'X** and postmultiply this inverse by the product **X'y**. We recall that the first column of **X** consists of unities. Accordingly, in the sample problem of Table 2.1, we have:

$$\mathbf{X} = \begin{array}{c} X_0 \ X_1 \ X_2 \\ \begin{bmatrix} 1 & 4 & 2 \\ 1 & 9 & 7 \\ \vdots & \vdots & \vdots \\ 1 & 1 & 3 \end{bmatrix} \end{array}$$

so that the cross product matrix in this case is:

$$\mathbf{X'X} = \overset{\mathbf{X'}}{\begin{bmatrix} 1 & 1 & \ldots & 1 \\ 4 & 9 & \ldots & 1 \\ 2 & 7 & \ldots & 3 \end{bmatrix}} \overset{\mathbf{X}}{\begin{bmatrix} 1 & 4 & 2 \\ 1 & 9 & 7 \\ \vdots & \vdots & \vdots \\ 1 & 1 & 3 \end{bmatrix}}$$

$$= \begin{bmatrix} 10 & 43 & 40 \\ 43 & 255 & 229 \\ 40 & 229 & 226 \end{bmatrix} = \begin{bmatrix} m & \sum X_1 & \sum X_2 \\ \sum X_1 & \sum X_1{}^2 & \sum X_1 X_2 \\ \sum X_2 & \sum X_1 X_2 & \sum X_2{}^2 \end{bmatrix}$$

Note that the first row (and column) of **X'X** consists, respectively, of m, the sample size, the sum of X_1, and the sum of X_2.

The inverse of **X'X** is found from standard methods (see Appendix B):

$$(\mathbf{X'X})^{-1} = \begin{bmatrix} 0.377 & -0.041 & -0.026 \\ -0.041 & 0.048 & -0.041 \\ -0.026 & -0.041 & 0.051 \end{bmatrix}$$

while the product **X'y** is:

$$\mathbf{X'y} = \overset{\mathbf{X'}}{\begin{bmatrix} 1 & 1 & \ldots & 1 \\ 4 & 9 & \ldots & 1 \\ 2 & 7 & \ldots & 3 \end{bmatrix}} \overset{\mathbf{y}}{\begin{bmatrix} 3 \\ 7 \\ \vdots \\ 2 \end{bmatrix}} = \begin{bmatrix} 43 \\ 247 \\ 232 \end{bmatrix}$$

so that:

$$\mathbf{b} = \begin{bmatrix} 0.247 \\ 0.493 \\ 0.484 \end{bmatrix} = \overset{(\mathbf{X'X})^{-1}}{\begin{bmatrix} 0.377 & -0.041 & -0.026 \\ -0.041 & 0.048 & -0.041 \\ -0.026 & -0.041 & 0.051 \end{bmatrix}} \overset{\mathbf{X'y}}{\begin{bmatrix} 43 \\ 247 \\ 232 \end{bmatrix}}$$

as found earlier.

In brief, the primary computation involved in solving for **b** is the inverse of **X'X**. A variety of procedures can be used to find the (regular) inverse of **X'X**. Typically, however, $(X'X)^{-1}$ will be found from computer packaged routines.

Other computational procedures for finding **b** are available, however, and we next turn to a discussion of these alternatives.

2.5 Computing Formulas for Multiple Regression

The preceding solution $\mathbf{b} = (X'X)^{-1} X'y$, where **X** is understood to be of order $m \times (n + 1)$, of the matrix equation is probably the most direct and transparent way to solve for the vector **b**. However, other procedures for finding the vector of regression parameters exist. One procedure utilizes the correlation matrix of predictors while another employs the covariance matrix. In either case one less normal equation is involved since the first column of unities in **X** is eliminated.

To illustrate, suppose that we were to consider the sample problem of Table 2.1 in terms of the *standardized* data: Y_s, X_{s1} and X_{s2}. If we do this, the estimating equation becomes:

$$\mathbf{b^*} = (X'_s X_s)^{-1} X'_s y_s$$

However, since each variable is standardized to zero mean and unit standard deviation, we have:

$$\mathbf{R} = \frac{1}{m-1}(X'_s X_s); \ \mathbf{r(y)} = \frac{1}{m-1} X'_s y_s$$

where **R** denotes the correlation matrix of predictors and **r(y)** denotes a vector of simple correlations between each predictor and the criterion variable.

In this case we can solve for a standardized form of **b** without the need for an intercept term. This standardized form is the vector of beta weights, **b***. The least squares solution assumes a particularly simple form:

$$\mathbf{b^*} = R^{-1} \mathbf{r(y)}$$

where R^{-1} denotes the inverse of the correlation matrix of n predictor variables and the vector **r(y)** denotes the simple correlations r_{yj} between the criterion and each of the n predictors, in turn.[18]

The squared multiple correlation coefficient is also found quite easily from the matrix equation:

$$R^2 = \mathbf{r'(y)} R^{-1} \mathbf{r(y)}$$

And, since $\mathbf{b}^*$ is computed from $\mathbf{R}^{-1}\mathbf{r}(\mathbf{y})$ we can, alternatively, express R^2 (in scalar notation) as:

$$R^2 = r_{y1}b_1^* + r_{y2}b_2^* + \cdots + r_{yn}b_n^*$$

where: $r_{y1}, r_{y2}, \ldots, r_{yn}$ are the simple correlations of the criterion with each of the n predictors.

Often the researcher will wish to express the multiple regression equation in terms of original variable scores. It is a rather simple matter to convert the b_j^*'s into b_j's, the raw score partial regression coefficients, and then to solve for the intercept term b_0. To show this we first write the regression equation in terms of the beta weights, as applied to standardized variables:

$$\frac{\hat{Y}_i - \overline{Y}}{s_y} = b_1^* \frac{(X_{i1} - \overline{X}_1)}{s_1} + b_2^* \frac{(X_{i2} - \overline{X}_2)}{s_2} + \cdots + b_n^* \frac{(X_{in} - \overline{X}_n)}{s_n}$$

Next, we multiply both sides of the preceding equation by s_y, the standard deviation of the criterion variable. This has the effect of transforming the b_j^*'s to raw score form (since $b_j = b_j^* s_y / s_j$):[19]

$$\hat{Y}_i - \overline{Y} = b_1^* \frac{s_y}{s_1}(X_{i1} - \overline{X}_1) + b_2^* \frac{s_y}{s_2}(X_{i2} - \overline{X}_2) + \cdots + b_n^* \frac{s_y}{s_n}(X_{in} - \overline{X}_n)$$

$$= b_1(X_{i1} - \overline{X}_1) + b_2(X_{i2} - \overline{X}_2) + \cdots + b_n(X_{in} - \overline{X}_n)$$

Finally, since $\overline{Y} = b_0 + b_1\overline{X}_1 + b_2\overline{X}_2 + \cdots + b_n\overline{X}_n$, we can add $\overline{Y}$ to both sides of the above equation and get:

$$\hat{Y}_i = b_0 + b_1 X_{i1} + b_2 X_{i2} + \cdots + b_n X_{in}$$

as desired. Thus, by means of simple algebra we can easily work from the solution utilizing beta weights to the raw score format. The relationship of b_j to b_j^* depends on the ratio of the standard deviation of the criterion to the standard deviation of the j-th predictor. In the case of a preliminary standardization of variables, these ratios, of course, are each equal to 1 and $b_j^* = b_j$.

2.5.1 Sample Calculations

Let us now return to the hypothetical data of Table 2.1. In analyzing these data via the correlation matrix approach, we first ask:

1. What is the estimating equation for predicting Y from the best linear composite of X_1 and X_2?

2. What proportion of variation in Y is accounted for by variation in X_1 and X_2?

These questions have already been answered in our preliminary (and more intuitive) discussion of multiple regression. Now, however, we apply the computing approach based on **R**, the correlation matrix. Results are shown in Table 2.6. As noted, we first compute the regression vector **b*** (the beta coefficients).[20] R^2 is next computed as the scalar product of **r(y)** and **b***. Then the standard deviations of each of the variables are used to solve for the b_j's. Finally, b_0 is found from the means of the variables and the preceding b_j's.

From the value of R^2 we verify earlier findings that approximately 78 per cent of the variation in Y is accounted for by variation in X_1 and X_2. The partial regression coefficients, b_1 and b_2, are 0.493 and 0.484, respectively, and b_0, the intercept, is 0.25.

Once the regression equation and the value of R^2 are computed, the researcher may wish to compute other quantities and run various significance tests. In particular, the following questions may be raised:

1. Is the regression equation as a whole significantly different from an equation with both partial regression coefficients equal to zero?

2. Does each predictor separately account for statistically significant variation in Y?

3. What does each predictor contribute in terms of total accounted-for variation in Y?

In sections 2.3 and 2.4 we considered these questions on an informal basis. We now work through the same set of calculations in terms of the matrix algebra

Table 2.6 Regression Calculations from Sample Data of Table 2.1

Correlations		Computing the **b*** Vector	

$$\mathbf{R} = \begin{array}{c} \\ X_1 \\ X_2 \end{array} \begin{array}{cc} X_1 & X_2 \\ \begin{bmatrix} 1.000 & 0.838 \\ 0.838 & 1.000 \end{bmatrix} \end{array}; \quad \mathbf{r(y)} = \begin{array}{c} r_{y1} \\ r_{y2} \end{array} \begin{bmatrix} 0.850 \\ 0.847 \end{bmatrix};$$

$$\mathbf{R}^{-1} \quad\quad \mathbf{r(y)} \quad\quad \mathbf{b^*}$$
$$\begin{bmatrix} 3.358 & -2.814 \\ -2.814 & 3.358 \end{bmatrix} \begin{bmatrix} 0.850 \\ 0.847 \end{bmatrix} = \begin{bmatrix} 0.471 \\ 0.452 \end{bmatrix}$$

$$R^2 = (0.850 \quad 0.847) \begin{bmatrix} 0.471 \\ 0.452 \end{bmatrix} = 0.783$$

Computing the b_j Coefficients

$$\overline{Y} = 4.3; s_y = 2.908 \qquad b_1 = \frac{0.471(2.908)}{2.791} = 0.493$$

$$\overline{X}_1 = 4.3; s_1 = 2.791 \qquad b_2 = \frac{0.452(2.908)}{2.708} = 0.484$$

$$\overline{X}_2 = 4.0; s_2 = 2.708 \qquad b_0 = 4.3 - 0.493(4.3) - 0.484(4.0)$$
$$= 0.25$$

Estimating Equation (Original Units)

$$\hat{Y} = 0.25 + 0.493X_1 + 0.484X_2$$

approach. Since tests of significance are involved, we now assume that the error term ϵ_i is normally distributed (in addition to the other assumptions that are part of the general linear model).

2.5.2 Testing the Significance of the Overall Equation

The first question of interest concerns the significance of the overall equation:

$$\hat{Y} = 0.25 + 0.493X_1 + 0.484X_2$$

As we indicated for the two-variable case, there are two ways of expressing the null hypothesis:

$$H_0: \beta_1 = \beta_2 = 0$$
$$H_0: R_p^2 = 0$$

where R_p^2 denotes the population squared multiple correlation.

The procedure for conducting the test of either hypothesis is the same and involves computation of:

1. Sum of squares due to regression.
2. Residual sum of squares.
3. Total sum of squares to be accounted for in the criterion variable.

One then constructs the following type of table:

Source	Sums of Squares	Degrees of Freedom	Mean Squares	F Ratio
Due to Regression	R^2SST	n	R^2SST$/n$	$\dfrac{R^2(m - n - 1)}{(1 - R^2)n}$
Residual	$(1 - R^2)$SST	$m - n - 1$	$[(1 - R^2)$SST$]/(m - n - 1)$	
Total	SST	$m - 1$		

The quantity SST represents the total sum of squares (expressed in terms of squared deviations from $\overline{Y}$) to be accounted for. This, in turn, can be partitioned into two additive components: (a) that due to regression, R^2SST and (b) the residual, $(1 - R^2)$SST, which was earlier denoted as SSE. The mean squares for each are found by dividing the appropriate sum of squares by its degrees of freedom. The F ratio represents the ratio of these two mean squares. (The common term SST drops out in the computation of the F ratio.)

As reviewed in Appendix B, the F distribution is the distribution followed by the ratio of two independent, unbiased estimates of the same normal population variance σ^2. If R_p^2 is zero, then the sample R^2 reflects only sampling error and the F ratio will tend to be equal to unity.

Most regression programs routinely go through the preceding calculations. To illustrate, Table 2.7 shows an analysis of the data of Table 2.1, as obtained from BMD-03R, a commonly used regression program in the University of California Biomedical Series (Dixon, 1973). We note that the (sample) R^2 is 0.783 and, hence $1 - R^2$ is 0.217. The sum of squares to be accounted for, as measured about $\overline{Y}$, is 76.1. This total is partitioned into the quantities:

Due to Regression (SSR)	$R^2\text{SST} =$	59.61
Residual (SSE)	$(1 - R^2)\text{SST} =$	16.49
Total	SST	76.10

where SSR denotes sum of squares for regression and SSE denotes sum of squares for residual, or error. The mean squares and F value are:

$$\text{MSR} = \frac{59.61}{2} = 29.805; \quad F = 12.652$$

$$\text{MSE} = \frac{16.49}{7} = 2.356$$

The tabular F ($\alpha = 0.05$), with 2 and 7 degrees of freedom, is only 4.74; hence, we reject the null hypothesis that $\beta_1 = \beta_2 = 0$ (or, equivalently, that $R_p^2 = 0$).[21]

In addition to the ANOVA table in Table 2.7, most regression programs compute the variance and standard error of the estimate (see Table 2.7). The former is simply the residual sum of squares, $(1 - R^2)\text{SST}$, divided by degrees of freedom; the latter is the square root of this quantity.

One other quantity in Table 2.7 that is of general interest is the adjusted R^2, computed as:

$$\boxed{R^2 \text{ (adjusted)} = 1 - (1 - R^2)\frac{m - 1}{m - n - 1}}$$

$$= 1 - \frac{[0.217(9)]}{7}$$

$$= 0.721$$

One can notice from the formula that as the number of predictor variables increases, the downward adjustment of R^2 becomes greater, holding sample size constant. Adjusted R^2 is usually computed to reflect the fact that the sample R^2 tends to capitalize on chance variation in the specific data set under analysis.[22] This transformation of R^2 is sometimes referred to as a *correction for shrinkage*.

Table 2.7 Summary Output of BMD-03R Regression Analysis of Sample Data from Table 2.1

Sample R^2	0.783	
Adjusted R^2	0.721	
Sample R (Unadjusted)	0.885	
SS Attributable to Regression	59.61	$R^2 \text{SST}$
SS of Deviations from Regression	16.49	$(1 - R^2)\text{SST}$
Variance of Estimate	2.356	$[(1 - R^2)\text{SST}]/(m - n - 1)$
Standard Error of Estimate	1.535	$\{[(1 - R^2)\text{SST}]/(m - n - 1)\}^{1/2}$
Intercept	0.247	

Analysis of Variance for the Multiple Linear Regression

Source of Variation	d.f.	SS	MS	F Ratio
Due to Regression (SSR)	2	59.61	29.805	12.652
Deviations About Regression (SSE)	7	16.49	2.356	
	9	76.10		

Additional Statistics

Variable Number	Mean	Standard Deviation	Regression Coefficient	Standard Error of Reg. Coeff.	Computed t Value	Partial Corr. Coeff.	Sum of Sq. Added	Prop. of Total Var.
X_1	4.3	2.791	0.493	0.336	1.466	0.485	55.013	0.723
X_2	4.0	2.708	0.484	0.346	1.397	0.467	4.597	0.060
Y	4.3	2.908					59.610	0.783

2.5.3 Testing the Significance
of Partial Regression Coefficients

While the preceding analysis has indicated that the overall regression is significant, it does not follow that *both* b_1 and b_2 contribute significantly to overall accounted-for variance. It may be the case that a simpler model involving only X_1 (or one involving only X_2) would be sufficient. Hence, we shall now want to examine the second question posed at the end of section 2.5.1.

Earlier, we approached the general problem of significance testing as a problem of comparing alternative models. In particular, we discussed the F ratio for testing whether the reduction in the sum of squares for error associated with a full model justified its acceptance over a restricted model. The general expression was:

$$F = \frac{\text{SSE}_r - \text{SSE}_f}{\text{SSE}_f} \cdot \frac{d_f}{d_r - d_f}$$

where SSE_r and SSE_f denote error sums of squares for restricted and full model, respectively, and d_r and d_f denote their respective degrees of freedom.

For example, in testing for the overall significance of R^2, we have:

$$F = \frac{76.10 - 16.49}{16.49} \cdot \frac{7}{2}$$

$$= 12.652$$

where SSE_r, SSE_f denote error sums of squares for the restricted and full model, in section 2.5.2. In this case 76.1 is the error sum of squares for the restricted model (in which only b_0 appears) while 16.49 is the error sum of squares for the full model that includes the parameters b_0, b_1 and b_2. The degrees of freedom for the full model are 7, since three parameters are being estimated; those for the restricted model are 9, since only one parameter (b_0) is being estimated.

It is useful to point out that the preceding models comparison formula can be also expressed in terms of R^2's.[23] The formula is:

$$F = \frac{R_f^2 - R_r^2}{1 - R_f^2} \cdot \frac{d_f}{d_r - d_f}$$

For example, in the preceding case, $R_f^2 = 0.783$ and $R_r^2 = 0$, since, in this latter case, b_1 and b_2 are both assumed to be zero. We then have:

$$F = \frac{0.783 - 0}{1 - 0.783} \cdot \frac{7}{2}$$

$$\cong 12.652$$

as desired.

The models comparison procedure can also be used to test the significance of each partial regression coefficient. For example, suppose we wished to test the statistical significance of b_2. If so, the full model contains b_0, b_1 and b_2 while the restricted model contains only b_0 and b_1. From section 2.2.3, we already know that the r^2 of Y regressed on X_1 alone is 0.723; this is the R^2 of the restricted model. From Table 2.7 we know that the R^2 based on *both* X_1 and X_2 is 0.783; this is the R^2 of the full model. With this information we set up the test:

$$F = \frac{R_f^2 - R_r^2}{1 - R_f^2} \cdot \frac{d_f}{d_r - d_f}$$

$$= \frac{0.783 - 0.723}{1 - 0.783} \cdot \frac{7}{1}$$

$$= 1.935$$

The tabular F ratio (alpha risk equal to 0.05) for 1 and 7 degrees of freedom in Table C.4 is 5.59; we conclude that b_2 is *not* significant at the 0.05 level. (We could also find the same result by employing the version of the models comparison procedure that is based on error sums of squares, as shown earlier.)

The models comparison test also gives results that are equivalent to the more traditional t test in which we first find the standard error of, say, b_2 and then compute the ratio of b_2 to its standard error. Since t tests (see Table 2.7) are part of the standard output of such packaged programs as BMD-03R, it is appropriate to show the formulas on which they are based:

$$t_j = \frac{b_j}{SE(b_j)}$$

$$SE(b_j) = \frac{s_y}{s_j} \sqrt{\frac{r^{jj}(1 - R^2)}{m - n - 1}}$$

where SE denotes the standard error and r^{jj} is the element of the j-th predictor in the matrix $\mathbf{R}^{-1}$ (see Table 2.6).

In the case of $b_2 = 0.484$, we have:

$$SE(b_2) = \frac{s_y}{s_2} \sqrt{\frac{r^{22}(1 - R^2)}{m - n - 1}}$$

$$= \frac{2.908}{2.708} \sqrt{\frac{3.358(0.217)}{7}}$$

$$= 0.346$$

$$t_2 = \frac{0.484}{0.346} = 1.397$$

$$t_2^2 = 1.952$$

As noted, $t_2 = 1.397$ is the same as that shown under the t value column in Table 2.7. Its square, $t_2^2 = 1.952$, is (within rounding error) the same value as found for F, above.

If use of the inverse element r^{jj} seems strange from an interpretation standpoint, it should be pointed out that the following relationship holds:

$$r^{jj} = \frac{1}{[1 - R^2_{j(n-1)}]}$$

where $R^2_{j(n-1)}$ is the squared multiple correlation between the j-th predictor and the remaining $n - 1$ predictors. Note, then, that as $R^2_{j(n-1)}$ approaches unity, r^{jj} gets very large. Since $SE(b_j)$ depends on (the square root of) r^{jj}, the more redundant X_j is with the remaining $n - 1$ predictors, the larger its standard error, and hence, the less reliable is its partial regression coefficient.

In the sample problem the addition of X_2 raised R^2 from 0.723 to only 0.783, not enough to be statistically significant. This is not surprising, given the fact (Table 2.3) that X_2's simple r^2 with X_1 is 0.702. That is, X_2 can be predicted reasonably well from X_1, or vice versa.

In summary, the models comparison test and the conventional t test yield *precisely the same test for the significance of b_j*, if we bear in mind that $F = t^2$ (in the case of one degree of freedom for the numerator in the F table). The comparison of models test puts into perspective the fact that what is being tested is whether the *addition* of the j-th predictor accounts for significant variation in Y when compared to a model that includes all variables *except the j-th*.[24]

It should be emphasized that separate tests of the b_j's are, in general, *not independent* since each test involves the choice between a model including all variables but the j-th (the restricted model) versus one with all variables including the j-th (the full model).

An additional value of the models-comparison procedure is its generality for testing the significance of *sets of predictors*. For example, suppose we had three rather than two predictors in our demonstration problem. Assume further that we wished to see if X_2 and X_3 added anything significant over the inclusion of only X_1. If this were the case, the comparison of models formula can be readily modified to:

$$F = \frac{R^2_{y123} - R^2_{y1}}{1 - R^2_{y123}} \cdot \frac{d_f}{d_r - d_f}$$

with degrees of freedom for the F ratio of $d_r - d_f = 2$ for numerator and $d_f = 6$ for denominator, respectively. In this case we examine the *increment in R^2* produced by the full model (Y on X_1, X_2, X_3) versus the R^2 produced by the restricted model (Y on X_1 only) relative to variance unaccounted by the full model.

2.5.4 Which Variable to Retain?

In Table 2.7 we note that the introduction of X_2 in addition to X_1 accounts for relatively little variance in Y (actually, an incremental R^2 of only 0.06). However, introducing X_1 into the regression equation before X_2 is largely an arbitrary decision. We might now ask: suppose X_2, rather than X_1, had been entered first? If such were the case, is X_1 statistically significant? From Table 2.7 the computed t value for X_1 is 1.466 (with 7 degrees of freedom). Not surprisingly, this is not significant at the 0.05 alpha level. Hence, *either* X_1 or X_2 could be dropped from the regression equation. If X_2 is entered first, followed by X_1, we would find that the only values that change in the whole set of output statistics in Table 2.7 are the last two columns. These become:

Variable No.	Sum of Squares Added	Proportion of Total Variance
X_1	5.065	0.067
X_2	54.545	0.716
	59.610	0.783

Since we already know that total R^2 is significant at the 0.05 level, but each of the individual t values is not significant, the model to be retained will involve *only one predictor*, X_1 or X_2.

Whether we elect to retain X_1 or X_2 (as long as each was significant in a bivariate regression with Y) would depend on considerations outside the regression analysis per se. The correlation between them is sufficiently high so that one or the other could be deleted.

To recapitulate, the predictors X_1 and X_2 are so highly correlated themselves that little would be lost if one of them were deleted. However, the order in which they enter the multiple regression equation does not affect such statistics as total R^2 or b_1 and b_2. What *is* affected is their respective contributions to total accounted-for sum of squares and to total R^2.

However, if either X_1 or X_2 is dropped entirely, in general it will be the case that the regression coefficient (and the intercept) for the retained predictor will change. *If predictors are correlated, the deletion of one (or more) from the regression will lead to different partial regression coefficients if the regression is recalculated on the basis of the retained predictors.* Only if the deleted predictors are uncorrelated with the retained set will we note no change in the recomputed partial regression coefficients.

2.5.5 Recapitulation

Section 2.5 has covered many of the same topics discussed on a more intuitive level in earlier sections. The major points of interest to the applied researcher are illustrated in Tables 2.5, 2.6, and 2.7.

Table 2.5 shows how beta weights, partial, and part correlation coefficients all involve simple transformations of the partial regression coefficient. This, in turn, is the simple regression coefficient that is obtained from a two-variable regression involving the original criterion variable regressed on a set of predictor-variable residuals.

Table 2.6 shows how the calculation of regression coefficients proceeds from a simple matrix equation involving the product of the inverse of **R**, the matrix of correlations among predictors, and $\mathbf{r(y)}$ a vector of simple correlations between Y and each X_j, in turn.

Table 2.7 provides a typical summary of the kind of output provided by packaged computer programs. A fair amount of space in section 2.5 was devoted to reviewing the various statistics computed by packaged programs. The understanding of these statistics is a prerequisite for general comprehension of multiple regression techniques.

In addition, the unifying idea of treating significance testing as comparisons of models received further elaboration in terms of the formula:

$$F = \frac{R_f^2 - R_r^2}{1 - R_f^2} \cdot \frac{d_f}{d_r - d_f}$$

as applied in testing for the significance of b_1 and b_2. While illustrated in the context of testing for the significance of R^2 and b_j, the preceding formula has wide applicability for testing the significance of sets of regression coefficients and related topics.

An excellent discussion of the models comparison approach has been presented by Cramer (1972). As he indicates, in the case of correlated predictors one could obtain the following outcomes, some of which appear to be contradictory:

1. R^2 and all b_j significant.
2. R^2 and some (but not all) b_j significant.
3. R^2 but none of the b_j significant.
4. All b_j significant but not R^2.
5. Some b_j significant, but not all, nor R^2.
6. Neither R^2 nor any b_j significant.

As cogently argued by Cramer, the first test to make is a test of the significance of the overall model, as evinced in R^2. If R^2 is not significant then conditions 4, 5 or 6 lead to *no differences in conclusions*. That is, assuming that the sample size is large enough for detecting true effects, in each of the cases 4, 5 and 6, one concludes that the regression equation is useless. Hence, one first conducts a test of R^2 in order to see if the analysis should go further. If overall R^2 is not significant, one need not bother to test individual coefficients.

If R^2 for the full model is significant, then one has a justification in looking for a simpler model. If case 1 prevails, there is no need for simplification. If case 2 prevails, one can make a series of model comparisons in which a subset of r

variables is considered as the restricted model, to be compared with a full model involving all n variables. (This, however, is different from discarding all variables whose individual t tests are not significant.) Finally, case 3, while improbable, could still occur if each predictor is contributing something, however slight, to overall accounted-for variance. As Cramer shows, it is possible that an overall test of R^2 is significant and yet no subset of regression coefficients (individual ones or combinations) achieves significance. Note, however, that section 2.2.5 has already indicated that $b_1^{(s)}$ is significant at the 0.05 level. Hence, we could employ the single predictor X_1 in a *bivariate* model, assuming that its inclusion is sensible from a content viewpoint.

In short, if R^2, representing the fit of the overall model is not significant, the researcher can end the matter then and there. If R^2 is significant one may search for a simpler model involving a subset of r predictors. If so, one should check to see if the full subset of $n - r$ deleted variables adds anything beyond a model utilizing only the r predictors.

2.6 Computer Programs for Multiple Regression

As would be supposed, a large number of packaged computer programs exist for conducting multiple regression analysis. However, most of these are organized around a relatively few rules for the selection of predictor variables. In large-scale regression problems involving ten or more predictors, it is not unusual to find that the predictors themselves are so highly correlated that a relatively few predictors represent most of the accounted-for variance in the criterion. Hence, many computer programs are concerned with the problem of selecting and ordering predictors to be introduced into the regression analysis.

We first comment on the characteristics of more or less standard (all variables) programs in which *all* candidate predictors are used in a single regression analysis. We then turn to programs that are concerned with the problem of sequential predictor selection.

2.6.1 The BMD-03R All-Variables Regression Program

The BMD-03R program (Dixon, 1973), as used in section 2.5, is one of the most widely disseminated programs for performing all-variables type regressions. The program can accommodate up to 49 predictors and a sample size of up to 99,999 cases. Moreover, regressions can be performed on subsamples of the data matrix; up to 28 different subsamples can be handled in a single run. This is usually done by entering the data in subsample blocks (up to a maximum of 28) and then combining various blocks in different subproblems in the computer run. A main advantage of subsample splitting is the flexibility it provides for running cross validations. One can split the data into halves, thirds (or whatever) and make separate regression runs for comparing parameter estimates obtained by the various subsamples as well as various aggregations of those subsamples.

Flexibility is also provided in the selection of the predictor (and criterion) variables for various subproblems in the master run. A table of residuals $(Y_i - \hat{Y}_i)$ can also be requested for further examination.

In any given subproblem, the variables and cases are selected by the user and an all-variables type regression is run. Each subproblem represents a particular combination of variables and cases; up to 99 such subproblems can be handled in any particular run. (We shall use this program again in analyzing some of the T.V. commercial data in section 2.7.)

2.6.2 Variable Selection Programs

Draper and Smith (1966) provide an extensive account of various procedures that can be employed in the sequential selection of predictor variables in multiple regression. Our account will be much briefer since the present purpose is to illustrate only one of these approaches in any detail. Figure 2.8 shows a number of procedures that have been advanced for selecting predictor variables.

All Possible Regressions   As the name suggests, this procedure involves the fitting of all possible regressions. With n predictors, each of which may or may not be included in the equation, one has 2^n possibilities ranging from fitting just the constant b_0 to an equation in which all predictors are entered.

The approach that is typically taken involves comparing overall fits given by the R^2 value for all combinations in which the number of predictors is fixed at some value $r \leqq n$. The winners in each subset are then selected and compared in terms of R^2 across subsets. Often the researcher plots residual mean squares

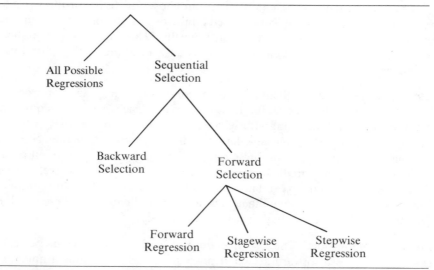

Figure 2.8 Illustrative Procedures for the Selection of Predictor Variables

or unaccounted-for variance $(1 - R^2)$ against the number of variables that are included as a guide to selection of the appropriate subset size. One then selects the "best" subset size in terms of R^2 and various substantive considerations.

The sequential selection procedures (to be described) are, in a sense, approximations to the all-possible-regressions approach. Such approximations are needed since for problems of realistic size it is not practical, even on computers, to run all possible regressions. (Most all-possible-regression programs are limited to ten or twelve predictors.)

Backward Selection   This class of procedures starts out with a full or all-variables regression. The program then tests to see what the drop in R^2 would be if each variable, in turn, were deleted from the regression; this is summarized by computing an F ratio for each predictor.[25] If the F ratio associated with the least important predictor is less than some predetermined critical value of F, that variable is deleted and a new regression with only $n - 1$ predictors is computed. The process is then continued until all remaining predictors, when considered to be the last variable entered, exceed the critical F ratio.

Forward Regression   One can, of course, proceed from the opposite end by entering predictors sequentially until some "best" equation is obtained. The forward regression procedure first computes all simple correlations of each predictor with the criterion. The predictor with the highest simple r^2 is entered into the equation. One then computes all partial r^2's and associated F ratios for the $n - 1$ remaining predictors. The next predictor to enter is the one whose squared partial correlation (conditioned on the variable already entered) is highest, assuming its associated F ratio exceeds some predetermined critical value for inclusion. (This rule is equivalent to choosing the largest F ratio, as noted earlier.)

The next variable to be entered is again chosen in terms of the size of its squared partial correlation (conditioned on the two predictors already entered), assuming that its F ratio exceeds the critical value. The process terminates when a candidate variable to be entered exhibits an F ratio below the critical value for inclusion.

Stagewise Regression   Stagewise regression is a forward selection procedure but one that is *not* based on a full least squares approach. The idea is quite simple, however. One starts out with selecting a predictor on the basis of the one exhibiting the largest simple correlation with the criterion. One then computes residuals from this equation. These residuals are regressed, in turn, on each of the $n - 1$ remaining predictors and the predictor most highly correlated with the residuals is chosen next and new residuals from the regression are computed. The process terminates when the next candidate predictor's correlation with the sequentially obtained residuals is not statistically significant.

Stepwise Regression   The last procedure to be described is stepwise regression. This is also a forward selection procedure, but commands a number of advantages over the two preceding approaches.

Stepwise regression starts out in the same way as the forward regression procedure. Its major difference is that predictor variables already entered in preceding selections can be deleted in subsequent selections. In the forward regression procedure, once a candidate predictor enters the regression equation, it stays in over all subsequent steps. In stepwise regression, at each stage in the process, each variable already entered is tested to see if it should be retained in the equation. This is done by assuming that each variable already in the equation is the last to enter.

It may turn out that some variable entered early in the selection process is, in fact, highly correlated with subsequently entered variables. If so, this variable may not pass the F ratio threshold associated with the last-to-be-entered assumption and may be deleted at a later stage. Thus, at each stage, *all* variables then included in the regression are checked for retention on the assumption that each was entered last.

2.6.3 The BMD-02R Stepwise Regression Program

The BMD-02R stepwise regression program (Dixon, 1973) is also widely distributed and we shall comment on it as a prototype of stepwise procedures. This program will accommodate up to 74 predictors and a sample size of up to 9,999 cases. Regression equations with or without an intercept can be computed. Variables can be subjected to various preliminary transformations (for example, logarithmic, positive square root, etc.) prior to analysis. Provision is also made for the printing of residuals, if desired by the user.

The user can set critical values for the F ratio for predictor variable inclusion and the F ratio for predictor variable retention. In addition, the user can specify in advance which, if any, predictors are to be considered as "forcing." Up to seven levels of forcing variables can be chosen, if desired. All variables at a higher level in the forcing hierarchy are considered for inclusion before any variable in the next layer of the hierarchy is examined.

For the sake of simplicity, however, let us assume that all variables are free variables of common status. The user is first asked to state an F ratio for inclusion, an F ratio for deletion and a tolerance level. The tolerance level is defined as the value $1 - R^2$ for the j-th predictor, when regressed on all predictors then in the equation. (As such, it is a measure of the lack of redundancy of X_j with predictors already in the equation.) The program's default values for these three control parameters are, respectively, 0.01, 0.005, and 0.001. (These are "loose" values in the sense of tending to include and retain all predictors.)

The program typically prints means, standard deviations, the covariance, and correlation matrices of all $n + 1$ variables (including the criterion). Then the following features of the program are implemented:

1. The predictor with the highest simple r^2 with the criterion (or, equivalently, the one with the highest F ratio) is entered, assuming it passes the F ratio control value to enter.

2. Any new predictor is added to the equation on the basis of its displaying the highest squared partial r (or, equivalently, the highest F value) with those predictors already in the equation, assuming that its F level and tolerance value exceed the control parameters.

3. A predictor can be deleted at any stage if its F value to remove drops below the control parameter. That is, at any given stage beyond the first, each predictor in the equation is tested as though it were the last to enter. If any F values are less than the F for retention the one with the lowest F value is deleted at that stage.

4. At each stage in the accretion of predictors a regression equation is estimated (including intercept), as well as a multiple R, and an ANOVA table for testing R^2. Also, the standard error of each predictor in the equation is computed as well as the F value associated with the predictor-retention test.

5. Also, at each stage, the partial r's of all predictors not in the equation, their tolerance levels, and F values to enter are computed and displayed.

6. At the end of the process, a summary table with R, R^2, incremental R^2, and the F values to enter, associated with the full sequence of predictors, is printed.

7. The BMD-02R program also provides a capability for listing and plotting residuals $Y_i - \hat{Y}_i$ versus specified predictors and versus $\hat{Y}$ itself.

After the regression run has been made, the user is free to go back to that step in the sequence in which R^2 and all individual F values for retention are significant. Since the equation for that subset of variables is already available, the program does not need to be re-run.

Of course, considerations other than significance levels may direct which equation is chosen. Moreover, there is no guarantee that the particular subset of r ($<n$) predictors chosen by stepwise regression is the best possible subset of size r, in the sense of providing the largest sample R^2. One would have to compute all possible combinations of n predictors taken r at a time to see if this is so.

We use BMD-02R in the course of analyzing some of the TV commercial data in the next section of the chapter.

2.7 Analyzing the TV Commercial Data

In section 2.5 we used an all-variables type of regression program (BMD-03R) to analyze the hypothetical data of Table 2.1. We now turn to a problem of more realistic size—the television commercial study, briefly introduced in Chapter 1. Appendix A provides a more detailed account of the study as well as the questionnaire and data.

One of the substantive questions of interest in the study concerned what the linear association might be between the criterion variable:

Y: Post-exposure purchase interest in the Alpha brand of radial tires (question C-2)

and the predictors:

X_1: Interest in the product class of radial tires, irrespective of brand (question A-1).

X_2: Whether Alpha was the brand selected in the respondent's last purchase of replacement tires (question A-2).

X_3: Pre-exposure interest in Alpha's brand of radial tires (question A-3).

In particular, one might theorize that a high post-exposure purchase interest in Alpha radials would depend, first, on interest in the product class of radial tires; second, on brand "loyalty," as evinced in selecting Alpha's brand (radial or otherwise) on the last purchase occasion; and third, on current purchase interest in Alpha's radial tire before being exposed to the TV commercials.

In terms of the regression model, we would like to include the above predictors in the order named. However, we might also be interested in whether the demographic variables of Part E of the questionnaire add anything to our descriptive model of post-exposure purchase interest. If such is the case, this additional information could be useful in helping to delineate various subgroups in the market that may be particularly attracted toward Alpha radials.

With the foregoing as background, we conduct two regression analyses of interest:

1. The sample of 252 respondents is first split (essentially randomly) into first and second halves. BMD-03R is then applied to each half independently. The results are then compared and double cross validation is performed.

2. As a demonstration of stepwise regression, the total sample is analyzed by the BMD-02R program.

Each of these analyses is discussed separately and the results are then compared. Table 2.8 lists the predictor variables that were employed in the preliminary runs.

2.7.1 Data Organization

As described above, two separate runs of BMD-03R were made on the first and second halves of the sample. The criterion variable Y is the post-exposure purchase interest rating (expressed on a 0–10 point scale) of Alpha radials, and the predictors are as indicated in Table 2.8. However, an additional aspect of the analysis is the fact that five of the predictor variables are dummies:

X_1: Is respondent interested in the product class of radial tires? If yes, coded 1; if no, coded 0.

X_2: Was Alpha the tire brand last purchased by respondent? If yes, coded 1; if no, coded 0.

X_7: Is respondent married? If yes, coded 1; if no, coded 0.

X_8: Is respondent's occupation professional or white collar? If yes, coded 1; if no, coded 0.

X_9: Is respondent's occupation blue collar? If yes, coded 1; if no, coded 0.

Table 2.8 shows the total-sample means on the criterion and each of the ten predictor variables while Appendix A provides additional detail on the marginal distributions of the demographic variables, X_4 through X_{10}.

The dummy variables X_1, X_2, and X_7 are each dichotomies and present no special problems. On the other hand, the dummy variables X_8 and X_9 are developed from an originally trichotomous classification:

| | Dummy Variables | |
Occupation Category	X_8	X_9
Professional or white collar	1	0
Blue collar	0	1
Other (including unemployed)	0	0

Note, then, that the third category does not appear explicitly in the regression. Its contribution "is buried" in the intercept term of the regression equation while the contributions of X_8 and X_9 will be expressed as *differential* contributions relative to the third category. That is, the intercept term is obtained when all predictors (including X_8 and X_9) assume the value zero.

Generally, however, we are interested in only relative effects compared to some (often arbitrary) base category. Any of the three classes of occupation could serve as this base category without affecting the relative contribution of the remaining two categories. That is, *differential* effects (plus or minus) are not influenced by which category serves as the reference class.

The proportion of respondents in the third category is easily obtained from the means of X_8 and X_9 (in Table 2.8) as $1.0 - (0.44 + 0.39) = 0.17$, since categories have been chosen to be exclusive and exhaustive. As such, a respondent must be assignable to one and only one of the three categories.

The possibilities for dummy-variable coding, and the procedures used to do it, go far beyond the simple cases described here. Chapter 3 explores this area much more thoroughly in the context of analysis of variance and covariance.

Table 2.8 Total-Sample Means on the Criterion and Ten Predictor Variables

Variable	Mean	Description of Variable
Y	6.14	Post-exposure rating of Alpha radials on purchase interest.
X_1	0.58	Proportion expressing interest in the product class of radial tires
X_2	0.19	Proportion choosing the Alpha brand on last tire purchase
X_3	5.45	Pre-exposure rating of Alpha radials on purchase interest
X_4	33.20	Respondent age (years)
X_5	3.50	Number of household family members, including respondent
X_6	13.80	Number of years of formal education
X_7	0.57	Proportion who are married
X_8	0.44	Proportion who are professional or white collar
X_9	0.39	Proportion who are blue collar
X_{10}	149.50	Annual household income before taxes (hundreds of dollars)

2.7.2 Split-Sample Analysis

We now turn to the results of the split-sample analysis, using the all-variables program, BMD-03R. As indicated, three runs were made initially:

1. A first-half analysis involving 126 respondents.
2. A second-half analysis involving 126 respondents.
3. A full-sample analysis involving 252 respondents.

Pertinent summary output of these three runs appears in Table 2.9. Squared multiple correlations (at the sample level) are 0.434, 0.338, and 0.345, respectively, for first-half, second-half, and total sample. All three F ratios due to regression are significant at the 0.05 alpha level or better.

Perhaps the most interesting finding, however, is the fact that the demographic variables (X_4 through X_{10}) add very little to accounted-for variance. Their cumulative contribution is 0.05, 0.02, and 0.01, respectively, for the first half, second half and total sample. Their relatively small effect—*after* the contributions of X_1, X_2, and X_3 are taken into account—is also shown in the lack of stability in even the signs of the regression coefficients between first and second halves of the sample.

On the other hand, the first three predictors: (a) general interest in radials; (b) Alpha is brand last purchased; and (c) pre-exposure interest rating in Alpha radials, show relatively strong and stable effects.[26] In addition, the signs of the regression coefficients (all positive) that are associated with the first three predictors are in accord with expectations. Predictor X_3, pre-exposure interest in Alpha radials, in particular displays the major contribution to variance accounted-for.

From a context standpoint, the opportunity to capitalize on respondent differences in post-exposure purchase interest in terms of demographics is virtually non-existent insofar as these data are concerned. That is, after the association of the seven demographic predictors with the first three predictors is taken into account, it is clear that the demographics add very little to R^2.

A confirmation of the preceding results was found by examining the total-sample results using the stepwise regression procedure, BMD-02R. In this case, no prior ordering of the predictors' variables is undertaken. Rather, predictor variables are added sequentially on the basis of their partial correlations with predictors already in the equation.

It is of interest to note that the first three predictors sequentially entered by BMD-02R were X_3, X_1, and X_2 (in that order) with R^2's of:

1. $0.291-X_3$ alone.
2. $0.331-X_3$ and X_1.
3. $0.337-X_3$, X_1 and X_2.

The next predictor to enter, X_4 (respondent age), contributed only 0.004 in accounted-for variance (as can also be verified from Table 2.9 in the BMD-03R total-sample run).

Thus, the results suggest that the first three predictors—particularly X_3 and X_1—are most influential in accounting for variance in the criterion variable. From a pragmatic standpoint, it does not seem worthwhile to consider any of the demographic variables.[27]

2.7.3 Double Cross Validation

Since the first and second halves of the sample appear to produce fairly stable results for predictors X_1, X_2, and X_3, one might next check to see whether these findings hold up under double cross validation. This was carried out by:

1. Re-running BMD-03R for the first and second halves of the sample, using *only* predictors X_1, X_2, and X_3.

2. Estimating second-half criterion values from the regression equation obtained from the first-half analysis.

3. Estimating first-half criterion values from the regression equation obtained from the second-half analysis.

4. Finding simple two-variable correlations between Y and $\hat{Y}$ in each case.

Summary results appear in the top portion of Table 2.10.

If we examine the signs of the partial regression coefficients, an anomaly is noted in the case of predictor X_2 (Alpha is last brand purchased). In the first-half analysis we find that b_2 is negative. Not only does this appear counter to expectations, but b_2 also differs in sign from its counterpart in Table 2.9. Thus, when the subset of demographic variables is excluded from the analysis—which, in the case of correlated predictors, will change the values of the remaining partial regression coefficients—we find in the case of X_2 that even the sign of b_2 changes in the first-half analysis, shown in Table 2.10.

An explanation of this anomaly is not hard to find. If we examine Table 2.9 we see that the variance accounted-for by X_2 is only 0.005 in the first-half analysis. It appears that the partial regression coefficient b_2 is unstable across replications.

This conjecture is supported by re-running the double cross validation, this time with only predictors X_1 and X_3. The results appear in the lower portion of Table 2.10. Not only do we find that the partial regression coefficients are stable in sign and value across separate halves of the sample, but the split-half predictions, as measured by the simple correlation between Y and $\hat{Y}$, are slightly higher if X_2 is excluded.

As a final cross-check on the significance of the excluded subset of predictors, namely X_2 and the set of seven demographic variables, we can see if the *total set* of eight excluded variables collectively adds anything to the regression results. In section 2.5.3 we considered the models comparison formula:

$$F = \frac{R_f^2 - R_r^2}{1 - R_f^2} \cdot \frac{d_f}{d_r - d_f}$$

In the present context we can modify this formula to test whether the subset

Table 2.9 Selected Output of BMD-03R Regressions

	First Half	Second Half	Total Sample
R^2 (Unadjusted)	0.434	0.338	0.345
F Ratio due to Regression	8.834	5.882	12.851
Incremental Variance Accounted-For by:			
X_1	⎡0.139	⎡0.062	⎡0.099
X_2	0.383 ⎢0.005	0.313 ⎢0.086	0.337 ⎢0.037
X_3	⎣0.239	⎣0.165	⎣0.201
X_4	0.037	0.001	0.004
X_5	0.000	0.000	0.000
X_6	0.004	0.001	0.000
X_7	0.000	0.006	0.000
X_8	0.004	0.005	0.002
X_9	0.003	0.005	0.000
X_{10}	0.005	0.006	0.005
Sign of Regression Coefficient for:			
X_1	+	+	+
X_2	+	+	+
X_3	+	+	+
X_4	−	−	−
X_5	−	−	−
X_6	+	−	−
X_7	−	+	+
X_8	+	+	+
X_9	−	+	−
X_{10}	+	+	+

of $n - p = 8$ predictors adds anything to accounted-for variance beyond the $p = 2$ predictors. The formula is:

$$F = \frac{R_{yn}^2 - R_{yp}^2}{1 - R_{yn}^2} \cdot \frac{m - n - 1}{n - p}$$

with $n - p$ degrees of freedom for numerator and $m - n - 1$ degrees of freedom for denominator.

In the present case, we know on a total-sample basis (see Table 2.9) that R_{yn}^2 is 0.345. From the stepwise regression results we found that X_1 and X_3 alone

Table 2.10 Selected Output from Double Cross Validations

	First Half	Second Half	Total Sample
	Predictors X_1, X_2, and X_3		
b_0 (intercept)	2.266	3.673	2.957
b_1	1.717	0.741	1.209
b_2	−0.127	1.100	0.585
b_3	0.481	0.384	0.434
R^2	0.383	0.313	0.337

$$r(Y_{\text{First}}; \hat{Y}_{\text{Second}}) = 0.583$$

$$r(Y_{\text{Second}}; \hat{Y}_{\text{First}}) = 0.523$$

	First Half	Second Half	Total Sample
	Predictors X_1 and X_3 Only		
b_0 (intercept)	2.264	3.685	2.961
b_1	1.707	0.698	1.223
b_3	0.478	0.429	0.452
R^2	0.383	0.290	0.331

$$r(Y_{\text{First}}; \hat{Y}_{\text{Second}}) = 0.603$$

$$r(Y_{\text{Second}}; \hat{Y}_{\text{First}}) = 0.527$$

resulted in an R^2 of 0.331. Thus, we can see if the subset of the $n - p = 8$ remaining predictors makes any significant contribution:

$$F = \frac{(0.345 - 0.331)}{(1 - 0.345)} \cdot \frac{241}{8}$$

$$= 0.644$$

As can be seen, the F ratio is even less than unity, indicating that the whole subset of the eight remaining variables (X_2 and the seven demographics) *makes no significant contribution over X_1 and X_3*.

2.7.4 Recapitulation

At this point it may be useful to summarize the empirical findings. We first found that the subset of seven demographic variables added very little to accounted-for variance in the criterion variable: post-exposure purchase interest in Alpha radials. This was borne out in all of the BMD-03R runs and in the BMD-02R stepwise regression applied to the total sample. Hence, the attempt to find demographic groups that might vary with regard to post-exposure purchase interest was not successful.

Second, upon double cross validation, only predictors X_1 (interest in the product class of radial tires) and X_3 (pre-exposure purchase interest in Alpha radials) maintained stability over the split-half predictions. These two predictors accounted for approximately 33 percent of the variation in the criterion variable on a total-sample basis. Moreover, the signs of the partial regression coefficients agreed with substantive expectations. By way of summary:

1. Post-exposure purchase interest in Alpha radials is higher if the respondent is interested in the product class of radial tires to begin with. If the respondent is interested in the product class, then the differential effect on post-exposure interest is 1.22 rating points on a 0–10 scale.

2. Post-exposure purchase interest in Alpha radials increases with increases in pre-exposure interest in Alpha radials. The partial regression coefficient b_3 (lower portion of Table 2.10) is 0.452; hence, for each rating point change in pre-exposure interest, post-exposure interest changes by 0.45 of a rating point on a 0–10 scale.

3. Somewhat surprisingly, predictor X_2 (brand last purchased was Alpha) does not contribute much to post-exposure purchase interest in the Alpha brand of radial tires.

The last finding is of interest from a content viewpoint. Apparently, previous purchases of the Alpha brand exhibit little effect on post-exposure purchase interest. Two possibilities as to why this might be the case come to mind: (a) of those respondents who chose the Alpha brand on their last tire purchase, about as many were later satisfied as dissatisfied with their choice or (b) the radial tire is a new enough product that previous brand "loyalty" (as applicable to non-radial tires, for the most part) is not relevant.

Although either situation could be at work insofar as some respondents are concerned, other evidence from the survey suggested that radials were perceived to be a significantly important enough innovation for the sample as a whole to upset previous brand loyalty patterns.

Perhaps the most interesting result of all of this is the fact that large-scale regressions (in this case, one involving ten predictors) are to be handled with caution. As found in the split-half analyses (and subsequent double cross validation) only two predictors, X_1 and X_3, made stable and interpretable contributions to post-exposure purchase interest. As banal as this may sound, it is no exaggeration to state that many empirical studies are reported without even rudimentary attempts to cross validate the findings on fresh data. Since all of the techniques that we shall be describing display a penchant for capitalizing on chance variation in the specific sample being analyzed, we cannot over-emphasize the need for prior specification of relationships and the employment of cross validation procedures.

2.8 Summary

The purpose of this chapter has been to introduce the reader to single criterion, multiple predictor association. The multiple regression model represents a prototype of this class of procedures, while the general linear model represents the organizing framework.

We first described regression from an intuitive point of view and presented concepts in terms of both the response surface and vector models. A small numerical example was used to illustrate the procedures.

We then turned to a discussion of the general linear model and a matrix formulation of the regression problem. Different types of solution techniques were presented and compared. Various statistics—multiple correlation, the partial regression coefficient, the partial and part correlation coefficients, standard errors—were described and computing formulas presented.

Attention was then directed to the television commercial data and illustrative application of all-variables regression and stepwise regression was shown. In addition, double cross validation was illustrated.

Review Questions

1. Consider the first three variables of the data set listed in question 3 of section 1.8:

Case	Language Aptitude Y	Analogical Reasoning X_1	Geometric Reasoning X_2
A	2	3	15
B	6	8	9
C	5	2	7
D	9	4	3
E	11	10	2
F	12	15	1
G	1	4	12
H	7	3	4

Assume that you wish to see if Y is linearly associated with X_1 and X_2.
   a. Compute the 3×3 product moment correlation matrix of Y, X_1 and X_2.
   b. Find the regression equation (in original units) of Y on X_1 and X_2.
   c. Find R^2 and the partial correlations $r_{y1.2}$ and $r_{y2.1}$.
   d. Test for the significance of R^2, b_1, and b_2 at the 0.05 alpha risk level.
   e. Using matrix methods, find SSR, SSE, and SST.

2. Refer to the data of Table 2.1 in the present chapter. Construct two new variables X_2^2 and $X_1 X_2^2$.
   a. Using some packaged computer program, regress Y on X_1, X_2, X_2^2, and $X_1 X_2^2$.
   b. Find part correlation coefficients for each of the four predictors.
   c. Find the standard error of each of the predictors and carry out individual t tests, using a 0.05 alpha risk level.
   d. Test a model that includes only X_1 and X_2 only versus one that includes all four predictors; use an alpha level of 0.05.

e. Test a model that includes X_2 only versus one that includes all four predictors; use an alpha level of 0.05.

f. What is the drop in R^2 if $X_1 X_2^2$ is omitted?

3. Consider the television commercial data of Appendix A.

a. Split the data into halves on the basis of odd versus even respondent numbers.

b. Perform separate *stepwise* regression runs of post-exposure purchase interest Y (question C-2) on:

X_1: Interest in the product class of radial tires (question A-1).

X_2: Whether Alpha was the brand selected in the respondent's last purchase of replacement tires (question A-2).

X_3: Pre-exposure interest in Alpha's brand of radial tires (question A-3).

c. How do the results of these split-half runs compare with those found in section 2.7.2?

d. Compute double cross validation correlations using the full regression equation of each half. Compare these results with the assignment of unit weights to each predictor variable and then computing the product moment correlation between $\hat{Y}$ and Y (unit weights) across the total sample.

e. Compute double cross validation correlations of Y on X_1 and X_3 and compare these with the results in part d.

4. Return to the data in Table 2.1 and dichotomize X_1 so that cases a, b, f, g, and i are coded 1 and the remaining five cases are coded zero.

a. Regress Y on the new X_1 and the original X_2.

b. How do R^2, b_1 and b_2 compare to the results in the chapter?

c. If the code value 1 is referred to as "high" extra-curricular activity and the code value 0 as "low" extra-curricular activity, how would you interpret b_1?

d. Suppose a new case is presented with X_1 coded as 1 (high) and X_2 is 8. What is $\hat{Y}$?

e. What happens to R^2, b_1, and b_2 under the coding $+1$ and -1 for 1 and 0, respectively?

5. From the data of Table 2.1 graph Y_s versus X_{s1} (i.e., the standardized form of the data) in a scatter plot.

a. Show that $r_{y1}^2 = b_{y \cdot 1}^* b_{1 \cdot y}^* = r_{1y}^2$.

b. Show that the slopes are equal and that a $45°$ line bisects these slopes.

c. If X_{si1} is $+2$ standardized units, $\hat{Y}_{si}$ will be closer to zero than X_{si1}. Under what circumstances will this "regression effect" disappear?

d. What happens to the slopes $b_{y \cdot 1}$ and $b_{1 \cdot y}$ if Y_d and X_{d1} are plotted? How does their difference in standard deviations affect the relationship between the slopes?

6. In the vector model, the correlation between a pair of variables is portrayed as the cosine of the angle separating them. Moreover, vector length is proportional to the standard deviation with the constant of proportionality being $\sqrt{m-1}$ (where m is the number of observations).

a. Show Y and X_1 as a pair of vectors in two dimensions using the data of Table 2.1.

b. What happens to the lengths of this pair of vectors in portraying the correlation between Y and X_1?

c. What is the angle separating Y and X_2 in the vector model?

d. Suppose that we have the matrix product $\mathbf{X}_d'\mathbf{X}_d$ where $\mathbf{X}_d$ is a mean corrected matrix of order $m \times 2$ ($m > 2$). How can the off-diagonal element of $\mathbf{X}_d'\mathbf{X}_d$ be expressed as a scalar product in terms of the standard deviations of X_1 and X_2, their correlation r_{12}, and the sample size m?

Notes

1. This is the so-called "classical" or fixed design model. More recent versions (Johnston, 1972) allow X_1 to be a random variable as well, subject to certain restrictions that are discussed later in the chapter.

2. Later we shall assume that the ϵ_i's are normally distributed (in order to make various significance tests). This assumption is also suggested in the plot of Figure 2.1.

3. In actuality the computing version of the formula for finding the slope b_1 is used (see Table 2.2).

4. The use of X_1 in a linear regression can do no worse than $\overline{Y}$. That is, even if b_1 turns out to be zero, the predictions are $\hat{Y}_i = b_0 = \overline{Y}$ which are the same as using the mean of the criterion values in the first place.

5. Some additional properties of the least squares regression line are useful to point out:

a. $\sum_{i=1}^{m} X_i e_i = \sum_{i=1}^{m} \hat{Y}_i e_i = 0$;

b. $\sum_{i=1}^{m} Y_i = \sum_{i=1}^{m} \hat{Y}_i$; and

c. the regression line goes through the point $(\overline{Y}, \overline{X})$.

(We already know, of course, that $\sum_{i=1}^{m} e_i = 0$ and that $\sum_{i=1}^{m} e_i^2$ is a minimum under least squares fitting procedures.)

6. The F statistic follows the F distribution of Table C.4 (Appendix C) provided that our assumptions about the error term ϵ_i are met. A description of the F test appears in Appendix B.

7. The constraints that result in the loss of two degrees of freedom are:

$$\sum_{i=1}^{m} e_i = 0 \text{ and } \sum_{i=1}^{m} X_i e_i = 0.$$

8. For example, the t value associated with a test of $b_1{}^{(s)}$ is $\sqrt{20.87} = 4.57$ and the F value found from the ratio of the mean-square for regression to the mean-square for error is 20.87. All three tests yield equivalent results.

9. Only if r^2 is equal to 1 (perfect correlation) will the regression slopes be each equal to 1 (assuming that both X_1 and X_2 are expressed in standardized X_s form). In this special case the *single regression slope* is characterized by a vector through the origin that makes a 45° angle (if X_1 and X_2 are positively related) or a 135° angle (if X_1 and X_2 are negatively related) with the horizontal axis.

10. One could also obtain the same value of (say) b_1, the partial regression coefficient for X_1, by regressing a set of Y-residuals (net of any linear association of Y with X_2) on a set of X_1-residuals (net of any linear association of X_1 with X_2). However, the extra step of finding the Y-residuals is not really needed.

11. This interpretation is fully analogous to the case of a single predictor variable, as illustrated in Figure 2.2.

12. This follows from elementary principles of matrix rank in which the rank of a matrix cannot exceed its smaller order.

13. Still another way to compute the (squared) part and partial correlations is in terms of the *difference in R^2* between a model that includes all predictors and one that includes all predictors except X_j. The squared *part* correlation is then defined as:

$$r^2_{yj(k,\ell,\dots)} = R^2_{yj,k,\ell,\dots} - R^2_{yk,\ell,\dots}$$

The squared partial correlation expresses this difference relatively in terms of variance still to be accounted for in the restricted model that excludes X_j:

$$r^2_{yj \cdot k,\ell,\dots} = \frac{R^2_{yj,k,\ell,\dots} - R^2_{yk,\ell,\dots}}{1 - R^2_{yk,\ell,\dots}}$$

14. The standardized regression coefficient b_j^* is often called a *beta weight*. However, these betas are not to be confused with the β_j's used to denote universe values of the partial regression coefficients. Furthermore, in the multiple regression case, standardizing the partial regression coefficient does *not* yield the partial correlation coefficient (unlike the two-variable case) unless the predictors are uncorrelated.

15. To illustrate, from Table 2.3 we can compute the standard deviation of the residuals of Y regressed on X_2 as:

$$\{\tfrac{1}{8}[(0.518)^2 + (-0.027)^2 + \cdots + (-1.391)^2]\}^{1/2} = 1.55$$

which is the same as $s_y \sqrt{1 - r^2_{y2}} = 2.908\sqrt{1 - 0.716} = 2.908(0.533) = 1.55$.

16. Linn and Werts (1969) also describe a part correlation coefficient in which the roles of Y and X_j are reversed. However, this measure has received little attention, as yet, in applied studies.

17. Although not essential to the general linear model, we shall also assume that the X_j are linearly independent; that is, we cannot predict perfectly some variable j from a linear combination of the remaining predictors. Throughout the chapter, we shall restrict discussion to a general linear model of full rank (i.e., one with an absence of linear dependencies).

18. The constant $(m - 1)$ cancels out when the inverse of $\mathbf{R}$ is post-multiplied by $\mathbf{r(y)}$.

19. Alternatively, if one works with the covariance matrix $\mathbf{C}$ (rather than $\mathbf{R}$) to begin with, the solution for the vector $\mathbf{b}$ is obtained directly as:

$$\mathbf{b} = \mathbf{C}^{-1}\mathbf{a(y)}$$

where $\mathbf{C}^{-1}$ is the inverse of the covariance matrix and $\mathbf{a(y)}$ represents the *covariance* vector (rather than a vector of correlations) between the criterion and each of the predictors in turn. R^2 in this format is computed as:

$$R^2 = \frac{\mathbf{a'(y)C^{-1}a(y)}}{s_y^2} = \frac{\mathbf{a'(y)b}}{s_y^2}$$

One still has to solve for b_0 as shown above.

20. In the 2×2 case, the inverse of **R** can be found quite simply by: (a) interchanging the main-diagonal elements of **R**; (b) affixing minus signs to the off-diagonal elements; and (c) multiplying this matrix (called the adjoint) by $1/|\mathbf{R}|$, the reciprocal of the determinant of **R**. To illustrate:

$$\mathbf{R}^{-1} = \frac{1}{1 - (-0.838)^2}\begin{bmatrix} 1 & -0.838 \\ -0.838 & 1 \end{bmatrix} = \begin{bmatrix} 3.358 & -2.814 \\ -2.814 & 3.358 \end{bmatrix}$$

21. We could use the models comparison test (as illustrated in section 2.2.5 for the two-variable case) here, as well. However, so as to adhere closely to the BMD-03R format, we first focus on the more traditional analysis of variance approach (and then discuss the models-comparison approach).

22. More precisely, even if the null hypothesis H_0: $R_p^2 =$ is true, $E(R^2) \neq 0$. Rather, $E(R^2) = \frac{n}{m-1}$ so that as n approaches m, the sample size, $E(R^2)$ approaches 1. The adjusted R^2 can actually decrease if a new predictor enters the regression equation since the increase in accounted-for sum of squares may be more than counterbalanced by the loss of a degree of freedom in the denominator ($m - n - 1$). In contrast, unadjusted (i.e., sample-based) R^2 can never decrease as a new predictor is introduced into the regression equation.

23. It can be shown by simple algebra that:

$$\frac{R_f^2 - R_r^2}{1 - R_f^2} = \frac{\text{SSE}_r - \text{SSE}_f}{\text{SEE}_f}$$

This can be seen by the substitution on the left-hand side of:

$$\frac{1 - \dfrac{\text{SSE}}{\text{SST}} - \left[1 - \dfrac{\text{SST}}{\text{SST}}\right]}{1 - \left[1 - \dfrac{\text{SSE}}{\text{SST}}\right]} = \frac{\dfrac{\text{SST} - \text{SSE}}{\text{SST}}}{\dfrac{\text{SSE}}{\text{SST}}}$$

But, since $\text{SST} \equiv \text{SSE}_r$ and $\text{SSE} \equiv \text{SSE}_f$, we have:

$$\frac{R_f^2 - R_r^2}{1 - R_f^2} = \frac{\text{SSE}_r - \text{SSE}_f}{\text{SSE}_f}$$

as desired.

24. It should also be pointed out that the *squared partial correlation* (see Table 2.7) is also related to F and, hence, to t^2. The relationship is:

$$F_j = \frac{r_{yj \cdot k,\ell,\dots}^2}{1 - r_{yj \cdot k,\ell,\dots}^2} \cdot (m - n - 1)$$

For example:

$$F_2 = \frac{r_{y2 \cdot 1}^2}{1 - r_{y2 \cdot 1}^2} \cdot (m - n - 1) = \frac{(0.467)^2}{1 - (0.467)^2} \cdot (10 - 3) = 1.952$$

We note that F_j (and, hence, t_j^2) is *monotonically* related to the squared partial correlation of Y and X_j. In this context F_j is often referred to as a "partial" F ratio. Many computer programs that select predictor variables sequentially (as described later in the chapter) utilize the largest F_j as a basis for predictor selection. As can be noted above, however, the squared partial correlation will rank the variables in exactly the same way.

25. As described in section 2.5.3, in this context the F value is often referred to as a "partial" F_j ratio. It is used to test whether X_j accounts for significant variance in Y beyond the variance accounted for by other predictors already in the equation. As recalled, $F_j = t_j^2$ and, furthermore, F_j is monotonically related to the squared partial correlation $r_{yj \cdot k, \ell, \dots}^2$. Either F_j or $r_{yj \cdot k, \ell, \dots}^2$ will rank the predictors the same way.

26. An exception to this—commented upon later—is noted in the case of predictor X_2. In the first-half analysis its contribution to accounted-for variance is only 0.005.

27. This intuitive observation was also supported by a models-comparison test to see if the demographics, as a group, provided a significant addition at the 0.05 alpha level; they did not.

CHAPTER 3

Univariate Analysis of Variance and Covariance

3.1 Introduction

In the preceding chapter's discussion of the general linear model, it was indicated that univariate analysis of variance (ANOVA) and analysis of covariance (ANCOVA) can be subsumed under this model by suitable modification of the predictor-variable matrix, $\mathbf{X}$.[1] Part of the motivation for the present chapter is to point out the nature of the formal similarities between multiple regression and analysis of variance and covariance, where both are considered as illustrations of the general linear model.

We first discuss these correspondences in the context of hypothetical data, similar to the procedure followed in the preceding chapter. Since the data set has been made intentionally small, the reader should be able to follow the computations and their rationale with little difficulty. And, to keep the discussion particularly simple, we emphasize the most basic case of one treatment variable (the so-called single-factor, or one-way design) and one covariate.

Multiple-factor designs, in which two or more treatment variables are changed simultaneously, constitute the bulk of experimental layouts in ANOVA and ANCOVA. While we do not delve deeply into these more complex designs, some of the essentials of factorial designs (a prototype of multiple-factor layouts) are discussed and illustrated numerically in the next section of the chapter.

We then turn to a discussion of some of the data obtained in the TV commercial study. While these are essentially survey-type data, for illustrative purposes we consider a subset of the data that is organized as a two-way ANOVA with equal cell frequencies. Analyses are carried out with and without covariates, so as to provide a real-world counterpart to earlier analyses involving hypothetical data.

The concluding section of the chapter is concerned with general matters of research strategy in testing alternative models in the ANOVA and ANCOVA contexts. In particular, we describe some of the problems involved in setting up significance tests when unequal numbers of cases appear across the various treatment combinations.

While it is formally true that ANOVA and ANCOVA are special cases of the general linear model, it is nonetheless useful to study them in their own right. First, ANOVA and ANCOVA provide major computational simplifications in the case of balanced designs, where an equal number of observations appear in each cell of an experimental layout. Second, their development underscores the virtues of experimental design in which the separate contribution of various treatment-variable levels to the response variable can be unambiguously measured. Third, in cases involving two or more treatment variables, we can study not only simple (or main) effects of the treatments but their interactions as well.

3.2 Preliminary Aspects of ANOVA

Historically, multiple regression has been used by applied researchers to test models and make predictions from naturalistic data, for example, sales of some product from field data collected on its relative price, perceived quality and distribution coverage, students' classroom performance from their scores on a set of aptitude and intelligence tests, and so on. In contrast, analysis of variance and covariance have emerged from the experimental tradition, particularly the biological, agricultural and psychological disciplines, where active intervention is often feasible in the process being examined. In this case emphasis centers on the relative performance of various treatment variables or experimental factors and, in particular, whether mean responses differ significantly across factor levels.[2]

But, from the standpoint of estimating **b**, the vector of parameters in the general linear model:

$$y = Xb + e$$

multiple regression, ANOVA and ANCOVA are formally the same. The basis of this similarity lies in the opportunity to use dummy variables in the design matrix **X**. As described in Chapter 1, if one has a categorical variable consisting of k treatment or factor levels, this nominally scaled variable can be represented as a set of coded (dummy) variables that assume a restricted set of values, such as $(0, 1)$ or $(-1, 1)$. As in Chapter 1, if the variable has k levels it can be recorded into $k - 1$ dummies with no information loss. This device—called reparameterization—not only retains all information in the original categorization, but can also guard against singularity in the design matrix **X** when the usual regression model (employing an intercept term) is fitted.[3]

As a simple example, suppose a researcher is interested in finding out whether sex of respondent (male versus female) and work status (full-time student, full-time employed, and all other) lead to different mean levels of endorsement of certain U.S. foreign policies, where degree of endorsement is expressed on a nine-point rating scale. If so, the following predictor-variable coding could be used:[4]

Sex	Dummy Variable X_1	Work Status	Dummy Variable X_2	X_3
Male	1	Full-time student	1	0
Female	0	Full-time employed	0	1
		All other	0	0

Notice in each case that knowledge of the values assumed on any subset of $k - 1$ dummies is sufficient to know the individual's state regarding any of the k categories. For example, if an individual's work status is not (1, 0) or (0, 1) it must be (0, 0), namely, the coding assigned to the category: all other.

The choice of which class in each categorization receives the coding of all zeroes is arbitrary. In the preceding example, when $X_1 = 0$, $X_2 = 0$ and $X_3 = 0$, the *combined* class—female and all other—is designated. As recalled from Chapter 2, the intercept b_0 estimates the value for Y when all X_j's are zero. Accordingly, in the above dummy-variable coding b_0 measures the mean response to the combined class, female-all other. The effects of other classes can then be expressed, *differentially,* as increases or decreases in the response variable mean as compared to mean response in the base class.

3.2.1 Research Questions in ANOVA

The objectives of ANOVA are broadly similar to those of multiple regression, once we get used to the idea of working with dummy-variable predictors. Again we use least squares to find a linear composite of the predictors whose values best match (in the sense of a minimum sum of squared deviations) those of the criterion variable. As in regression analysis, the criterion variable is assumed to be at least intervally scaled.[5] The major research questions in a typical ANOVA are:

1. Can the response variable Y be adequately represented by a linear composite of the dummy-valued predictors X_j, X_k, etc.?

2. If so, how good is the fit between the Y_i and the $\hat{Y}_i$ that are computed by the linear model?

3. Is the relationship between Y and the overall linear model statistically significant?

4. Which treatment variables and which levels within treatment variables[6] appear to be statistically significant in accounting for variation in the response variable?

5. In the case of two or more treatment variables, do we find significant interactions among the treatment-level responses?

In contrast to multiple regression, however, ANOVA typically focuses on questions 3 through 5, the *significance testing* part, more so than on parameter estimation.

There are three main types of ANOVA models that are employed in applied research:

1. The *fixed-effects* model assumes that the treatment levels under study comprise the full population of interest to the researcher. The vector of parameters being estimated is not considered to be a random variable.

2. The *components of variance or random-effects* model assumes that the researcher has selected a random sample of treatment levels from some large universe with an unknown mean and variance.[7] The parameters estimated by the ANOVA procedure are considered to be random variables.

3. The *mixed-effects* model assumes that the design includes one or more factors of each kind—fixed and random.

This chapter deals primarily with the fixed-effects model. It is this model that represents a special case of the general linear model, introduced in Chapter 2. However, even the components of variance and mixed models use the same computational procedures that are discussed here. It is the way that error terms are handled and interpretations made that differs across the models.

As a case in point, the researcher might be interested in how employees of a large department store react to various types of customer complaints (where customers displaying different kinds of complaints are impersonated by professional actors). If the experiment is conducted with four (considered as replicates) salespersons each from three departments—say, sportswear, furniture, and cosmetics—a fixed-effects model would yield conclusions that concerned possible differences among these, and only these, departments. On the other hand, if the researcher is interested in making inferences about some average courtesy or efficiency level of *all* salespersons in the store, he might set up an experiment in which salespersons are selected randomly from the full set of departments. In this case, his interest would center on the total store's population of sales personnel.

While components of variance and mixed models are employed from time to time, it is fair to say that most behavioral and administrative science applications employ the more familiar fixed-effects model in which inferences are confined to the specific factor levels under study.

3.2.2 A Numerical Example of Single-Factor ANOVA

In keeping with our practice of working with miniature numerical examples, suppose we consider a small educational experiment in which three groups of four students each are exposed to a one-hour lecture (presented at three different times) by the same instructor. The instructor selects the groups at random and covers essentially the same content. The three lectures vary only with respect to the *type of visual* aids that the instructor uses in presenting the material.

The three levels of this treatment variable are:

Type of Visual Aid	Dummy Variable X_1	X_2
1. Blackboard	0	0
2. Transparencies (with Overhead Projector)	1	0
3. Handout (Multilithed) Material	0	1

Prior to hearing their specific lecture, each of the 12 students is given a written test on the general area from which the lecture is to be selected. Two days after the lecture, each student is given a shorter test directed at the specific material of the lecture. (Students are not told that this second test is going to be given.)

Let us assume that the instructor is interested in the extent to which the various kinds of visual aids used in his experiment influence student comprehension of the material covered in the lecture. Although students are assigned to the three groups at random, the instructor is also concerned about the influence that prior knowledge of the material may have on the results of the experiment. Accordingly, his collection of pre-exposure information (via the first test) can be used to adjust the post-exposure test results for possible differences in prior knowledge of the subject area and, in so doing, increase the precision of the experiment.

Table 3.1 shows the (hypothetical) data for this little experiment. The response variable Y denotes the number of correct answers on the short (post-exposure) test. The covariate X_3 indicates the (coded) score for the pre-exposure test; higher scores denote better performance. (We discuss this variable later in the chapter.)

Analysis of variance is essentially concerned with *tests of mean response.* For example, are the three mean response scores of 5.5, 6.5 and 7.5 in Table 3.1 significantly different? As we recall from basic statistics, running separate *t* tests between all possible pairs of the three groups' mean responses would not be a suitable procedure since the tests are not independent. Rather than three separate, two group *t* tests—with an attendant inflation of the alpha risk—we would prefer conducting one *F* test to see if *any* of the three group means differ. Our hypotheses are:

$$H_0: \mu_1 = \mu_2 = \mu_3$$

$$H_1: \text{Not all } \mu\text{'s are equal}$$

where each μ denotes the population *mean response,* namely the average number of correct answers on the post-exposure test.[8]

Table 3.1 Hypothetical Data Involving One Three-Level Treatment Variable and One Covariate

Student	Response (Post-Exposure Test Score) Y	Dummy Variables		Covariate (Pre-Exposure Test Score) X_3
		X_1	X_2	
a	4	0	0	2.2
b	5	0	0	2.9
c	6	0	0	4.3
d	7	0	0	5.2
1. Blackboard $\overline{Y}_1 = 5.5$				$\overline{X}_{31} = 3.65$
e	5	1	0	4.8
f	6	1	0	6.4
g	7	1	0	6.7
h	8	1	0	8.1
2. Transparencies $\overline{Y}_2 = 6.5$				$\overline{X}_{32} = 6.5$
i	6	0	1	8.3
j	7	0	1	9.2
k	8	0	1	9.8
l	9	0	1	11.1
3. Handouts $\overline{Y}_3 = 7.5$				$\overline{X}_{33} = 9.6$

3.3 Alternative Approaches to ANOVA

Before discussing the reasons why ANOVA and dummy-variable regression lead to the same kinds of tests, we demonstrate their numerical correspondence in terms of the sample problem. Table 3.2 shows the type of computer results that one commonly gets from a single-factor ANOVA. First, the descriptive results, $\overline{Y}_1$, $\overline{Y}_2$, $\overline{Y}_3$, appear. These are followed by a summary table of the quantities:

1. Among-groups sum of squares (SSA)
2. Within-groups sum of squares (SSW)
3. Total sum of squares (SST)

along with associated degrees of freedom, the among and within-groups mean squares (MSA and MSW) and the F ratio.

Without delving into details at this point we note that:

$$SSA = 8; \quad d.f. = 2; \quad MSA = 4$$
$$SSW = 15; \quad d.f. = 9; \quad MSW = 1.67$$
$$SST = 23; \quad d.f. = 11$$

and the F ratio is $4/1.67 = 2.4$, which, with 2 and 9 degrees of freedom, is *not significant* at the 0.05 level.

Next, let us take a look at the multiple regression results, also summarized in Table 3.2. First, we examine the ANOVA summary table obtained from the BMD-03R regression program. When this summary is compared to that obtained from the BMD-01V ANOVA program we find that the two are identical. Note further that the squared multiple correlation is:

$$R^2 = \frac{\text{SSR}}{\text{SST}}$$

$$= \frac{8}{23} = 0.348$$

The numerator of this expression, involving the sum of squares due to regression, is identical to the among-groups sum of squares SSA, while the denominator is identical to the total sum of squares SST obtained in the ANOVA run.

Next, let us examine the partial regression coefficients in Table 3.2. These are 1.0 and 2.0 for b_1 and b_2, respectively. The intercept is 5.5. Hence the regression equation is:

$$\hat{Y} = 5.5 + 1.0X_1 + 2.0X_2$$

As we know, the intercept of 5.5 is the mean response when both X_1 and X_2 are zero. Since we coded blackboard by the dummy-variable designation (0, 0), the intercept is simply the mean response for this variety of visual aid. The first partial regression coefficient shows the differential response when X_1 is at level 1. As recalled, we used the coding (1, 0) for the transparencies; therefore, the transparencies variety of visual aid contributes 1 test point to response above the reference level of blackboard. Similarly, under coding conditions (0, 1) for handouts, we find an average incremental response of 2 test points above the reference level of blackboard.

We conclude that the average response to blackboard is 5.5; that to transparencies is 6.5, and that to handouts is 7.5. A check of the ANOVA results shows, indeed, that the average responses to the three varieties of visual aids are precisely these numbers.

In summary, the multiple regression approach produces a set of results that are in full accord with those obtained from ANOVA. In particular, the ANOVA table provides, in the context of regression, a test of *among-group differences*. The partial regression coefficients measure the *differential response* of each treatment level vis-a-vis the base class. Moreover, the average response to the base class itself is given by the intercept b_0.[9]

In the next two sections of the chapter we shall consider an alternative formulation of the single-factor ANOVA model and the regression model that is analogous to this formulation.

Table 3.2 A Comparison of Single-Factor ANOVA and Dummy-Variable Regression

Single Factor ANOVA (BMD-01V)

| | Treatment Group | | | | | | | |
	Black-board	Trans-parencies	Hand-outs	Source	Sums of Squares	d.f.	Mean Squares	F
Sample Size	4	4	4	Among Groups	8	2	4.0	2.4*
Mean	5.5	6.5	7.5	Within Groups	15	9	1.67	
				Total	23	11		

Multiple Regression (BMD-03R)

$$R^2 = 0.348$$
$$\text{Intercept} = 5.5$$

ANOVA Table for Linear Regression

Source	d.f.	Sums of Squares	Mean Squares	F
Due to Regression	2	8	4.0	2.4*
Deviations About Regression	9	15	1.67	
	11	23		

Variable	Mean	Standard Deviation	Partial Reg. Coeff.	Standard Error of Reg. Coeff.	t Value	Partial Corr. Coeff.
X_1	0.333	0.492	1.000	0.913	1.095	0.343
X_2	0.333	0.492	2.000	0.913	2.191	0.590
Y	6.500	1.446				

*Not significant at the 0.05 level.

3.3.1 The Traditional ANOVA Approach

The adaptation of multiple regression models to ANOVA computations has been of relatively recent origin. A decade or so back, virtually all of the elementary textbooks provided detailed procedures for obtaining the necessary sums of squares and mean squares of various types of ANOVA designs, as illustrated in Table 3.2. The practice is still widespread. In the single-factor case, the *traditional* ANOVA model is commonly expressed as follows:

$$\boxed{Y_{ij} = \mu + \tau_j + \epsilon_{ij}}$$

where μ is a constant denoting the overall mean response and τ_j is the differential effect of the j-th level ($j = 1, 2, \ldots, n$) of the treatment variable. The error term ϵ_{ij} ($i = 1, 2, \ldots, m_j; j = 1, 2, \ldots, n$) is assumed to be NID $(0, \sigma^2)$;

that is, *normally and independently distributed with a mean of zero and a constant variance of* σ^2.

Moreover, it is customary to introduce one further condition regarding the τ_j, namely:

$$\sum_{j=1}^{n} \tau_j = 0$$

In this way of looking at things each τ_j is expressed as a *deviation* about the overall mean μ. In the case of the sample problem our earlier statement of the hypotheses:

$$H_0: \mu_1 = \mu_2 = \mu_3$$

H_1: Not all μ's are equal

is now transformed to the (equivalent) representation:

$$H_0: \tau_1 = \tau_2 = \tau_3 = 0$$

H_1: Not all of the τ's
are equal to zero

This formulation of the model is simply a different (but equivalent) parameterization. Subsequently, we shall show how the analogous regression model is modified to reflect this particular formulation.

In keeping with the traditional view of the ANOVA, we seek a decomposition of the total sum of squares SST (as measured about the *overall* mean of the response variable) into two parts:

1. Squared deviations of each individual treatment-level mean about the overall mean.

2. Squared deviations of each observation about its respective treatment-level mean.

This decomposition of squared deviations can be expressed as:

$$\underset{\text{[Total]}}{\sum_{j=1}^{n} \sum_{i=1}^{m_j} (Y_{ij} - \overline{Y}..)^2} = \underset{\text{[Among]}}{\sum_{j=1}^{n} m_j(\overline{Y}._j - \overline{Y}..)^2} + \underset{\text{[Within]}}{\sum_{j=1}^{n} \sum_{i=1}^{m_j} (Y_{ij} - \overline{Y}._j)^2}$$

where the dot-type subscript indicates that the variable so designated has been summed over in the computations. For example $\overline{Y}..$ denotes the overall mean while $\overline{Y}._j$ denotes the mean of the j-th treatment level, as computed after summing over the m_j observations comprising that treatment level. In the sample problem $m_j = 4$ for $j = 1, 2, 3$, since we have four replicates for each of the three treatment levels.

We can continue to use abbreviations similar to those employed in designating various sums of squares in regression analysis. In the present case we have:

$$SST = SSA + SSW$$

where SST denotes the total sum of squares of individual observations about the overall mean, SSA denotes the sum of squares due to differences among treatment-level means (as measured about the overall mean), and SSW denotes the (pooled) sum of squares within each treatment level, as measured from each of the individual treatment-level means.

Table 3.3 illustrates how one would traditionally go about conducting a one-way ANOVA. First, the data of Table 3.1 are shown in the conventional tabular form. As noted, the individual treatment-level means are:

$$\overline{Y}_1 = 5.5; \quad \overline{Y}_2 = 6.5; \quad \overline{Y}_3 = 7.5$$

while $\overline{Y}..$, the overall mean, is 6.5. The three sums of squares, described above, are computed as shown in the table, leading to the same quantities noted in the BMD-01V results of Table 3.2.

The degrees of freedom and mean squares of Table 3.3 are also shown in Table 3.2.[10] However, it is the last column of Table 3.3, labeled Expected Mean Square, that is of theoretical interest. First, we observe that the expectation of the mean square for the within-groups variation is:

$$E[SSW/(m - n)] = \sigma^2$$

The mean square for *within-levels* variation is an (unbiased) estimator of $\text{var}(\epsilon_{ij}) = \sigma^2$.

The *among-levels* mean square is also of interest:

$$E[SSA/(n - 1)] = \sigma^2 + \frac{1}{n - 1} \sum_{j=1}^{n} m_j \tau_j^2$$

If we examine the first term on the right, we note that it is simply σ^2 all over again. Therefore, *if the second term is zero,* we have another estimate (indeed, an independent estimate) of $\text{var}(\epsilon_{ij}) = \sigma^2$.

If the null hypothesis is *not* true, however, at least one τ_j is not equal to zero. In this case, the second term:

$$\frac{\sum_{j=1}^{n} m_j \tau_j^2}{n - 1}$$

Table 3.3 Parameter Estimates Obtained from Single-Factor ANOVA

Observation	Factor Levels		
	1	2	3
1	4	5	6
2	5	6	7
3	6	7	8
4	7	8	9
$\bar{Y}_{.j}$	5.5	6.5	7.5

$$\bar{Y}_{..} = 6.5$$

Source of Variation	Sums of Squares	d.f.	Mean Squares	Expected Mean Square
Among Levels	SSA $= 8$	$n - 1 = 2$	$\dfrac{\text{SSA}}{n-1} = 4$	$\sigma^2 + \dfrac{1}{n-1}\displaystyle\sum_{j=1}^{n} m_j \tau_j^2$
Within Levels	SSW $= 15$	$m - n = 9$	$\dfrac{\text{SSW}}{m-n} = 1.67$	σ^2
Total	SST $= 23$	$m - 1 = 11$		

$$\text{SSA} = \sum_{j=1}^{n} m_j (\bar{Y}_{.j} - \bar{Y}_{..})^2 = 4(5.5 - 6.5)^2 + 4(6.5 - 6.5)^2 + 4(7.5 - 6.5)^2$$
$$= 8$$

$$\text{SSW} = \sum_{j=1}^{n} \sum_{i=1}^{m_j} (Y_{ij} - \bar{Y}_{.j})^2 = (4 - 5.5)^2 + (5 - 5.5)^2 + \cdots + (8 - 7.5)^2 + (9 - 7.5)^2$$
$$= 15$$

$$\text{SST} = \sum_{j=1}^{n} \sum_{i=1}^{m_j} (Y_{ij} - \bar{Y}_{..})^2 = (4 - 6.5)^2 + (5 - 6.5)^2 + \cdots + (8 - 6.5)^2 + (9 - 6.5)^2$$
$$= 23$$

will be greater than zero. This term—often called a *noncentrality* parameter—reflects systematic differences in treatment-level means.[11] The larger this term, the greater the F ratio and the greater the tendency to reject the null hypothesis.

In essence, then, this is what ANOVA is about. Differences in treatment-level means introduce an additional (positive) term that increases the F ratio, tending to lead to rejection of the null hypothesis. (In terms of a multiple regression involving dummy variables, this will be tantamount to testing whether *any* of the partial regression coefficients b_j differ from zero.)

3.3.2 ANOVA via Multiple Regression

In Chapter 2 we placed considerable emphasis on the use of model comparison tests in multiple regression. These same ideas apply to ANOVA. As we recall, in model comparison tests we set up a model—conventionally called a *full* model—that contains various parameters of interest. We then find the error sum of squares associated with this model. Next, a *restricted* model is set up in which one or more of the parameters in the full model are missing. We find the error sum of squares associated with the restricted model. These two separate error sums of squares are then compared to see if the additional parameter(s) leads to a significant reduction in error.

In Table 3.1's *regression approach* to our sample problem we first set up the dummy-variable model:

$$Y_{ij} = \beta_0 + \beta_1 X_{i1} + \beta_2 X_{i2} + \epsilon_{ij}$$

in which we let $X_{ij} = 1$ if the observation came from the j-th treatment level and 0, otherwise. We let $j = 1$ denote the second category of transparencies and $j = 2$ denote the third category of handouts. As recalled, β_0 denotes the intercept term representing the effect due to the base category of blackboard. Observations in the base category were coded $(0, 0)$ for $j = 1$ and $j = 2$, respectively.

The BMD-03R regression results of Table 3.2 also reflected this type of coding in the intercept term ($b_0 = 5.5$) and in the partial regression coefficients:

$$b_1 = 1.0; \, b_2 = 2.0$$

Our current objective is to point out how the regression model of Table 3.1 can be readily transformed into the traditional ANOVA model:

$$Y_{ij} = \mu + \tau_j + \epsilon_{ij}$$

described in the preceding section. Continuing to let μ denote the overall mean of the data and τ_j denote the deviation from the overall mean associated with the j-th treatment ($j = 1, 2, 3$), we now have the regression model correspondences:

$$\beta_0 = \mu + \tau_1$$

$$\beta_1 = \tau_2 - \tau_1 = \tau_2 - \beta_0 + \mu$$

$$\beta_2 = \tau_3 - \tau_1 = \tau_3 - \beta_0 + \mu$$

We let the total-sample mean $\overline{Y}..$ estimate μ and t_j estimate τ_j. Then, by simple algebra and the parameter estimates obtained from the original dummy-variable coding used in the BMD-03R computations of Table 3.2, we have:

$$t_1 = b_0 - \overline{Y}.. \qquad = 5.5 - 6.5 \qquad = -1$$

$$t_2 = b_0 + b_1 - \overline{Y}.. = 5.5 + 1 - 6.5 = \quad 0$$

$$t_3 = b_0 + b_2 - \overline{Y}.. = 5.5 + 2 - 6.5 = \quad 1$$

In brief, then, we have merely changed our point of reference from which the effects are measured differentially. However, the two models are completely equivalent. Moreover, we can still use the sum of squares for deviations from regression in Table 3.2 to characterize the within-groups (or error) sum of squares in the ANOVA formulation:

$$\mathrm{SSE}_f = \mathrm{SSW} = \sum_{j=1}^{n} \sum_{i=1}^{m_j} (Y_{ij} - \overline{Y}_{\cdot j})^2 = 15$$

where SSE_f denotes the sum of squares for the *full* regression model. That is, SSE_f in the regression model is simply SSW in the ANOVA model. In terms of the regression approach, SSE_f is the residual sum of squares found after fitting all three regression parameters, b_0, b_1, and b_2.

Under the *restricted* regression model we assume that $b_1 = b_2 = 0$. In the regression context the appropriate sum of squares was computed around the overall mean of the criterion variable. Similarly, in the ANOVA context, we assume that all $\tau_j = 0$, so that $\overline{Y}..$ is the relevant estimate of all Y_{ij}, leading to:

$$\mathrm{SSE}_r = \mathrm{SST} = \sum_{j=1}^{n} \sum_{i=1}^{m_j} (Y_{ij} - \overline{Y}..)^2 = 23$$

where SSE_r denotes the sum of squares for the restricted model. As recalled from Chapter 2, the statistic for the models comparison test is:

$$F = \frac{\mathrm{SSE}_r - \mathrm{SSE}_f}{\mathrm{SSE}_f} \cdot \frac{d_f}{d_r - d_f} = \frac{8}{15} \cdot \frac{9}{2} = 2.4$$

These results have already been observed in the BMD-03R regression output of Table 3.2.

3.3.3 Concluding Remarks

This section of the chapter has tried to show—in the simplest (single-factor) case—the similarities between the traditional sum of squares decomposition that constitutes the computational apparatus of ANOVA and the multiple regression approach. Dummy-variable coding is the device that produces the compatibility between the two procedures.

Various forms of dummy-variable coding—some of which are directly analogous to the traditional ANOVA model—are available and we consider one of these alternatives in a later section of the chapter. However, none of these alternatives alters the appropriate sums of squares, SSR, SSE, and SST in the regression model. Their differences are confined to the individual regression parameters, b_0, b_1, b_2, etc., and to statistics based on these parameters.[12] The main thing to remember is that a particular system of dummy-variable coding leads to a particular set of regression parameters which should be made explicit in the interpretation of the regression results. As noted, $b_0 = 5.5$ could be interpreted as either the mean response for blackboard or, alternatively, as the overall mean ($\overline{Y}.. = 6.5$) less the treatment effect $t_1 = -1$ for blackboard.

In the beginning of section 3.2.1 we listed a set of four research questions dealing with fitting and testing a linear model to a set of response data where, in the present context, the design matrix of the linear model consists of dummy-valued predictors that denote various treatment levels. As has been illustrated, ANOVA is essentially a procedure for testing whether the *mean values of these responses differ across treatment levels.*

One question has not been considered as yet: if the treatment levels differ significantly in mean response, which *specific* levels are contributing to this general significance? That is, if

$$H_0: \mu_1 = \mu_2 = \mu_3$$

$$H_0: \tau_1 = \tau_2 = \tau_3 = 0$$

is rejected, which treatment levels are different and how are they different? This question, in turn, involves consideration of a series of detailed tests within the whole treatment and is part of the topic of *multiple comparisons.* (We defer detailed discussion of this topic until Chapter 5.)

3.4 Preliminary Aspects of ANCOVA

Up to this point, we have ignored the last column in Table 3.1 of the sample problem. This column, representing each student's pre-exposure test score, was called a *covariate*. It is now time to introduce it into the analysis.

3.4.1 The Objectives of ANCOVA

Earlier, we indicated that one of the primary purposes of covariance analysis was to *increase the precision* of an experiment by removing possible sources of variance in the criterion variable that are attributable to factors not being controlled by the researcher. If these influences can be removed statistically, the residual error can be decreased, resulting in a more sensitive experiment.

The potential value of covariance analysis lies in the existence of a substantial correlation between the covariate and the response variable—in the sample problem, pre-exposure versus post-exposure test scores. A rule of thumb (Cochran, 1957) suggests that the absolute value of this correlation should exceed 0.3, although the results will depend on the error variance without covariate adjustment and the degrees of freedom associated with the error variance.

Ideally, it would be nice to have the students take the pre-exposure test, find their scores, and then make up test groups of subjects that vary by pre-exposure scores. If two such groups: (a) low pre-exposure scores and (b) high pre-exposure scores, were set up, one could select an equal number of subjects randomly from each group for exposure to each of the three visual aid treatments levels. This rather simple idea of subdividing test units into more homogeneous subgroups is called *blocking,* and represents a common practice in experimentation. Blocking provides the advantage of being able to cope with departures in simple (e.g., linear) functional forms that relate Y to X and also permits the researcher to examine the possibility of block versus treatment-level interaction.[13]

Still, occasions arise in which it is just not practical to utilize blocking. For example, too many blocks might have to be made up to cover all treatment levels or the test units themselves may come in indivisible groups such as already-formed student classes. Moreover, covariance—unlike blocking—can be used *after the fact* in cases where it can be safely assumed that covariate values have not been altered by the treatments. For these (and related) reasons, covariance analysis is a useful topic to discuss.

Still looking at things intuitively, suppose we return to Table 3.1 and examine the behavior of the covariate, pre-exposure test score. First, we observe that post-exposure score does, indeed, tend to vary with it. Suppose we try a very simple thing, namely subtracting pre-exposure test score from post-exposure test score, giving us a set of *difference* scores:

Student	Difference $(Y - X_3)$	Student	Difference $(Y - X_3)$
a	1.8	g	0.3
b	2.1	h	−0.1
c	1.7	i	−2.3
d	1.8	j	−2.2
e	0.2	k	−1.8
f	−0.4	l	−2.1

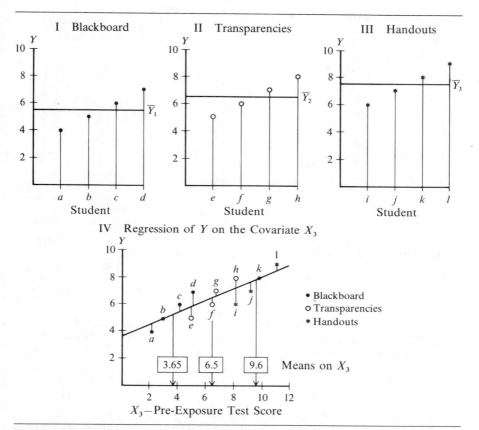

Figure 3.1 Comparing Variation in Response in Original Terms Versus Residuals from Regression of Y on the Covariate X_3

Intuitively, it appears that the within-groups variation might be decreased by means of this device since, in a sense, each subject is serving as his own control.

One could analyze these difference (or change) scores via ANOVA in just the same way that has already been described in section 3.3. (As we shall see later, analyzing change scores via ANOVA is, indeed, a type of ANCOVA under a restricted set of conditions.)

Let us plot Y, the response variable in Table 3.1, versus X_3, the covariate, and compare the variation in *residuals* from the regression line with variation in the original responses (without the introduction of X_3). Figure 3.1 shows this comparison. We see that introduction of the covariate leads to a set of $(Y_i - \hat{Y}_i)$ residuals in panel IV that seem to display less variability than the original values of Y, when we examine each treatment group separately in Panels I, II, and III of the figure.

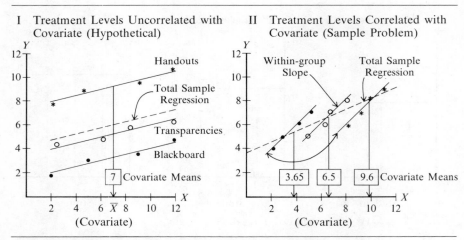

Figure 3.2 The Two Principal Types of Covariance Situations

Still, there is one disturbing note about Panel IV. If we imagine projecting the points onto the covariate (horizontal) axis, we see that the three treatment levels occupy different portions of the axis. Their means on the covariate:

$$\overline{X}_{31} = 3.65; \ \overline{X}_{32} = 6.5; \ \overline{X}_{33} = 9.6$$

are not equal. That is, *the treatment levels are correlated with the covariate.*[14]

Figure 3.2 shows the two basic situations that can occur regarding treatment levels versus covariate. Panel I illustrates (hypothetically) the case in which all treatment levels exhibit the *same mean* on the covariate, namely $\overline{X} = 7$. Since the slope of each within-group regression is the same *and* since their covariate means are equal, the slope of the total-sample regression is the same as the within-group slopes. In this case, the responses differ *only in terms of intercept* as we go across the three treatment levels. It is these differences in intercept that represent the effects of the treatment levels.

Panel II, however, shows the actual case for our sample problem. While within-group regression slopes appear to be equal, the treatment level means on the covariate are not. Since the *total-sample* regression line reflects differences among the covariate means as well as differences among the within-groups regression slopes, the total sample regression has a slope that is different from (actually less than) the slopes of the individual groups.

What this means in terms of the sample problem of Table 3.1 is that, despite random selection procedures, it just so happened that the treatment groups differed with regard to their mean values on the covariate. That is, the average pre-exposure test score was highest for treatment group 3 (handouts), next highest for group 2 (transparencies), and lowest for group 1 (blackboard).

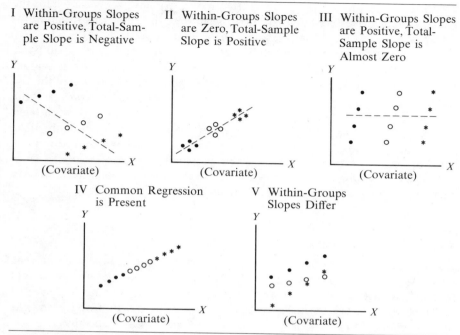

I Within-Groups Slopes are Positive, Total-Sample Slope is Negative

II Within-Groups Slopes are Zero, Total-Sample Slope is Positive

III Within-Groups Slopes are Positive, Total-Sample Slope is Almost Zero

IV Common Regression is Present

V Within-Groups Slopes Differ

Figure 3.3 Illustrations of Situations in Which Treatment Levels are Correlated with the Covariate

In experimental design situations in which test units are selected randomly, there will be no systematic bias operating but it may happen on occasion that the covariate means still differ, strictly as a reflection of sampling error. In non-experimental situations where test units cannot be sampled at random, high and systematic correlations may exist between the treatment levels and the covariate.

In an exaggerated way, Figure 3.3 shows some of the strange situations that can arise when treatment levels are correlated with the covariate. In Panel I we have a case in which the within-groups slopes are highly correlated positively and yet the total sample regression slope is negative.

Panel II shows a case of within-groups regression slopes that are zero and yet the total sample regression is positive. Panel III shows the converse in which high within-groups association is present while the total sample regression slope is virtually zero.

Panel IV shows a rather bizarre situation in which the treatment-level means not only differ on the covariate but all observations fall on a single (total sample) regression line. In this case removal of the influence of the covariate would remove all treatment-level effects as well since the residuals $(Y_i - \hat{Y}_i)$ would all be zero after covariate adjustment.

Panel V shows something that is a bit different from what we have been discussing so far. Up to now we have been assuming that all within-groups slopes are the same (up to sampling error). In Panel V we have the case where the within-groups slopes are unequal. As we shall describe later, this situation involves an interaction between covariate and treatment levels.[15]

As we shall discuss in more detail later, if the covariate means are all equal (within the range of experimental error), ANCOVA presents no difficulties whatsoever. In this case its purpose is strictly to *increase the precision* of the significance test regarding the equality of treatment-level effects. However, if the covariate means are unequal (i.e., the covariate is correlated with the treatment levels), ANCOVA can still be performed but the results are more ambiguous since the adjustment for differences in the covariate means can also remove some of the effects of the treatment levels along with it. Indeed, as Panel IV of Figure 3.3 shows in exaggerated fashion, removal of the covariate effect on Y can also remove *all* of the treatment-level effects.

3.4.2 The ANCOVA Model

Having described covariance at an introductory level, it is time to discuss the basic ANCOVA model. While the particular system for coding treatment levels is not critical, we shall continue to use the original coding of Table 3.1 in which the first treatment level (blackboard) receives a $(0, 0)$ code and the second and third treatment levels are coded $(1, 0)$ and $(0, 1)$ respectively.

We recall that the traditional ANOVA model considered in section 3.3 was:

$$Y_{ij} = \mu + \tau_j + \epsilon_{ij}$$

where μ denotes the overall mean response and τ_j $(j = 1, 2, \ldots, n)$ denotes the differential effect of treatment j on Y, subject to the condition:[16]

$$\sum_{j=1}^{n} \tau_j = 0$$

As before, ϵ_{ij} denotes the error term.

In ANCOVA, one augments the preceding model by the explicit introduction of another term X_{ij} representing the covariate:[17]

$$Y_{ij} = \mu + \tau_j + \beta(X_{ij} - \overline{X}..) + \epsilon_{ij}$$

The variable X_{ij} denotes the value of observation i $(i = 1, 2, \ldots, m_j)$ on treatment j and β denotes the pooled *within-groups* slope. (In the case of Panel I of Figure 3.2 this is also the total sample regression slope.) The X_{ij} are assumed to be fixed values while the ϵ_{ij} are random. As before, the ϵ_{ij} are assumed to be NID $(0, \sigma^2)$.

One way of thinking about covariance is to compare its error term directly with that of the basic single-factor ANOVA model:

$$Y_{ij} = \mu + \tau_j + \epsilon_{ij}$$

Since the only random portion of Y_{ij} is contributed by ϵ_{ij} and since $\sum_{j=1}^{n} \tau_j = 0$, we have a set of Y_{ij} that are also NID $(0, \sigma^2)$. Next, if we write the covariance model as:

$$Y_{ij} = \mu + \tau_j + \epsilon_{ij}^* + \beta(X_{ij} - \overline{X}..)$$

we can view $\beta(X_{ij} - \overline{X}..)$ as representing that part of Y that is accounted for by changes in X, while ϵ_{ij}^* is the *residual* error that still remains. In this side-by-side comparison of the two models we can write:

$$\epsilon_{ij} = \epsilon_{ij}^* + \beta(X_{ij} - \overline{X}..)$$

Then, assuming that the two components of ϵ_{ij} are independent, we have:

$$\sigma^2(\epsilon) = \sigma^2(\epsilon^*) + \sigma^2 \text{ (due to } X)$$

which asserts that $\sigma^2(\epsilon^*)$, *after* covariance adjustment, is no more than $\sigma^2(\epsilon)$, the error variance in the usual ANOVA context that ignores the covariate X.

Given the presence of the term $(X_{ij} - \overline{X}..)$ and assuming that $\beta \neq 0$, we no longer have the case of a single mean response to each treatment level. Rather, the response within treatment level will vary with changes in X_{ij}. However, when $(X_{ij} - \overline{X}..)$ is zero, the *intercept* for the j-th treatment level measures $\mu + \tau_j$.

Since the assumption is made in the ANCOVA model that all of the individual treatment-level slopes are equal, *differences in intercept terms measure the effect of various treatment levels.* These differences remain constant over variations in the values of X_{ij}. Figure 3.4 shows the basic idea that is involved here for a hypothetical case of three treatment levels. As shown, differences in treatment-level means are characterized by differences in intercepts. Clearly, the slopes of the individual treatment-level regressions must be equal to justify the assumption of a *constancy of intercept differences* throughout the range of the X_{ij}'s.

We can also note from Figure 3.4 that if the null hypothesis:

$$H_0: \tau_1 = \tau_2 = \tau_3 = 0$$

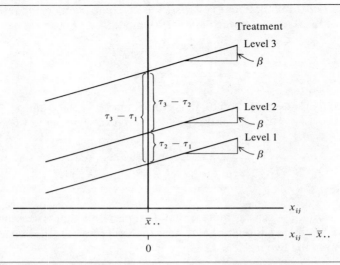

Figure 3.4 Treatment-Level Regressions in the ANCOVA Model (Hypothetical Situation)

is true, all treatment-level regression lines must be the same so that all intercept differences are zero.[18]

Translation of the basic ANCOVA model into the *regression format* of the general linear model is straightforward. Since X_1 and X_2 have earlier been described as dummy-valued variables, let us differentiate these from the covariate by making a simple notational change. We shall let:

$$W_1 = \left.\begin{array}{c} 1 \\ 0 \end{array}\right\} \begin{array}{l} \text{if observation is from treatment level 2} \\ \text{if otherwise.} \end{array}$$

$$W_2 = \left.\begin{array}{c} 1 \\ 0 \end{array}\right\} \begin{array}{l} \text{if observation is from treatment level 3} \\ \text{if otherwise.} \end{array}$$

while X continues to denote the covariate. With this notational change, we have the model:

$$Y_{ij} = \beta_0 + \beta_1 W_{i1} + \beta_2 W_{i2} + \beta_3 (X_{ij} - \overline{X}..) + \epsilon_{ij}$$

where we now let ϵ_{ij} denote the error term for *this* model. In view of our procedure that codes treatment level 1 as (0, 0), we have the following correspondences between the regression formulation and the traditional ANCOVA model:

$$\beta_0 = \mu + \tau_1$$

$$\beta_1 = \tau_2 - \tau_1$$

$$\beta_2 = \tau_3 - \tau_1$$

$$\beta_3 = \beta$$

As before, tests concerning treatment effects in the traditional ANCOVA model can be translated into tests of various regression models, a point that is illustrated in the next section.

What has just been described above is the *basic* ANCOVA model, in either traditional or regression format. It should be reiterated, however, that a fundamental assumption of the model—namely that all individual treatment-level slopes are equal—must hold before the main test of equality of intercept is applied. As noted in Panel V of Figure 3.3, this assumption may not hold; if it does not, the analysis is stopped. Hence, in the next sections we shall consider a more elaborate version of the basic ANCOVA model, one that permits testing individual treatment-level slopes *before* the main test of intercept equality is considered.

3.5 Alternative Approaches to ANCOVA

Having formulated the basic ANCOVA model, both traditionally and in terms of the regression format, it is useful to demonstrate that the two approaches yield the same numerical results.

3.5.1 Traditional ANCOVA

Long before the advent of computers, procedures were developed for ANCOVA that were suited to desk calculators; these traditional methods still appear in most introductory books on the subject. Traditional ANCOVA works on the basis of an adjustment that is made to the basic ANOVA model. That is, we begin the computations with the regular ANOVA model; the quantities of this model are then adjusted for the covariate and ANOVA is performed on these adjusted values.

In Table 3.2 we showed how ANOVA decomposes the total sum of squares around $\overline{Y}$ into among-groups and within-groups portions. Adapting our earlier notation to the present context, we have the relationship:

$$\boxed{SSA(Y) + SSW(Y) = SST(Y)}$$

What is done with the criterion variable Y can also be done with the covariate X and the cross-product term YX.

Table 3.4 shows the initial calculations required to develop the sums of squares and cross products *before* removing the covariate effect. For example:

$$\text{SST}(Y) = \sum_{j=1}^{n} \sum_{i=1}^{m_j} (Y_{ij} - \overline{Y}..)^2 = \sum_{j=1}^{n} \sum_{i=1}^{m_j} Y_{ij}^2 - \left(\sum_{j=1}^{n} \sum_{i=1}^{m_j} Y_{ij}\right)^2 \Big/ m$$

$$= 530 - (78)^2/12 = 23$$

$$\text{SST}(X) = \sum_{j=1}^{n} \sum_{i=1}^{m_j} (X_{ij} - \overline{X}..)^2 = \sum_{j=1}^{n} \sum_{i=1}^{m_j} X_{ij}^2 - \left(\sum_{j=1}^{n} \sum_{i=1}^{m_j} X_{ij}\right)^2 \Big/ m$$

$$= 606.06 - (79)^2/12 = 85.98$$

$$\text{SST}(YX) = \sum_{j=1}^{n} \sum_{i=1}^{m_j} (Y_{ij} - \overline{Y}..)(X_{ij} - \overline{X}..)$$

$$= \sum_{j=1}^{n} \sum_{i=1}^{m_j} Y_{ij}X_{ij} - \left(\sum_{j=1}^{n} \sum_{i=1}^{m_j} Y_{ij} \sum_{j=1}^{n} \sum_{i=1}^{m_j} X_{ij}\right) \Big/ m$$

$$= 552.1 - (78)(79)/12$$

$$= 38.6$$

Similarly, we could find the other quantities of interest by means of the analogous formulas shown in Table 3.4 (and the computational equivalences shown above).

Having found the unadjusted sums of squares and cross products in Table 3.4, we can next adjust these to obtain the quantities that are used in the ANCOVA test. In effect, adjustments are found by regressing Y on X—either in terms of the total sample, the among-groups means or the within-groups deviations. Illustrating the procedure for the sum of squares involving the total sample, we have:

$$\text{SST(Adj. } Y) = \text{SST}(Y) - b_T\text{SST}(YX)$$

$$= \text{SST}(Y) - \frac{[\text{SST}(YX)]^2}{\text{SST}(X)}$$

$$= 23 - \frac{(38.6)^2}{85.98} = 5.67$$

In other words SST(Adj.Y) is simply the squared residuals obtained from regressing Y on X in terms of the total sample.

Table 3.4 Computational Summary of ANCOVA—Sample Data from Table 3.1

Preliminary Calculations	Group 1 (Black-board)	Group 2 (Trans-parencies)	Group 3 (Hand-outs)	Total
m_j	4	4	4	12
$\sum Y$	22	26	30	78
$\sum X$	14.6	26	38.4	79
$\sum Y^2$	126	174	230	530
$\sum XY$	85.5	174.1	292.5	552.1
$\sum X^2$	58.78	174.50	372.78	606.06

Sums of Squares and Cross Products—Before Covariate Adjustment

Source	Y	X	YX	d.f.
Treatments	8	70.85	23.8	2
Error	15	15.13	14.8	9
Total	23	85.98	38.6	11

Formulas

Total:

$$\text{SST}(Y) = \sum_{j=1}^{n} \sum_{i=1}^{m_j} (Y_{ij} - \overline{Y}..)^2$$

$$\text{SST}(X) = \sum_{j=1}^{n} \sum_{i=1}^{m_j} (X_{ij} - \overline{X}..)^2$$

$$\text{SST}(YX) = \sum_{j=1}^{n} \sum_{i=1}^{m_j} (Y_{ij} - \overline{Y}..)(X_{ij} - \overline{X}..)$$

Among:

$$\text{SSA}(Y) = \sum_{j=1}^{n} m_j(\overline{Y}._j - \overline{Y}..)^2$$

$$\text{SSA}(X) = \sum_{j=1}^{n} m_j(\overline{X}._j - \overline{X}..)^2$$

$$\text{SSA}(YX) = \sum_{j=1}^{n} m_j(\overline{Y}._j - \overline{Y}..)(\overline{X}._j - \overline{X}..)$$

Within:

$$\text{SSW}(Y) = \sum_{j=1}^{n} \sum_{i=1}^{m_j} (Y_{ij} - \overline{Y}._j)^2$$

$$\text{SSW}(X) = \sum_{j=1}^{n} \sum_{i=1}^{m_j} (X_{ij} - \overline{X}._j)^2$$

$$\text{SSW}(YX) = \sum_{j=1}^{n} \sum_{i=1}^{m_j} (Y_{ij} - \overline{Y}._j)(X_{ij} - \overline{X}._j)$$

What can be done for the SST(Y) can also be done for SSW(Y):

$$\text{SSW(Adj. } Y) = \text{SSW}(Y) - b_w\text{SSW}(YX)$$

$$= \text{SSW}(Y) - \frac{[\text{SSW}(YX)]^2}{\text{SSW}(X)}$$

$$= 15 - \frac{(14.8)^2}{15.13} = 0.52$$

Finally, we can obtain the quantity SSA(Adj. Y), by subtraction, as 5.15.[19] These quantities appear in Table 3.5. Note that one additional degree of freedom is lost via the estimation of the (pooled) within-groups regression slope. This slope can be easily obtained from Table 3.4 as:

$$b_w = \frac{14.8}{15.13} = 0.98$$

On the other hand, the total sample regression slope is:

$$b_T = \frac{38.6}{85.98} = 0.45$$

as also found from Table 3.4.[20]

However, before proceeding with the main test of the equality of covariate-adjusted treatment level effects, we conduct the prior test of equality of individual treatment-level slopes. From Table 3.5 we note that these slopes are quite similar:

Group 1 (Blackboard):	$b_1 = 0.95$
Group 2 (Transparencies):	$b_2 = 0.93$
Group 3 (Handouts):	$b_3 = 1.09$

The pooled within-groups slope is $b_w = 0.98$. The associated F ratio is only 0.47 and, hence, we accept the null hypothesis of equality of individual regression slopes.[21]

The main test of ANCOVA—namely a test of the equality of covariate-adjusted treatment level effects—is then carried out in Table 3.5. The adjusted treatment mean square is 2.575; that for error is 0.065. The associated F ratio of 39.38, with 2 and 8 degrees of freedom, is highly significant at the 0.05 level. We conclude that treatment-level effects differ, once the influence of the covariate is removed.

This result is in sharp contrast to the ANOVA results of section 3.3.1 where the null hypothesis of equality of treatment-level effects was accepted. Thus, *after* covariate adjustment, the treatment-level effects are found to be significantly different.

Table 3.5 Traditional ANCOVA via Adjustment of Sums of Squares and Cross Products

Source	Adjusted Sums of Squares	Adjusted d.f.	Mean Squares	F
Treatments	5.15	2	2.575	39.38
Error	0.52	8	0.065	
	5.67	10		

Individual Regression Slopes

Source	Sum of Squares before Covariate Adjustment			
	Y	X	YX	Slope
Group 1 (Blackboard)	5	5.49	5.2	0.95
Group 2 (Transparencies)	5	5.50	5.1	0.93
Group 3 (Handouts)	5	4.14	4.5	1.09
Total	15	15.13	14.8	$0.98 = b_w$ (Pooled Slope)

Sum of Squared Deviations from Individual Regressions

$$[5 - (5.2)^2/5.49] + [5 - (5.1)^2/5.5] + [5 - (4.5)^2/4.14] = 0.45$$

Squared Deviations from Pooled Within-Groups Regression

$$[15 - (14.8)^2/15.13] = 0.52$$

Commonality of Individual Slopes: $F = \dfrac{0.07}{0.45} \cdot \dfrac{6}{2} = 0.47$

3.5.2 ANCOVA Via Multiple Regression

Application of the traditional ANCOVA approach in Table 3.5 has resulted in two principal findings:

1. Acceptance of the null hypothesis ($F = 0.47$) that the three within-treatment level regression slopes are equal. Acceptance of this hypothesis is a precondition for carrying out the main test of covariate adjusted treatment differences.

2. Rejection of the null hypothesis ($F = 39.38$) that the covariate-adjusted treatment level effects are equal. This is the main test of interest.

We now want to show that both of these tests can be conducted within the models comparison approach of the regression model by the incorporation of dummy variable predictors.

First, we set up the *full* regression model:

$$Y_{ij} = \beta_0 + \beta_1 W_{i1} + \beta_2 W_{i2} + \beta_3(X_{ij} - \overline{X}..)$$
$$+ \beta_4 W_{i1}(X_{ij} - \overline{X}..) + \beta_5 W_{i2}(X_{ij} - \overline{X}..) + \epsilon_{ij}$$

The only difference between this formulation and the regression model described at the end of section 3.4 is that we have introduced two new cross-product terms and their associated partial regression coefficients, β_4 and β_5.

Table 3.6 Variables Setup for Multiple Regression of Table 3.1 Data

Observation	Y	W_1	W_2	X_d	W_1X_d	W_2X_d
a	4	0	0	−4.4	0	0
b	5	0	0	−3.7	0	0
c	6	0	0	−2.3	0	0
d	7	0	0	−1.4	0	0
e	5	1	0	−1.8	−1.8	0
f	6	1	0	−0.2	−0.2	0
g	7	1	0	0.1	0.1	0
h	8	1	0	1.5	1.5	0
i	6	0	1	1.7	0	1.7
j	7	0	1	2.6	0	2.6
k	8	0	1	3.2	0	3.2
l	9	0	1	4.5	0	4.5

This complication is necessary for developing the *full* ANCOVA model— one that permits a test of common treatment-level slopes before conducting the principal test of equality of treatment-level effects. Since W_1 and W_2 are zero-one dummies, β_4 and β_5 are estimating *differential slopes* for treatment level 2 compared to level 1 and treatment level 3 compared to level 1. Estimates of the six parameters are given by: b_0, b_1, b_2, b_3, b_4 and b_5, the partial regression coefficients.[22] If we find that the null hypothesis $\beta_4 = \beta_5 = 0$ holds, then we have no evidence of different slopes within treatment levels.

At this point, we are ready to apply the regression approach to the sample problem data in Table 3.1. The data values to be read into the multiple regression program (BMD-03R) appear in Table 3.6.

In actuality, several regressions on the same data are needed for making the various hypothesis tests:

1. Y on W_1, W_2, X_d, W_1X_d, and W_2X_d.[23]
2. Y on W_1, W_2 and X_d.
3. Y on X_d.
4. Y on W_1 and W_2.

Table 3.7 shows the R^2 values found from each of these regressions.

As noted earlier, two principal research questions are present in any ANCOVA application:

1. Are the individual treatment-level regression slopes equal?
2. If the hypothesis of equal slopes is accepted, do the treatment-level means differ after covariate adjustment?

In section 3.2 (and in Chapter 2 as well), extensive use was made of the models-comparison approach in which we find the quantity:

$$F = \frac{SSE_r - SSE_f}{SSE_f} \cdot \frac{d_f}{d_r - d_f}$$

and then compare sums of squares for restricted and full models. As can be seen from Table 3.7, a variety of model comparisons can be set up for carrying out the required tests.

First of all, let us consider whether the treatment-level slopes are equal. In this case the full model computes all six parameters $b_0, b_1, \ldots, b_5$; its R^2 is 0.980 while R^2 for a restricted model that omits the cross-product terms W_1X_d and W_2X_d is 0.977, a difference of only 0.003. The total sum of squares about $\overline{Y}$ to be accounted for is SST = 23.

If we multiply 0.003 by SST = 23 we get 0.07 as the incremental sum of squares. The error sum of squares for the full model (as also shown in Table 3.7) is 0.45. Taking note of the appropriate degrees of freedom[24] for each model, we have:

$$F = \frac{SSE_r - SSE_f}{SSE_f} \cdot \frac{d_f}{d_r - d_f}$$

$$= \frac{(0.980 - 0.977)23}{0.45} \cdot \frac{6}{2} = 0.47$$

Since $F = 0.47 < 1$, the result is clearly not significant and we conclude that all treatment-level slopes are equal.[25]

Having accepted the null hypothesis of equal treatment-level regression slopes, we can proceed to the second (and main) comparison of models involving

Table 3.7 Applying the BMD-03R Multiple Regression Model to the Data of Table 3.6

Regression Run	R^2
Y on X_d	0.754
Y on W_1, W_2, X_d	0.977
Y on W_1, W_2	0.348
Y on $W_1, W_2, X_d, W_1X_d, W_2X_d$	0.980

Regression-Type ANCOVA (from BMD-03R)

Source	d.f.	Sums of Squares		Mean Squares
Covariance Adjustment	1	0.754(23)	= 17.33	17.33
Adjusted Group Means	2	(0.977 − 0.754)(23)	= 5.15	2.57
Difference in Group Regressions	2	(0.980 − 0.977)(23)	= 0.07	0.04
Error	6	(1.000 − 0.980)(23)	= 0.45	0.08
Total	11		23.00	

the treatment-level means. In this case the full model includes W_1, W_2 and X_d while the restricted model includes only X_d (and in each case the intercept).[26] We have:

$$F = \frac{(0.977 - 0.754)23}{(0.45 + 0.07)} \cdot \frac{8}{2} = 39.38$$

The source entitled Adjusted Group Means in Table 3.7 represents the effect of treatment differences in intercept levels, after taking into account the within-groups regression of Y on X. This difference in R^2 is $0.977 - 0.754 = 0.223$ and is significant. The accounted-for sum of squares is:

$$0.223(23) = 5.15$$

This quantity represents the difference in the accounted-for sum of squares between a model that includes the covariate *and* the treatments versus one that includes only the covariate.

The denominator in the first term of the F ratio represents the pooled error sum of squares of $23(1.000 - 0.977) = 0.52$ based on the present full model (since we have already accepted the null hypothesis that all individual treatment-level slopes are equal). As such, we are now dealing with the basic covariance model that assumes at the outset that these slopes are equal.

Most importantly, the F value of 39.38, with 2 and 8 degrees of freedom, is highly significant beyond the 0.05 level.[27] Inclusion of the covariate has made the test of treatment-level mean differences more sensitive so that the null hypothesis of equality of treatment-level means is now rejected.

In addition to this main result we could also test to see whether the covariance adjustment is significant (it is overwhelmingly so). Furthermore, although not shown in Table 3.7, the estimate of the pooled within-groups regression slope is $b_3 = 0.98$, as represented in the expression:

$$\boxed{\hat{Y}_{ij} = b_0 + b_1 W_{i1} + b_2 W_{i2} + b_3(X_{ij} - \overline{X}..)}$$

The estimate of the total-sample regression slope is $b_T = 0.45$, as represented in the expression:

$$\boxed{\hat{Y}_{ij} = b_0 + b_T(X_{ij} - \overline{X}..)}$$

in which the dummy variables W_1 and W_2 are ignored.[28]

In short, the traditional ANCOVA and the regression approaches yield precisely the same results. To recapitulate:

1. The full ANCOVA approach first tests to see if the within-groups slopes of all treatment levels are equal. If not, the analysis is terminated.

2. If so, one proceeds to conduct the main test involving the equality of covariate-adjusted treatment-level effects.

3. In the regression approach, testing for the equality of within-groups slopes is equivalent to a test of *interaction* between the covariate and the treatment-level dummy variables.

4. In the regression approach the main test is equivalent to a test of a model that includes both treatment-level dummies *and* the covariate versus one that includes only the covariate.

If the treatment variable and the covariate are uncorrelated, removal of the covariate presents no problem. However, if the treatment variable and the covariate *are* correlated (as was the case in the sample problem), the results are more ambiguous. Removal of the covariate effect also can remove some of the effect of the treatment variable, simply because the covariate is credited with any association that it shares with the treatment variable.

3.5.3 Concluding Remarks

Now that we have illustrated the compatibility of the traditional and regression approaches to ANCOVA, we should consider the various ways in which the technique can be used.

Many researchers restrict their use of covariance to an experimental situation where the continuous variable (X in this case) always plays the role of covariate. However, there is no necessity to do so. As Blalock (1960) has suggested, either the categorical or the continuous variable, depending on the situation, can play the role of covariate.

For example, in Panel I of Figure 3.2 we could assume that our interest is primarily focused on the variable X and now let the categorical treatment variable become the covariate. If we were to adjust for the effect of treatment intercept differences (by subtracting out each intercept), the points would all fall on a single regression line (the dotted line in Panel I). Y could then be estimated perfectly from changes in X, once the "covariate" (represented by the qualitative treatment variable) has been partialed out. As such, precision has increased dramatically.

However, it is also relevant to point out that occasions can arise in which covariance adjustment can actually decrease precision. For example, suppose the covariate (e.g., X, the continuous variable) is highly related to the treatment variable and *not* highly related to the response variable Y. Covariance adjustment in this case could remove a large part of the treatment effect in the course of partialing out the covariate, thereby reducing precision.

Before leaving the topic of single-factor covariance analysis some additional points should be mentioned:

1. ANCOVA can be rather straightforwardly extended to deal with two or more treatments (each at two or more levels) and two or more covariates.

2. The functional form of the covariate term is not required to be linear. For example, functions of the form $Y_{ij} = \mu + \tau_j + \beta_1(X_{ij} - \overline{X}..) + \beta_2(X_{ij} - \overline{X}..)^2 + \epsilon_{ij}$ could be used. Note, however, that linearity in the parameters *is* maintained.

3. The practice of using difference or change scores (of the type involving post-exposure test score minus pre-exposure test score) in place of ANCOVA reduces precision *unless the within-groups slope is close to unity*. In this latter case we have: $Y_{ij} = \mu + \tau_j + X_{ij} + \epsilon_{ij}$ or $Y_{ij} - X_{ij} = \mu + \tau_j + \epsilon_{ij}$ if we take X_{ij} in original rather than mean-corrected form. However, the ANOVA of difference scores offers little advantage in the age of rapid calculation by computers.[29]

Perhaps the most controversial aspect of ANCOVA concerns what situations are appropriate for using it. Evans and Anastasio (1968) describe three prototypical contexts in which ANCOVA might be considered:

1. Individuals are assigned to treatment levels at random; the covariate is unaffected by the treatment variable.

2. Intact groups are used (e.g., various student classes) but treatment levels are assigned at random to the groups; the covariate is unaffected by the treatment variable.

3. Intact groups and treatment levels occur together in a natural fashion.

In the first case no controversy arises and, within sampling variability, the treatment-level means on the covariate should be equal.

In the second case Evans and Anastasio suggest caution in the application of ANCOVA. Finally, they feel that ANCOVA is definitely inappropriate in case three. Instances of the third case are more prevalent in the analysis of survey data where the "treatment" may merely be a classificatory variable rather than representing the result of experimental intervention. While we later illustrate this type of application in the context of the TV commercial data, the reader should be alerted to the possibility that covariance adjustment can remove possible classificatory effects if the classification variable is highly correlated with the covariate. If the covariate is credited with *all* accounted-for variance that it shares with the classificatory variable, the independent effect of the latter may be minuscule.

There is nothing *intrinsically* wrong with this procedure for apportionment of shared variance but substantive interpretation of the results may be less than clear. (This topic is considered more generally in Chapter 5.)

3.6 Analyzing TV Commercial Data

Up to this point our emphasis has been strictly on the application of ANOVA and ANCOVA to single-factor designs. This class of design was characterized by the presence of one factor or treatment variable, expressed in terms of some number of levels. Multiple-factor designs extend the purview of ANOVA and ANCOVA to cases in which we have two or more factors, each at two or more levels.

If all combinations of the factor levels are included in the design, this situation is usually called a *full factorial*. (If, for a variety of possible reasons, it is not possible or practical to include all combinations, the term *incomplete factorial* is often used to designate the situation.) Multiple-factor designs possess a number of advantages over conducting a series of single-factor studies—primarily, higher efficiency and the opportunity to measure interaction effects.

Here, we shall emphasize complete factorial designs, since the nice algebraic properties of traditional multiple-factor ANOVA and ANCOVA are lost when the design has missing observations.[30] We briefly describe the simplest of multiple-factor designs, namely the two-factor model, and then show the computerized results of applying ANOVA and ANCOVA to the TV commercial data. Then, in section 3.7 we continue our discussion of the two-factor model (including such aspects as the interpretation of interaction effects) and show how the regression approach compares to traditional ANOVA, when applied to this class of designs.

3.6.1 The Two-Factor ANOVA Model

Analysis of variance for the two-factor design with an equal number of replications per cell proceeds analogously to the single factor case. Table 3.8 shows the traditional ANOVA estimators and decomposition of sums of squares for the two-factor design with r replications. This table is analogous to Table 3.3 for the single-factor case. The two-factor model is:

$$Y_{ijk} = \mu + \alpha_i + \beta_j + (\alpha\beta)_{ij} + \epsilon_{ijk}$$

subject to the constraints:

$$\sum_{i=1}^{a} \alpha_i = \sum_{j=1}^{b} \beta_j = \sum_{i=1}^{a} (\alpha\beta)_{ij} = \sum_{j=1}^{b} (\alpha\beta)_{ij} = 0$$

and where:

$$\epsilon_{ijk} \text{ is NID}(0, \sigma^2)$$

for $i = 1, 2, \ldots, a; j = 1, 2, \ldots, b; k = 1, 2, \ldots, r; m = abr$.

We do not delve into the computational forms for obtaining the various sums of squares in Table 3.8 for the reason that nowadays virtually all ANOVA problems that are multiple-factor in nature are analyzed via computer programs. Accordingly, a specific ANOVA summary table for the two-factor orthogonal design is presented in the context of an actual set of data drawn from the TV commercial study.

Table 3.8 Point Estimates and Decomposition of Sums of Squares for the Two-Factor ANOVA Design

Parameter	Estimator
μ	$\overline{Y}...$
α	$\overline{Y}_i.. - \overline{Y}...$
β	$\overline{Y}._j. - \overline{Y}...$
$(\alpha\beta)_{ij}$	$\overline{Y}_{ij}. - \overline{Y}_i.. - \overline{Y}._j. + \overline{Y}...$

Among-Treatments Sum of Squares

$$SSA = SS(\alpha) + SS(\beta) + SS(\alpha\beta) = r \sum_{i=1}^{a} \sum_{j=1}^{b} (\overline{Y}_{ij}. - \overline{Y}...)^2$$

$$SS(\alpha) = rb \sum_{i=1}^{a} (\overline{Y}_i.. - \overline{Y}...)^2$$

$$SS(\beta) = ra \sum_{j=1}^{b} (\overline{Y}._j. - \overline{Y}...)^2$$

$$SS(\alpha\beta) = r \sum_{i=1}^{a} \sum_{j=1}^{b} (\overline{Y}_{ij}. - \overline{Y}_i.. - \overline{Y}._j. + \overline{Y}...)^2$$

Within-Treatments Sum of Squares

$$SSW = \sum_{i=1}^{a} \sum_{j=1}^{b} \sum_{k=1}^{r} (Y_{ijk} - \overline{Y}_{ij}.)^2$$

Total Sum of Squares

$$SST = \sum_{i=1}^{a} \sum_{j=1}^{b} \sum_{k=1}^{r} (Y_{ijk} - \overline{Y}...)^2$$

3.6.2 The TV Commercial Study

The preceding examples, employing artificial data, have served to illustrate the principles, and limitations, of analysis of variance and covariance, including their relationship to multiple regression and the general linear model. However, it is now time to consider a problem of realistic size. In the television commercial study data were available on:

1. Whether or not Alpha was the last tire brand purchased (A-2).
2. Pre-exposure interest in the Alpha brand of radial tires (A-3).
3. Post-exposure (following the commercial) interest in the Alpha brand of radial tires (C-2).
4. Demographic variables (Part E).

These same variables, along with A-1, were used in the multiple regression analyses of the preceding chapter. However, this time we would like to illustrate the application of ANOVA and ANCOVA. We continue to let post-exposure interest in Alpha (variable C-2) be the response variable.

For illustrative purposes we let A-2 and A-3 be "treatment" variables (expressed categorically) and the seven variables of Part E of the questionnaire serve as covariates. Variable A-2 is already expressed categorically but A-3 is not. However, the A-3 ratings can be categorized into "high" versus "low" classes by (arbitrarily) classifying a respondent as high if his pre-exposure rating of Alpha on question A-3 is 8, 9 or 10 and low if his rating is 0, 1, or 2.

Since our interest in demonstrating ANOVA and ANCOVA via computer procedures is confined to the balanced-design case, we desire to have an equal number of respondents in each of the following cells:

Brand Last Purchased was Alpha (A-2)	Pre-Exposure Interest in Alpha (A-3)		Total
	High	Low	
Yes	17	17	34
No	17	17	34
Total	34	34	68

To achieve equal cell sizes, we had to search the data to find an equal number of replicates for each of the two categorial variables. The search led to 17 respondents in each of the four cells above.

A number of computer programs are available for conducting analysis of variance and covariance in cases where orthogonal designs are involved. In particular, BMD-12V (part of the UCLA Biomedical package) is quite versatile in the sense of being able to handle both univariate and multivariate analysis of variance and covariance.

3.6.3 The BMD-12V Analysis

This program is set up for orthogonal designs in which an equal number of cases appear in each cell of the factor design matrix. While the program is designed for both univariate and multivariate analysis of variance, here we confine our application to univariate ANOVA and ANCOVA. Two analyses were run— one with and one without the seven covariates (demographic variables).

Table 3.9 shows the summary table for the 2 × 2 ANOVA case. Also shown are the mean responses for each of the 2 × 2 cells. As can be noted from the large F value of 13.79, pre-exposure interest (high versus low) leads to significant differences in post-exposure response; the tabular F value (in Table C.4) for an alpha risk of 0.05 is only 4.0. As can be seen from the table of mean responses, the difference in rating points is $7.68 - 5.32 = 2.36$.

Table 3.9 Two-Way Analysis of Variance of Post-Exposure Interest in the Alpha Brand

Source	Sums of Squares	d.f.	Mean Squares	F
Alpha was Last Brand Purchased	5.88	1	5.88	
Pre-Exposure Interest in Alpha Radials	94.12	1	94.12	13.79
Interaction	0.06	1	0.06	
Residual	436.93	64	6.83	

Mean Responses on Post Exposure Interest

	Pre-Exposure Interest in Alpha Radials		
Alpha was Last Brand Purchased	High	Low	Marginal Means
Yes	8.00	5.59	6.79
No	7.35	5.06	6.21
Marginal Means	7.68	5.32	6.50

On the other hand, the effect due to whether or not Alpha represented the last tire brand purchased, is not statistically significant (F value < 1) and neither is the interaction term. As noted from the table of means, this main-effect difference in rating points of $6.79 - 6.21 = 0.58$ is rather small.

Table 3.10 shows the comparison of ANCOVA in which the demographics serve as covariates. Also shown are the cell means adjusted for all seven of the

Table 3.10 Two-Way Analysis of Covariance of Post-Exposure Interest in Alpha Brand

Source	Sums of Squares	d.f.	Mean Squares	F
Alpha was Last Brand Purchased	11.47	1	11.47	1.72
Pre-Exposure Interest in Alpha Radials	64.84	1	64.84	9.71
Interaction	0.82	1	0.82	
Covariates	56.19	7	8.03	1.20
Residual	380.74	57	6.68	

Mean Responses Adjusted for All Covariates

	Pre-Exposure Interest in Alpha Radials		
Alpha was Last Brand Purchased	High	Low	Marginal Means
Yes	8.07	5.79	6.93
No	6.99	5.15	6.07
Marginal Means	7.53	5.47	6.50

covariates. In this case the effect of the seven covariates is removed via a pooled within-groups *multiple* regression. If we compare this table to the ANOVA summary in Table 3.9, we see that pre-exposure interest remains significant, although the covariate-adjusted F value is actually lower than its unadjusted counterpart.[31] Alpha as last brand purchased continues to be non-significant, although its F value now exceeds unity. The seven covariates, even as a group, do not reach significance.

The covariate-adjusted mean responses differ somewhat from their unadjusted counterparts but in this analysis the differences are rather minor. About all that can be said is that pre-exposure interest continues to be important, even after allowing for the possible influence of demographics. As observed in the preceding chapter, however, the demographic variables did not account for much variance in the related multiple regressions in which post-exposure interest in Alpha radials served as the criterion variable.

Hence, the present results—based on only 68 out of 252 responses—generally support earlier findings in which we noted that the demographics played no important role in accounting for criterion variance. Still, it should be borne in mind, insofar as the ANOVA and ANCOVA illustrations are concerned, that the original data *are* survey data and that our use of equal size samples in each cell is strictly for expository purposes and does not represent the real state of affairs.

3.7 Regression Approaches to the Analysis of Multiple-Factor Designs

As briefly indicated in section 3.6.1, one of the most useful designs in applied research is the full factorial, in which we examine responses to all combinations of two or more factors, each at two or more levels. Just as the regression approach served us well in the analysis of single-factor ANOVA and ANCOVA designs, in this part of the chapter we shall show how regression can also be used to analyze multiple-factor designs. While we illustrate the procedure for the simplest (two-factor) case, it turns out that *any type* of multiple-factor design can be analyzed by the regression model, with results equivalent to those obtained from traditional ANOVA and ANCOVA.

We first apply the regression technique to a two-factor design that has an equal number of replicates per cell. The concluding part of this section describes, more briefly, some strategic approaches to the analysis of designs in which there are unequal numbers of replicates across cells.

In earlier discussions of the regression approach to ANOVA and ANCOVA, we emphasized the system of dummy-variable coding in which k categories are coded into a set of $k - 1$ zero-one dummies, one category being assigned $k - 1$ zeros. Another form of coding—one that is directly responsive to the usual types of ANOVA restrictions—is now discussed.

3.7.1 The Two-Factor Model

As briefly outlined in section 3.6.1, the ANOVA model for the two-factor design can be written as follows:

$$Y_{ijk} = \mu + \alpha_i + \beta_j + (\alpha\beta)_{ij} + \epsilon_{ijk}$$

where Y_{ijk} denotes the response, μ is the overall mean, α_i is the effect due to level i ($i = 1, 2, \ldots, a$) of the first factor and β_j is the effect due to level j ($j = 1, 2, \ldots, b$) of the second factor. The interaction term is denoted by $(\alpha\beta)_{ij}$ and $\epsilon_{ijk} = $ NID $(0, \sigma^2)$ denotes the error term with $k = 1, 2, \ldots, r$. In addition, the following restrictions are usually imposed:

$$\sum_{i=1}^{a} \alpha_i = \sum_{i=1}^{b} \beta_j = \sum_{i=1}^{a} (\alpha\beta)_{ij} = \sum_{j=1}^{b} (\alpha\beta)_{ij} = 0$$

Let us examine a form of dummy-variable coding that is appropriate for the zero-sum restrictions, just noted.

To be specific, suppose that the first factor consists of two levels and the second factor consists of three levels. Assuming that the regression program automatically computes a constant (and, hence, we always have an extra column in the design matrix that consists of all $+1$'s), we can use the dummy-variable coding of Table 3.11.

Considering the two-level factor A, we set up a design in which $+1$ is used if A_2 is the level. If A_1 is the level, -1 is used. In the case of the second factor,

Table 3.11 Dummy-Variable Coding for Two-Factor Design

Cell	Response						
(Below)	Y	Intercept	A_2	B_2	B_3	A_2B_2	A_2B_3
a_2, b_2	94; 97; 93	1	1	1	0	1	0
a_2, b_3	87; 90; 90	1	1	0	1	0	1
a_2, b_1	91; 90; 87	1	1	-1	-1	-1	-1
a_1, b_2	85; 88; 93	1	-1	1	0	-1	0
a_1, b_3	77; 81; 78	1	-1	0	1	0	-1
a_1, b_1	70; 75; 79	1	-1	-1	-1	1	1

Factorial Layout

	b_1	b_2	b_3	Total
a_1	70; 75; 79	85; 88; 93	77; 81; 78	726
a_2	91; 90; 87	94; 97; 93	87; 90; 90	819
Total	492	550	503	1,545

we set up two dummies, B_2 and B_3. If B_2 is present a coding of $(1, 0)$ is used. If B_3 is present a coding of $(0, 1)$ is used while if B_1 is present, a coding of $(-1, -1)$ is used. Thus, for a k-level factor we set up $k - 1$ dummies, reserving the designation of all -1's for one of the k categories. Codings for the interaction terms A_2B_2 and A_2B_3 are found by simply multiplying the dummy-variable entries in the appropriate main-effect columns. For example, the first entry of A_2B_2 is $1 \times 1 = 1$ while the first entry of A_2B_3 is $1 \times 0 = 0$.[32]

The response data and factorial layout for this 2×3 factorial (with three replications per cell) also appears in Table 3.11. Our purpose now is to analyze these response data as a factorial experiment using one of the standard programs for doing this type of ANOVA. We shall then apply the regression approach to the ANOVA of factorial designs and compare the results.

3.7.2 BMD-02V Analysis of the Two-Factor Design

The BMD-02V program is another program of the UCLA Biomedical package and is particularly versatile for analyzing data based on factorial designs. This program will take up to eight different factors and up to 999 levels per factor (assuming an upper limit of 18,000 on the product of factor levels). Complete and balanced data are required, however. Since this program develops its own predictor coding, there is no need to use the dummy-variable system of Table 3.11; however, that coding is used later in the regression approach. Applying this program to the response data of Table 3.11 leads to the ANOVA summary shown in Table 3.12.

As can be seen from Table 3.12 both of the main effects are significant at the 0.05 level. (The interaction reaches significance only at the 0.1 level, however.) Figure 3.5 shows a plot of the mean response to factor B at each of the two levels of factor A. Interaction would be indicated by a significant *lack of parallelism* between the line segments making up A_1 and A_2. As noted, the sample-data difference in mean response between A_2 and A_1 at level B_1 is somewhat greater than at level B_2. However, the departures from parallelism are not great enough to lead to a significant interaction at the 0.05 level. (More will be said on the topic of interaction in Chapter 5.) However, for now we wish to see if the ANOVA of a two-factor design can be carried out via multiple regression.

Table 3.12 ANOVA Summary from BMD-02V

Source	d.f.	Sums of Squares	Mean Squares	F	Probability
A	1	480.50	480.50	54.73	< 0.05
B	2	316.33	158.17	18.01	< 0.05
$A \times B$ Interaction	2	56.33	28.17	3.21	
Residual	12	105.33	8.78		
Total	17	958.49			

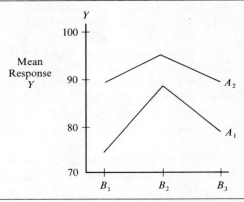

Figure 3.5 Plot of Interaction Effect — Mean Response to Factor B at Two Levels of Factor A

3.7.3 Two-Factor ANOVA via Multiple Regression

To apply regression analysis to the data of Table 3.11, we use the dummy-variable coding illustrated in that table. As noted earlier, each column sums to zero and the cross-product columns A_2B_2 and A_2B_3 receive dummy-variable coding that represents the cross-multiplication of the relevant dummy variables. The following regressions (BMD-03R) were carried out:

1. Y on A_2, B_2, B_3, A_2B_2, A_2B_3.
2. Y on A_2.
3. Y on B_2, B_3.
4. Y on A_2B_2, A_2B_3.

Table 3.13 shows a summary of the regression results.

If we compare Table 3.13 to Table 3.12, we see that *all sums of squares for A, B and A $\times$ B* are reproduced via the dummy-variable regression. However, separate F ratios for A, B, and $A \times B$ are only computed when these variables represent the predictor set. In addition, a full model is computed for which the F ratio is 19.44, with 5 and 12 degrees of freedom. This value shows that the full model is, indeed, significant.

Separate ANOVA tables for the various restricted models:

$$Y \text{ on } A_2$$
$$Y \text{ on } B_2, B_3$$
$$Y \text{ on } A_2B_2, A_2B_3$$

are easily set up from the column "Sums of Squares Added." Hence, all tests made in Table 3.12 can be made here as well.

The regression coefficients of Table 3.13 also provide useful information. Given the way the predictor variables were coded, we have:

Table 3.13 Regression Approach to ANOVA of Factorial Designs

Variable	Mean	Standard Deviation	Reg. Coeff.	Sums of Squares Added	Proportion of Total Variance
A_2	0	1.03	5.17	480.50	0.50
B_2	0	0.84	5.83	280.33⎱ 316.33	0.29⎱ 0.33
B_3	0	0.84	−2.00	36.00⎰	0.04⎰
A_2B_2	0	0.84	−2.17	56.33	0.06
A_2B_3	0	0.84	0	0	
Y	85.83	7.51			
Intercept	85.83				

ANOVA for the Multiple Linear Regression (Full Model)

Source	d.f.	Sums of Squares	Mean Squares	F
Due to Regresion	5	853.17	170.63	19.44
Deviations About Regression	12	105.33	8.78	
Total	17	958.50		

Separate ANOVA's (Restricted Models)

Source	d.f.	Sums of Squares	Mean Squares	F
Y on A_2	1	480.50	480.50	54.73
Y on B_2, B_3	2	316.33	158.17	18.01
Y on A_2B_2, A_2B_3	2	56.33	28.17	3.21
Residual	12	105.33	8.78	

$$A_1 = -5.17 \quad B_1 = -3.83 \quad A_2B_2 = -2.17 \quad A_1B_1 = -2.17$$

$$A_2 = 5.17 \quad B_2 = 5.83 \quad A_2B_3 = 0 \quad A_1B_2 = 2.17$$

$$B_3 = -2.00 \quad A_2B_1 = 2.17 \quad A_1B_3 = 0$$

The coefficients for the interaction terms follow from the restriction that rows and columns sum to zero:

$$\begin{array}{c c c c} & B_1 & B_2 & B_3 \\ A_1 & -2.17 & 2.17 & 0 \\ A_2 & 2.17 & -2.17 & 0 \end{array}$$

Since the overall mean $\overline{Y}$ is 85.83, the various regression coefficients can be used to estimate main effects and interactions. For example, to estimate the averaged response of the (2, 3) cell in the factorial layout of Table 3.11, we have:

$$\hat{Y}_{ij} = \overline{Y} + A_2 + B_3 + A_2B_3$$
$$= 85.83 + 5.17 - 2.00 + 0$$
$$= 89$$

which necessarily equals the actual averaged cell value of $(87 + 90 + 90)/3$ $= 89$ in Table 3.11.

In summary, the regression approach to ANOVA (and ANCOVA as well) is just as appropriate for multiple-factor designs as it is for single-factor designs. Moreover, as long as the design is orthogonal, no ambiguity is introduced in terms of how one goes about applying the regression model.

Things are not so simple in the case of nonorthogonal designs. However, even here, the regression model can be applied, whereas the simplicity of traditional ANOVA and ANCOVA models is lost. Since the analysis of nonorthogonal designs is both lengthy and complex, we confine our discussion to introductory concepts only.

3.7.4 Regression Approaches to Nonorthogonal Designs

Throughout Chapter 2 (and this chapter as well) we have stressed the idea of viewing significance tests as tests of alternative models. This point of view is particularly attractive in the case of nonorthogonal ANOVA (and ANCOVA) designs involving unequal numbers of observations per cell.[33] One of the most persuasive arguments for models comparison testing is presented by Appelbaum and Cramer (1974).

The strategy recommended by Appelbaum and Cramer can be illustrated in terms of the two-factor design already discussed. Suppose we set up the following alternative models:

1. $Y_{ijk} = \mu + \alpha_i + \beta_j + (\alpha\beta)_{ij} + \epsilon_{ijk}$
2. $Y_{ijk} = \mu + \alpha_i + \beta_j + \epsilon_{ijk}$
3. $Y_{ijk} = \mu + \beta_j + \epsilon_{ijk}$
4. $Y_{ijk} = \mu + \alpha_i + \epsilon_{ijk}$
5. $Y_{ijk} = \mu + \epsilon_{ijk}$

In the orthogonal case one could regard the first model as providing *all* of the parameter estimates required in the other models. However, in the case of nonorthogonal designs the various predictors are correlated and, in general, different estimates of the parameters will be obtained in each regression run. These correlated estimates make matters considerably more complex.

Appelbaum and Cramer advocate the following sequence of model comparisons, applicable to either orthogonal or nonorthogonal designs:

1. Begin with a test of $A \times B$ interaction by comparing the full model (model 1) with a restricted model (model 2) that allows for only main effects. If the null hypothesis of no interaction is rejected, one would usually stop the analysis at this point and examine the nature of the interaction.

2. However, suppose the interaction effect is not significant and one wishes to press on and test for main effects. To test for an A effect, models 2 and 3 are compared. To test for a B effect, models 2 and 4 are compared. These are referred to as *elimination tests*.[34] If both A and B are significant, model 2 is adopted. If only one is significant, one adopts the specific model (model 3 or model 4) appropriate for that effect. If neither effect is significant one proceeds to the next step.

3. Tests comparing models 3 and 5 and models 4 and 5 are referred to, respectively, as *B ignoring A* and *A ignoring B*.[35] If neither test is significant, then model 5 is accepted. If both tests are significant one can retain either α_i or β_j, but not both; moreover the choice of which one to retain is indeterminate. If only one of the tests is significant, the model related to that effect alone is retained.

Various patterns (Appelbaum and Cramer, 1974) that can arise in the two-factor case are shown below:

Pattern of Results

Test	1	2	3	4	5	6	7
A eliminating B	S	S	NS	NS	NS	NS	NS
B eliminating A	S	NS	S	NS	NS	NS	NS
A ignoring B	—	—	—	NS	S	S	NS
B ignoring A	—	—	—	NS	S	NS	S

where S denotes a significant and NS denotes a nonsignificant result. Pattern 1 shows that both A and B are required under the appropriate elimination tests in which model 2 is compared with models 3 and 4, respectively. Patterns 2 and 3 indicate that only one of the effects (A or B) need be retained. Pattern 4 shows that neither main effect is needed; that is, model 5 is appropriate. These are the standard patterns and represent the *only possible patterns in the case of orthogonal designs.*

Patterns 5, 6, and 7 can apply in the case of nonorthogonal designs. Pattern 5 shows serious confounding of A and B effects in which only one of these can be retained but the choice is not determinate. Patterns 6 and 7 also show highly confounded situations although the choice of which factor to include is now determinate—A in the case of pattern 6 and B in the case of pattern 7. Still, the researcher should be wary if either of these situations arises since the elimination tests are not significant.

The strategy of Appelbaum and Cramer can be extended to three-factor and higher-order designs, albeit at some increase in complexity. As the authors indicate, their procedure applies when one does not have a prior basis for ordering certain factors in the design. It would seem that the Appelbaum and Cramer approach has much to recommend it as a logical procedure for model comparison testing. Other strategies for analyzing nonorthogonal designs are available, as described by Overall and Klett (1972). Only in the case of orthogonal designs do all of the strategies lead to the same results.

However, perhaps the main conclusion to be reached from all of this discussion is that orthogonal designs provide a simplicity of interpretation that is highly desirable. Fortunately, the regression approach can be applied with the same facility to either the orthogonal or nonorthogonal case. The main point to keep in mind is the need to be explicit in the formulation of what models are being compared. This is particularly important in nonorthgonal designs.

3.8 Summary

This chapter has been concerned with the relationship of the general linear model to ANOVA and ANCOVA. In particular, we illustrated how regression approaches to analysis of variance and covariance could be carried out in cases involving: (a) a single-factor ANOVA design; (b) a single-factor design with one covariate; (c) a two-factor design with several covariates; and (d) a factorial design. In addition, a number of packaged programs for performing ANOVA and ANCOVA were illustrated and their results compared to those obtained from multiple regression.

We pointed out some of the dangers associated with the utilization of ANCOVA in cases where treatments differ in terms of covariate means. Moreover, we also described some of the problems that can arise in the analysis of nonorthogonal designs via regression approaches to ANOVA and ANCOVA.

Review Questions

1. Consider the following experimental design data consisting of one treatment (at four levels) and one covariate:

	Case	Y	W_1	W_2	W_3	Covariate X
Level 1	a	2	0	0	0	1
	b	6	0	0	0	3
	c	5	0	0	0	2
Level 2	d	4	1	0	0	1
	e	7	1	0	0	2
	f	9	1	0	0	4
	g	8	1	0	0	4
Level 3	h	6	0	1	0	2
	i	8	0	1	0	4
	j	10	0	1	0	8
Level 4	k	12	0	0	1	3
	l	14	0	0	1	10

The columns W_1, W_2, and W_3 refer to the dummy-variable coding of the four-level treatment variable. Assume that you wish to test whether the four treatment-level mean responses are significantly different, ignoring the covariate:

a. Write the single-factor ANOVA model for the present problem.

b. Perform a traditional ANOVA analysis for the four-level treatment variable; use an alpha level of 0.05.

c. Perform a regression analysis in which W_1, W_2, and W_3 are dummy-variable regressors and compare your results with those of part b.

d. Change the coding of the treatment variable to: level 1 $\Rightarrow$ 2, 2, 2; level 2 $\Rightarrow$ 3, 2, 2; level 3 $\Rightarrow$ 2, 3, 2; level 4 $\Rightarrow$ 2, 2, 3, and repeat the regression run. Compare your results with those of part c.

2. An experiment was conducted involving a single treatment with three levels, leading to the results:

$$\text{Sample size: } m_1 = 25 \qquad m_2 = 25 \qquad m_3 = 25$$

$$\text{Response: } \quad \overline{Y}_{.1} = 64.6 \qquad \overline{Y}_{.2} = 76.7 \qquad \overline{Y}_{.3} = 85.6$$

$$\text{SSA} = 7,182.67$$
$$\text{SSW} = 1,642.83$$
$$\text{SST} = 8,825.60$$

a. Compute the three separate t tests using the sum of squares for within-groups variation as a common measure of variability and $\text{SSW}/(m_j - 1)$ as a common measure of a given sample's variance. Use an alpha level of 0.01.

b. Set up an ANOVA test of the difference among the three treatment-level means and compare your results with those of part a.

3. Return to the data of question 1 but now consider the covariate X.

a. Write out the single-factor ANCOVA model for the present problem.

b. Perform a traditional ANCOVA analysis of the four-level treatment variable with one covariate; use an alpha level of 0.05.

c. Carry out the counterpart regression approach and compare your results with those of part b.

d. Conduct a separate F test to see if the four treatment groups differ on the covariate X.

4. Continuing to use the data of question 1, now assume equal cell sizes in which treatment level 1 consists of cases a, b, c; level 2 consists of cases d, e, f; level 3 consists of g, h, i; and level 4 consists of cases j, k, l.

a. Using the coding: level 1 $\Rightarrow$ -1, -1, -1; level 2 $\Rightarrow$ 1, 0, 0; level 3 $\Rightarrow$ 0, 1, 0; and level 4 $\Rightarrow$ 0, 0, 1, conduct a single-factor ANOVA via regression; let alpha equal 0.05.

b. Next, introduce the covariate X and run an ANCOVA via regression.

c. Subdivide X into three equal-size blocks in which the codes are:

Block 1 $\Rightarrow$ -1, -1; cases a, b, c, d

Block 2 $\Rightarrow$ 1, 0; cases e, f, g, h

Block 3 $\Rightarrow$ 0, 1; cases i, j, k, l

Using this coding conduct a two-factor ANOVA and compare the results of the blocking procedure to the ANCOVA results in the preceding part.

5. Consider the TV commercial data in Appendix A. Let A-1, interest in the product class of radial tires, be a two-level treatment variable and A-3, pre-exposure interest in Alpha radials, denote a covariate.

   a. Use the regression approach to ANCOVA on the whole sample ($m = 252$) to see if significant differences on C-2, post-exposure interest, exist after covariate adjustment; use an alpha level of 0.1.

   b. Divide the sample into halves (based on odd-numbered versus even-numbered respondents) and carry out ANCOVAs on separate halves; continue to use an alpha level of 0.1.

   c. Reverse the roles of A-1 and A-3 by treating A-1 as a covariate. At an alpha level of 0.05, is A-3 significantly correlated with C-2, after the influence of A-1 is removed?

6. Consider the two-factor design data of Table 3.11. However, now assume that the second observation (88) in A_1B_2 and the third observation (90) in A_2B_3 are missing.

   a. Using the models-comparison approach of Appelbaum and Cramer, make tests of the models illustrated in section 3.7.4; use an alpha level of 0.01.

   b. Suppose the researcher has a prior interest in the main effect of factor A, followed by the interaction effect $A \times B$. Set up models-comparison tests using this prespecified order.

   c. Compare the results of A eliminating B with A ignoring B.

Notes

1. This statement is correct only if we restrict discussion to models of the fixed-effects variety where the treatment or factor levels constitute the complete universe of interest to the researcher.

2. The terms *treatment* and *factor* are used here interchangeably and refer to some design variable that is being manipulated by the researcher. Specific realizations of a treatment or factor are conventionally called *levels* even though the treatment may be categorical, such as four varieties of fertilizer in an agronomy experiment.

3. In general, however, different reparameterizations lead to different hypotheses to be tested, a point that is illustrated later on.

4. ANOVA is often used to analyze survey or observational data, although its power lies primarily in *experimental* contexts where treatments can be manipulated. (In nonexperimental contexts factors are often referred to as *classificatory* variables.)

5. The scale properties of the criterion variable are prescribed from the standpoint of substantive interpretation of the results and are not computational requirements of the technique.

6. The term *multiple comparisons* is often used to refer to the detailed examination of within-factor effects, *after* establishing that variation due to the factor, as a whole, is significant.

7. The term *components of variance* arises from the fact that in this model the variance of the variable has two components. One component represents the variance in the distribution of Y while the second reflects variance in the universe from which the treatment levels have been selected.

8. Estimates of the three means μ_1, μ_2, and μ_3 are the three response-variable means, $\overline{Y}_1$, $\overline{Y}_2$, and $\overline{Y}_3$, for blackboard, transparencies, and handouts, respectively.

9. Moreover, none of these correspondences depends on the fact that an equal number of observations was employed for each treatment variable. However, equal sample sizes provide the most efficient estimates and in the case of two or more factors their advantages are even more pronounced, as will later be discussed in the context of multiple-factor ANOVA.

10. Insofar as degrees of freedom for SST are concerned, we have $m - 1$ since the constraint equation $\sum_{j=1}^{n} \sum_{i=1}^{m_j} (Y_{ij} - \overline{Y}..) = 0$ must be satisfied. Degrees of freedom for SSA are $n - 1$ since the constraint equation $\sum_{j=1}^{n} m_j (\overline{Y}._j - \overline{Y}..) = 0$ must be satisfied. Finally, the degrees of freedom for SSW are $m - n$ since each treatment-level sum of squares has to satisfy the constraint $\sum_{i=1}^{m_j} (Y_{ij} - \overline{Y}._j) = 0$. Since this is true for every treatment level, we have $(m_1 - 1) + (m_2 - 1) + \cdots + (m_n - 1) = m - n$ degrees of freedom for SSW.

11. The noncentrality parameter is not a variance in the sense of a statistic related to a random variable. (Recall that the treatment levels are fixed.) Rather, it is a measure of the extent to which the treatment-level deviations about the overall mean depart from zero.

12. A variety of different dummy-variable coding systems (and their main properties) are described in the book by Kerlinger and Pedhazur (1973).

13. By *interaction* generally is meant that the pattern of response to changes in the levels of one treatment variable depends on the level of some other treatment variable.

14. In general the product moment correlation of a continuous variable with a zero-one variable is given by the point biserial coefficient:

$$r_{pb} = \frac{\overline{X}_3^{(1\text{-code})} - \overline{X}_3^{(0\text{-code})}}{s_{x_3}} \sqrt{\frac{m_1 m_0}{m(m-1)}}$$

where:

$\overline{X}_3^{(1\text{-code})}$ denotes the mean on X_3 of those observations coded 1

$\overline{X}_3^{(0\text{-code})}$ denotes the mean on X_3 of those observations coded 0

s_{x_3} denotes the estimated standard deviation of X_3

m_1 denotes the number of cases coded 1

m_0 denotes the number of cases coded 0

m denotes the total number of cases

If the mean on X_3 for the 1-coded subgroup differs from that for the 0-coded subgroup, the variables are correlated.

15. If covariate slopes differ across treatment levels, one usually stops the analysis at this point (since one of the assumptions of covariance analysis is violated) and examines covariate relationships separately within each treatment-level group (see Kerlinger and Pedhazur, 1973).

16. More generally, the condition is $\sum_{j=1}^{n} m_j \tau_j = 0$, where m_j denotes the number of observations in the j-th treatment level. In the sample problem, $m_1 = m_2 = m_3 = 4$; naturally, the special case involving equal sample sizes can be expected to hold.

17. One is not restricted to working with a single covariate. The topic of multiple covariates is discussed at a later point in the chapter.

18. If β should turn out to be zero, then all regression lines in Figure 3.4 are horizontal and parallel and we have a situation that is identical to ANOVA (after allowing for loss of a single degree of freedom used for estimating β).

19. Note that SSA (Adj. Y) is *not* obtained in the same manner as the other two quantities but, rather, by subtraction. The quantity SSA (Adj. Y) = 5.15 represents a *difference* between the accounted-for sums of squares of a model that includes covariate and treatments versus one that includes only the covariate. One finds this quantity by subtraction because, in practice, the treatment levels and covariate may be correlated; as such, their separate sums of squares do *not* add up to the total.

20. One could also compute the among-groups regression slope $b_A = \frac{23.8}{70.85} = 0.34$ from Table 3.4. Ordinarily, however, no major interest is attached to this specific regression slope.

21. Of further interest is the fact that the pooled within-groups regression slope is simply a weighted average of the three individual slopes, using quantities obtained from the expression:

$$\sum_{i=1}^{m_j} (X_{ij} - \overline{X}_{\cdot j})^2$$

as weights (see Table 3.5). To illustrate:

$$b_w = \frac{[5.49(0.95) + 5.50(0.93) + 4.14(1.09)]}{15.13}$$

$$= 0.98$$

22. In the parlance of multiple-factor ANOVA, b_4 and b_5 are estimates of interaction effects between treatment levels and the covariate.

23. We use X_d to denote deviations from the overall covariate mean; that is $X_{dij} = X_{ij} - \overline{X}_{\cdot\cdot}$. Incidentally, the model could easily be formulated in terms of the original X_{ij} instead of deviations about the overall mean. If so, $\overline{X}_{\cdot\cdot}$ is absorbed by the intercept b_0; the interpretation of b_0 is, of course, changed thereby.

24. The degrees of freedom for the full model are $12 - 6 = 6$ since six parameters (including the intercept) are being estimated. The degrees of freedom for the restricted model are $12 - 4 = 8$ since only $b_0, b_1, b_2,$ and b_3 are being estimated.

25. Note that this is the same ($F = 0.47$) value found under the traditional ANCOVA approach.

26. The degrees of freedom for the full model are $12 - 4 = 8$ since four parameters ($b_0, b_1, b_2,$ and b_3) are being estimated. In the restricted model only b_0 and b_3 are being estimated so degrees of freedom are $12 - 2 = 10$; hence, $10 - 8 = 2$ for $d_r - d_f$.

27. Note that this is the same ($F = 39.38$) value found under the traditional ANCOVA approach.

28. The coefficient b_w (within-groups regression slope) is routinely found from the BMD-03R regression involving the two dummies and the covariate. The total sample coefficient b_T is found from the simple regression of Y on $(X_{ij} - \overline{X}_{\cdot\cdot})$ only.

29. Another approach to ANCOVA involves applying ordinary ANOVA to a set of residuals obtained by first regressing Y on the covariate. If appropriate adjustments in degrees of freedom are made and, more importantly, if the covariate is uncorrelated with the treatment levels (so that $b_T = b_w$), then one gets the same results as obtained from ANCOVA. Moreover, a legitimate ANOVA of change scores requires all of these conditions as well as the additional one that $b_T = 1$.

30. A variety of special procedures (Myers, 1972) have been proposed for handling cases in which an unequal number of observations appear across cells. Still, the simple algebraic properties of ANOVA are not maintained if the design is multiple-factor. In the case of single-factor designs no problem arises since the average of each effect does not require summing over the levels of some other factor.

31. As indicated in section 3.5, application of ANCOVA can actually lower the treatment F value if the covariates are correlated with the treatment levels. This happens because the covariates are credited with any variance in the response that they share with the treatment levels.

32. Note that the cross product of any column of dummy variables *across* factors (not within factor) is zero. Note also that the cross product of either interaction column with any main effect column is also zero (but not between the two interaction columns). Finally, note that all columns display a cross product of zero with the intercept column. This type of design produces *orthogonal* (uncorrelated) estimates of each main effect and the overall two-factor interaction. Orthogonal designs permit the uncorrelated estimation of each principal effect. Balanced multiple factor designs (i.e., those with an equal number of replicates across cells) lead to orthogonal estimates under appropriate dummy coding. Nonorthogonal designs (e.g., those with missing observations) lead to correlated effects; in this case, simple ANOVA models cannot be used.

33. Actually, main-effect orthogonal designs can be produced with weaker conditions than an equal number of replicates per cell (see Winer, 1971).

34. An elimination test is one in which the restricted model, for example, A eliminating B, is compared to a full model that includes B in order to see if a particular factor—in this case factor B—can be eliminated.

35. Ordinarily these are not proper tests unless model 3 or model 4 is, indeed, a correct model. In the case of orthogonal designs we would get the same results as obtained from the elimination tests; in general, however, nonorthogonal designs would produce different results.

CHAPTER 4

Two-Group Discrimination and Classification

4.1 Introduction

Cases arise in the analysis of single criterion, multiple predictor data where the criterion variable is dichotomous rather than continuous. For example, one might wish to study how consumers classified as brand-loyal versus brand-switcher differ in terms of shopping characteristics. Or, one might want to predict a new student as being either successful or unsuccessful in a program of graduate study on the basis of a set of evaluation scores obtained from achievement tests and personal interviews. Two-group linear discriminant analysis represents the prototype of a growing number of procedures that have been developed to assist the researcher in a variety of discrimination and classification problems.

We first describe the two-group discrimination problem from an informal and intuitive point of view, in the context of a small problem involving artificial data. Various "rules" for assigning observations to classes are described and discussion centers on the earliest procedure of this kind to be proposed—R. A. Fisher's linear discriminant function.

After describing other approaches to two-group discrimination, we discuss questions of statistical significance as related to testing for the equality of co-variance matrices and centroids. This, in turn, leads to the question of assigning cases to groups, and, in particular, to the incorporation of discriminant functions in general classification procedures based on statistical decision theory.

The last section of the chapter discusses various computer algorithms for performing two-group discrimination and the interpretive aspects of discriminant analysis as applied to a portion of the TV commercial data.

4.2 Fundamentals of Two-Group Discrimination

The basic idea behind two-group discriminant analysis is to reduce what may originally be a large set of multiple (and correlated) measurements on a set of persons or objects, to a single linear composite with values that maximally distinguish between members of the two groups. In so doing we transform a set of multivariate profiles into a set of univariate data. If a large number of predictor variables are involved, the reduction in problem complexity can be sizable.

Under certain distributional assumptions about the predictor variables we can also test to see whether the vectors of means (i.e., the centroids) of the two groups are equal. And, if we establish that they are different, we can use other aspects of discriminant analysis to assign new multivariate profiles to the two groups. This is done by computing a discriminant score for each profile and then comparing that score to a numerical cutoff value, computed from the calibration sample used to construct the linear composite. If the score is on one side of the cutoff value, we assign the profile to the first group; if the score is on the other side of the cutoff, the profile is classified as a member of the second group. (If the score equals the cutoff, one can randomly assign it to one of the two groups.)

To motivate the discussion let us assume that a researcher in the field of political science is interested in whether support for an ecology-oriented piece of legislation is related to legislators' attitudes toward energy conservation and state government intervention in certain types of corporate practices.

Table 4.1 shows a set of artificial data assumed to be developed from interviews with ten state legislators. Group 1 denotes those who oppose the bill while group 2 denotes those who support the bill. The criterion variable Y receives a dummy-value code of 1, indicating opposition, or 0, indicating support. Predictor X_1 denotes an attitude score related to the need for energy conservation and X_2 denotes an attitude score regarding the need for state government intervention in the control of various corporate practices. Higher scores represent higher degrees of the attitude in question.

We note from Table 4.1 that supporters of the bill tend to display higher attitude scores on energy conservation and government intervention than those who oppose the bill. If we express each attitude score as a deviation from its respective overall sample mean, group 1's centroid consists of negative deviations while group 2's centroid consists of positive deviations. Since sample sizes are equal, the (equally-weighted) sum of the two centroid vectors equals the null vector:

$$\begin{bmatrix} -2.5 \\ -1 \end{bmatrix} + \begin{bmatrix} 2.5 \\ 1 \end{bmatrix} = \begin{bmatrix} 0 \\ 0 \end{bmatrix}$$

Table 4.1 Illustrative Data for Two-Group Discriminant Analysis

	Criterion Y	Attitude Toward Energy Conservation X_1	Attitude Toward State Government Intervention X_2	Deviations from Overall Centroid	
				X_{d1}	X_{d2}
	1	2	4	−4.5	−1.4
Group 1	1	3	2	−3.5	−3.4
(Oppose the	1	4	5	−2.5	−0.4
Legislation)	1	5	4	−1.5	−1.4
	1	6	7	−0.5	1.6
	0	7	6	0.5	0.6
Group 2	0	8	4	1.5	−1.4
(Support the	0	9	7	2.5	1.6
Legislation)	0	10	6	3.5	0.6
	0	11	9	4.5	3.6
Total-Sample Centroid	$\overline{X}._{j}.$	6.5	5.4	0	0
Group Means $\begin{cases} \overline{X}._{j1} \\ \overline{X}._{j2} \end{cases}$		4	4.4	−2.5	−1
		9	6.4	2.5	1

Note also in Table 4.1 that we use a third subscript to denote group membership and the dot notation of Chapter 3 to denote variables that are summed over in the computation of specific means of interest.

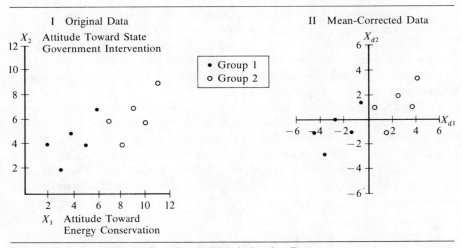

Figure 4.1 Scatter Plots of Two-Group Discrimination Data

Figure 4.1 shows scatter plots of both the original and mean-corrected attitude scores. The only effect of transforming the original attitude scores to mean-corrected form is to translate the origin of the configuration to the centroid, as shown in Panel II of the figure.

Examination of the two scatter plots in Figure 4.1 suggests that the scores are positively correlated on a within-groups as well as a total-sample basis. Moreover, it is seen that the group means are more widely separated on X_1 (their difference is 5 score points) than on X_2 (their difference is only 2 score points).

4.2.1 Research Objectives of Two-Group Discriminant Analysis

With the foregoing as background, the purposes of two-group discriminant analysis can be summarized, in the context of this problem, as follows:

1. Suppose one wished to develop a set of linear composite scores, one for each legislator, that would exhibit the property of maximizing between-groups to within-groups variability on the new composite. How should this set of scores be developed?

2. Are the two groups significantly different with respect to their means on the original variables X_1 and X_2, when these two predictor variables are considered jointly?

3. Suppose we had some new legislator's scores on X_1 and X_2. The legislator is known to be from one of the two groups but it is not known which specific one. How could we use his scores to assign the legislator to group 1 versus group 2?

The first of these questions represents a problem of dimensional reduction. Here we attempt to find a new axis (say, in the configuration of Panel II of Figure 4.1) so that projections of the points onto that axis (or linear composite) would exhibit the property of maximizing the separation between the group means relative to the within-groups variation.

The second question concerns statistical inference. Are the two groups of legislators significantly different with respect to their centroid locations in, say, the X_{d1}, X_{d2} plane of Panel II? (This is the type of problem in which one might employ Student's t-test, were only one variable involved.)

The third question refers to prediction—assigning a new legislator profile to one of the two groups on the basis of statistics developed from legislators who comprise the calibration sample. We consider all of these questions in due course.

4.2.2 Some Geometric Considerations in Two-Group Discrimination

Still by way of preliminaries, let us focus attention on X_1 and X_2 in mean-corrected form in Table 4.1. (It turns out that all of the arguments described here hold for the original attitude data as well.) The centroids of the two legislative groups are:

$$\begin{array}{cc} \text{Group 1} & \text{Group 2} \\ \begin{bmatrix} \overline{X}_{d11} \\ \overline{X}_{d21} \end{bmatrix} = \begin{bmatrix} -2.5 \\ -1 \end{bmatrix}; & \begin{bmatrix} \overline{X}_{d12} \\ \overline{X}_{d22} \end{bmatrix} = \begin{bmatrix} 2.5 \\ 1 \end{bmatrix} \end{array}$$

Intuitively, we might think that one way to develop a discriminant axis would be to draw a straight line, through the origin in Panel II of Figure 4.1, connecting the two subgroup centroids. One could then project each of the points onto this line and assign each point to the closer of the two centroids.

However, before examining the consequences of this particular rule, let us consider some general background to the problem. As stated earlier, in linear discriminant analysis we seek a new axis (or linear composite) whose point projections display maximal separation between the two groups relative to their within-groups variability.[1]

Figure 4.2 shows some hypothetical axes of interest. Here we assume large samples of observations whose scatter plots approximate the form of various concentration ellipses.[2] For purposes of illustration assume that the concentration ellipses in the various panels of Figure 4.2 each enclose 95 percent of the observations.

Let us consider Panel I first. We note that the two groups differ considerably on X_1 but very little on X_2. Hence, if we elected to use one of the *original* axes as the discriminant axis, choice of X_1 would be appropriate. If X_1 is chosen, the degree of overlap in the two unidimensional distributions is relatively small compared to the case where the points are all projected onto X_2; in this latter case the distributions overlap so much that the overall-sample distribution does not even suggest bimodality. Clearly, if for some reason we had to discard one of the variables we would prefer it to be X_2, given its relative inability to discriminate between the two groups.

Panel II illustrates the case that still other axes (i.e., linear composites of both X_1 and X_2) might be employed to provide discrimination. In each case the points are projected onto some axis whose position, relative to the original axes of the space, can be described by a set of direction cosines relative to the original axes.

Now let us examine, via Panel III of Figure 4.2, the discrimination rule mentioned earlier, namely, to connect the centroids of the two groups—assumed here to be of equal size—by a line through the origin.[3] Panel III illustrates a rather unusual case where the two groups are each characterized by: (a) equal variances on X_1 and X_2 and (b) a zero within-groups covariance between X_1 and X_2. That is, the within-groups covariance matrix is a scalar matrix.[4] Intuitively, it seems that point projections on a line through the origin connecting the centroids of groups A and B would be a reasonable approach to maximizing separation between group means relative to pooled within-groups variability. As would also be suggested intuitively, it can be shown that any point assigned to a group by this procedure would also be assigned to the *same* group on the basis of the smaller (ordinary Euclidean) distance between that point and the centroid of each of the two groups in the original two-dimensional space.

In the illustration appearing in Panel III we also observe that the between-groups difference in means on X_1 is the same as that on X_2. (However, this as-

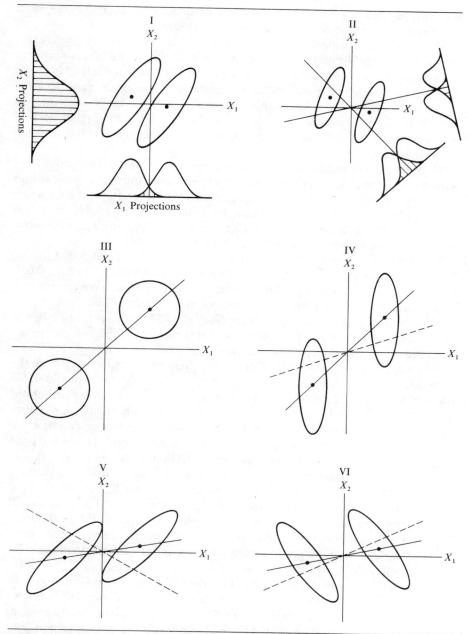

Figure 4.2 Some Geometrical Aspects of Two-Group Discrimination

sumption is not necessary for this particular assignment rule to be applied.) Hence, the line connecting the two centroids forms a 45° angle with X_1 and each predictor gets equal weight in the formation of the linear composite. In this special case the linear composite itself can be located by direction cosines of 0.707 and -0.707 with respect to X_1 and X_2, respectively.

However, what happens if the within-groups variability is *not* representable by a scalar matrix? In particular, let us first assume that X_1 and X_2 continue to be uncorrelated within groups but that the variance of X_2 is considerably larger than that of X_1. This type of situation is depicted in Panel IV. We note that the concentration ellipses have axes that are parallel to the original axes (i.e., X_1 and X_2 do not covary within each group), but the variance of X_2 considerably exceeds that of X_1. In this case the covariance matrix for each group is diagonal rather than scalar.

Under this set of conditions, the reader can convince himself that projections onto the dotted line in Panel IV would lead to a wider separation between groups than the solid line connecting the two centroids. Moreover, in this new case we observe from the angle that the dotted line makes with the original axes that X_1 — the variable with the *smaller* variance — would receive more weight in forming the linear composite. This seems intuitively reasonable.

Panel V shows a case in which each covariance matrix indicates that X_1 and X_2 covary on a within-groups basis. Assume now that the variance of X_1 equals the variance of X_2 within each group. Each of the two (equal) covariance matrices, then, consists of positive off-diagonal elements and, under this set of conditions, the concentration ellipses are tilted at a 45° angle, as noted in Panel IV. Finally, we have drawn the ellipses so that the means of the two groups are more widely separated on X_1 than they are on X_2.

If one applied the earlier rule of projecting points onto a line connecting the centroids of the two groups, it is easy to see from Panel V that a considerable amount of overlap would take place between the two (univariate) distributions obtained from this particular linear composite. On the other hand, if one projects the points onto the dotted line shown in Panel V, the groups can be separated perfectly. What this tells us to do is find a linear composite in which one takes a weighted *difference* between X_1 and X_2. As our intuition might suggest, this type of rule appears reasonable, so as to avoid a type of "double counting" that reflects the positive within-groups covariance between X_1 and X_2.[5]

If X_1 and X_2 are negatively correlated on a within-groups basis, we would want to develop a rule that considers their weighted *sum*, rather than difference. This idea is shown in Panel VI. We see that the dotted line does, indeed, result in wider separation of the group means on the point projections than the solid line connecting the two centroids.

While our comments regarding the last three panels of Figure 4.2 may appear plausible, we still do not know whether the dotted lines in Panels IV, V and VI are the very best that we can do with respect to a two-group discrimination. What we can say at this point is that a rule which involves point projection onto a line connecting the two group centroids appears intuitively reasonable *only* when X_1 and X_2 exhibit concentration ellipses in the form of scalar covariance matrices (see Panel III of Figure 4.2). *If the variances are unequal or the covariances are non-zero, other linear composites would appear to result in projections with less overlap in the resulting unidimensional distributions.*[6]

4.3 Formal Approaches to Two-Group Discrimination

We can now introduce more formal procedures for dealing with two-group discrimination. For expository purposes we continue to work with the two predictor variables in the sample problem of Table 4.1, although everything that is said here applies to three or more predictors, as well.

In line with the intuitive discussion presented earlier, we shall describe two interrelated, and essentially equivalent, ways of developing two-group linear discrimination:

1. Fisher's method (Fisher, 1936).

2. Mahalanobis' D^2 method (Mahalanobis, 1936).

Each procedure is described in turn, after presenting some preliminary notation and computations.

4.3.1 Background Notation and Computations

Before discussing the first formal approach for carrying out two-group discrimination, a set of preliminary operations for computing mean-corrected sums of squares and cross product (SSCP) and covariance matrices is outlined. We shall want to compute the following SSCP matrices:

1. Total (mean-corrected) sums of squares and cross products matrix, denoted **T**.

2. Pooled within-groups (mean-corrected) sums of squares and cross products matrix, denoted **W**.

3. Between-groups (mean-corrected) sums of squares and cross products matrix, denoted **A**.

While the procedure is illustrated for the case of $n = 2$ predictors, and $G = 2$ groups, it should be mentioned that the same format can be followed for $n > 2$ or $G > 2$, as will be seen in subsequent chapters.

Table 4.2 shows the matrix algebra details of finding **W**, **A**, and **T**. Also shown are the various numerical results corresponding to the intermediate matrices of interest. As noted, the key transformation devices consist of the matrix **H** and various unit vectors that can be used for carrying out the desired manipulations. We note at the end that the within-groups and between-groups matrices sum to the total-sample matrix.

The next step is to compute the pooled within-groups covariance matrix $\mathbf{C}_w$ from **W** as follows:

$$\mathbf{C}_w = \frac{\mathbf{W}}{\sum\limits_{g=1}^{G} (m_g - 1)} = \frac{\mathbf{W}}{10 - 2} = \frac{1}{8}\begin{bmatrix} 20 & 16 \\ 16 & 26.4 \end{bmatrix}$$

$$= \begin{bmatrix} 2.5 & 2 \\ 2 & 3.3 \end{bmatrix}$$

where m_g denotes the number of cases in group g ($g = 1, 2, \ldots, G$). In this problem $m_1 = m_2 = 5$. Notice that $\mathbf{C}_w$ is proportional to $\mathbf{W}$.

In the special case of two groups, certain simplifications are possible in the computation of the between-groups SSCP matrix, $\mathbf{A}$. These simplifications can be derived via simple algebra and result in the following formula for the special case of two groups:

$$\mathbf{A} = \frac{m_1 m_2}{m_1 + m_2}\, \mathbf{dd'}$$

where $\mathbf{d}$ denotes the difference vector between group centroids. In terms of the sample-problem data of Table 4.1 we have the difference vector:[7]

$$\mathbf{d} = \begin{bmatrix} -2.5 \\ -1 \end{bmatrix} - \begin{bmatrix} 2.5 \\ 1 \end{bmatrix} = \begin{bmatrix} -5 \\ -2 \end{bmatrix}$$

Since $m_1 = m_2 = 5$, we then get:

$$\mathbf{A} = \frac{25}{10}\begin{bmatrix} -5 \\ -2 \end{bmatrix}[-5 \quad -2]$$

$$= \frac{25}{10}\begin{bmatrix} 25 & 10 \\ 10 & 4 \end{bmatrix}$$

$$= \begin{bmatrix} 62.5 & 25 \\ 25 & 10 \end{bmatrix}$$

which, as can be seen, agrees with that shown (via the general procedure) in Table 4.2.

4.3.2 Fisher's Discriminant Function

R. A. Fisher, the eminent British statistician, introduced the linear discriminant function in the mid-thirties. At this point we shall want to develop the formal apparatus for computing Fisher's discriminant function and compare this approach with some of the intuitive considerations described earlier in connection with Figure 4.2.

To set the stage for developing significance tests for the discriminant function, let us assume that we have two random samples of observations on a set of continuous variables from two multivariate normal populations, with assumed common covariance matrices, whose centroids may differ.

We assume that an estimate of the common population covariance matrix can be obtained by pooling the two samples' covariance matrices. We have called this pooled within-groups covariance matrix $\mathbf{C}_w$ and have noted that it is proportional to $\mathbf{W}$, the pooled within-group SSCP matrix. In two-group discriminant analysis we seek a set of weights, expressed by the vector:

Table 4.2 Calculation of SSCP Matrices from Sample Data of Table 4.1

Define

$$\mathbf{H} = \begin{bmatrix} 1 & 0 \\ 1 & 0 \\ 1 & 0 \\ 1 & 0 \\ 1 & 0 \\ 0 & 1 \\ 0 & 1 \\ 0 & 1 \\ 0 & 1 \\ 0 & 1 \end{bmatrix}$$

Data Matrix

$$\mathbf{X} = \begin{bmatrix} 2 & 4 \\ 3 & 2 \\ 4 & 5 \\ 5 & 4 \\ 6 & 7 \\ 7 & 6 \\ 8 & 4 \\ 9 & 7 \\ 10 & 6 \\ 11 & 9 \end{bmatrix}$$

Then, find

$$\overline{\mathbf{X}}_g = (\mathbf{H}'\mathbf{H})^{-1}\mathbf{H}'\mathbf{X} = \begin{bmatrix} 4 & 4.4 \\ 9 & 6.4 \end{bmatrix} \begin{matrix} \text{Group 1 Centroid} \\ \text{Group 2 Centroid} \end{matrix}$$

$$\overline{\mathbf{x}}' = \tfrac{1}{10}\mathbf{1}'\mathbf{X} = (6.5 \quad 5.4) \text{ Total-sample Centroid}$$

where $\mathbf{1}'$ is a 1×10 unit vector.

Matrix of Within-Groups Deviations	Matrix of Between-Groups Deviations	Matrix of Total-Sample Deviations
$\mathbf{P} = \mathbf{X} - \mathbf{H}\overline{\mathbf{X}}_g$	$\mathbf{Q} = \mathbf{H}\overline{\mathbf{X}}_g - \mathbf{1}\overline{\mathbf{x}}'$	$\mathbf{R} = \mathbf{X} - \mathbf{1}\overline{\mathbf{x}}'$

$$\mathbf{P} = \begin{bmatrix} -2 & -0.4 \\ -1 & -2.4 \\ 0 & 0.6 \\ 1 & -0.4 \\ 2 & 2.6 \\ -2 & -0.4 \\ -1 & -2.4 \\ 0 & 0.6 \\ 1 & -0.4 \\ 2 & 2.6 \end{bmatrix} \quad \mathbf{Q} = \begin{bmatrix} -2.5 & -1 \\ -2.5 & -1 \\ -2.5 & -1 \\ -2.5 & -1 \\ -2.5 & -1 \\ 2.5 & 1 \\ 2.5 & 1 \\ 2.5 & 1 \\ 2.5 & 1 \\ 2.5 & 1 \end{bmatrix} \quad \mathbf{R} = \begin{bmatrix} -4.5 & -1.4 \\ -3.5 & -3.4 \\ -2.5 & -0.4 \\ -1.5 & -1.4 \\ -0.5 & 1.6 \\ 0.5 & 0.6 \\ 1.5 & -1.4 \\ 2.5 & 1.6 \\ 3.5 & 0.6 \\ 4.5 & 3.6 \end{bmatrix}$$

where $\mathbf{1}$ is a 10×1 unit vector.

Within-Groups SSCP Matrix $\mathbf{W} = \mathbf{P}'\mathbf{P}$	Between-Groups SSCP Matrix $\mathbf{A} = \mathbf{Q}'\mathbf{Q}$	Total-Sample SSCP Matrix $\mathbf{T} = \mathbf{R}'\mathbf{R}$
$\mathbf{W} = \begin{bmatrix} 20 & 16 \\ 16 & 26.4 \end{bmatrix}$	$\mathbf{A} = \begin{bmatrix} 62.6 & 25 \\ 25 & 10 \end{bmatrix}$	$\mathbf{T} = \begin{bmatrix} 82.5 & 41 \\ 41 & 36.4 \end{bmatrix}$

$$\boxed{\mathbf{T} = \mathbf{W} + \mathbf{A}}$$

$$\mathbf{k} = \begin{bmatrix} k_1 \\ k_2 \\ \vdots \\ k_n \end{bmatrix}$$

from which the following linear composite **t**—called a *discriminant axis*—can be obtained:

$$\mathbf{t} = \mathbf{X}_d \mathbf{k} = \begin{bmatrix} X_{d11} & X_{d12} & \cdots & X_{d1n} \\ X_{d21} & X_{d22} & \cdots & X_{d2n} \\ \vdots & \vdots & \vdots & \vdots \\ X_{dm1} & X_{dm2} & \cdots & X_{dmn} \end{bmatrix} \begin{bmatrix} k_1 \\ k_2 \\ \vdots \\ k_n \end{bmatrix} = \begin{bmatrix} t_1 \\ t_2 \\ \vdots \\ t_m \end{bmatrix}$$

where $\mathbf{X}_d$ is the $m \times n$ matrix of mean-corrected scores on the original predictors.[8]

Fisher's linear discriminant function is then defined as:

$$t_i = \mathbf{x}'_{di}\mathbf{k} = x_{di1}k_1 + x_{di2}k_2 + \cdots + x_{din}k_n$$

$$(\text{for } i = 1, 2, \ldots, m).$$

That is, each t_i denotes the *score* of the i-th person (in the case of the data of Table 4.1) on the discriminant axis **t**.

The mean of the linear composite **t** is found by substituting the centroid of $\mathbf{X}_d$ in the discriminant function:

$$\bar{t} = \bar{\mathbf{x}}'_d\mathbf{k} = (0, 0, \ldots, 0) \begin{bmatrix} k_1 \\ k_2 \\ \vdots \\ k_n \end{bmatrix} = 0$$

That is, $\bar{t} = 0$ since the matrix $\mathbf{X}_d$ is already expressed in mean-corrected form. The variance of **t** (as recalled from section 1.5) is given by:

$$\frac{\mathbf{t}'\mathbf{t}}{(m-1)} = \frac{1}{m-1}[\mathbf{k}'\mathbf{X}'_d\mathbf{X}_d\mathbf{k}]$$

which also can be broken down into a between-groups and a within-groups part.

In order to see how this happens, let us backtrack a little. First, from section 1.5 of Chapter 1, we know that the following relationship holds for the sum of squared deviations about the mean (which is zero) of $\mathbf{t}$:

$$\mathbf{t't} = \mathbf{k'X'_dX_dk} = \mathbf{k'Tk} = \mathbf{k'(A + W)k} = \mathbf{k'Ak} + \mathbf{k'Wk}$$

The expression on the right shows how the decomposition of $\mathbf{t't}$ consists of two scalars, $\mathbf{k'Ak}$ and $\mathbf{k'Wk}$, one based on the between-groups SSCP matrix of $\mathbf{X}_d$ and the other based on the pooled within-groups SSCP matrix of $\mathbf{X}_d$.

If we can find the vector $\mathbf{k}$ that maximizes between-groups variation relative to within-groups variation, we have the desired discriminant function. First, we set up the ratio:

$$\lambda = \frac{\mathbf{k'Ak}}{\mathbf{k'Wk}}$$

Finding a vector $\mathbf{k}$ that maximizes the ratio λ sounds, of course, like a problem in the calculus. Appendix B shows the procedure for differentiating λ with respect to $\mathbf{k}$ and setting the partial derivatives equal to the $\mathbf{0}$ vector. The resulting matrix equation leads to a vector $\mathbf{k}$ that maximizes the scalar λ in the above ratio. This matrix equation is:

$$(\mathbf{W^{-1}A} - \lambda\mathbf{I})\mathbf{k} = \mathbf{0}$$

where $\mathbf{I}$ is the identity matrix. (Details on the solution of the general eigenstructure equation above are provided in Chapter 7, covering multiple discriminant analysis.)

Fortunately, in the special case of two groups certain simplifications are possible. We have already defined $\mathbf{A}$, the between-groups, SSCP matrix, as:

$$\mathbf{A} = \frac{m_1m_2}{m_1 + m_2}\,\mathbf{dd'}$$

The expression $m_1m_2/(m_1 + m_2)$ will be a constant for any specific problem; accordingly, we can set it equal to the scalar C. In the special case of two groups, then, we have the substitution:

$$[(\mathbf{W^{-1}}C\mathbf{dd'}) - \lambda\mathbf{I}]\mathbf{k} = \mathbf{0}$$

Since C is a scalar, it can be used as a premultiplier, giving us:

$$[C\mathbf{W^{-1}(dd')} - \lambda\mathbf{I}]\mathbf{k} = \mathbf{0}$$

Postmultiplication of the bracketed expression by $\mathbf{k}$, followed by transposing, gives us:

$$\lambda\mathbf{k} = C\mathbf{W}^{-1}(\mathbf{dd}')\mathbf{k}$$

And, by regrouping, we have:

$$\lambda\mathbf{k} = C(\mathbf{W}^{-1}\mathbf{d})(\mathbf{d}'\mathbf{k})$$

Since $\mathbf{d}'\mathbf{k}$ is a scalar and λ is also a scalar we can assemble these (and C) into a single composite scalar:

$$C^* = C(\mathbf{d}'\mathbf{k})/\lambda$$

giving us the formula for finding the vector of discriminant weights:

$$\mathbf{k} = C^*\mathbf{W}^{-1}\mathbf{d}$$

However, since $\mathbf{C}_w$ is proportional to $\mathbf{W}$, $\mathbf{C}_w^{-1}$ will be proportional to $\mathbf{W}^{-1}$, and we could just as easily write:

$$\boxed{\mathbf{k} \propto \mathbf{C}_w^{-1}\mathbf{d}}$$

for $\mathbf{k}$ the desired vector of weights. The symbol $\propto$ denotes "is proportional to" and represents a general constant of proportionality to take note of the fact that the values of $\mathbf{k}$ are unique only up to a scale transformation.[9]

The vector of weights in Fisher's discriminant function is usually computed from the inverse of the pooled within-groups covariance matrix. However, as illustrated above, we would just as readily use $\mathbf{W}^{-1}$ since:

$$\boxed{\mathbf{C}_w^{-1} = (m - 2)\mathbf{W}^{-1}}$$

and we noted that $\mathbf{k}$ is defined only up to a constant of proportionality anyway.

Very often the vector $\mathbf{k}$, leading to the desired linear composite $\mathbf{t}$, is normalized to unit sum of squares (i.e., $\mathbf{k}'\mathbf{k} = 1$). Table 4.3 shows the calculations for finding the discriminant weights, based on the sample problem of Table 4.1. First, we note that Table 4.2 provides $\mathbf{W}$, the pooled within-groups SSCP matrix which we divide by $(m_1 + m_2 - 2)$ to obtain:

$$\mathbf{C}_w = \begin{bmatrix} 2.5 & 2 \\ 2 & 3.3 \end{bmatrix}$$

We then find $\mathbf{C}_w^{-1}$ and $\mathbf{k}$. We can normalize the components of $\mathbf{k}$ to unit sum of squares, thus expressing $\mathbf{k}$ as a set of direction cosines:

Table 4.3 Computations for Finding the Linear Discriminant Weights

Covariance Matrix (computed from Table 4.2)

$$\mathbf{W} = \begin{bmatrix} 20 & 16 \\ 16 & 26.4 \end{bmatrix}; \qquad \mathbf{C}_w = \frac{1}{8} \begin{bmatrix} 20 & 16 \\ 16 & 26.4 \end{bmatrix}$$

$$= \begin{bmatrix} 2.5 & 2 \\ 2 & 3.3 \end{bmatrix}$$

Inverse of Within-Groups Covariance Matrix

$$|\mathbf{C}_w| = 4.25; \; \mathbf{C}_w^{-1} = \frac{1}{4.25} \begin{bmatrix} 3.3 & -2 \\ -2 & 2.5 \end{bmatrix} = \begin{bmatrix} 0.776 & -0.471 \\ -0.471 & 0.588 \end{bmatrix}$$

Difference Vector of Centroids

$$\mathbf{d} = \begin{bmatrix} -2.5 \\ -1 \end{bmatrix} - \begin{bmatrix} 2.5 \\ 1 \end{bmatrix} = \begin{bmatrix} -5 \\ -2 \end{bmatrix}$$

Discriminant Weights

$$\mathbf{k} = \mathbf{C}_w^{-1}\mathbf{d} = \begin{bmatrix} 0.776 & -0.471 \\ -0.471 & 0.588 \end{bmatrix} \begin{bmatrix} -5 \\ -2 \end{bmatrix} = \begin{bmatrix} -2.94 \\ 1.18 \end{bmatrix}$$

$$\mathbf{k} \text{ (normalized)} = \begin{bmatrix} -0.93 \\ 0.37 \end{bmatrix}$$

$$\mathbf{k} \text{ (normalized)} = \begin{bmatrix} -0.93 \\ 0.37 \end{bmatrix}$$

Figure 4.3 reproduces Panel II of Figure 4.1. The discriminant axis is then plotted in. As noted, the centroids of groups 1 and 2 are well separated on the discriminant axis $\mathbf{t}$, computed above.

Notice also that the direction of the discriminant axis is similar to that of Panel V in Figure 4.2. Since the variance of predictor X_1 is less than that of X_2 (2.5 versus 3.3) the function gives more (absolute) weight to X_1, the predictor with the smaller variance. Since the covariance between X_1 and X_2 is positive, the linear composite (axis) entails a weighted *difference* between X_1 and X_2 to eliminate the "double counting" effect.

Suppose we next consider values on the linear composite—that is, the discriminant scores—obtained by substituting the *mean* vectors of groups 1 and 2 in the discriminant function:

$$\bar{t}_1 = \mathbf{k}'\bar{\mathbf{x}}_{d1}$$

$$\bar{t}_2 = \mathbf{k}'\bar{\mathbf{x}}_{d2}$$

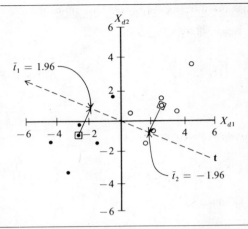

Figure 4.3 Discriminant Axis Plotted in Deviations-from-Centroid Space

If so, we have:

$$\bar{t}_1 = (-0.93 \quad 0.37)\begin{bmatrix} -2.5 \\ -1 \end{bmatrix} = 1.96; \; \bar{t}_2 = (-0.93 \quad 0.37)\begin{bmatrix} 2.5 \\ 1 \end{bmatrix} = -1.96$$

Next, we can compute all ten scores on the discriminant axis **t** and find the deviations around the respective group means of $\bar{t}_1$ and $\bar{t}_2$:

Group 1	t_{i1}	$t_{i1} - \bar{t}_{.1}$	Group 2	t_{i2}	$t_{i2} - \bar{t}_{.2}$
1	3.67	1.71	1	−0.24	1.72
2	2.00	0.04	2	−1.91	0.05
3	2.18	0.22	3	−1.73	0.23
4	0.88	−1.08	4	−3.03	−1.07
5	1.06	−0.90	5	−2.85	−0.89

Then it is a simple matter to find the between-groups and within-groups squared deviations, first with regard to $\bar{t}$ and then with regard to $\bar{t}_1$ and $\bar{t}_2$:

$$\text{SSA} = 5(1.96 - 0)^2 + 5(-1.96 - 0)^2 = 38.4$$

$$\text{SSW} = (1.71)^2 + (0.04)^2 + \cdots + (-1.07)^2 + (-0.89)^2 = 9.9$$

We note that these quantities are, indeed, equal to:

$$\mathbf{k'Ak} = (-0.93 \quad 0.37)\begin{bmatrix} 62.5 & 25 \\ 25 & 10 \end{bmatrix}\begin{bmatrix} -0.93 \\ 0.37 \end{bmatrix} = 38.4$$

$$\mathbf{k'Wk} = (-0.93 \quad 0.37)\begin{bmatrix} 20 & 16 \\ 16 & 26.4 \end{bmatrix}\begin{bmatrix} -0.93 \\ 0.37 \end{bmatrix} = 9.9$$

as asserted earlier.

Finally, we can compute a cutoff point, t_c, halfway between the mean discriminant scores of groups 1 and 2:

$$t_c = \frac{(\bar{t}_1 + \bar{t}_2)}{2} = \frac{(1.96 - 1.96)}{2}$$
$$= 0$$

and assign the i-th case to group 1 if $t_i > t_c$; otherwise, it is assigned to group 2.

As can be seen from the numerical values of t_i, shown above, no classification errors are made under this assignment rule. In the case of equal sample sizes the cutoff point will always equal zero (if the $\mathbf{X}_d$ matrix is used), since $\bar{t}_1 = -\bar{t}_2$.

4.3.3 Re-Examination of the Relationships in Figure 4.2

Now that we have provided a somewhat more formal introduction to two-group discrimination, let us examine the vector of weights in Fisher's discriminant function:

$$\mathbf{k} = \mathbf{C}_w^{-1}\mathbf{d}$$

in the context of the hypothetical cases shown in Figure 4.2. First, if $\mathbf{C}_w$, the pooled within-groups covariance matrix, is a scalar matrix (as would be the case in Panel III of Figure 4.2), then its inverse would also be a scalar matrix—with a diagonal consisting of the inverse elements of the original scalar matrix—and the elements of $\mathbf{k}$ would depend on those of $\mathbf{d}$ alone. Hence, we can now see that a discriminant axis that connects the centroids, where $\mathbf{d}$ is normalized to yield a set of direction cosines, is not only intuitively plausible but formally correct in the case of situations like those depicted in Panel III. If the elements of $\mathbf{d}$ are not equal then the discriminant axis will be tilted in the direction of the dimension of larger difference between the group means. If the sample sizes of groups 1 and 2 are unequal, then the centroid of the larger sample will lie closer to the origin but, otherwise, the preceding comments apply in just the same way.[10]

If $\mathbf{C}_w$ is diagonal, then $\mathbf{C}_w^{-1}$ is also diagonal and its elements are, again, the reciprocals of those of $\mathbf{C}_w$; hence, the effect of this situation would be to weight each axis (X_1 versus X_2 in the two-predictor case) *inversely* by the variance of X_1 and X_2. This agrees with our intuitive remarks regarding Panel IV of Figure 4.2.

If $\mathbf{C}_w$ has off-diagonal elements not equal to zero and major diagonal elements equal to each other, and if the off-diagonal elements are positive, then one finds a difference-type score. This was illustrated by Panel V in Figure 4.2. The sample problem (with unequal diagonal elements) is of this type.

However, there is an additional point of interest. As we know, in the sample problem, the difference vector of centroids is:

$$\mathbf{d} = \begin{bmatrix} -5 \\ -2 \end{bmatrix}$$

while the (normalized) discriminant weights are:

$$\mathbf{k} = \begin{bmatrix} -0.93 \\ 0.37 \end{bmatrix}$$

Note that in each vector separately, the absolute value of the first component of the vector is about 2.5 times the second component.

This relationship need not hold generally. To illustrate what we mean, let us now assume that we have a difference vector between group means of:

$$\mathbf{d} = \begin{bmatrix} 5 \\ 1 \end{bmatrix}$$

and that the pooled within-groups covariance matrix is:

$$\mathbf{C}_w = \begin{bmatrix} 10 & 8 \\ 8 & 10 \end{bmatrix}$$

That is, we assume (for convenience) that the variances of X_1 and X_2 are equal and that their covariance is highly positive. Then, in this case we have:

$$\mathbf{C}_w^{-1} = \frac{1}{36} \begin{bmatrix} 10 & -8 \\ -8 & 10 \end{bmatrix} = \begin{bmatrix} 0.278 & -0.222 \\ -0.222 & 0.278 \end{bmatrix}$$

and the vector $\mathbf{k}$ of discriminant weights is:

$$\mathbf{k} = \begin{bmatrix} 0.278 & -0.222 \\ -0.222 & 0.278 \end{bmatrix} \begin{bmatrix} 5 \\ 1 \end{bmatrix} = \begin{bmatrix} 1.17 \\ -0.832 \end{bmatrix}$$

Notice now that the absolute value of the ratio of the first component to the second component drops from 5:1 (in the $\mathbf{d}$ vector) to only 1.4:1 (in the $\mathbf{k}$ vector), *solely as a consequence of the high correlation between predictors*. Thus, the high positive covariation between X_1 and X_2 has two effects:

1. The absolute values of the weights assigned to X_1 and X_2 are *more nearly equal* than those given by $\mathbf{d}$ alone.

2. The resulting linear composite involves a weighted *difference* score between the two variables (as was earlier pointed out in the intuitive discussion of Panel V of Figure 4.2).

Similar arguments underlie the example of Panel VI, although in this case we would have a weighted sum since X_1 and X_2 are negatively correlated on a within-groups basis.

In summary, Fisher's discriminant *function* is defined as:

$$t_i = \mathbf{x}'_{di}\mathbf{k}$$

When $\mathbf{k}$ is applied to the full data matrix $\mathbf{X}_d$ of mean-corrected data one finds the linear composite:

$$\mathbf{t} = \mathbf{X}_d\mathbf{k}$$

which is the discriminant *axis* that maximizes the ratio of between-groups to within-groups variability.

Cases are assigned to groups 1 or 2 on the basis of whether $t_i > t_c = (\bar{t}_1 + \bar{t}_2)/2$. If so, t_i is assigned to group 1. If otherwise, it is assigned to group 2. This assumes, of course, that $\bar{t}_1 > \bar{t}_2$, which is the case in the sample problem.

4.3.4 Mahalanobis' D^2

Use of either set of discriminant weights:

$$\mathbf{k} = \begin{bmatrix} -2.94 \\ 1.18 \end{bmatrix} \quad \text{or} \quad \mathbf{k} \text{ (normalized)} = \begin{bmatrix} -0.93 \\ 0.37 \end{bmatrix}$$

as computed in Table 4.3, entailed a reduction of a two-dimensional configuration of points to a single axis $\mathbf{t}$, the discriminant axis shown in Figure 4.3. Indeed, $\mathbf{k}$ (normalized) can be viewed as a set of direction cosines that define the location of the axis while the individual t_i's denote scores of the profiles' orthogonal projections on the axis.

If the dimensionality of the (mean-corrected) profiles space is high, a sizable reduction takes place since only one discriminant axis need be found for two groups. Regardless of how high the original dimensionality of the profiles is, we still have:

$$\mathbf{t} = \mathbf{X}_d\mathbf{k}$$

as the *single* discriminant axis (in the case of two groups).

Cases arise, however, in which we may wish to assign points to groups on the basis of the *original number of dimensions* in which the profiles are characterized. For example, in Panel III of Figure 4.2, a rather natural classification procedure would be to measure the respective distance (or possibly the squared distance) of each point to the centroid of group 1 and the centroid of group 2. One could then assign the point to that group whose centroid was the closer.

While this rule seems eminently reasonable in the case of Panel III, in which the covariance matrices of each group show equal variance on each axis and no covariation between X_1 and X_2, the presence of either unequal variances or covariance between predictors (as in Panels IV, V, and VI, for example) suggests the need for some kind of adjustment of "statistical distance" to regular (Euclidean) distance.

Two routes can be followed in order to bring about this adjustment. First, one could generalize the basic idea of squared Euclidean distance to handle covariance matrices that are not scalar matrices. Or, one could perform a preliminary transformation of the (mean-corrected) data matrix that produces a set of scores whose pooled within-groups covariance matrix *is* a scalar matrix. Then one could assign cases in this *transformed* space on the basis of nearest centroid computed via ordinary squared Euclidean distance. Here, we confine our discussion to the first procedure. (The concept of a preliminary transformation to squared Euclidean distance is discussed in Chapter 7, in the context of multiple discriminant analysis.)

Mahalanobis' D^2 (named after its developer) represents a generalization of the traditional squared Euclidean distance formula that adjusts for pooled within-groups covariance matrices that are *not* scalar matrices. Assuming that we have two mean-corrected centroids, h and i, in a space of n dimensions, their Mahalanobis' D^2 is defined as:

$$D^2_{hi} = (\mathbf{x}_h - \mathbf{x}_i)'\mathbf{C}_w^{-1}(\mathbf{x}_h - \mathbf{x}_i)$$

where $\mathbf{C}_w^{-1}$ denotes the inverse of the pooled within-groups covariance matrix.

Mahalanobis' D^2 is a generalization of the familiar (squared) Euclidean distance d^2_{hi}:

$$d^2_{hi} = (\mathbf{x}_h - \mathbf{x}_i)'(\mathbf{x}_h - \mathbf{x}_i)$$

$$= (X_{h1} - X_{i1}, X_{h2} - X_{i2}, \ldots, X_{hn} - X_{in}) \begin{bmatrix} X_{h1} - X_{i1} \\ X_{h2} - X_{i2} \\ \vdots \\ X_{hn} - X_{in} \end{bmatrix}$$

$$= \sum_{j=1}^{n} (X_{hj} - X_{ij})^2$$

$$= \mathbf{d}'\mathbf{d}$$

Note that in the case of d^2_{hi}, $\mathbf{C}_w = \mathbf{C}_w^{-1} = \mathbf{I}$, an identity matrix. The idea of squared Euclidean distance is illustrated in Panel III in Figure 4.2; we see that the concentration ellipses are circles.

However, suppose we have the case of Panel IV in which the pooled within-groups covariance matrix is diagonal, for example:

$$\mathbf{C}_w = \begin{bmatrix} 1 & 0 \\ 0 & 2 \end{bmatrix}$$

Mahalanobis' D^2 is:

$$D^2 = \mathbf{d}'\mathbf{C}_w^{-1}\mathbf{d} = \mathbf{d}' \overset{\mathbf{C}_w^{-1}}{\begin{bmatrix} 1 & 0 \\ 0 & 1/2 \end{bmatrix}} \mathbf{d}$$

Note that D^2 shrinks the second axis to adjust for the initial difference in variances. Of course, the sample problem of Table 4.1 is more like Panel V of Figure 4.2 in that the predictors not only have unequal variances but are correlated as well.

In Table 4.3 we calculated $\mathbf{C}_w^{-1}$ for the sample data and we also know what the difference vector between the two group centroids is. Accordingly, we can compute Mahalanobis' D^2 between the two group centroids in the sample problem as:

$$D^2 = \mathbf{d}'\mathbf{C}_w^{-1}\mathbf{d}$$

$$= (-5 \quad -2) \begin{bmatrix} 0.776 & -0.471 \\ -0.471 & 0.588 \end{bmatrix} \begin{bmatrix} -5 \\ -2 \end{bmatrix} = 12.35$$

where, as recalled, the group centroids are, respectively:

$$\bar{\mathbf{x}}_{d1} = \begin{bmatrix} -2.5 \\ -1 \end{bmatrix}; \bar{\mathbf{x}}_{d2} = \begin{bmatrix} 2.5 \\ 1 \end{bmatrix}$$

Similarly, we could compute D^2 between each individual point and the centroid of each group. For example, the profile of the first case in group 1 is $(-4.5, -1.4)$. Its Mahalanobis' D^2 from the centroids of groups 1 and 2, respectively, is:

Group 1

$$D_{11}^2 = (-2.0 \quad -0.4) \begin{bmatrix} 0.776 & -0.471 \\ -0.471 & 0.588 \end{bmatrix} \begin{bmatrix} -2.0 \\ -0.4 \end{bmatrix} = 2.5$$

Group 2

$$D_{12}^2 = (-7 \quad -2.4) \begin{bmatrix} 0.776 & -0.471 \\ -0.471 & 0.588 \end{bmatrix} \begin{bmatrix} -7 \\ -2.4 \end{bmatrix} = 25.6$$

In this case the first profile would be assigned (correctly) to group 1 since its D^2 of 2.5 is the smaller of the two.

One could also compute D^2's for the remaining nine points from the group 1 and group 2 centroids, respectively. Assignment on the basis of which centroid has the smaller D^2 would proceed as above. As the reader may verify, all ten profiles would be correctly classified under this rule.

A further point of interest is: what is the relationship of Mahalanobis' D^2 to Fisher's discriminant function? As it turns out, *both procedures produce exactly the same point assignments*. To see why this is so, let us reconsider the assignment rule that was followed in the case of Mahalanobis' D^2. This rule involves assigning some profile $\mathbf{x}_i$ to group 1 if:

$$(\mathbf{x}_i - \overline{\mathbf{x}}_1)'\mathbf{C}_w^{-1}(\mathbf{x}_i - \overline{\mathbf{x}}_1) < (\mathbf{x}_i - \overline{\mathbf{x}}_2)'\mathbf{C}_w^{-1}(\mathbf{x}_i - \overline{\mathbf{x}}_2)$$

which (continuing to denote $\mathbf{d} = \overline{\mathbf{x}}_1 - \overline{\mathbf{x}}_2$) can be rewritten as:

$$(\mathbf{x}_i'\mathbf{C}_w^{-1}\mathbf{d}) > \tfrac{1}{2}(\overline{\mathbf{x}}_1'\mathbf{C}_w^{-1}\overline{\mathbf{x}}_1 - \overline{\mathbf{x}}_2'\mathbf{C}_w^{-1}\overline{\mathbf{x}}_2)$$

This says that we assign $\mathbf{x}_i$ to group 1 if its score $(\mathbf{x}_i'\mathbf{C}_w^{-1}\mathbf{d})$ exceeds some constant, as computed from the term on the right of the inequality.

But the term on the left is proportional to $\mathbf{x}_i'\mathbf{k}$ since, as recalled:

$$\mathbf{k} \propto \mathbf{C}_w^{-1}\mathbf{d}$$

Furthermore, the cutoff value on the right of the equality behaves just as t_c does in the discriminant axis computations. The above rule will always produce group assignments that are *identical* to those obtained from Fisher's discriminant function.[11]

Of additional interest is the relationship of Mahalanobis' D^2 between group centroids in the $\mathbf{X}_d$ space to the pooled within-groups variance that is computed from $\mathbf{t}$, the single discriminant axis. In terms of the sample problem:

$$D^2 = \overset{\mathbf{d}'}{(-5 \quad -2)} \overset{\mathbf{C}_w^{-1}}{\begin{bmatrix} 0.776 & -0.471 \\ -0.471 & 0.588 \end{bmatrix}} \overset{\mathbf{d}}{\begin{bmatrix} -5 \\ -2 \end{bmatrix}}$$

$$= 12.35$$

This quantity is the Mahalanobis' D^2 between group centroids in the original $\mathbf{X}_d$ space. However, if one finds discriminant scores via the *non-normalized version* of $\mathbf{k}$ (as computed in Table 4.3):

$$\mathbf{k} = \begin{bmatrix} -2.94 \\ 1.18 \end{bmatrix}$$

and then computes their *pooled within-groups variance* along the single discriminant axis $\mathbf{t}$, that number will be equal to D^2.

In summary, Fisher's discriminant function and Mahalanobis' D^2 are equivalent insofar as group assignments are concerned. Their chief difference lies in the fact that Mahalanobis' D^2 is computed in the *original space* of the predictors while the discriminant approach, in effect, collapses that space to a *single axis* (in the case of two groups), denoted by **t**. In this sense the discriminant function provides a more parsimonious model. Moreover, as the number of predictors increase, the single axis **t** can represent a considerable reduction in original dimensionality.

4.3.5 Other Formulations of the Discriminant Function

At this point we have discussed two approaches to two-group discrimination:

1. Fisher's discriminant function, defined as:

$$\boxed{t_i = \mathbf{x}'_{di}\mathbf{k}}$$

where profiles are assigned on the basis of which side of the cutoff value t_c they lie.

2. Mahalanobis' D^2, in which profiles are assigned to groups on the basis of which D^2 is smaller:

$$D^2_{i1} = (\mathbf{x}_i - \bar{\mathbf{x}}_1)'\mathbf{C}_w^{-1}(\mathbf{x}_i - \bar{\mathbf{x}}_1) \quad \text{or}$$

$$D^2_{i2} = (\mathbf{x}_i - \bar{\mathbf{x}}_2)'\mathbf{C}_w^{-1}(\mathbf{x}_i - \bar{\mathbf{x}}_2)$$

We indicated that both procedures yield identical profile assignments.

Still other discriminant-type formulations are possible. For example, discriminant weights can be computed from:

1. The pooled within-groups SSCP, covariance, or correlation matrices.

2. The total-sample SSCP, covariance, or correlation matrices.

3. A multiple regression approach in which the dummy variable in Table 4.1 serves as a criterion and X_1 and X_2 are predictor variables.

In the case of the within-groups SSCP matrices versus their counterpart covariance matrices, it is already clear that these produce discriminant weights that are the same up to a constant of proportionality. Since **k** is determined only up to a scale transformation, it does not matter whether SSCP or covariance matrices are used.

Use of the total-sample SSCP matrix to obtain discriminant weights is illustrated in Table 4.4. As noted, we obtain the same vector **k** of normalized discriminant weights.[12] This will also be true of the total-sample covariance matrix.

In the case of the within-groups total-sample correlation matrices, it turns out that the discriminant weights are *not* the same up to a proportionality constant. However, if we call these new weights $\mathbf{k}_s$, then $\mathbf{k}_s$ (representing the *standardized* data case of correlation matrices) is related to **k** by a diagonal matrix of predictor variable standard deviations which we can call **S**.

Table 4.4 Computing Discriminant Weights from the Inverse of the Total-Sample SSCP Matrix

Total-Sample SSCP Matrix, **T** (from Table 4.2)

$$\mathbf{T} = \begin{bmatrix} 82.5 & 41 \\ 41 & 36.4 \end{bmatrix}$$

Computing the Inverse of **T**

$$|\mathbf{T}| = 1322; \quad \mathbf{T}^{-1} = \frac{1}{1322} \begin{bmatrix} 36.4 & -41 \\ -41 & 82.5 \end{bmatrix}$$

$$\mathbf{T}^{-1} = \begin{bmatrix} 0.028 & -0.031 \\ -0.031 & 0.062 \end{bmatrix}$$

Solving for **k**

$$\mathbf{k} = \mathbf{T}^{-1}\mathbf{d} = \begin{bmatrix} 0.028 & -0.031 \\ -0.031 & 0.062 \end{bmatrix} \begin{bmatrix} -5 \\ -2 \end{bmatrix} = \begin{bmatrix} -0.078 \\ 0.031 \end{bmatrix}$$

$$\mathbf{k} \text{ (normalized)} = \begin{bmatrix} -0.93 \\ 0.37 \end{bmatrix}$$

To illustrate, let us take the case involving $\mathbf{R}_T$, the total-sample correlation matrix versus $\mathbf{C}_T$, the total-sample covariance matrix. We continue to let **k** denote the discriminant weights, as obtained from $\mathbf{C}_T$.

If **X**, the original data matrix, is first standardized to mean zero and unit standard deviation, then the correlation matrix **R** is simply:

$$\boxed{\mathbf{R} = \mathbf{X}_s'\mathbf{X}_s/(m-1)}$$

If we then compute the standardized weights vector $\mathbf{k}_s$, which we know to be proportional to $\mathbf{R}^{-1}\mathbf{d}_s$, then:

$$\boxed{\mathbf{k}_s = \mathbf{S}\mathbf{k}}$$

where **S** is the diagonal matrix of total-sample standard deviations of the predictors. That is, $\mathbf{k}_s$ denotes a set of weights that are applied to *standard scores* $\mathbf{X}_s$. As such, to obtain the vector **k** (weights applied to mean-corrected scores only), one needs to multiply $\mathbf{k}_s$ by $\mathbf{S}^{-1}$, the inverse of the matrix of standard deviations:

$$\boxed{\mathbf{k} = \mathbf{S}^{-1}\mathbf{k}_s}$$

In the case of the sample problem data, we can find:[13]

$$\mathbf{k}_s = \mathbf{R}_T^{-1}\mathbf{d}_s$$

$$= \overset{\mathbf{R}_T^{-1}}{\begin{bmatrix} 2.272 & -1.700 \\ -1.700 & 2.272 \end{bmatrix}} \overset{\mathbf{d}_s}{\begin{bmatrix} -1.65 \\ -0.99 \end{bmatrix}} = \begin{bmatrix} -2.07 \\ 0.55 \end{bmatrix}$$

However, the matrix of standard deviations of the predictors is:

$$\mathbf{S} = \begin{bmatrix} 3.028 & 0 \\ 0 & 2.011 \end{bmatrix}$$

Hence, we find $\mathbf{k}$ from:

$$\mathbf{k} = \mathbf{S}^{-1}\mathbf{k}_s = \begin{bmatrix} 1/3.028 & 0 \\ 0 & 1/2.011 \end{bmatrix} \begin{bmatrix} -2.07 \\ 0.55 \end{bmatrix} = \begin{bmatrix} -0.68 \\ 0.27 \end{bmatrix}$$

$$\mathbf{k} \text{ (normalized)} = \begin{bmatrix} -0.93 \\ 0.37 \end{bmatrix}$$

Note that this is the same result as found earlier in Table 4.3.

Perhaps the most interesting relationship, however, is that the partial regression coefficients obtained by regressing the dummy variable Y (of Table 4.1) on predictors X_1 and X_2 are also equal to the entries of $\mathbf{k}$, up to a proportionality transformation.[14]

In the sample problem the regression of Y on X_1 and X_2 turns out to be:

$$\hat{Y} = -0.32 - 0.19X_1 + 0.08X_2; \quad R^2 = 0.79$$

(We can ignore the intercept term of -0.32 since the discriminant function utilizes a covariance matrix obtained from the mean-corrected data.) If the partial regression weights $b_1 = -0.19$ and $b_2 = 0.08$ are normalized, we obtain:

$$\mathbf{k} = \begin{bmatrix} -0.93 \\ 0.37 \end{bmatrix}$$

as desired.[15] Furthermore, as shown by Cramer (1975), the F test value and individual t values obtained from a multiple regression on the 0-1 criterion variable are appropriate for testing the full model and individual predictors, respectively.

In short, we have almost an embarassment of riches insofar as computational alternatives for computing discriminant weights are concerned. We shall emphasize Fisher's function:

$$t_i = \mathbf{x}_{di}'\mathbf{k}$$

in most of what follows, although the reader should be alert to the various procedures that can be used to compute discriminant weights, particularly in terms of various computer programs that have been developed for discriminant analysis.

Both Fisher's discriminant function and multiple regression applied to a binary-valued criterion variable are spatial reduction procedures in the sense that n-dimensional profiles are reduced to a set of scores along a single axis **t**. Mahanlanobis' D^2 procedure assigns profiles to groups in terms of the original space. However, as illustrated, the assignments are identical to those found from the discriminant function.

4.4 Significance Testing

Up to this point we have been concerned primarily with ways of computing a linear discriminant axis and assigning profiles to groups. A companion question (and one that was introduced at the beginning of the chapter) concerns whether the centroids of the two groups differ significantly.[16] This is the type of problem in which one would usually employ a t test, were only a single variate involved. In the sample problem of Table 4.1, however, the two t tests—one for testing if universes 1 and 2 differ with respect to their means on X_1, the other for testing for a significant difference on X_2—are not independent.

If we reject the null hypothesis that the universe means of groups 1 and 2 are equal on X_1, and if X_1 and X_2 are highly correlated, chances are that the second t test would reject the second null hypothesis as well. In the case of correlated variates, a series of t tests leads to a confusion of alpha risks since each alpha risk is not independent of the others.

4.4.1 Hotelling's T^2 Statistic

Student's t statistic for testing the difference between the universe means of two independent samples (variances unknown but assumed equal) was generalized for the case of multivariate data by Hotelling (1931). Hotelling's T^2 test employs the pooled within-groups covariance matrix $\mathbf{C}_w$ in the following formula:

$$T^2 = \frac{m_1 m_2}{m_1 + m_2} \cdot \mathbf{d}'\mathbf{C}_w^{-1}\mathbf{d}$$

where m_1, m_2 and **d** are defined as before. Hotelling showed that with n predictors and $m = m_1 + m_2$ observations, the following relationship holds:

$$\frac{(m - n - 1)}{n(m - 2)} \cdot T^2 \sim F_{[n, m-n-1]}$$

where $\sim$ means "is distributed as." That is, one can compute T^2 and then convert T^2 to the well-known F distribution with n degrees of freedom in the numerator and $m - n - 1$ degrees of freedom in the denominator.

By way of analogy, recall that one can test for the difference between the means of two independent samples (variances unknown but assumed equal) by:

$$t = \frac{\overline{X}_1 - \overline{X}_2}{\sqrt{\dfrac{s_w^2}{m_1} + \dfrac{s_w^2}{m_2}}}$$

where s_w^2 represents the pooled variances:

$$s_w^2 = \frac{\displaystyle\sum_{i=1}^{m_1}(X_{i1} - \overline{X}_1)^2 + \sum_{i=1}^{m_2}(X_{i2} - \overline{X}_2)^2}{m_1 + m_2 - 2}$$

The multivariate analogue merely replaces s_w^2 by $\mathbf{C}_w$ and the scalar quantity $\overline{X}_1 - \overline{X}_2$ by a vector of centroid differences, $\mathbf{d}$.

Thus, were we to square t, we would have:

$$t^2 = \frac{(\overline{X}_1 - \overline{X}_2)^2}{\dfrac{s_w^2}{m_1} + \dfrac{s_w^2}{m_2}}$$

$$= \frac{m_1 m_2}{m_1 + m_2}(\overline{X}_1 - \overline{X}_2)(s_w^2)^{-1}(\overline{X}_1 - \overline{X}_2)$$

and the analogy to T^2 becomes even more transparent.

In terms of the sample problem of Table 4.1, we have:

$$T^2 = \frac{5(5)}{10}\overset{\mathbf{d'}}{\overbrace{(-5 \quad -2)}}\overset{\mathbf{C}_w^{-1}}{\overbrace{\begin{bmatrix} 0.776 & -0.471 \\ -0.471 & 0.588 \end{bmatrix}}}\overset{\mathbf{d}}{\overbrace{\begin{bmatrix} -5 \\ -2 \end{bmatrix}}}$$

$$= 2.5(12.35)$$

$$T^2 = 30.9$$

$$F_{[2,7]} = \frac{(m - n - 1)}{n(m - 2)}T^2 = \frac{10 - 2 - 1}{2(10 - 2)}(30.9)$$

$$F_{[2,7]} = 13.5$$

The tabular value of F at the $\alpha = 0.05$ level (see Table C.4) is 4.74, with 2 and 7 degrees of freedom. Hence, the null hypothesis of equality of group centroids is rejected.

4.4.2 The Relationship of T^2 to Mahalanobis' D^2

If we return to the formula for Hotelling's T^2:

$$T^2 = \frac{m_1 m_2}{m_1 + m_2} \cdot \mathbf{d'C}_w^{-1}\mathbf{d}$$

we note that in any given problem T^2 will be proportional to Mahalanobis' D^2 where the latter is defined as:

$$D^2 = \mathbf{d'C}_w^{-1}\mathbf{d}$$

By means of simple algebra, Mahalanobis' D^2 can, in turn, be related to the F ratio in the following way:

$$\frac{m_1 m_2 (m_1 + m_2 - n - 1)}{n(m_1 + m_2)(m_1 + m_2 - 2)} \cdot D^2 \sim F_{[n, m-n-1]}$$

Mahalanobis' D^2 measure and, hence, T^2 can also be related to R^2, the squared multiple correlation coefficient computed from the multiple regression approach involving a binary-valued criterion variable. The relationship is as follows:

$$R^2 = \Phi/(1 + \Phi)$$

where:

$$\Phi = \frac{m_1 m_2}{(m_1 + m_2)(m_1 + m_2 - 2)} \cdot D^2$$

Furthermore, F can also be related to Φ (and, hence, to R^2, D^2 and T^2), as follows:

$$F = \frac{(m_1 + m_2 - n - 1)}{n} \cdot \Phi$$

In the two-group case it is relatively easy, then, to move back and forth across the four measures, T^2, D^2, F, and R^2. (Moreover, the quantity Φ can be shown to be interpretable in its own right as the ratio of the between-groups to the pooled within-groups sum of squares of the computed discriminant scores.)

In terms of the sample problem of Table 4.1, we found that $D^2 = 12.35$. When a dummy-variable regression of Y or X_1 and X_2 was carried out, an R^2 of 0.79 was obtained. From section 4.3.2 we also recall that:

$$SSA = 38.4; \quad SSW = 9.9$$

Thus, we can find the relationships:

$$\Phi = \frac{m_1 m_2}{(m_1 + m_2)(m_1 + m_2 - 2)} \cdot D^2 = \frac{SSA}{SSW}$$

$$= \frac{5(5)}{10(8)} \cdot 12.35 = \frac{38.4}{9.9}$$

$$= 3.86$$

Moreover:

$$R^2 = \frac{\Phi}{1 + \Phi} = \frac{3.86}{4.86}$$

$$= 0.79$$

as indicated above.

4.4.3 Testing for the Equality of Covariance Matrices

Up to this point we have been *assuming* that the two within-group covariance matrices in the sample problem were equal, analogous to the usual assumption of equality of variances in the t test. If one wishes to test this assertion, prior to computing the discriminant function or conducting Hotelling's T^2 test, then another test is needed. Bartlett (1947) has provided a chi square approximation statistic for testing the equality of two or more covariance matrices *prior* to computing discriminant functions. More recently, Box (1949) has developed a more sophisticated (but more complex) procedure for testing the equality of two or more covariance matrices that is based on an F test.

Let us assume a sample of G groups and n predictors. Also, we assume G separate covariance matrices $\mathbf{C}_g(g = 1, 2, \ldots, G)$ and a pooled within-groups covariance matrix $\mathbf{C}_w$ with a total number of observations $\sum_{g=1}^{G} m_g = m$. If so, the following null hypothesis can be tested:

$$\boxed{\sum_1 = \sum_2 = \cdots = \sum_g = \cdots = \sum_G}$$

where Σ_g is the universe covariance matrix for the g-th group. We shall illustrate the test for equality of covariance matrices via the simpler chi square statistic, as developed by Bartlett. (Note that the test is designed for $G \geq 2$ groups; hence, it will also be applicable to multiple discrimination problems considered in Chapter 7.)

Since the two within-groups covariance matrices in the sample problem of Table 4.1 happen to be exactly equal, to make the test somewhat more interesting, let us illustrate it with the following covariance matrices:

$$\mathbf{C}_1 = \begin{bmatrix} 3 & 2 \\ 2 & 4 \end{bmatrix}; \mathbf{C}_2 = \begin{bmatrix} 3 & 2.5 \\ 2.5 & 3 \end{bmatrix}$$

where we shall assume that $m_1 = 15$ and $m_2 = 20$.

The matrices $\mathbf{C}_1$ and $\mathbf{C}_2$ are pooled as follows to obtain $\mathbf{C}_w$:

$$\mathbf{C}_w = \frac{\left\{ (15 - 1)\begin{bmatrix} 3 & 2 \\ 2 & 4 \end{bmatrix} + (20 - 1)\begin{bmatrix} 3 & 2.5 \\ 2.5 & 5 \end{bmatrix} \right\}}{(35 - 2)}$$

$$= \begin{bmatrix} 3 & 2.3 \\ 2.3 & 4.6 \end{bmatrix}$$

Table 4.5 shows the test calculations, based on Bartlett's chi square approximation.[17] In this illustration, the computed chi square approximation is only 0.31, which is clearly non-significant. We accept the null hypothesis that:

$$\Sigma_1 = \Sigma_2$$

However, should the null hypothesis of covariance matrix equality be rejected, then, strictly speaking, a linear discriminant function is not appropriate.

If the covariance matrices are unequal, a bias occurs in the test for equality of centroids. In the two-group case it turns out that the null hypothesis in the T^2 test is accepted more frequently when the covariance matrices are unequal. Moreover, in terms of classification errors, use of a linear discriminant function in the case of unequal covariance matrices tends to assign too many observations to the group with the larger covariance matrix, since this matrix is contributing most to the pooled within-groups covariance.[18]

As Cooley and Lohnes (1971) indicate, many researchers do not bother to test for covariance matrix equality on the presumption that Hotelling's T^2 test of centroid equality is fairly robust and that with fairly large sample sizes it would be quite easy to reject the null hypothesis of covariance matrix equality anyway without doing much violence to the subsequent test of centroid equality. However, the reader may be well advised to run this preliminary test until more evidence has been assembled on the "robustness" question (Gilbert, 1969).

In summary, Bartlett's test, or versions based on the F approximation (Timm,

Table 4.5 Bartlett's Chi Square Approximation for Testing the Equality of Two or More Covariance Matrices

General Formula

$$B = (m - G) \ln |C_w| - \sum_{g=1}^{G} (m_g - 1) \ln |C_g|$$

Illustrative Problem

$$C_1 = \begin{bmatrix} 3 & 2 \\ 2 & 4 \end{bmatrix}; \qquad C_2 = \begin{bmatrix} 3 & 2.5 \\ 2.5 & 5 \end{bmatrix}; \qquad C_w = \begin{bmatrix} 3 & 2.3 \\ 2.3 & 4.6 \end{bmatrix}$$

$$m_1 = 15 \qquad m_2 = 20 \qquad m = 35$$

$$|C_1| = 8.00 \qquad |C_2| = 8.75 \qquad |C_w| = 8.51$$

$$B = (35 - 2) \ln 8.51 - [(15 - 1) \ln 8.00 + (20 - 1) \ln 8.75]$$
$$= 70.66 - [14(2.08) + 19(2.17)]$$
$$= 0.31$$

Degrees of Freedom

$$\text{d.f.} = \tfrac{1}{2}[(G - 1)(n)(n + 1)] = \tfrac{1}{2}[(2 - 1)(2)(2 + 1)] = 3$$

Test

$B = 0.31$ is approximately distributed as χ^2 with 3 degrees of freedom. The null hypothesis of equal covariance matrices is accepted for $\alpha = 0.05$.

1975), would usually be employed to test for the equality of covariance matrices. Following acceptance of the null hypothesis, one would then go on to compute T^2 and find the associated F ratio for testing the null hypothesis of equality of universe centroids. If this hypothesis is rejected one can then proceed to employ the discriminant function in various classification tasks.

4.5 Decision Theory and Classification Procedures

Up to this point we have discussed in some detail:

1. Various ways of computing discriminant functions.

2. Significance testing of the equality of covariance matrices followed by a test of centroid equality.

The third problem—that of assigning profiles to groups—has been described more briefly in terms of:

1. Computing a cutoff point $t_c = (\bar{t}_1 + \bar{t}_2)/2$ halfway between the mean discriminant scores, t_1 and t_2, of groups 1 and 2 and then assigning profiles to group 1 or 2 on the basis of which side of t_c the profile falls.

2. Assigning profiles to the group with the closer centroid (in the case of Mahalanobis' D^2 method).

Both procedures yield identical assignments.[19] Moreover, both approaches assume that the prior probabilities of a profile belonging to each group are equal and that the costs of misclassification are also equal.

It is now time to relax these more stringent assumptions by considering two-group discriminant analysis in the broader context of *statistical decision theory*. Statistical decision theory provides a general model for classification that can take unequal prior probabilities and unequal costs of misclassification into account. By *prior* probability is meant the probability that a randomly selected case belongs to each of the two groups *before* its profile is actually observed. As such, prior probabilities refer to the incidence with which members of each group appear in the population and are sometimes called *base rates*. In terms of the sample problem, let us define a set of prior probabilities:

$$q_1$$
$$q_2 = 1 - q_1$$

that a randomly chosen legislator belongs to group 1 and group 2, respectively. These prior probabilities may be estimated from the sample, previous studies, general experience in the substantive field, or by other means.

We next define a set of conditional probabilities:

$$P(\mathbf{x}|1)$$
$$P(\mathbf{x}|2)$$

of obtaining a specific profile vector $\mathbf{x}$, given that the legislator comes from group 1 and group 2, respectively.

We can then define another type of conditional probability, called a *posterior* probability:

$$P(1|\mathbf{x})$$
$$P(2|\mathbf{x})$$

that group 1 or 2, respectively, is the group represented by the profile vector $\mathbf{x}$. This probability is called *posterior* because it incorporates information about group 1 versus group 2 that is provided by examination of the profile $\mathbf{x}$. What this whole line of discussion says so far is:

1. If I know nothing about the specific characteristics of the profile $\mathbf{x}$, the prior probability that the observation is from group 1 is q_1.

2. However, having observed **x**, the posterior probability that group 1 is the group from which **x** emanated is $P(1|\mathbf{x})$.

With the preceding comments as background, we wish to know how the posterior probability $P(1|\mathbf{x})$ can be computed from the prior probabilities q_1 and q_2 and the conditional probabilities $P(\mathbf{x}|1)$ and $P(\mathbf{x}|2)$. Bayes' theorem[20] computes this probability as follows:

$$P(1|\mathbf{x}) = \frac{q_1 \cdot P(\mathbf{x}|1)}{q_1 \cdot P(\mathbf{x}|1) + q_2 \cdot P(\mathbf{x}|2)}$$

What Bayes' formula tells us is that the posterior probability $P(1|\mathbf{x})$ is the ratio of the joint probability of group 1 being present *and* observing **x** to the total probability of **x** occurring, as given by the expression in the denominator. Similarly, to obtain $P(2|\mathbf{x})$ we replace the present numerator by the product $q_2 \cdot P(\mathbf{x}|2)$.

4.5.1 Bayesian Classification Procedures

Up to this point in our discussion of assignment procedures using Fisher's linear discriminant function or Mahalanobis' D^2 measure, we have always been assuming equal prior probabilities across groups. That is, without knowledge of the profile vector **x** we would just as likely classify the observation as being from group 1 as from group 2.

Under this assumption we always set the cutoff point t_c midway between the mean discriminant scores of group 1 versus group 2. Thus:

$$t_c = (\bar{t}_1 + \bar{t}_2)/2$$

represents the simple cutoff point that we have been using so far. Now we consider more elaborately constructed cutoff points.

As background, let $f_1(\mathbf{x})$ and $f_2(\mathbf{x})$ denote the conditional probability functions of observing profile **x**, given group 1 and group 2, respectively. Then a Bayes' classification procedure classifies **x** into group 1 if:

$$P(1|\mathbf{x}) > P(2|\mathbf{x})$$

or, more directly from Bayes' theorem, we classify **x** into group 1 if:

$$\Omega = \frac{q_1 f_1(\mathbf{x})}{q_2 f_2(\mathbf{x})} > 1$$

and into group 2 if $\Omega < 1$. (We make a random assignment if $\Omega = 1$.) This rule has the effect of assigning each observation to the group whose posterior probability is the higher. Also, the procedure will *minimize the expected probability of misclassification if all probabilities are known.*

The import of all of this is that the simple assignment rules based on Fisher's linear discriminant function require modification if prior probabilities over groups 1 and 2 are unequal. Our intuition would suggest that we should want to move the cutting point on the discriminant axis from being halfway between the mean discriminant scores of groups 1 and 2 to a point closer to the *less frequently occurring group.* In this way the probability of misclassifying members of the more frequently occurring group is reduced.

Computation of the new cutting point employs the *original* (non-normalized) discriminant vector from Table 4.3:[21]

$$\mathbf{k} = \mathbf{C}_w^{-1}\mathbf{d} = \begin{bmatrix} -2.94 \\ 1.18 \end{bmatrix}$$

The Bayes' rule that utilizes prior probabilities is:

Assign $\mathbf{x}_i$ to group 1 if:

$$t_i = \mathbf{x}_i'\mathbf{k} > (\bar{t}_1 + \bar{t}_2)/2 + ln\,(q_2/q_1)$$

Assign $\mathbf{x}_i$ to group 2 if:

$$t_i = \mathbf{x}_i'\mathbf{k} < (\bar{t}_1 + \bar{t}_2)/2 + ln\,(q_2/q_1)$$

where *ln* denotes the natural logarithm.

Moreover, we can generalize the preceding rule one step further by also incorporating the costs of misclassification. For example, we can let $C(2\,|\,1)$ denote the conditional cost of erroneously assigning an $\mathbf{x}$ that is really from group 1 to group 2 and $C(1\,|\,2)$ as the conditional cost of erroneously assigning a group 2 type to group 1.

If this additional aspect is considered, we obtain the more general Bayes' rule:[22]

Assign $\mathbf{x}_i$ to group 1 if:

$$t_i = \mathbf{x}_i'\mathbf{k} > \frac{\bar{t}_1 + \bar{t}_2}{2} + ln\left[\frac{q_2 C(1\,|\,2)}{q_1 C(2\,|\,1)}\right]$$

Assign $\mathbf{x}_i$ to group 2 if:

$$t_i = \mathbf{x}_i'\mathbf{k} < \frac{\bar{t}_1 + \bar{t}_2}{2} + ln\left[\frac{q_2 C(1\,|\,2)}{q_1 C(2\,|\,1)}\right]$$

4.5.2 Illustrating a Bayesian Classification Procedure

Suppose we illustrate the generalized Bayes' rule by assuming the following additional values in the sample problem of Table 4.1:

$$q_1 = 0.1; \; C(2|1) = 15$$

$$q_2 = 0.9; \; C(1|2) = 5$$

If we return to the first solution (Table 4.3) for **k**, we find:

$$\mathbf{k} = \begin{bmatrix} -2.94 \\ 1.18 \end{bmatrix}; \; \bar{\mathbf{x}}_1 = \begin{bmatrix} -2.5 \\ -1 \end{bmatrix}; \; \bar{\mathbf{x}}_2 = \begin{bmatrix} 2.5 \\ 1 \end{bmatrix}$$

Next we can find $\bar{t}_1 = \bar{\mathbf{x}}_1'\mathbf{k}$ and $\bar{t}_2 = \bar{\mathbf{x}}_2'\mathbf{k}$ as:

$$\bar{t}_1 = (-2.5 \quad -1)\begin{bmatrix} -2.94 \\ 1.18 \end{bmatrix} = 6.17; \; \bar{t}_2 = (2.5 \quad 1)\begin{bmatrix} -2.94 \\ 1.18 \end{bmatrix} = -6.17$$

Thus, groups 1 and 2 are $6.17 + 6.17 = 12.34$ units apart on the discriminant axis.[23]

We next compute:

$$ln\left[\frac{0.9(5)}{0.1(15)}\right] = ln3 = 1.10$$

In the *absence* of knowledge about either prior probabilities or costs of misclassification, the cutoff value would, of course, be:

$$t_c = (\bar{t}_1 + \bar{t}_2)/2 = (6.17 - 6.17)/2 = 0$$

If $t_i > 0$, we assign the case to group 1; if $t_i < 0$, we assign the case to group 2. But, given the above information and computations, *the generalized Bayes' rule* states:

Assign **x** to group 1 if:

$$t_i = \mathbf{x}'\mathbf{k} > (0 + 1.10)$$

Assign **x** to group 2 if:

$$t_i = \mathbf{x}'\mathbf{k} < (0 + 1.10)$$

Notice, in this case, that the prior probability associated with group 2 is so large, even considering classification costs that are more serious if a group 1 case were misclassified, that the cutting point is moved 1.1 units closer to the mean discriminant score ($\bar{t}_1 = 6.17$) on group 1. In this way fewer real group 2 cases are erroneously assigned as group 1 cases.

Instances also arise in which the researcher would like to estimate the posterior probabilities themselves.[24] Assuming that the observations arise from known multivariate normal populations with common covariance matrix, the estimated posterior probability of, say, group 1, given $\mathbf{x}$ is:

$$P(1|\mathbf{x}) = \cfrac{1}{1 + \cfrac{q_2}{q_1} \exp\left\{-\mathbf{x}'\mathbf{k} + \cfrac{\bar{t}_1 + \bar{t}_2}{2}\right\}}$$

where exp (y) denotes e^y and the remaining notation is as before.[25] For example, if we take the third (mean-corrected) individual observation in group 1 of Table 4.1:

$$\mathbf{x}_{31} = \begin{bmatrix} -2.5 \\ -0.4 \end{bmatrix}$$

we can find the estimated posterior probability that it is from group 1 as:

$$P(1|\mathbf{x}_{31}) = \cfrac{1}{1 + \cfrac{0.9}{0.1} \exp\left\{-\begin{pmatrix} -2.5 & -0.4 \end{pmatrix} \begin{bmatrix} -2.94 \\ 1.18 \end{bmatrix} + \cfrac{6.17 - 6.17}{2}\right\}}$$

$$= \frac{1}{1 + 9 \exp(-6.88 + 0)}$$

$$= \frac{1}{1.009252}$$

$$= 0.99$$

The posterior probability of its being from group 2 is $1 - 0.99 = 0.01$. Thus, in the sample problem, the posterior probability is quite high that $\mathbf{x}_{31}$ is from group 1, despite the 0.9 prior probability that a *randomly* chosen case is from group 2.

Before leaving the topic of classification procedures, it should be emphasized that the parameters q_g, and $P(\mathbf{x}|g)$ are assumed to be known. In practice, however, we use various sample statistics. (For example, the population Mahalanobis D^2 is estimated from the counterpart sample quantity.) The effect of employing sample estimates of universe parameters is to *underestimate the actual probability of misclassification*, thus introducing a bias in the estimate.

To produce less biased estimates, one could apply cross validation in which each group is divided into two subsamples. The cases in the first subsample of each group are used to compute the function, and members of the second subsample are classified by that function.

Another alternative is to hold out one observation from group 1 and compute the discriminant function on the basis of the rest of the total sample. The function is used to classify the hold-out case. One then proceeds to the second observation in group 1 and so on until each member of group 1 has been classified. The proportion misclassified estimates $P(2|1)$. The same idea is then applied to the second group to obtain an estimate of $P(1|2)$. While the computational burden is high, Lachenbruch and Mickey (1968) indicate that the bias is virtually zero.

Finally, in empirical applications the researcher may not want to assign each and every case to group 1 or group 2. That is, he may want to suspend judgment on those cases that involve a fairly low probability of belonging to either group. Procedures are available (Van de Geer, 1971) for dealing with this problem.

4.6 Two-Group Discrimination via Computer Program

A large number of discriminant analysis programs are available to the interested researcher. Many of these programs have been designed to conduct linear discrimination for more than two groups. We reserve discussion of multiple discriminant programs—although they can readily handle two-group discrimination as a special case—until a subsequent chapter.

Here we confine our attention to one program, the BMD-04M program (Dixon, 1973) that is designed specifically for two groups. We process the data of Table 4.1 as a way to demonstrate the nature of the program's algorithm and output.

4.6.1 The BMD-04M Two-Group Discriminant Analysis Program

The BMD-04M program is designed for two-group discrimination involving up to 25 predictors and sample sizes in each group up to 300 cases. The computation of discriminant weights uses the $\mathbf{W}$ matrix (i.e., the pooled within-groups SSCP matrix) rather than $\mathbf{C}_w$. As such, the discriminant weights differ by a constant of proportionality from those based on $\mathbf{C}_w$:[26]

$$\mathbf{k(W)} = \frac{1}{(m-2)} \mathbf{k(C}_w)$$

BMD-04M was applied to the sample-problem data of Table 4.1; results appear in Table 4.6.

We note that the weights vector $\mathbf{k}$, based on the $\mathbf{W}$ (rather than the $\mathbf{C}_w$) matrix is:

$$\mathbf{k} = \begin{bmatrix} -0.37 \\ 0.15 \end{bmatrix}$$

Table 4.6 Application of BMD-04M to Sample Problem of Table 4.1

Variable	Sample Mean on Group 1	Sample Mean on Group 2	Difference Vector **d**
1	4.0	9.0	−5.0
2	4.4	6.4	−2.0

$$\mathbf{W} = \begin{bmatrix} 20 & 16 \\ 16 & 26.4 \end{bmatrix}; \qquad |\mathbf{W}| = 272; \qquad \mathbf{W}^{-1} = \begin{bmatrix} 0.10 & -0.06 \\ -0.06 & 0.07 \end{bmatrix}$$

$$\mathbf{k} = \mathbf{W}^{-1}\mathbf{d} = \begin{bmatrix} -0.37 \\ 0.15 \end{bmatrix}; \qquad \text{Mahalanobis' } D^2 = 12.35; \qquad F_{[2,7]} = 13.51$$

Discriminant Score Statistics

Group Number	Sample Size	Mean	Variance	Standard Deviation
1	5	−0.82	0.19	0.44
2	5	−2.37	0.19	0.44

Cutoff Point: $(-0.82 - 2.37)/2 = -1.60$

Score Rank	First-Group Values	Second-Group Values	First-Group Item No.	Second-Group Item No.
1	−0.15		1	
2	−0.74		3	
3	−0.81		2	
4	−1.18		5	
5	−1.25		4	
6		−1.69		1
7		−2.28		3
8		−2.35		2
9		−2.72		5
10		−2.79		4

As noted above, the entries of this **k** vector are 1/8 of the respective entries computed for **k** in Table 4.3 (prior to normalization), where $\mathbf{C}_w$ was used. The program also computes Mahalanobis' D^2 and the accompanying T^2 test. (The computed $F_{[2,7]} = 13.51$ is significant at the 0.05 level.)

The program then computes a discriminant score for each observation by means of:

$$t_i = \sum_{j=1}^{n} k_j X_{ij}$$

as applied to the *original* ratings of Table 4.1. For example, to obtain t_{11}, the first discriminant score of group 1, we have:

$$t_{11} = -0.37(2) + 0.15(4)$$
$$= -0.15$$

Notice that the group 1 scores are arrayed in order of the size of the discriminant score: cases 1, 3, 2, 5, 4, and similarly so for the group 2 scores in Table 4.6. We also see that no overlap in scores is found. For example, the score for item 4 of group 1 is -1.25 which is still closer to the group 1 mean score of -0.82 than it is to the group 2 mean score of -2.37.

The program also computes the within-groups variance and standard deviation of the discriminant scores. However, the discriminant weights obtained from BMD-04M would have to be rescaled to:

$$(m - 2)\mathbf{k} = \mathbf{k}(\mathbf{C}_w)$$

$$8\begin{bmatrix} -0.37 \\ 0.15 \end{bmatrix} \cong \begin{bmatrix} -2.94 \\ 1.18 \end{bmatrix} = \mathbf{k}(\mathbf{C}_w)$$

in order to yield appropriate values for the Bayesian classification rules, described in section 4.5.1.

4.6.2 Discriminant Analysis of the TV Commercial Data

To round out the discussion of classical discriminant methods for the two-group case, we return to the television commercial ratings, already discussed in Chapters 2 and 3. This time we focus on the substantive question of the relationship between:

Group 1 respondents who select the Alpha brand of radial tires in Part D of the questionnaire ($m_1 = 78$) versus

Group 2 respondents who do not select the Alpha brand ($m_2 = 174$).

and the predictor variables:

1. A-2 Whether Alpha was the brand selected in respondent's last purchase of replacement tires.

2. A-3 Pre-exposure interest in Alpha radial tires.

3. C-1 Post-exposure believability of the Alpha commercial.

4. C-2 Post-exposure interest in Alpha radial tires.

A two-group discriminant analysis was run with the four variables above serving as predictors.[27]

Table 4.7 shows summary output for the analysis. The group means indicate the direction of predictor-variable differences that might be expected. Alpha brand selectors are more likely to be past purchasers of the Alpha brand (35 percent versus 11 percent for non-selectors) and generally rate it higher on pre-exposure and post-exposure interest and on believability of the commercial.

Table 4.7 Summary Output of Two-Group Discriminant Analysis—Alpha Brand Selectors Versus Others

	Sample Means			Standardized Weights Based on	
Predictor	Alpha Selectors	Others	Discriminant Weights	C_w	C_t
A-2	0.36	0.11	1.20	0.45	0.47
A-3	7.21	4.67	0.15	0.45	0.48
C-1	8.05	6.20	0.18	0.44	0.47
C-2	7.73	5.41	0.07	0.19	0.21
Mean Discriminant Score	0.80	−0.36			
	$m_1 = 78$	$m_2 = 174$			

Classification Matrix

		Predicted by Function		
		Alpha Selectors	Others	Totals
Actual	Alpha Selectors	52	26	78
	Others	43	131	174
	Totals	95	157	252

Pooled Within-Groups Covariance Matrix, C_w

	A-2	A-3	C-1	C-2
A-2	0.143			
A-3	0.196	9.053		
C-1	0.002	1.679	5.983	
C-2	0.133	3.891	4.134	7.599

$$F_{[4, 247]} = 17.98$$

The pooled within-groups covariance matrix shows that the four predictors tend to be positively associated.

The computed F value of 17.98, with 4 and 247 degrees of freedom, is highly significant.[28] However, the classification matrix suggests that only (52 + 131)/252, or 73 percent of the sample is correctly classified. Hence, the separation effected by the discriminant function is not at all good from a practical point of view.[29]

Of considerable interest to the applied researcher is the question of the "relative importance" of predictor variables in the usual case where the predictors are intercorrelated themselves. As we shall show in Chapter 5, there is no unambiguous answer to this question in the case of correlated predictors. However, one aspect of the problem—specification of a common unit by which discriminant coefficients can be expressed—is relatively simple to solve.

If we examine the original discriminant weights in Table 4.7, it should be remembered that they are to be applied to original scores expressed in mean-corrected form.[30] As such, the absolute value of the discriminant coefficients is dependent on the standard deviation of the particular predictor variable entering into the computation of the pooled within-groups covariance matrix $\mathbf{C}_w$.[31] However, if one were to transform the predictor variable to unit standard deviation prior to running the discriminant analysis, we know from section 4.3.5 that the following expression results:

$$\mathbf{k}_s \propto \mathbf{Sk}$$

where, as indicated earlier, $\mathbf{S}$ is the diagonal matrix of total-sample standard deviations.

Often the researcher wants to make statements about the discriminant weights in terms of a standardized unit. If so, he should use the vector $\mathbf{k}_s$ which, in turn, can easily be computed from $\mathbf{k}$ and the matrix of total-sample predictor standard deviations. These standard deviations, in turn, are computed as the square roots of the diagonal entries of $\mathbf{C}_t$, the total-sample covariance matrix.

Alternatively, the original discriminant weights can be standardized by means of the standard deviations of $\mathbf{C}_w$, the pooled within-groups covariance matrix. All that needs to be done is to find the square roots of the diagonal entries of $\mathbf{C}_w$ in Table 4.7 and multiply the respective discriminant weights by each of these standard deviations.

For contrast, both standardization procedures appear in Table 4.7. Both transformations yield quite similar results. The major point of interest is that the weight of 1.20 for predictor A-2 is adjusted downwards to reflect its relatively small standard deviation. We note that the first three predictors all exhibit about the same standardized weight while C-2 displays a standardized weight that is somewhat less than half of each of the others.

Both standardization procedures are used in practice. However, since the standard deviations based on $\mathbf{C}_t$ reflect between-groups as well as within-groups variation, a somewhat stronger case might be made for employing the square roots of the diagonal entries of $\mathbf{C}_w$.

It should also be mentioned that our use of the computer program (BMD-07M) to conduct the analysis of Table 4.7 assumed equal prior probabilities and equal costs of misclassification. Of course, it is possible to relax these requirements if procedures based on section 4.5.1 are employed.

4.6.3 Tests of Predictor Variables

In section 2.5 of Chapter 2 we illustrated how tests of the significance of one or more predictor variables could be carried out by the examination of a pair of R^2's for full and restricted models, respectively. More specifically, we have the relationship:

$$\frac{R_f^2 - R_r^2}{1 - R_f^2} \cdot \frac{d_f}{d_r - d_f} \sim F_{[d_r - d_f, d_f]}$$

where R_f^2 denotes the R^2 for some full model and R_r^2 denotes the R^2 for some restricted model; d_f and d_r are their respective degrees of freedom. The test is whether the additional predictors in the full model add anything over the predictors of the restricted model.

In an analogous way one might want to test in two-group discriminant analysis if a subset of p predictors ($p < n$) can do essentially as good a job as all n predictors. Rao (1952) has provided such a test, also based on the F statistic:

$$Q = \frac{m - n - 1}{n - p} \cdot \frac{C(D_n^2 - D_p^2)}{1 + CD_p^2}$$

where C, in turn, is computed as:

$$C = \frac{m_1 m_2}{m(m - 2)}$$

The statistic $Q \sim F_{[n-p, m-n-1]}$, where, as before, $m_1 + m_2 = m$, the total sample size. The quantity D_n^2 denotes the Mahalanobis D^2 for the full set of n predictors while D_p^2 denotes the Mahalanobis D^2 for the subset of p predictors in the restricted model.

As a special case, we can consider testing for the significance of a single predictor via the formulation:

$$Q = \frac{m - n - 1}{1} \cdot \frac{C(D_n^2 - D_{n-1}^2)}{1 + CD_{n-1}^2}$$

where $n - p = 1$ (since only 1 predictor is held out in the restricted model) and $Q \sim F_{[1, m-n-1]}$. This test is analogous to Student's t test, as used in computing the significance of *individual* partial regression coefficients. Furthermore, it is subject to the same kinds of caveats regarding interpretation that were described in Chapter 2 in the context of single-predictor regression coefficient tests.

In the TV commercial data of Table 4.7 Mahalanobis' D^2, as based on all four predictors, was $D_n^2 = 1.35$. However, if only the first two predictors are employed $D_p^2 = 0.99$. We can next compute Q to see if the second two predictors (C-1 and C-2) add anything beyond the contributions of A-2 and A-3.

First, we compute:

$$C = \frac{78(174)}{252(250)} = 0.2154$$

We then go on to compute Q as follows:

$$Q = \frac{252 - 4 - 1}{4 - 2} \cdot \frac{0.2154(1.35 - 0.99)}{1 + 0.2154(0.99)}$$

$$= 7.89$$

The statistic $F_{[2,247]} = 7.89$ is highly significant at the 0.01 level, indicating that predictor variables C-1 and C-2, as a pair, *do* contribute to between-groups separation beyond the collective contributions of A-2 and A-3.

Rao's test of discriminant coefficients is an extremely useful procedure and plays an important role in stepwise discrimination procedures that are discussed in a later chapter. It is completely analogous to the F and t tests obtained from a regression analysis performed on a 0, 1 criterion variable.

4.6.4 Additional Problems in Classification

Before concluding our analysis of the T.V. commercial data, let us return to the classification matrix in Table 4.7. We have already indicated that 73 percent of the total sample was correctly classified. Notice, however, that the group sizes differ markedly—only 78 Alpha selectors versus 174 others. Had we elected to classify everyone as an "other," on the basis that this is the most frequently occurring group, we would have assigned 174/252 or 69 percent of the cases correctly. This "hit" ratio is not much lower than 73 percent.

Morrison (1969) points out some of the difficulties involved in the interpretation of classification matrices when group sizes are unequal, as in the case here. A chance model that Morrison proposes is based on the assumption that the researcher would still try to identify Alpha selectors if he had no discriminant model—there being 31 percent Alpha selectors and 69 percent others. Morrison's proposed chance model would yield the probability of correct assignment as:

$$P(\text{Correct}) = P(\text{Correct} | \text{Classified Alpha}) \cdot P(\text{Classified Alpha})$$
$$+ P(\text{Correct} | \text{Classified Others}) \cdot P(\text{Classified Others})$$

Letting P denote the true proportion of Alpha selectors and α denote the proportion classified as Alpha selectors, we have:

$$P(\text{Correct}) = \alpha P + (1 - \alpha)(1 - P)$$

If we let $P = \alpha = 0.31$, we have:

$$= \alpha^2 + (1 - \alpha)^2$$
$$= (-0.31)^2 + (0.69)^2$$
$$= 0.096 + 0.476$$
$$= 0.572 \text{ or } 57\%$$

However, if the two groups are of equal size we shall always find that:

$$P(\text{Correct}) = 0.5$$

regardless of the value of α.

The classification matrix of Table 4.7 would, of course, be more realistic if the function were adjusted for differences in prior probabilities or costs of misclassification. We do not carry out this step but merely indicate that the results would generally differ from those obtained here.

What is brought out in the example of Table 4.7 is the importance—if emphasis is placed on case assignment rather than dimensional reduction—of using equal-sized groups, if at all possible. This not only has the virtue of estimating each group's centroid with the same precision but, as noted, leads to less equivocal results about the value of classifications based on the discriminant function versus some alternative model. As was pointed out, if one is interested solely in improving the probability based on assignment of all cases to the modal group, then in the problem of Table 4.7 the function would have to do better than a classification accuracy of 69 percent.

Thus, it goes almost without saying that one can find highly significant differences between group centroids and yet find that the discriminant function does not do very well at all in classifying profiles. As already noted, this phenomenon is just another illustration of the *difference between statistical and operational significance.*

In closing it should be noted that other procedures are available for classifying multivariate profiles. For example, Kendall (1966) has proposed a method for two-group classification that requires only ordinal data for the various predictors. Overall and Klett (1972) have discussed classification procedures for cases where all of the predictor variables are nominally scaled. In short, the field of multivariate profile classification contains a variety of special-purpose techniques that can augment the more basic tools discussed here.

4.7 Summary

This chapter has been concerned with linear models of two-group discrimination and classification. After an intuitive introduction to the subject we described two procedures: (a) Fisher's linear discriminant function and (b) Mahalanobis' D^2 for performing two-group discrimination. Other procedures were briefly mentioned as well.

After showing how discriminant techniques can be developed under various approaches, we discussed the problems of significance testing and classification, including Bayesian methods. Some of the techniques described earlier were then applied to a portion of the TV commercial ratings data, and some questions about the interpretive aspects of discriminant analysis—the relative importance of predictors, classification accuracy, and cross validation—were raised and discussed.

Review Questions

1. Consider the following hypothetical data obtained by a political science researcher:

Group		Y	X_1: Age	X_2: Guns Owned
1		1	29	0
1		1	37	1
1	Legislators Who Support Stricter Gun Control Bill	1	45	3
1		1	28	0
1		1	31	0
1		1	40	0
2		0	47	1
2		0	52	3
2		0	61	5
2		0	58	6
2	Legislators Who Oppose Stricter Gun Control Bill	0	38	0
2		0	55	4
2		0	44	3
2		0	42	2
2		0	48	3

a. Compute Fisher's discriminant axis **t** for the above data, as expressed in original terms.

b. Compute Fisher's discriminant axis **t** for the mean-corrected and standardized versions of the above data.

c. Apply Bartlett's test for covariance matrix equality. Then, compute Mahalanobis' D^2 and its associated F ratio; test the null hypothesis of no centroid differences. (Use an alpha level of 0.05 in both cases.)

d. Use Fisher's function to classify each profile, where the cutoff is t_c $(\bar{t}_1 + \bar{t}_2)/2$, in terms of the original data.

2. Referring to the same data of question 1:

a. Regress Y on X_1 and X_2 and compare the partial regression weights with the discriminant weights in question 1a.

b. Find the sample R^2 and transform this to D^2 and F, via the methods of section 4.4.2.

c. Compute the SSCP matrices **T** and **W**. Find their inverses and solve for discriminant weights for comparison with those of question 1a.

3. Referring to the same data of question 1:

a. Assume prior probabilities of 0.4 (support) and 0.6 (oppose) and misclassification costs of $C(1|2) = 2$; $C(2|1) = 1$. Find the Bayes' rule that minimizes estimated expected cost of misclassification.

b. What are the respective estimated posterior probabilities of group 1 versus group 2, given that case 3 (in group 1) is observed?

4. Refer to the TV commercial data analysis in Table 4.7. Denoting an Alpha selector as 1 and an other brand selector as 0, regress Y on A-2, A-3, C-1, and C-2, using the total-sample data in Appendix A.

   a. What is the sample R^2 and how do the partial regression coefficients compare with the discriminant weights of Table 4.7?

   b. Find the transformation of R^2 to Φ and thence to T^2, D^2, and F, as described in section 4.4.2. Compare this F value with that obtained from the discriminant analysis procedure.

   c. Test, via the incremental R^2 procedure, whether the second two regression predictors are significant, using an alpha level of 0.05. Compare your results with those in section 4.6.3 in which Rao's test is used.

5. In the special case of two groups the between-groups covariance matrix $\mathbf{A}$ is given by:

$$\mathbf{A} = \frac{m_1 m_2}{m_1 + m_2} \mathbf{dd'}$$

where $\mathbf{d}$ is the difference vector of centroids:

$$\mathbf{d} = \bar{\mathbf{x}}_1 - \bar{\mathbf{x}}_2$$

and m_1 and m_2 denote the sample sizes of groups 1 and 2, respectively (see section 4.3.1).

In this special case the general entry a_{jk} of $\mathbf{A}$ can be written as:

$$a_{jk} = m_1(\bar{X}_{j1} - \bar{X}_{j}.)(\bar{X}_{k1} - \bar{X}_{k}.) + m_2(\bar{X}_{j2} - \bar{X}_{j}.)(\bar{X}_{k2} - \bar{X}_{k}.)$$

with overall mean on the j-th variable, for example, given by:

$$\bar{X}_{j}. = \frac{m_1 \bar{X}_{j1} + m_2 \bar{X}_{j2}}{m_1 + m_2}$$

Show algebraically that the expression:

$$\mathbf{A} = \frac{m_1 m_2}{m_1 + m_2} \mathbf{dd'}$$

holds in the special case of two groups.

Notes

1. Variability can be measured either as a sum of squared deviations or (proportionally) as a variance.

2. Concentration ellipses are discussed in more detail in Chapter 6. Any two points

falling on a given concentration ellipse are assumed to have the same probability of occurring. As will be discussed later, we assume here that the observations are drawn from multivariate normal distributions with common within-groups covariance matrices but (possibly) different centroids.

3. In the case of equal sized groups, each group centroid will be equally distant from the origin. If group sizes are unequal the centroid of the larger group will be closer to the origin.

4. By *scalar matrix* is meant a diagonal matrix whose diagonal entries are all equal to the same (positive) constant. (An identity matrix is a special case in which the diagonal entries are all 1's.)

5. Panel V shows rather dramatically that the solid line connecting the two centroids does considerably poorer than the perfect separation effected by projecting the points onto the dotted line.

6. Again, we see that finding some "best" linear composite of the predictors represents the objective of two-group discriminant analysis, fully in accord with the motivation underlying regression, ANOVA and ANCOVA.

7. While the difference vector **d** is computed here by taking the difference between subgroup centroids of the $\mathbf{X}_d$ (mean-corrected) matrix, the same difference vector would have been obtained from **X**, the matrix of original scores.

8. While the problem is developed here in terms of mean-corrected data, the same formula is obtained for the case in which **k** is applied to original-score data.

9. This same idea follows from decomposition of the more general eigenstructure equation in which eigenvectors are unique only up to a scale transformation, a reflection, or both.

10. This suggests that an underlying assumption of assignments based on Fisher's linear discriminant function is that the *prior* probability (before observing the profile) of a case belonging to group 1 is equal to that of its belonging to group 2 (i.e., 0.5 for each group).

11. Note, then, that no information about group assignment is lost by reducing the original m-dimensional problem to a one-dimensional discriminant analysis (as long as only two groups are involved).

12. The reason why this is so relates to the fact that the eigenstructure equation:

$$(\mathbf{T}^{-1}\mathbf{A} - \mu\mathbf{I})\mathbf{k} = \mathbf{0}$$

has the same eigenvector **k** as:

$$(\mathbf{W}^{-1}\mathbf{A} - \lambda\mathbf{I})\mathbf{k} = \mathbf{0}$$

but where the associated eigenvalue μ is related to the eigenvalue λ by $\mu = \lambda/(1 + \lambda)$.

13. $\mathbf{R}_T^{-1}$ is computed (via the adjoint matrix procedure) from:

$$\mathbf{R} = \begin{bmatrix} 1.000 & 0.748 \\ 0.748 & 1.000 \end{bmatrix}$$

R, in turn, can be computed from **T** in Table 4.2. The standardized difference vector is obtained from the entries:

$$\frac{1}{3.028}(-5) = -1.65$$

$$\frac{1}{2.011}(-2) = -0.99$$

where 3.028 and 2.011 are the total-sample standard deviations of X_1 and X_2, respectively. These are also computed from $\mathbf{T}$ in Table 4.2.

14. This is so because the vector of covariances:

$$\mathbf{a}(\mathbf{y}) = \begin{bmatrix} -1.39 \\ -0.56 \end{bmatrix}$$

of Y with X_1 and X_2, respectively, is *proportional* to the difference vector:

$$\mathbf{d} = \begin{bmatrix} -5 \\ -2 \end{bmatrix}$$

Next, recall from Chapter 2 that $\mathbf{b}$, the vector of partial regression coefficients is obtained from:

$$\mathbf{b} = \mathbf{C}_t^{-1}\mathbf{a}(\mathbf{y})$$

Then, remembering that $\mathbf{C}_t$ yields discriminant weights that are proportional to $\mathbf{C}_w$, we note that the correspondence is complete.

15. However, the two-group discriminant model is *not* a special case of the general univariate linear model. For example, the assumption of constant variance in the error term does not hold in the case of a dummy-valued criterion variable.

16. Ordinarily the significance testing phase would be undertaken *before* considering the assignment of profiles to groups. Our reason for discussing assignment rules first is based on the fact that they are closely related to procedures for computing discriminant functions.

17. In the general formula shown at the top of Table 4.5, *ln* denotes the natural logarithm. The vertical lines $|\mathbf{C}_w|$ denote the determinant of $\mathbf{C}_w$, the pooled within-groups covariance matrix.

18. The quadratic discriminant function (Van de Geer, 1971) could be used in this instance.

19. One could also assign cases to groups via the multiple regression formula. First, one substitutes the predictor-variable centroids for group 1 and group 2, respectively, and finds the appropriate $\overline{Y}$ value, each conditioned on the respective centroid. Call these predicted values:

$$\hat{Y}^{(1)} \text{ and } \hat{Y}^{(2)}$$

A new case is assigned to group 1 if its $\hat{Y}$ value is closer to $\hat{Y}^{(1)}$ than it is to $\hat{Y}^{(2)}$; otherwise, it is assigned to group 2.

20. So named after a British clergyman, Thomas Bayes, who proposed the theorem in the middle 1700's.

21. For all that follows in this section it is important to keep in mind that a *specific* $\mathbf{k}$ vector, namely $\mathbf{k} = \mathbf{C}_w^{-1}\mathbf{d}$, is being used. In general, other procedures (e.g., multiple regression) that find some vector proportional to $\mathbf{k}$ are *not* appropriate for this phase of the analysis.

22. Note that if $[q_2 C(1|2)/q_1 C(2|1)] < 1$, then the natural logarithm is negative, and, hence, the cutting point will be moved closer to $\bar{t}_2$. This is as it should be since a ratio of less than 1 implies either prior probabilities or costs of misclassification that favor group 1. If the converse holds, the cutting point is, of course, moved closer to $\bar{t}_1$, since the opposite situation prevails.

23. Note also that 12.34 (within rounding error) is also equal to Mahalanobis' D^2, as computed in section 4.3.4.

24. In the case of computing posterior probabilities, misclassification costs are not relevant.

25. This formula is a special case of a formula for finding posterior probabilities involving multiple (three or more) groups:

$$P(g|\mathbf{x}) = \frac{q_g\,e^{t_g}}{\displaystyle\sum_{g=1}^{G} q_g\,e^{t_g}}$$

where:

$$t_g = \mathbf{x}'\mathbf{C}_w^{-1}\overline{\mathbf{x}}_g - \frac{1}{2}\overline{\mathbf{x}}_g'\mathbf{C}_w^{-1}\overline{\mathbf{x}}_g + \ln q_g$$

for $g = 1, 2, \ldots, G$. Multiple discriminant analysis is described in Chapter 7.

26. In particular the vector $\mathbf{k}(\mathbf{W})$ should be adjusted to $\mathbf{k}(\mathbf{C}_w)$ *before* applying the Bayesian-type assignment rules discussed in the preceding section.

27. Actually, another discriminant program, BMD-07M (details of which are discussed in Chapter 7) was used here. Prior to this computer run, a test of the equality of co-variance matrices was made; the null hypothesis was accepted at the 0.01 level.

28. The F value of 17.98 was obtained from Hotelling's T^2 test, as described in section 4.4.1.

29. This is another illustration of the distinction between statistical significance and operational significance. While the results are highly significant statistically, operational significance of the discriminant function (in correctly assigning new cases) is not at all high.

30. Moreover, the discriminant weights are scaled by BMD-07M so that $\mathbf{k}'\mathbf{C}_w\mathbf{k} = 1$. However, this is not essential to the argument that follows.

31. This would also be true if $\mathbf{W}$ were used or if the scaling were based on $\mathbf{k}'\mathbf{k} = 1$, rather than $\mathbf{k}'\mathbf{C}_w\mathbf{k} = 1$.

CHAPTER **5**

Other Aspects of Single Criterion, Multiple Predictor Association

5.1 Introduction

Chapters 2 through 4 have dealt with the major topics of single criterion, multiple predictor association: multiple regression, analysis of variance and covariance, and two-group discriminant analysis. While these are the primary methods used for this class of problems, they are not the only ones. Accordingly, the first section of this chapter discusses another approach to the portrayal of association between certain types of single criterion, multiple predictor data.

The technique is called Automatic Interaction Detection (AID) and is appropriate for handling a single criterion variable that is either intervally scaled or dichotomous. While the original predictor variables may be nominal, ordinal, or interval-scaled, they are transformed to categorical variables before the analysis is begun. AID is especially designed for searching out relationships in really large samples of data, on the order of 1,000 cases or more. It often serves as a precursor to, rather than a substitute for, other techniques such as dummy-variable multiple regression.

The next principal section of the chapter deals with the topic of data transformation prior to the application of linear models. As discussed in previous chapters, linear models—that is, models that are linear in the parameters—are extremely versatile. This versatility is enhanced when the researcher can undertake various transformations of the criterion and/or predictor variables to recast an orginally nonlinear representation into a linear one.

190

We discuss this topic from two points of view. First, the use of prespecified functional forms, in which the original data are transformed according to specific linearizing functions, is described and illustrated numerically. We then turn to a discussion of non-specified functions—the only requirement here is that they be monotonic—that can be applied to the criterion variable in order to optimize its fit to some relatively simple (e.g., additive) model. We illustrate this technique in the context of some of the television commercial data.

The next set of topics pertain to ANOVA and ANCOVA. As recalled in Chapter 3, various tests were run to appraise the statistical significance of a set of treatment levels but we did not probe into the question of *which* levels of a factor were contributing to its overall significance. Tests involving specific levels of a factor are called *multiple comparisons* and are taken up in this section. A related subject—the employment of Hays' omega squared as a descriptive measure of factor importance—is also discussed here.

Introduction to Hays' omega squared measure sets the stage for discussion of a substantive question that is common to all multivariate techniques: how can we measure the relative importance of predictor variables in accounting for variation in the criterion variable? It turns out that in the usual case of correlated predictor variables, this question has no unambiguous answer. Accordingly, what is discussed are some of the general problems of multicollinearity (i.e., high intervariable correlation) and what may be done about it. We conclude the chapter with a brief discussion of various "importance" measures that have been proposed.

In summary, this chapter is something of a miscellany of subjects related to Part II. However, the topics that have been selected for discussion should be useful in rounding out the more basic material described in Chapters 2 through 4.

5.2 Automatic Interaction Detection

One of the most interesting problems in multivariate analysis is the phenomenon of interaction in which the response to changes in the level of one predictor variable depends on the level of some other predictor (or predictors). When interaction effects exist the simple additive property of individual predictor-variable contributions to changes in the criterion no longer holds.

The distinction between interaction effects and intercorrelated predictors is shown in Figure 5.1. Using an illustration drawn from public opinion research, assume that respondents are asked to rate the quality of some local municipality's bus service. Data are also collected on the respondent's extent of past usage of the service and whether he (or she) is working or not.

Next, suppose we wish to see if respondents' ratings of Y, the quality of the service (higher values indicating higher quality), are related to X_1, past usage of the service. However, we also suspect that the relationship between Y and X_1 might be dependent on whether the respondent is working or not (which can be denoted by a dummy variable X_2).

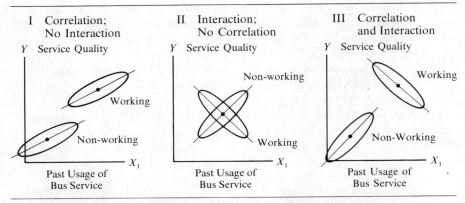

Figure 5.1 Illustrations of Correlation and Interaction (Hypothetical Data)

In Panel I we note the "classic" case of correlation between X_1 and X_2 but no interaction. We observe that Y is positively associated with changes in X_1 and that the slopes of the regression lines do not depend on the values of X_2 (working versus non-working). Rather, it is the intercept that differs between the two cases, in the sense that respondents who are working report higher quality of service on the average than those who are not working. The correlation between X_1 and X_2 is indicated by the fact that average past usage by working respondents exceeds that of non-working respondents.[1]

Panel II shows the case of interaction without correlation between X_1 and X_2. Here we see that across the total sample of working and nonworking respondents, Y does not depend on X_1 (i.e., the total-sample slope is horizontal). However, *within* the separate levels of X_2, Y increases with increases in X_1, given that the respondent is not working, while it decreases with increases in X_1, given that the respondent is working. The average past usage is the same for both working and non-working respondents, showing no correlation between X_1 and X_2.

Panel III shows the combined case of correlation and interaction. Across the total sample Y is positively associated with X_1. However, within the separate levels of X_2 we see that Y increases with increases in X_1, given that the respondent is not working, but decreases with X_1, given that the respondent is working. The positive association between Y and X_1, at the total-sample level, results primarily from the fact that *average* past usage is considerably higher for working respondents than it is for non-working respondents.

Our discussion of ANOVA and ANCOVA in Chapter 3 was concerned with experiments in which departures from parallelism of slopes (i.e., interaction) could be detected and the contributions of interaction effects isolated from those of main effects by the employment of orthogonal designs in which the predictor variables were designed to be uncorrelated.

However, in the analysis of observational and survey data one often finds cases in which the predictors are correlated, as well as interactive. Survey data is the type of data in which the analyst would ordinarily use multiple regression, which, as we know, is a type of additive model. If the researcher suspects interaction-type effects, it is a common procedure to add cross-product terms to the regression model. Unfortunately, cross-product terms express only one form of interaction (a multiplicative one) between the interval-scaled predictors. Furthermore, the cross products will generally be highly correlated with the original predictors. Moreover, one may have not only two-variable but three-variable or higher-way interactions to contend with as well.[2]

If only a few interactions are believed to exist, the researcher could introduce cross-product terms for only those predictors. A major problem, of course, is that in exploratory analyses of observational and survey data one does not ordinarily know *which* predictors are interactive (and how they are interactive). *Automatic Interaction Detection* (Sonquist and Morgan, 1964), abbreviated as AID, has been designed to help the researcher uncover interactions for subsequent analysis via regression or other types of models.

5.2.1 Basic Concepts of AID

As motivation for introducing AID, suppose a researcher has collected data for some $m = 200$ respondents regarding Y, X_1, and X_2 in the problem illustrated by Figure 5.1. However, now let us also assume that X_1, extent of past usage of the bus service, has been discretized into nine (graded) levels. This classification, along with the two levels of X_2, gives us $9 \times 2 = 18$ possible groups by which a respondent can be classified on the two predictors. We shall let Y_{ig} denote the criterion value on the i-th observation ($i = 1, 2, \ldots, m_g$) of the g-th group ($g = 1, 2, \ldots, 18$). Moreover, we shall employ the dot notation of Chapter 3 to indicate the index being summed over in computing various means.

After collecting the survey data (in which the 200 cases are well spread out across the 18 categories), the researcher decides to compute a separate mean value on the criterion variable Y for respondents falling into each of the 18 cells. If so, he can then compute a measure, called eta squared (η^2), that is analogous to R^2 in the context of multiple regression:[3]

$$\eta^2 = \frac{\sum_{g=1}^{18} m_g (\overline{Y}_{\cdot g} - \overline{Y}_{\cdot\cdot})^2}{\sum_{i=1}^{200} (Y_i \cdot - \overline{Y}_{\cdot\cdot})^2}$$

The numerator of this expression is the among-groups sum of squares across the separate cells that is found after splitting the sample into the aforementioned

18 groups and computing *separate* criterion means for each group. The denominator is the total sum of squares to be accounted for in Y. To illustrate, if one knows the overall $\overline{Y}..$ of the data and predicts each individual Y_i as $\overline{Y}..$, then the squared error is the denominator of η^2:

$$\sum_{i=1}^{200} (Y_i. - \overline{Y}..)^2 = \sum_{i=1}^{200} Y_i.^2 - 200\, \overline{Y}?.$$

The numerator of η^2 considers the group designation. In this case we are interested in the among-groups error sum of squares:

$$\sum_{g=1}^{18} m_g(\overline{Y}._g - \overline{Y}..)^2 = \sum_{g=1}^{18} m_g \overline{Y}?._g - 200\, \overline{Y}?.$$

where we try to predict each subgroup mean $\overline{Y}._g$ as $\overline{Y}..$, the overall mean. Each of these squared deviations is weighted by m_g, the number of cases in the g-th group. Thus, η^2 is simply a ratio of the among-groups to the total-group sum of squares. If the within-groups variation is 0, then $\eta^2 = 1$; if the among-groups variation is 0, then $\eta^2 = 0$.

What can be done with 18 groups can, of course, be done with a smaller (or larger) number of groups. For example, suppose we decided to develop only two (rather than 18) groups. If one knows the mean of each of the two groups and one also knows which group each individual case falls in, the numerator of η^2 is simply:

$$m_1 \overline{Y}?_1 + m_2 \overline{Y}?_2 - m\overline{Y}?.$$

Assuming that the means differ across the two groups, this expression will be positive—that is, $m_1 \overline{Y}?_1 + m_2 \overline{Y}?_2$ will exceed $m\overline{Y}?.$ and, hence, greater homogeneity is achieved by splitting the sample into the two groups than by maintaining one overall group.

The problem, then, is to find a classification in which criterion-variable means differ appreciably across cells. AID is a sequential branching technique that systematically exploits the preceding notions.

First of all, assume that we have either an interval-scaled or dichotomous (binary-valued) criterion variable and several predictor variables, all expressed as categorical. Suppose we then decide to split the total sample of cases into *two* groups using each predictor, in turn, as the basis of splitting. To be specific, assume we continue to work with the nine levels of X_1. If we restrict the splitting to maintain the natural order of X_1, that is: category 1 versus categories 2, 3, . . . , 9 or category 1 and 2 versus categories 3, 4, . . . , 9 and so on, we have eight distinct ways to form two groups. And, in general, with a G-state predictor, we shall have $G - 1$ possible partitions into two groups that maintain a natural order. (As will be noted later, other options available in AID do not restrict the partitions to those maintaining some natural order.)

What AID does is to examine all possible splits—in this case eight—of the sample on the basis of X_1 alone. It records the between-groups sum of squares around the subgroup means of Y (the numerator of eta squared) for each split. It then proceeds to predictor X_2 which, in our example, contains only two categories, and splits on that variable alone. If n different predictors are involved it considers all possible two-group splits on *each* of the n different predictors.

At this point the total sample would then be split on that predictor whose between-groups sum of squares is largest.[4] AID next looks at the resulting group with the larger within-group sum of squares and starts the process of trying binary splits all over again on this group, and so on, until one of various stopping criteria, to be described below, is reached.

5.2.2 An Example

The detailed aspects of AID are most easily explained by numerical example. In Chapter 4 we used two-group discriminant analysis to examine the association between alpha brand selectors (in part D of the television commercial study) and all others. The predictors were:

A-2: Alpha was brand selected on last tire purchase.

A-3: Pre-exposure interest rating of the Alpha radial tire.

C-1: Post-exposure believability rating of the Alpha commercial.

C-2: Post-exposure interest rating of the Alpha radial tire.

We recall that the classification matrix obtained from the two-group discriminant analysis indicated only fair discrimination. Variable A-3 appeared to be the most important of the predictors.

Our present objective is to use the same predictors in the AID algorithm. The criterion variable Y is again expressed as a binary $(0, 1)$ variable with an overall sample mean of $78/252 = 0.31$ and a total sample sum of squares (SST) of 53.86. That is, 31 percent of the sample selected the Alpha radial tire brand as their most preferred in Part D of the questionnaire. Our interest now is in using AID to partition the total sample of 252 respondents into successive sets of subgroups in a way that leads to a maximum between-groups sum of squares (in the criterion variable Y) at each level of partitioning. Table 5.1 summarizes some of the computer output obtained from AID.

For example, suppose we were to split the sample on the basis of predictor A-2. This predictor has only two levels to begin with—hence, the two subgroups are required to be Alpha purchasers versus others. Table 5.1 shows that 48 respondents reported purchase of the Alpha brand while 204 respondents did not. If the sample is split on predictor A-2, the between-groups sum of squares on the criterion variable (SSA) is 4.45, compared to a total sum of squares of 53.86.

We could then try various splits on predictor A-3. Since there are 11 possible ratings—0 through 10—there are 10 possible binary partitionings of the sample, assuming that we wish to maintain "monotonicity" across the 11 categories.

Table 5.1 Initial AID Splits of Total Sample on Each Predictor

Predictor	Categories	Number of Cases	Between-Groups Sum of Squares SSA	Ratio of Between-Groups to Total-Group Sum of Squares SSA/SST
A-2	Alpha purchasers versus Others	48 204	4.45	0.082
A-3	Ratings 0–8 versus Ratings 9–10	206 46	8.39	0.156
C-1	Ratings 0–8 versus Ratings 9–10	199 53	9.17*	0.170
C-2	Ratings 0–8 versus Ratings 9–10	197 55	6.70	0.124

*The total sample is first split on this predictor since its between-groups sum of squares (SSA) is the highest of all predictors.

By this is meant that the natural order of the (discretized) variable is maintained. For example, the partition 0, 1 versus 2 through 10 would be appropriate but not 0, 3 versus 2, 4, 5, . . . , 10. Of the 10 possible two-group splits on A-3, the one resulting in the largest SSA involves the ratings 0 through 8 versus 9 and 10 and is shown in Table 5.1.

Similarly, we can test out each possible partition for the (assumed monotonic) predictors C-1 and C-2, in each case recording the two-group split that leads to the highest between-groups sum of squares on the binary criterion variable. The best splits—taking each predictor singly—all appear in Table 5.1.

We note from the table that if we split the total sample on predictor C-1 into those giving the Alpha commercial high ratings of 9–10 versus those rating the commercial less high (in the range of 0–8), the between-groups sum of squares (SSA) is 9.17. This is the maximum between-groups sum of squares that can be achieved at this stage. Hence, the total sample is split into two groups of 53 "high believability" raters and 199 others on the basis of predictor C-1. Group 2, those expressing "high believability" in the commercial, also indicate high acceptance of the Alpha brand in that $\overline{Y}_2$ for this group is 0.68 while $\overline{Y}_3$ for the other group of 199 respondents is only 0.21. This is shown in the tree diagram of Figure 5.2.

The program proceeds to split group 3, the group with the higher sum of squares—this time on predictor A-3—and so on, until one obtains the complete tree diagram of Figure 5.2. If we examine the extreme cases, we see that:

1. Group 10, consisting of only 15 respondents, shows the highest criterion mean ($\overline{Y}_{10} = 0.93$). This group expresses high believability in the Alpha radial tire commercial *and* its members are all past purchasers of the Alpha brand.

2. Group 9, consisting of 38 respondents, shows the lowest criterion mean ($\overline{Y}_9 = 0.03$). This group:
   a. Does not express high believability in the Alpha commercial.
   b. Does not express high pre-exposure interest in Alpha.

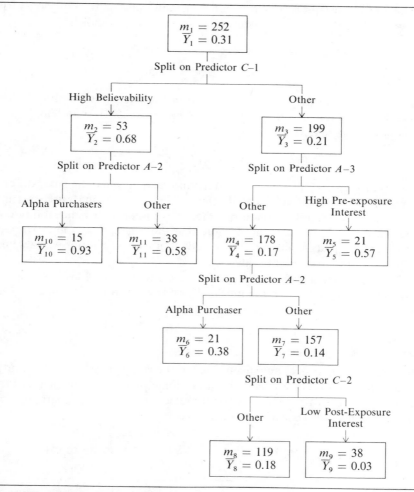

Figure 5.2 Tree Diagram from AID Analysis

c. Is not made up of past purchasers of Alpha.
d. Expresses low post-exposure interest in the Alpha brand.

Due to the small initial sample size of 252 respondents it is difficult to assess the "interactive" aspects of this illustrative application. For example, since group 10 has only 15 respondents in it, we are not able to tell reliably what the effects of further splits would be.

However, one simple check can be carried out. We can find out whether the change in the criterion variable between Alpha purchasers versus purchasers of other brands (predictor A-2) varies appreciably between high believability and other believability levels (predictor C-1). As noted from Figure 5.2, in the case of the *high believability* group ($m_2 = 53$), the difference in the criterion-variable mean for Alpha purchasers versus purchasers of other brands is:

$$\overline{Y}_{10} - \overline{Y}_{11}$$

$$0.93 - 0.58 = 0.35$$

Although not shown in Figure 5.2, we can find the corresponding criterion-variable means for the other believability group of 199 respondents. The corresponding difference is:

$$0.42 - 0.17 = 0.25$$

This result suggests the presence of some interaction, since the latter difference is smaller. That is, the differential effect of being an Alpha purchaser is less under lower believability conditions than it is under high believability conditions. However, the reader should note that the interaction is not extreme. In particular, being an Alpha purchaser leads to a higher criterion-variable mean *regardless* of the believability category.

The results appear to be consistent from a content point of view. At this stage, however, let us go through the mechanics of the algorithm in somewhat more detail.

5.2.3 The AID Algorithm

The objective of AID, at any given stage in the splitting process, is to partition the group into two classes according to a single predictor X_j which, in turn, is expressed categorically.[5] The criterion for selecting X_j—and that particular partition of X_j (into two classes)—is to maximize:

$$
m_1 \overline{Y}_{\cdot 1}^2 + m_2 \overline{Y}_{\cdot 2}^2 - m \overline{Y}_{\cdot}^2 = \text{SSA, the between-groups} \\ \text{sum of squares,}
$$

where:

$$\overline{Y}_{\cdot\cdot} = \frac{m_1 \overline{Y}_{\cdot 1} + m_2 \overline{Y}_{\cdot 2}}{m}$$

and $m_1 + m_2 = m$, the total group size.

At the beginning of the binary splitting process, the total sample is considered as the first group to be split. At any subsequent stage that unsplit group g with the largest total sum of squares:

$$\text{SST}_g = \sum_{i=1}^{m_g} Y_{ig}^2 - m_g \overline{Y}_{\cdot g}^2$$

is chosen for the next split.

A number of stopping criteria are built into the program with parameters that can be chosen by the researcher:

1. Minimum group size and total number of groups—the researcher can specify that no group be split with a sample size of less than m^*, where m^* is an integer. In the sample problem of Figure 5.2 m^* was set at 15 in order to guard against extremely small sample sizes. One can also fix the maximum total number of groups to be formed in a single AID run. This parameter was set at $G = 15$ in the sample problem.

2. Minimum total sum of squares—the researcher may specify that no group be split whose total sum of squares is less than some prespecified fraction P of the SST of the original input sample. That is:

$$\boxed{\text{Do not split if SST}_g < P(\text{SST})}$$

The parameter P was set at 0.01 in the sample problem.

3. Minimum between-groups sum of squares—the researcher may specify that no group be split if the best split at any given stage results in a between-groups sum of squares that is less than some prespecified fraction Q of the SST of the input sample. That is:

$$\boxed{\text{Do not split if SSA}_g < Q(\text{SST}); \text{ where } Q < P}$$

The parameter Q was set at 0.005 in the sample problem.

The preceding parameters are often reset from run to run, depending upon whether the analyst desires a "loose" (everything in) or "tight" view of the relationships.[6]

A few other characteristics of the AID procedure are of interest:

1. As indicated earlier, if a predictor X_j has a natural order, this order can be preserved if the sample is split into two groups on X_j. For example, if X_j is characterized by eleven classes (as was the case in predictors A-3, C-1, and C-2 in the sample problem) then by choice of the *monotonic option* all binary splits preserve the order.

2. Other predictors (e.g., those dealing with occupation, religion, etc.) may be left *"free"*; in this case the binary split does not need to maintain some natural order in X_j.

3. The program provides a capability for taking interval-scaled predictors and splitting them into some prespecified number of classes, usually two to fifteen. (Recoding can range up to 32 classes for each predictor, however.)

4. In a single run of AID, up to 63 predictors can be employed and up to 89 binary partitions can be performed.

In addition to the above characteristics, more recent versions of AID permit the analyst to incorporate a (single) covariate in the analysis. Other recently incorporated features enable the researcher to prespecify certain data splits, force a type of symmetry in the tree diagram, start out with a tree that is partially specified, and pursue a "look ahead" strategy that examines a sequence of splits (Sonquist, Baker and Morgan, 1971).

These extensions go well beyond the scope of this introduction. The Institute of Social Research at the University of Michigan has been active in the development of a number of special procedures (Morgan and Messenger, 1973) for analyzing survey data, in addition to the AID technique.

5.2.4 Some Concluding Caveats

As has already been discussed in preceding chapters, all multivariate techniques show some penchant for capitalizing on the specific characteristics of the sample at hand. We saw in Chapter 2 that cross-validation, and double cross-validation as well, represented an important part of multiple regression, when used in exploratory analyses.

The need for replicate analyses is particularly acute in the case of AID. Unlike regression models, AID provides no way to examine the statistical significance of the fitted model. AID should be used with extreme caution, particularly on small samples. AID's developers suggest sample sizes of at least 800–1,000 cases. Sample sizes of only 200 or 300 are quite likely to lead to data splits that do not hold up under cross-validation.

AID's tendency to capitalize on chance variation was examined in conjunction with the sample problem of Figure 5.2. This time two separate AID runs were made. First, the group of $252 - 78 = 174$ non-Alpha brand selectors (in Part D) was split randomly in halves. Two groups were then made up, each consisting of the same (78) Alpha selectors in Part D and a group of 87 non-Alpha selectors.

In the first of these two groups the predictors split the sample in the following order: C-1; A-3; C-2; A-2. In the second of the two groups the predictors split the sample in the following order: A-3; C-2; A-2; C-1. While this difference in splitting order is due cause for concern, in both sub-samples the combination of:

1. High believability in the Alpha commercial (C-1)
2. Past purchasers of the Alpha tire brand (A-2).

was sufficient to partition the sample into Alpha selectors versus others (Part D) reasonably well. The replications differed in terms of the order in which the predictors entered more than anything else.

Despite the stability of results in this illustration, the user is well advised to apply the AID technique with caution. J. A. Sonquist, one of the developers of AID, has stressed this point and also the potential value of AID as a *preliminary procedure* to a more structured approach, such as dummy-variable regression with selected cross product terms.

As for the "interactive" aspects of AID, it should be clear that as the tree diagram departs from symmetry, interaction is at work. Unfortunately, "departure from symmetry" is not a well-defined phrase. Moreover, additive data is a sufficient, but not necessary, condition for symmetric tree diagrams. It is fair to say that AID provides only *diagnostic* information on interaction—rather than actual parameter estimates—for later incorporation in regression-type models. It also is fair to say that appliers of AID have often ignored the advice of Sonquist to use AID in a *preliminary* data search prior to applying regression-type models with selected interaction terms. Quite probably, the simple form of AID output (see Figure 5.2) has contributed to applied researchers' tendencies to employ AID by itself, rather than as a preliminary technique to other methods, such as multiple regression or two-group discriminant analysis.

In summary, AID basically performs a series of one-way ANOVA's on a set of predictors for the purpose of successively splitting a sample into more and more homogeneous subsamples insofar as the criterion variable is concerned. Each splitting involves two subgroups, leading to the type of tree representation shown in Figure 5.2. Since different sequences of predictors (and different partitions of the predictor variables as well) can occur, AID can handle very general kinds of interactions.[7]

However, it should be reiterated that AID is designed for the preliminary analysis of really large samples of 800–1,000 cases. Furthermore, the use of AID generally benefits from a subsequent analysis in which the predictors revealed by AID are entered, along with various cross products, into a dummy-variable regression program. Finally, wherever practical, both steps (AID followed by dummy-variable regression) of the analysis should be cross validated, given the high tendency for AID to capitalize on chance variations in the data.

5.3 Prespecified Transformations to Linearity

The high versatility of linear models, illustrated in the preceding chapters, is enhanced even further by the possibility of a preliminary data transformation. Two basic strategies are available for data transformation prior to linear model fitting. First, we could use some prespecified function (e.g., logarithmic) to transform either the criterion variable or the predictors before applying a linear model. Alternatively, we could adopt a strategy that finds an arbitrary monotonic (order preserving) function of the criterion variable that is maximally correlated with a linear function of the predictors. In this section of the chapter we discuss the first of these two strategies.

The use of prespecified functions can be further classified into: (a) transformation of the predictors only; (b) transformation of the criterion variable only; and (c) transformations of both. Here we concentrate on (a) and (c) and defer discussion of (b) to section 5.4, in the context of best-fitting monotonic functions.

5.3.1 Transformation of Polynominal Functions

The first set of functions of interest to us are those that are already linear in the parameters, but not in the variables. In this case, the transformation simply involves the substitution of new variables for the original ones. Polynomials are illustrations of this class of functions.

For example, consider the following function:

$$Y = \beta_0 + \beta_1 X + \beta_2 X^2 + \epsilon$$

The presence of the X^2 term shows that the equation is nonlinear in the variables. However, let us set $Z_1 = X^2$, with the following result:

$$Y = \beta_0 + \beta_1 X_1 + \beta_2 Z_1 + \epsilon$$

and we see that the new function *is* linear in the variables.

Similarly, ignoring the error term, suppose we have the function:

$$Y = \beta_0 + \beta_1 X_1 + \beta_2 X_1^2 + \beta_3 X_1 X_2$$

By setting $Z_1 = X_1^2$ and $Z_2 = X_1 X_2$, we have:

$$Y = \beta_0 + \beta_1 X_1 + \beta_2 Z_1 + \beta_3 Z_2$$

In general, *any polynomial*—second degree, third degree, etc.—can be handled by the same procedure.

It should be emphasized, however, that powers and cross products will tend to be highly correlated with their correspondent linear terms. This condition, known as multicollinearity, can render the partial regression coefficients highly instable over cross-validation. (Some aspects of this problem are discussed later in the chapter.)

5.3.2 Other Transformations of the Predictor Variables

In addition to the polynominal models illustrated above, a number of other types of transformations of the predictor variables can be made to achieve linearity in the variables. (For simplicity of presentation the error term is ignored.)

Reciprocal Function:

$$Y = \beta_0 + \beta_1 \left[\frac{1}{X_1} \right] + \beta_2 \left[\frac{1}{X_2} \right]$$

However, let $Z_1 = 1/X_1$, $Z_2 = 1/X_2$ and write:

$$Y = \beta_0 + \beta_1 Z_1 + \beta_2 Z_2$$

Square Root Function:

$$Y = \beta_0 + \beta_1 X_1^{1/2} + \beta_2 X_2^{1/2}$$

However, let $Z_1 = X_1^{1/2}$, $Z_2 = X_2^{1/2}$ and write:

$$Y = \beta_0 + \beta_1 Z_1 + \beta_2 Z_2$$

Logarithmic Function:

$$Y = \beta_0 + \beta_1 \ln X_1 + \beta_2 \ln X_2$$

However, let $Z_1 = \ln X_1$, $Z_2 = \ln X_2$ and write:

$$Y = \beta_0 + \beta_1 Z_1 + \beta_2 Z_2$$

Again, no new principles are involved and we see that application of the linear regression model, following the preliminary transformation, is quite straight-forward.[8]

5.3.3 Transformation of Criterion and Predictor Variables

A number of other transformations are possible in which both the criterion and the predictors are involved, prior to application of the linear model:

Multiplicative Model:

$$Y = \beta_0 X_1^{\beta_1} X_2^{\beta_2} X_3^{\beta_3} \epsilon$$

If we specify the error ϵ to have a mean of 1 and some finite variance $\sigma^2(\epsilon)$, then we can transform the preceding model, by taking common logarithms of both sides, to the expression:

$$\log Y = \log \beta_0 + \beta_1 \log X_1 + \beta_2 \log X_2 + \beta_3 \log X_3 + \log \epsilon$$

This same general idea can be extended to still other models. Ignoring the error term, some of those possibilities are:

Exponential Model:

$$Y = e^{[\beta_0 + \beta_1 X_1 + \beta_2 X_2]}$$

By taking natural logarithms of both sides, we obtain:

$$ln\ Y = \beta_0 + \beta_1 Z_1 + \beta_2 Z_2$$

Reciprocal Model:

$$Y = \frac{1}{\beta_0 + \beta_1 X_1 + \beta_2 X_2}$$

By taking reciprocals of both sides, we obtain:

$$\frac{1}{Y} = \beta_0 + \beta_1 Z_1 + \beta_2 Z_2$$

Still other transformations are possible which are not detailed here.

In short, a wide variety of functions can be transformed so as to be linear in the parameters. This flexibility serves to extend the value of linear models across a wide variety of contexts.[9]

5.3.4 Dummy-Variable and Piecewise Regression

As discussed in Chapter 2, still another way to cope with the problem of non-linearity is to use dummy-variable regression. For example, we may believe that Y is related to predictor X_j in a nonlinear fashion, but are not sure what the specific nature of the function is. What can be done in cases like this is to discretize the predictor variable into some specified number of (say k) classes and then to encode these into $k - 1$ dummy variables to be used in a conventional multiple regression.

What we obtain from this procedure is $\hat{Y}$ for each level of the predictor. Each value of $\hat{Y}$ can then be plotted against the respective midpoint of each class of X_j in order to obtain some idea of what the functional relationship is. However, as recalled from Chapter 3, discretizing interval-scaled predictors uses up degrees of freedom rather rapidly. Hence, if we have any good idea of the nature of the function (e.g., quadratic in X_j, linear in $X_j X_k$) it would be preferable to consider the specific function. In some cases, however, this information will not be available and the dummy-variable approach does offer a high degree of flexibility (albeit with the loss of degrees of freedom).

A variation on this theme is to use dummy variables in the context of *piecewise linear regression*. To illustrate, suppose we return to the example portrayed in Figure 5.1. Let us now assume that we are interested only in the relationship between Y, quality of the bus service, and X_1, past usage of the service. Moreover, we shall assume that X_1 is expressed as a continuous variable (as it was originally, prior to being discretized for the AID application).

In the present case, however, we assume that the relationship shown in Panel I of Figure 5.3 is observed. This scatter plot shows that Y appears to exhibit one

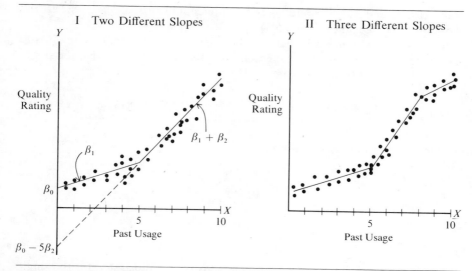

Figure 5.3 Illustrations of Piecewise Linear Regression

slope when X is between 0 and 5 years and a different slope when X is between 5 and 10 years.

One might want to fit a quadratic function to the data portrayed in Panel I. Alternatively, however, suppose we consider the possibility of fitting a piecewise linear function. This can be done quite simply by setting up the following linear model:

$$Y_i = \beta_0 + \beta_1 X_{i1} + \beta_2(X_{i1} - 5)X_{i2} + \epsilon_i$$

in which we define the dummy variable:

$$X_{i2} = \left. \begin{matrix} 1 \\ 0 \end{matrix} \right\} \begin{matrix} \text{if } X_{i1} > 5 \\ \text{if } X_{i1} \leqq 5 \end{matrix}$$

Note that if $X_1 \leqq 5$, then $X_2 = 0$; hence, we have:

$$Y_i = \beta_0 + \beta_1 X_{i1}; X_1 \leqq 5$$

However, if $X_1 > 5$, then $X_2 = 1$ and we have:

$$Y_i = (\beta_0 - 5\beta_2) + (\beta_1 + \beta_2)X_{i1}; X_1 > 5$$

In this case β_0 and $\beta_0 - 5\beta_2$ are the two Y intercepts and β_1 and $\beta_1 + \beta_2$ are the two slopes (see Panel I of Figure 5.3).

As Panel II of Figure 5.3 suggests, there is no need to confine ourselves to one change in slope if the data suggest otherwise. In the case of the three slopes suggested by Panel II, we have the model:

$$Y_i = \beta_0 + \beta_1 X_{i1} + \beta_2(X_{i1} - 5)X_{i2} + \beta_3(X_{i1} - 8)X_{i3} + \epsilon_i$$

in which we now define two dummies:

$$X_{i2} = \begin{matrix} 1 \\ 0 \end{matrix} \begin{cases} \text{if } X_{i1} > 5 \\ \text{if } X_{i1} \leq 5 \end{cases}$$

$$X_{i3} = \begin{matrix} 1 \\ 0 \end{matrix} \begin{cases} \text{if } X_{i1} > 8 \\ \text{if } X_{i1} \leq 8 \end{cases}$$

We could then proceed in a manner analogous to that illustrated for the single changes in slope to fit the three linear pieces in Panel II.

Neter and Wasserman (1974) show a variety of examples of the use of dummy variables in piecewise linear regression, including the possibility of a change in intercept at the point where the slope also changes. For our more limited purposes, however, piecewise regression is seen to be another case in which dummy variables can be employed to increase the versatility of the linear model.

5.3.5 Clues for Specifying the Transformation

A key problem in the use of prespecified transformations to linearity is the choice of what transformation to apply. Sometimes prior theory will guide the choice of the transformation. In other cases one may try generalized polynominals of, say, the second degree, and then examine the incremental variance accounted-for by various squared terms and cross products.

If the computer program prints out residuals $(Y_i - \hat{Y}_i)$, one should always plot these residuals against the fitted $\hat{Y}_i$ values. Ideally, we would like these residuals to suggest no dependence on $\hat{Y}$, assuming that the error term is considered to be normally and independently distributed with constant variance.

Figure 5.4 shows a few of the common problems that may be encountered in the model fitting process. Panel I shows a U-shaped plot; this is the kind of plot that could arise if one is fitting only a linear term when a quadratic variable is more appropriate. Panel II indicates that the dispersion of the residuals increases with $\hat{Y}$. This suggests that the model is violating the ordinary least squares assumption about the constancy of error variance. (Sometimes use of log Y or $Y^{1/2}$ will alleviate this condition.)

Panel III shows a situation in which we have an outlier—a point that shows an extremely poor fit of $\hat{Y}$ to Y. The researcher might wish to check this par-

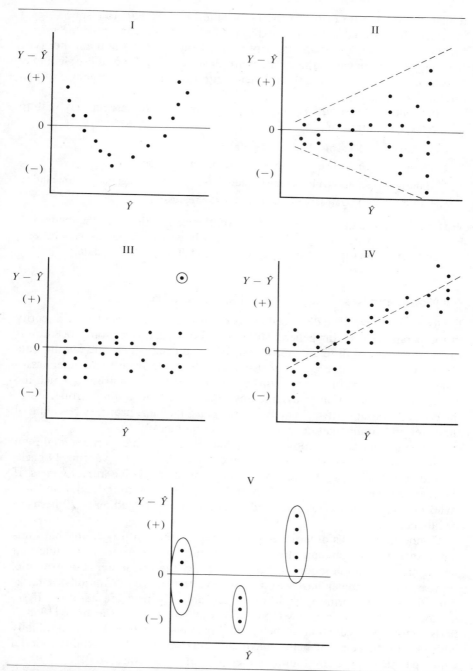

Figure 5.4 Some Common Fitting Problems Revealed by Plotting the Residuals ($Y - \hat{Y}$) Versus the Fitted Values $\hat{Y}$

ticular observation for errors in coding or other sources. Panel IV shows a case in which the residuals appear to exhibit a linear trend. (One possible reason for this is the omission of an intercept term.) Panel V shows still another type of situation—one in which the $\hat{Y}$ values are clustered. This kind of problem may be alleviated by discretizing the predictor into three classes via the use of two dummy variables.

Other patterns of residuals can occur. Moreover, the researcher may wish to undertake other kinds of plots:

1. Plotting $Y - \hat{Y}$ against X in the regression.
2. Plotting $Y - \hat{Y}$ in histogram form (see Daniel and Wood, 1971).
3. If the observations involve a time sequence, plotting $Y - \hat{Y}$ against the time index.
4. Plotting $Y - \hat{Y}$ against X's that are not in the regression.

We do not explore these matters further, on the assumption that the reader has been presented with enough illustrations to show how residuals can be employed to examine the reasons underlying lack of fit of model to data.

5.4 Monotonic Rescaling of the Criterion Variable

Up to this point we have discussed *prespecified* functions for achieving linearity in the parameters. If this general strategy is followed, a common procedure is to try a number of different functional forms and see which one fits the data "best" in terms of some global measure like R^2. Aside from the dangers associated with this type of strategy (Birnbaum, 1973), cases can arise in which the researcher does not wish to assume anything more than ordinal properties in the criterion variable. If so, he may wish to consider some arbitrary rescaling of the criterion variable that need only obey monotonicity.

To be more specific, suppose a researcher had set up an experimental design in which a subject was asked to rank a set of package designs—varying by background color and style of printing—in terms of their relative attractiveness. If the researcher employed four different "levels" of background color and five styles of printing, a 4×5 factorial design would result, leading to 20 items to be ranked.

Monotonic analysis of variance is the name given to a procedure that finds a non-specified (other than monotonic) function of the criterion variable that best agrees with some model that *is* specified by the researcher. For example, suppose the researcher has reason to believe that the contributions of background color and printing style to judged attractiveness are additive. If so, monotonic ANOVA—called MONANOVA by Kruskal and Carmone (1968)— finds a monotonic function of the original response variable that is "optimally additive" in the sense of being maximally correlated with an additive model whose parameters are also found by the technique (Kruskal, 1965).

Applications of MONANOVA have taken place in the context of conjoint measurement (Krantz and Tversky, 1971) and functional measurement (Anderson, 1970)—two closely related developments in mathematical psychology. While we do not pursue these substantive areas here, some discussion of the technical basis of MONANOVA is in order before showing how the procedure can be applied.

5.4.1 Basic Concepts of MONANOVA

We first assume that the researcher has developed some type of factorial design that, with suitable coding, can be characterized by a set of dummy variables. If we consider the problem as just another regression problem, we could then find a set of parameters

$$\beta_1, \beta_2, \ldots, \beta_n$$

corresponding to partial regression coefficients. However, in the context of ANOVA, the β's could be considered as row and column effects of a main effects, additive model.

The original ordinal response variable can be denoted by $Y_i (i = 1, 2, \ldots, m)$ while we can let $Z_i = f(Y_i)$ be the monotonically transformed observations to which the main effects, additive model is to be applied. Moreover, we can let $z_i(\beta)$ denote a set of criterion-variable predictions that are computed by the dummy-regression model.

A suitable measure of the departure of fit of the $z_i(\beta)$ from the Z_i is given by:

$$\sum_{i=1}^{m} [Z_i - z_i(\beta)]^2$$

which Kruskal (1965) normalizes through division by the scale factor:

$$\sum_{i=1}^{m} [z_i(\beta) - \bar{z}(\beta)]^2$$

where $\bar{z}(\beta)$ is the mean of the $z_i(\beta)$. Kruskal's fit measure, called *stress*, is then defined as the square root:

$$S = \sqrt{\frac{\sum_{i=1}^{m} [Z_i - z_i(\beta)]^2}{\sum_{i=1}^{m} [z_i(\beta) - \bar{z}(\beta)]^2}}$$

The objective of MONANOVA is to find a set of β's and a set of Z's that minimize S, subject to obeying the monotonic constraints in the criterion variable:

$$Y_i < Y_{i*} \text{ implies } Z_i \leqq Z_{i*}$$

MONANOVA proceeds iteratively to find the set of β's and the set of Z's that minimize stress. To illustrate, one could start by assuming that the original Y_i's were really interval-scaled. One could regress this criterion variable on the set of dummy variables and find the first set of β's and the predicted criterion-values $z_i(\beta)$. MONANOVA then adjusts these predicted values as little as possible to obtain the first set of Z_i's that *are* monotonic with the Y_i's. Then this set of Z_i's is regressed on the dummy variables and a new set of β's and $z_i(\beta)$ are obtained, and so on. (Actually, MONANOVA uses a gradient method of solution. However, the preceding method accomplishes precisely the same result and is intuitively simpler to describe.)

Thus, for a fixed set of Z's, one can find a set of β's that minimize stress. One can then find a new set of Z's that are closer to the $z_i(\beta)$ obtained on the previous iteration and solve for a new set of β's and so on, subject to maintaining the monotonicity constraints.[10] MONANOVA also provides a plot of the Y_i's versus the fitted values $z_i(\beta)$ and a plot of the Z_i's versus the fitted values $z_i(\beta)$ as a guide to the identification of a specific functional form that could be later used to transform the original criterion variable. It should be mentioned, however, that there are no tests of significance associated with the employment of MONANOVA. In this respect, it is like the AID procedure.

Carroll (in Green and Wind, 1973) shows how this monotonic ANOVA procedure can be generalized to deal with interval-scaled predictors and/or non-additive ANOVA models, thus extending its scope to monotonic multiple regression and monotonic ANCOVA.[11] We do not pursue these generalizations here. Rather, our objective is to show how MONANOVA can be used in the analysis of behavioral data.

5.4.2 An Application of MONANOVA

In Part B of the television commercial survey (see Appendix A), respondents were asked to rate a set of 25 hypothetical tire purchase options on a 5-point scale, ranging from highly unlikely to buy (coded 1) to highly likely to buy (coded 5). Each purchase option consisted of a four-element profile involving:

1. Advertised tread mileage: 30,000; 40,000; 50,000; 60,000; and 70,000 miles.
2. Brand: Alpha; Beta; Delta; Gamma; Epsilon.
3. Price per tire: $40; $55; $70; $85; $100.
4. Driving time to tire store: 10; 20; 30; 40; 50 minutes.

Since each of the four factors above could appear at each of five levels, a 5^4 factorial design of 625 possible combinations was involved.

On the assumption that a main effects, additive model would apply, only 25 profile descriptions, to be used as stimuli, were made up according to a graeco-latin square design, as shown in Figure A.3 in Appendix A. We note from the figure that each level of factor 3 (price) and factor 4 (driving time in minutes) appears once and only once in each row and each column. Furthermore, each pair of factor 3 and factor 4 levels appears exactly once in the whole figure. The graeco-latin square effects a considerable reduction in treatment combinations (from 625 to only 25) but does assume that a non-interactive model is appropriate.

Each of the 25 combinations in Figure A.3 was printed on a 3 × 5 card which served as a tire purchase description for evaluation by the respondent. Each respondent then rated each card description on the 1–5 likelihood-of-buying scale.

Analysis of the Part B questionnaire responses to the 25 hypothetical purchase options was preceded by application of factor and cluster analysis—the objective of which was to obtain clusters of respondents who exhibited high within-cluster similarity in terms of their responses to this part of the questionnaire. Subsequent chapters discuss these procedures, but here we assume that this has all taken place and we have developed three clusters of respondents:

$$Cluster\ 1;\ m_1\ =\ 62$$
$$Cluster\ 2;\ m_2\ =\ 85$$
$$Cluster\ 3;\ m_3\ =\ \underline{93}$$
$$240\ respondents$$

Only 240 of the original 252 respondents were analyzed this way. The 12 respondents who selected no brand in Part D of the questionnaire were excluded since interest centered on whether brand selectors differed systematically in terms of their utilities for each of the four factors making up the hypothetical tire purchase options.

As Table 5.2 bears out, no association was found between cluster membership and brand selection.[12] In particular, Alpha brand selectors (in Part D of the questionnaire) did not show any major differences in their profile evaluations from purchasers of other brands. We conclude that the brand which a respondent chooses in Part D does not appear to be highly dependent on the respondent's utilities derived for the four tire-purchase factors used in part B of the questionnaire.

However, the clusters themselves may differ with respect to the evaluation of the hypothetical tire purchase options—at least the clusters were formed on the basis of possible heterogeneity in the sample (albeit heterogeneity that does not appear to be related to respondent brand choice in Part D of the questionnaire). Accordingly, responses to the set of 25 profiles were averaged within each of the three clusters and the rankings of the three sets of averaged ratings were separately analyzed by MONANOVA.

Table 5.2 Distribution of Cluster Membership (Based on Part B) Across Brand-Selection in Part D*

Brand Selected in Part D	Clusters Obtained from Part B			Totals
	1	2	3	
Beta	5	17	14	36
Gamma	9	3	12	24
Delta	29	37	36	102
Alpha	19	28	31	78
Totals	62	85	93	240

*Chi square is 9.70; not significant at $\alpha = 0.05$ level with 6 degrees of freedom.

Figure 5.5 shows a comparison of the results of the three separate MONANOVA computer runs. By way of illustration, suppose we examine the results for cluster 1 first. (As observed from Table 5.2 there were 62 respondents in this cluster.) Panel I of Figure 5.5 shows that anticipated tread mileage was important to the members of cluster 1, with utility increasing monotonically with mileage. Insofar as utility for brands is concerned, Panel II indicates that cluster 1 respondents most like the beta and delta brands, although their utility function does not vary markedly across the five brands.

Panel III shows that price of tire is, indeed, an important factor for cluster 1 respondents; as might be expected, utility declines monotonically with increases in price per tire. Panel IV indicates that driving time from home to tire store is not an important factor at all; although utility generally declines with increases in driving time, the decline is hardly noticeable.

Cluster 2 and cluster 3 respondents also show results that are very similar to those of the first cluster—despite our attempts to develop the most "heterogeneous" subgroups possible. While we note that cluster 2 shows a greater utility range for tread mileage than does cluster 1, in general the results are quite similar across the three clusters. This lack of high heterogeneity in response may account for our failure to find an association between brand choice in Part D of the questionnaire and tire profile evaluation. It seems clear that tread mileage and price are the two most important factors *across all three clusters of respondents.*

As indicated earlier, the MONANOVA algorithm attempts iteratively to minimize a badness-of-fit function, called stress. This function has a lower limit of zero and an upper limit of unity. Since a main effects, additive model is being fitted in the present application, the stress values can be checked to see how well the additive model accounts for the data, once the best monotonically transformed values $Z_i = f(Y_i)$ are found. The stress values found in the application were: 0.02, 0.08, and 0.04, respectively, for clusters 1, 2, and 3.

These values of stress are generally considered to be good fits. However, as a further check on the adequacy of an additive model, the fitted z_i were rank

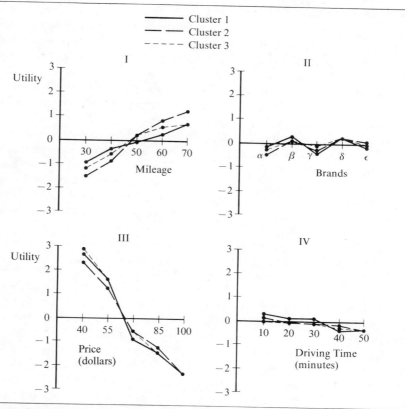

Figure 5.5 Plots of Utility Functions Obtained from MONANOVA

correlated with the original Y_i values. The rank correlations were: 0.99, 0.97, and 0.98, respectively, for clusters 1, 2, and 3.

In summary, we found that two factors—tread mileage and price—were most important to all three respondent clusters. Furthermore, the utility functions across the three respondent groups were quite similar, despite attempts to find heterogeneity in the original Part B response data. Finally, from a methodological viewpoint, a main-effects, additive model represented a good fit to the Part B profile evaluation data.

As illustrated, monotonic ANOVA provides a way to transform a criterion variable to optimize the fit between the monotonically transformed values and some hypothesized linear model *without* prior specification of the transformation. As Kruskal indicates, one can use MONANOVA as a preliminary device to see whether *any* monotonic function of the criterion is desirable. One can then choose a specific function, if deemed desirable, after examining the program's preliminary monotonic rescaling of the criterion variable.[13]

5.4.3 Concluding Comments

Sections 5.3 and 5.4 exhibit a common purpose, namely to show how versatile the linear model is, once the researcher is able to find a satisfactory transformation of the criterion variable, predictor variables, or both. The analysis of residuals—as illustrated in Figure 5.4—often provides clues regarding transformations that can render the data suitable for representation by linear models. Where possible, prior theory about the content area should be used to provide a similar guide.

However, occasions arise in which current theory is unable to suggest a specific transformation. If so, various functions—square root, logarithmic, and so on—may be tried out on an *ad hoc* basis. The researcher can then utilize some overall goodness of fit criterion like R^2 to compare models. More appropriately, if one of the models is a special case of another, one can apply the models comparison test of Chapter 2 to compare variance accounted-for under some full versus restricted model.

Selection of the "best" model becomes increasingly difficult as the number of predictors increase. For example, if only one predictor is involved, the plotting of Y and X is a simple matter and can provide direct hints regarding the selection of an appropriate functional relationship. If ten predictors are involved, the situation becomes much more cumbersome and uncertain. Fortunately, linear models are often very good approximations to more complex nonlinear models, particularly if disordinal interactions are absent (Green, 1973).

Dummy variables and piecewise linear approximation provide still more flexibility in dealing with nonlinearities (at the expense of losing degrees of freedom). Finally, the newer nonmetric methods like MONANOVA enable the researcher to deal with a criterion variable that is only rank ordered to begin with. Extensions by Carroll (discussed in Green and Wind, 1973) allow the researcher to carry out nonmetric multiple regression and nonmetric ANCOVA on rank ordered criterion variables as well. Carroll has further generalized these methods to deal with linear model fitting to a criterion variable that is only nominally scaled.

In short, there are a myriad of available ways to transform data prior to linear model fitting. The problem is less one of finding some type of transformation than knowing which one to choose. Aside from the examination of residual plots and the sequential testing of models with additional parameters, the researcher often has little to go on in choosing the "best" transformation. Generally, the principle of Occam's razor is a good one to follow—choose the simplest transformation that "works." (However, it is even better to have substantive theory guide the selection process.)

5.5 Some Additional Topics in the Analysis of Variance

In the discussion of ANOVA and ANCOVA in Chapter 3, our attention was focused primarily on tests of the *overall* significance of various factor effects. For example, in single-factor ANOVA, we have the usual hypotheses:

$$H_0: \mu_1 = \mu_2 = \cdots = \mu_K$$
$$H_1: \text{At least one } \mu_k \text{ differs}$$
$$\text{from the rest}$$

Suppose that we reject H_0 and conclude that at least one treatment level effect differs from the rest.

The topic of *multiple comparisons* is relevant for investigating specific factor level effects, once the researcher has concluded that H_0 is to be rejected. This problem, in turn, involves two subproblems:

1. What procedures can be used to find interval estimates of separate factor level effects?

2. Since the confidence levels related to these separate estimates will ordinarily not be independent, what procedures can be used to establish an *overall* confidence level regarding the full set of separate interval estimates?

The first problem involves multiple comparison estimation and the construction of confidence intervals around various factor level effects or contrasts.[14] We discuss this problem first.

The second problem involves the development of a confidence coefficient with regard to the whole set of separate (and interdependent) interval estimates that are computed in examining the fine structure of a set of factor level effects. In this section we emphasize the *Bonferroni approach,* one of several procedures for developing this overall confidence percentage.

The concluding topic of this section describes Hays' omega squared measure. Omega squared is a descriptive measure of variance accounted-for in the context of analysis of variance. It plays the same role in ANOVA as R^2 plays in multiple regression. Increasingly, researchers who use experimental designs in their work are becoming interested in measuring the strength of relationships as well as their statistical significance. Omega square is one of a variety of measures that are useful for this purpose.

5.5.1 Multiple Comparisons

The topic of multiple comparison tests can be most easily described by example. Consider the 2×3 factorial design in Table 5.3. The data in this table originally appeared in Table 3.11 and the accompanying ANOVA summary appeared in Table 3.12 of Chapter 3. As recalled, both main effects were significant at the 0.05 alpha level while the interaction term was significant only at the 0.1 level.

Let us adopt 0.05 as a relatively stringent alpha risk and thus conclude that the interaction effect is *not* significant. This simplification is made in order to justify more detailed examination of levels within main effects. Moreover, insofar as this example is concerned, the line segments in the interaction plot do not depart markedly from parallelism (see Figure 3.5 in Chapter 3), suggesting that the interaction is not too serious.

Table 5.3 Illustrative Response Data from Factorial Design

Factor Level	B_1	Original Data B_2	B_3	Total	Means
A_1	70; 75; 79	85; 88; 93	77; 81; 78	726	80.67
A_2	91; 90; 87	94; 97; 93	87; 90; 90	819	91.00
Total	492	550	503	1,545	
Means	82.0	91.67	83.83	85.83	

Source	d.f.	ANOVA Sums of Squares	Mean Squares	F
A	1	480.50	480.50	54.73
B	2	316.33	158.17	18.01
$A \times B$ Interaction	2	56.33	28.17	3.21
Residual	12	105.33	8.78	
	17	958.49		

However, if the interaction term in a two-factor design is highly significant, one would ordinarily not investigate the main effects in detail, particularly if the researcher is not able to find a monotonic function of the original responses that can "remove" the interaction. Here, we assume that the interaction is not important and detailed examination of main effects is justified. For expository purposes, we shall discuss multiple comparisons involving only the three-level factor, factor B.

We examine the three sample means corresponding to levels 1, 2, and 3 of factor B:

$$\overline{Y}(B_1) = 82.0; \quad \overline{Y}(B_2) = 91.67; \quad \overline{Y}(B_3) = 83.83$$

While the F ratio ($F = 18.01$) of Table 5.3 has already established that the universe means are not equal, we may still be interested in certain details. For example, may we reasonably assume that $\mu(B_1) = \mu(B_3)$ but neither of these universe means equals $\mu(B_2)$?

Once the researcher has established that the null hypothesis for some factor is to be rejected—say, that for factor B in Table 5.3—he may wish to carry out one or more of the following tasks:

1. Construct confidence intervals for the individual factor level means: $\mu(B_1)$; $\mu(B_2)$; $\mu(B_3)$.

2. Construct a confidence interval for the difference between two factor level means, such as $\mu(B_2) - \mu(B_1)$.

3. Construct a confidence interval for some other contrast,[15] such as $\mu(B_2) - [\mu(B_1) + \mu(B_3)/2]$.

The same general procedure is used for all three of these problems. Essentially what we shall be doing is constructing a confidence interval related to either a single mean or to a difference between two means. The notion of *contrast* merely generalizes the idea to cover a weighted difference with each term of the difference involving a linear composite of separate means.

As in the construction of any confidence interval about a mean, we need a point estimator of that mean and some standard error measure. As was discussed in Chapter 3, an unbiased point estimator of some universe mean μ_j is $\overline{Y}_j$, the sample mean. Moreover, an unbiased estimate of the variance of a sampling distribution of some $\overline{Y}_j$ of interest is provided by the residual mean square obtained from ANOVA divided by the number of observations on which the mean is based:

$$s^2(\overline{Y}_j) = \frac{\text{MSE}}{m_j}$$

where m_j denotes the number of observations used in computing $\overline{Y}_j$.

In the sample problem of Table 5.3 MSE $= 8.78$. This being the case, let us examine three cases of interest:

1. Constructing a 95 percent confidence interval for $\mu(B_1)$.

2. Constructing a 95 percent confidence interval for the difference between $\mu(B_2)$ and $\mu(B_1)$.

3. Constructing a 95 percent confidence interval for the contrast of $\mu(B_2)$ versus the mean of $\mu(B_1)$ and $\mu(B_3)$.

Each set of computations is handled similarly.

To illustrate, suppose we wished to find a 95 percent confidence interval around $\overline{Y}(B_1) = 82.0$. We note from Table 5.3 that the sample size $m(B_1) = 6$; and MSE $= 8.78$. An unbiased estimate of the variance about $\overline{Y}(B_1)$ is then computed as follows:

$$s^2[\overline{Y}(B_1)] = \frac{\text{MSE}}{m(B_1)} = \frac{8.78}{6} = 1.463$$

$$s = 1.21$$

The associated confidence interval based on Student's t distribution is computed as:[16]

$$\overline{Y}(B_1) - t\{1 - \alpha/2; \text{d.f. (resid.)}\}s[(\overline{Y}(B_1)] \leq$$

$$\mu(B_1) \leq$$

$$\overline{Y}(B_1) + t\{1 - \alpha/2; \text{d.f. (resid.)}\}s[\overline{Y}(B_1)]$$

Since $t\{1 - \alpha/2; \text{d.f. (residual)}\} = t\{0.975; 12\} = 2.179$, we have:

$$82.0 - 2.179(1.21) \leqq \mu(B_1) \leqq 82.0 + 2.179(1.21)$$

$$\boxed{79.36 \leqq \mu(B_1) \leqq 84.64}$$

as the confidence interval of interest.

We note that constructing the confidence interval is a straightforward application from univariate estimation, as reviewed in Appendix B. In particular, we use a level of t that involves the sum of the areas in both tails. For this reason $\alpha = 0.05$ is divided by 2, in a manner analogous to applying a two-tailed significance test.

Estimating a confidence interval around the difference between $\overline{Y}(B_2)$ and $\overline{Y}(B_1)$ proceeds analogously. First the unbiased point estimate of their difference is:

$$D = \overline{Y}(B_2) - \overline{Y}(B_1) = 91.67 - 82.0$$
$$= 9.67$$

Since $\overline{Y}(B_2)$ and $\overline{Y}(B_1)$ are independent estimators, the variance of D, their difference, is estimated by:

$$s^2(D) = \frac{\text{MSE}}{m(B_2)} + \frac{\text{MSE}}{m(B_1)}$$

$$= 8.78 \left[\frac{1}{6} + \frac{1}{6} \right]$$

$$= 2.926; \; s(D) = 1.711$$

The associated confidence interval is then computed as:

$$D - t\{1 - \alpha/2; \text{d.f. (resid.)}\}s(D) \leqq$$
$$[\mu(B_2) - \mu(B_1)] \leqq$$
$$D + t\{1 - \alpha/2; \text{d.f. (resid.)}\}s(D)$$

In this case, we have:

$$9.67 - 2.179(1.711) \leqq [\mu(B_2) - \mu(B_1)] \leqq 9.67 + 2.179(1.711)$$

or:

$$\boxed{5.94 \leqq [\mu(B_2) - \mu(B_1)] \leqq 17.13}$$

The third problem, involving the contrast $\mu(B_2) - [\mu(B_1) + \mu(B_3)]/2$ entails a somewhat different approach (Neter and Wasserman, 1974). First, we can denote the estimator of a contrast by the symbol L, defined as:

$$L = \sum_{j=1}^{J} c_j \overline{Y}_j$$

where we impose the restriction that $\sum_{j=1}^{J} c_j = 0$. Given that the $\overline{Y}_j$ are assumed to be independent, the variance of L is estimated by:

$$s^2(L) = \text{MSE} \sum_{j=1}^{J} c_j^2/m_j$$

In terms of the sample problem, we first find L, the value of the linear contrast, as:

$$L = \overline{Y}(B_2) - [\overline{Y}(B_1) + \overline{Y}(B_3)]/2$$
$$= 1(91.67) - 0.5(82.0) - 0.5(83.83)$$
$$= 8.755$$

with estimated variance:

$$s^2(L) = \text{MSE} \sum_{j=1}^{J} c_j^2/m_j$$
$$= 8.78\left[\frac{1}{6} + \frac{0.25}{6} + \frac{0.25}{6}\right]$$
$$= 8.78(0.25) = 2.195$$
$$s(L) = 1.482$$

The 95 percent confidence interval associated with this contrast is then found from:

$$L - t\{1 - \alpha/2; \text{d.f. (resid.)}\}s(L) \leq$$
$$\{\mu(B_2) - [\mu(B_1) + \mu(B_3)]/2\} \leq$$
$$L + t\{1 - \alpha/2; \text{d.f. (resid.)}\}s(L)$$

In this case we have:

$$8.755 - 2.179(1.482) \leq \{\mu(B_2) - [\mu(B_1) + \mu(B_3)]/2\}$$
$$\leq 8.755 + 2.179(1.482)$$

or:

$$5.53 \leqq \{\mu(B_2) - [\mu(B_1) + \mu(B_3)]/2\} \leqq 11.98$$

Furthermore, as might be surmised, the comparison involving the difference between *two* means is a special case of the linear contrast in which we have:

$$D = L = 1(91.67) - 1(82.0)$$

and:

$$s^2(D) = s^2(L) = \text{MSE}\left[\frac{(1)^2}{6} + \frac{(-1)^2}{6}\right]$$

as illustrated earlier.

We conclude from the above interval estimations that $\mu(B_2)$ exceeds $\mu(B_1)$ and also that $\mu(B_2)$ exceeds the average of $\mu(B_1)$ and $\mu(B_3)$. The reader can further verify that there is sufficient evidence to conclude that $\mu(B_2)$ exceeds $\mu(B_3)$ but $\mu(B_3)$ cannot be confidently assumed to exceed $\mu(B_1)$. In these latter two cases, the 95 percent confidence intervals are, respectively:

$$4.11 \leqq [\mu(B_2) - \mu(B_3)] \leqq 11.57$$

$$-1.90 \leqq [\mu(B_3) - \mu(B_1)] \leqq 5.56$$

Notice that the first of these confidence intervals does not include 0 while the second interval does. Thus, while the confidence interval approach is primarily used for interval estimation, it can also be employed to test hypotheses of the type illustrated here.

The concept of linear contrasts is much more extensive than this introduction might suggest.[17] For our purposes, however, let us summarize results for the three pairs of factor level differences:

$$5.94 \leqq [\mu(B_2) - \mu(B_1)] \leqq 17.13$$

$$4.11 \leqq [\mu(B_2) - \mu(B_3)] \leqq 11.57$$

$$-1.90 \leqq [\mu(B_3) - \mu(B_1)] \leqq 5.56$$

Each of these intervals has been *separately* constructed on the basis of a confidence coefficient of 95 percent. Since the tests are not independent, we seek some procedure that will enable us to attach a level of confidence to the full set of (three) statements.

5.5.2 The Bonferroni Method

A number of approaches are available for developing a global confidence co-efficient for a set of interdependent multiple comparisons.[18] However, only one of these methods—a technique due to Bonferroni (Neter and Wasserman, 1974)—is described here. The Bonferroni method is both simple to describe and easy to use.

To see how the Bonferroni method is derived, assume that we have found two separate confidence intervals. We shall let E_1 denote the event that the first confidence interval does *not* cover the parameter value of interest and E_2 the event that the second confidence interval does *not* cover its parameter value of interest. In terms of the preceding problem, these parameters could be:

$$D_1 = \mu(B_2) - \mu(B_1)$$

$$D_2 = \mu(B_2) - \mu(B_3)$$

Given that the confidence coefficient $1 - \alpha$ is being used in each case, α denotes the probability that E_1 is true and α also denotes the probability that E_2 is true:

$$\boxed{P(E_1) = P(E_2) = \alpha}$$

(In the sample problem α was set at 0.05.) From elementary probability theory,[19] we have:

$$P(E_1 \cup E_2) = P(E_1) + P(E_2) - P(E_1 \cap E_2)$$

Next, subtracting both sides from 1, we have:

$$1 - P(E_1 \cup E_2) = 1 - P(E_1) - P(E_2) + P(E_1 \cap E_2)$$

However, this equation, in turn, can be written as:

$$1 - P(E_1 \cup E_2) = P(\overline{E_1 \cup E_2}) = P(\overline{E}_1 \cap \overline{E}_2)$$

where the expression $P(\overline{E}_1 \cap \overline{E}_2)$ is the probability that *both* confidence intervals cover the parameter of interest. We can then write:

$$P(\overline{E}_1 \cap \overline{E}_2) = 1 - P(E_1) - P(E_2) + P(E_1 \cap E_2)$$

and, since our choice of α will be such that $P(E_1 \cap E_2) \geqq 0$, we find the Bonferroni inequality to be:

$$\boxed{P(\overline{E}_1 \cap \overline{E}_2) \geqq 1 - P(E_1) - P(E_2)}$$

which, in terms of the sample problem, can be expressed as:

$$P(\overline{E}_1 \cap \overline{E}_2) \geqq 1 - \alpha - \alpha = 1 - 2\alpha$$

Hence, if each separate statement carries a 95 percent confidence coefficient, then the global confidence is 90 percent that *both* intervals cover their respective parameter values.[20]

The Bonferroni inequality represents a lower bound on the global confidence coefficient and, in this sense, is a conservative approach.[21] Moreover, it can be readily extended to more than two confidence interval statements (each with confidence coefficient $1 - \alpha$) as follows:

$$P\left[\bigcap_{p=1}^{r} \overline{E}_p \right] \leqq 1 - r\alpha$$

where one has r separate interval estimates, each with confidence level[22] $1 - \alpha$.

In terms of the sample problem in the preceding section, we can illustratively consider three intervals, each based on a $1 - \alpha = 0.95$ confidence coefficient:

$$D_1: \quad 5.94 \leqq [\mu(B_2) - \mu(B_1)] \leqq 17.13$$

$$D_2: \quad 4.11 \leqq [\mu(B_2) - \mu(B_3)] \leqq 11.57$$

$$D_3: \quad -1.90 \leqq [\mu(B_3) - \mu(B_1)] \leqq 5.56$$

The lower bound for the global confidence coefficient is then:

$$P(\overline{E}_1 \cap \overline{E}_2 \cap \overline{E}_3) \geqq 1 - 3(0.05) = 0.85$$

Hence, we can assert with a confidence level of at least 0.85 that all three intervals cover their respective parameter values.

The Bonferroni method can be applied to cases in which the selected confidence coefficient differs from interval to interval. For example, if we have three intervals with respective coefficients of:

$$1 - \alpha_1 = 0.99; \ 1 - \alpha_2 = 0.95; \ 1 - \alpha_3 = 0.90$$

then the Bonferroni approach yields a global coefficient of at least:

$$1 - 0.01 - 0.05 - 0.1 = 0.84$$

Naturally, as the number of different intervals increase, the global coefficient becomes lower and lower. Hence, to maintain a specified overall confidence, the various individual intervals may become too wide for practical value. Therefore, the Bonferroni approach is best suited for cases in which a *relatively small number of confidence intervals are of interest*.

Some other aspects of the Bonferroni method are useful to point out:

1. Unlike some alternative procedures for making multiple comparisons, the Bonferroni method can be used for either equal or unequal sample sizes within factor level(s).

2. The Bonferroni method is particularly well suited when the number of confidence intervals being computed is less than or equal to the number of factor levels used in the design.

3. The Bonferroni method requires the researcher to state in advance of seeing the experimental results the confidence intervals that are of interest to him.[23]

In summary, the Bonferroni method provides a simple and direct way to develop a *family-type confidence coefficient* when the number of different intervals being computed is relatively small and specifiable in advance of seeing the data. As a rule of thumb, the researcher should compute no more separate confidence intervals than the number of levels of the factor under study.

The Bonferroni procedure can be used for either single factor or multiple factor designs, with sample sizes that are equal or unequal across factor levels. In the case of multiple factor designs, such as the two-factor example of Table 5.3, the Bonferroni approach can be used to develop a global confidence coefficient *across* intervals originally computed within either the A factor or the B factor. No distinction need be made concerning the source of these intervals.

In two factor and higher-way designs, it should be reiterated that one should first check for significant interaction effects prior to the examination of main effects. If important interaction effects are observed one would not ordinarily proceed to obtain main-effect multiple comparisons for those factors that are involved in the interactions.

Rather, what would be done is to examine the *individual* treatment means themselves. For example, in the sample problem of Table 5.3, a total of $2 \times 3 = 6$ treatment means exist, if we consider the design as a single factor experiment. If the interaction term were highly significant (and not removable by some suitable monotonic transformation), the Bonferroni method could still be used. However, in this case the comparisons would involve all or some subset of the *six individual treatment means.*

In summary, the Bonferroni method is highly flexible for handling multiple comparisons involving single factor levels, multiple factor levels, or interactions. As long as the number of comparisons under study are relatively small the Bonferroni approach can lead to an overall confidence coefficient regarding a set of intervals that individually are still narrow enough to be useful. Moreover, the Bonferroni coefficient is both conceptually simple and easy to work with computationally.

5.5.3 Hays' Omega Squared Measure

In Chapter 2's discussion of multiple regression, R^2 played an important role as a summary measure of the adequacy of fit of some overall model to the data.

Furthermore, other measures, such as beta weights and partial correlation co-efficients were described as aids to assessing the separate contributions of various predictor variables to variation in the criterion. In contrast, Chapter 3's discussion of ANOVA and ANCOVA emphasized significance testing rather than descriptive measures of goodness to fit.

Sections 5.5.1 and 5.5.2 of this chapter, however, have focused on estimation problems in analysis of variance. In this same spirit we now discuss the use of the omega squared coefficient as a goodness of fit measure. Omega squared's role in ANOVA is analogous to R^2 in the context of multiple regression.

By way of introduction, the reader may recall discussion of the correlation ratio η^2, defined in section 5.2.1 as:

$$\eta^2 = \frac{\text{SSA}}{\text{SST}} = 1 - \frac{\text{SSW}}{\text{SST}}$$

where SSA, SSW, and SST are, respectively, the among-groups, within-groups, and total-group sums of squares. The correlation ratio can be defined as the proportionate reduction in squared error when the criterion is estimated by the mean of the group within which the observation falls rather than by the total-sample mean.

If several ways of classifying the observations are available, as would be the case in a factorial design, the correlation ratio is easily modified. Table 5.4 shows eta squares for each effect—A, B, and the $A \times B$ interaction—in the sample problem of Table 5.3, as well as the total correlation ratio of 0.89. That is, 89 percent of the variation in the response variable is accounted for by the various group designations.

As easy as the correlation ratio is to compute, it provides no simple connection to *universe* values. Accordingly, early attempts (Kelley, 1935) were made to improve on the sample-based η^2 by making it "unbiased." Later, Hays' omega squared (Hays and Winkler, 1971) was designed to provide unbiased estimators of the terms in the expression:

$$\omega^2 = \frac{\sigma_y^2 - \sigma_{y|x}^2}{\sigma_y^2}$$

where σ_y^2 is defined as the marginal distribution of Y, without taking levels of X into consideration and $\sigma_{y|x}^2$ is the conditional distribution of Y, given knowledge of the group in X that the observation belongs to.

Hays developed unbiased estimators of both σ_y^2 and $\sigma_{y|x}^2$, leading to an estimator of ω^2 that was denoted $\hat{\omega}^2$.[24] In the single factor case, Hays' estimated omega squared is defined as:

Table 5.4 Eta Squared and Omega Squared Measures for the Sample Problem of Table 5.3

Source	Sums of Squares	Eta Squared		d.f.	Omega Squared	
A	480.50	0.50		1	0.49	
B	316.33	0.33	0.89	2	0.31	0.88
A × B Interaction	56.33	0.06		2	0.08	
Residual	105.33			12		
	958.49			17		

$$\hat{\omega}^2 = \frac{SSA - [(J - 1) \, MSW]}{SST + MSW}$$

where MSW denotes the within-groups mean square and J denotes the number of levels over which the single factor is classified.

Hays' measure is readily extended to multiple factor designs. For example, in the case of the B effect in the sample problem of Table 5.4, we have:

$$\hat{\omega}^2 = \frac{SS(B) - [\text{d.f.} \, (B) \cdot MSE]}{SST + MSE}$$

where d.f.$(B) = 2$ are the degrees of freedom associated with the B main effect and $MSE = 105.33/12 = 8.78$ is the mean square for the residual (or error) term. Hence:

$$\hat{\omega}^2(B) = \frac{316.33 - [2(8.78])}{958.49 + 8.78}$$

$$= 0.31$$

Estimated values of omega squared are computed for each effect, A, B, and $A \times B$, in Table 5.4. Collectively they account for 0.88 of the variation in the response variable.[25]

A comparison of effects in the table indicates that factor A appears most important, followed by B and then by their interaction. However, in factorial designs care should always be taken in discussing relative importance. A *highly* significant interaction (and one not removable by an appropriate monotonic transformation) makes the measurement of main effect importance irrelevant. Rather, it is the interaction itself that is the principal finding.

Omega squared is quite easy to compute and all of the values needed for its calculation are obtained in the course of conducting an ANOVA. Since ANOVA is often used in orthogonal designs (where the factors are not correlated), the separate $\hat{\omega}^2$ contributions are not ambiguous in the way separate contributions to R^2 can be in the counterpart case of multiple regression involving correlated predictors.

This is not to say that $\hat{\omega}^2$ is without limitations. As Glass and Hakstian (1969) argue, the whole question of the relative importance of factors depends on the adequacy with which the researcher has sampled the domains of interest in setting up his factor levels. The problem of domain sampling becomes particularly acute when the factors are factorially complex and differ in kind as well as in degree. While these caveats are not to be taken lightly, $\hat{\omega}^2$ can still be useful in assessing the operational significance of factors as an adjunct to the statistical significance tests that are part of the ANOVA routine.[26]

In summary, the three subtopics of section 5.5—multiple comparisons, the Bonferroni method for computing global confidence coefficients, and omega squared—all focus on the estimation of experimental effects, as opposed to the usual emphasis in ANOVA on significance testing. While the testing of various models is central to experimentation, the *interpretation* of experimental results can be considerably helped by the estimation procedures described in section 5.5. As such, their utilization is complementary to the more traditional emphasis on significance testing in analysis of variance.

5.6 Multicollinearity and Predictor Variable Importance

In section 2.5 of Chapter 2 we described the formula for computing the standard error of a partial regression coefficient b_j:

$$\text{SE}(b_j) = \frac{s_y}{s_j} \sqrt{\frac{r^{jj}(1 - R^2)}{m - n - 1}}$$

where, as recalled, r^{jj} is the diagonal element of the j-th predictor in $\mathbf{R}^{-1}$, the inverse of the matrix of predictor-variable correlations.[27] As also recalled,

$$r^{jj} = \frac{1}{[1 - R^2_{j(n-1)}]}$$

where $R^2_{j(n-1)}$ is the squared multiple correlation between X_j and the remaining $n - 1$ variables. As X_j becomes increasingly correlated with the other predictors, the standard error of b_j gets larger and larger.

Large standard errors make the partial regression coefficients unstable from sample to sample. The general problem of highly correlated predictors is called *multicollinearity*. This problem can be present in any kind of multivariate problem—regression, two-group discrimination, AID—where the predictors are not designed in advance to be orthogonal.[28]

5.6.1 The Problem of Multicollinearity

Any situation in which data are collected on a *naturalistic* basis without experimenter intervention or control, has the potential for exhibiting multicollinearity among two or more predictor variables. If "high" multicollinearity is present, various difficulties arise. As Johnston (1972) points out, any of the following problems may occur:

1. Reduced precision in estimating the coefficients of predictive functions and increased difficulty in disentangling the separate effects of each predictor variable on the criterion variable, may take place.

2. Predictor variables may be dropped incorrectly (perhaps mechanically so in stepwise regression procedures) because of high standard errors.

3. Estimation of regression (or discriminant) coefficients may become highly sensitive to the specific sample; addition or deletion of a few observations may produce marked differences in the values of the regression coefficients, including changes in algebraic sign.

Unless one is dealing with experimental design data, it is almost always the case that predictor variables will be correlated to some degree. The question is: how much multicollinearity can be tolerated without seriously affecting the results? Unfortunately there is no simple answer to this question.

The study of multicollinearity in data analysis evolves around two major problems: (a) how can it be detected, and (b) what can be done about it?

What constitutes "serious" multicollinearity is ambiguous. Some researchers have adopted various rules of thumb, e.g., any pair of predictor variables correlating above a certain level, such as 0.9, are "collinear" and one of them should be omitted. While looking at simple correlations between pairs of predictors has merit, it can miss more subtle relationships involving three or more predictors.

Rules of thumb can also be applied to *multiple* correlations between each predictor and all other predictors. One rule states that no predictor should have a multiple correlation with the remaining predictors that exceeds that of the criterion variable with the full set of predictors.

Another test for multicollinearity provided in many regression programs involves examining the determinant of the correlation matrix of predictors. As the determinant approaches zero, extreme multicollinearity is the case; as it approaches unity, the predictor variables are exhibiting mutual uncorrelatedness. Unfortunately it is not clear what a "reasonable" value of the determinant should be in concluding that multicollinearity is not serious.

5.6.2 Procedures for Coping with Multicollinearity

Essentially there are three basic procedures for dealing with multicollinearity: (a) ignore it; (b) delete one or more of the "offending" predictors; or (c) transform the set of predictor variables into a new set of predictor-variable combinations that are mutually uncorrelated.

Ignoring multicollinearity need not be as cavalier as it might sound. First, one can have multicollinearity in the predictor variables and still have strong enough effects that the estimating coefficients remain reasonably stable. Moreover, multicollinearity may be prominent in only a subset of the predictors, a subset that may not contribute much to accounted-for variance anyway. A prudent procedure in checking one's predictor set for multicollinearity is to examine the standard errors of the regression coefficients (which will tend to be large in the case of high multicollinearity). Second, one may randomly drop some subset of the cases (perhaps 20 percent or so), rerun the regression, and then check to see if the signs and relative sizes of the regression coefficients are stable. Third, a number of recently developed regression routines incorporate checks for serious multicollinearity; if the program does not indicate this condition, the researcher can generally assume that the problem is not acute.

If multicollinearity is "severe" one rather simple procedure is to drop one or more predictor variables that represent the major offenders. Usually, because of their high intercorrelations with the retained predictors, the overall fit will not markedly change. (Pragmatically, if a particular pair of predictors are highly collinear, one would retain that member of the pair whose measurement reliability and/or theoretical importance is higher in the substantive problem under study.)

Methods also exist, e.g., principal components analysis, for transforming the original set of predictors to a mutually uncorrelated set of linear combinations. If these components (linear composites) are interpretable in themselves, the researcher may use *these* in his regression or discriminant analysis rather than the original variables. If *all* components are retained, the predictive accuracy will be precisely the same as that obtained from the original set of predictors.[29]

5.6.3 The Relative Importance of Predictor Variables

In the case of correlated predictors—whether or not multicollinearity is a problem—there is always some *ambiguity* about the relative importance of predictors. This is true whatever the measure used—squared betas, squared partial correlations, squared part correlations, or squared simple correlations of each predictor with the criterion.

In Chapter 2 we had occasion to utilize two multiple regression programs, the BMD-03R program and the BMD-02R stepwise regression program. Each of these programs develops its own "importance" measure—at least measures that are often used by researchers to ascribe relative importance to each of a set of predictors. It is of interest to note that in the case of correlated predictors these measures will, in general, *not* agree.

The BMD-03R program utilizes the increase in accounted-for variance associated with: (a) a model that includes all *preceding* predictors in the *prespecified order stated by the researcher,* except the j-th, versus (b) a model that includes the j-th as well.

This relationship can be expressed generally for n predictors as:

$$R^2 = r_{y1}^2 + r_{y2(1)}^2 + r_{y3(1,2)}^2 + \cdots + r_{yn(1,2,\ldots,n-1)}^2$$

where, for example, the second term of the equation is equal to the squared correlation of Y with the residuals found after first regressing X_2 on X_1 (i.e., the squared coefficient of part correlation).[30] Similar remarks pertain to the remaining terms.

It follows, then, that the proportion of accounted-for variance measure in BMD-03R will depend on the *order* with which the analyst sets up his predictor variables. That is, although total R^2 will remain the same for any given set of predictors, the individual contributions to accounted-for variance will differ in BMD-03R, as the order in which the predictors appear differs. A prudent procedure is to select as the first subset of predictors in the analysis those variables of most theoretical interest to the researcher.

As recalled from Chapter 2, BMD-02R operates on the basis of adding predictor variables sequentially. However, the sequence is *not* based on the (possibly artibrary) order in which the researcher arranges his list of predictors. Rather, in BMD-02R that predictor with the highest simple (squared) correlation with the criterion variable is added first. New predictors are added on the basis of the relative size of their partial F ratios (or, equivalently, on the basis of their squared partial correlations) with the criterion variable. At each step the program computes the proportion of accounted-for variance for all predictors in the equation. For any given equation these contributions—like the BMD-03R case—will sum to the total R^2 associated with that equation. Also like BMD-03R, the contribution of the last predictor is constrained only by the variables already in the equation. The procedures differ in the sense that the contribution in BMD-03R depends on a *prespecified* order, while that in BMD-02R depends on what the j-th predictor adds to variance already accounted for by other predictors previously entered.

In the case of correlated predictors, there is no unambiguous measure of relative importance. The measure that is used will reflect the way the researcher interprets the question. As noted by Gorsuch (1973), three major strategies are open to the researcher.

1. Overlapping variance can be attributed to one of the variables, as the variables are entered sequentially. This is what happens in BMD-03R and other procedures in which variables are introduced according to an order specified in advance by the researcher. It is also true of BMD-02R, once the ordering of variables has been determined by the program.

2. The overlapping variance can be distributed among the predictors; this is the case in importance measures like the squared beta, squared partial correlation, or squared simple correlation of each predictor, in turn, with the criterion.

3. The overlapping variance can be absorbed into new, derived variables that are uncorrelated with the original variables.

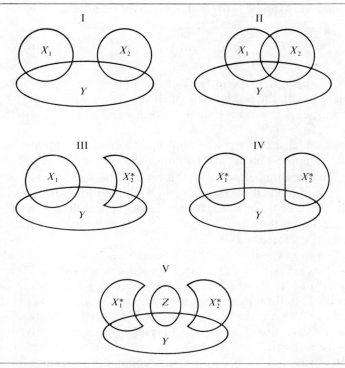

Figure 5.6 Procedures for Dealing with Overlapping Variance

Figure 5.6 shows the various types of situations examined by Gorsuch. In Panel I, no problem arises; although X_1 and X_2 both contribute to variance in Y, their contributions are separate. This is the type of situation that arises in orthogonal designs and, as pointed out earlier, leads to unambiguous allocation of criterion-variable variance across the predictors.[31]

Panel II, however, is the more usual situation encountered in multiple regression and other instances in which the predictors exhibit multicollinearity. Here we note that X_1 and X_2 each contribute to accounted-for variance in Y but they also share some of this accounted-for variance.

Panel III illustrates the case in which X_1 is credited with not only its unique variance but also all that it shares with X_2. Panel IV splits—often in a rather arbitrary way—the shared variance between X_1 and X_2. Panel V shows the case (illustrated by Gorsuch) in which a new composite variable Z is created to represent shared variance in X_1 and X_2.[32]

No firm recommendations can be made regarding which strategy to adopt in selecting an importance measure in the correlated predictors situation; obviously, the strategy will depend on the researcher's objectives. A prudent procedure is to examine more than one such measure to see if the relative importances in a given problem are sensitive to the measure being used.

5.7 Summary

This chapter has been concerned with several topics related to single criterion, multiple predictor association—Automatic Interaction Detection, variable transformation, multiple comparisons, omega squared, multicollinearity, and the relative importance of predictor variables.

The AID procedure was described first and illustrated with data drawn from the television commercial study. In carrying out this application the topic of cross-validation was also described briefly and illustrated with the same data bank.

We then turned to a discussion of prespecified functions that can be used to transform data prior to the application of linear models. A variety of such functions were illustrated and the analysis of residuals was also briefly described as a way to suggest possible data transformations. Monotonic ANOVA was then discussed and illustrated numerically with data obtained from the television commercial study.

The next major topic dealt with three interrelated aspects of ANOVA and ANCOVA—multiple comparisons, overall confidence coefficients, and Hays' omega squared measure. These topics collectively emphasize the parameter estimation phase of analysis of variance as a counterpart to significance testing in applied research.

The last section of the chapter discussed a pervasive problem in multivariate analysis—multicollinearity and determining the relative importance of predictor variables in accounting for variance in the criterion variable. The conclusion is that in the case of correlated predictors, there is no unequivocal answer to the "relative importance" question.

In short, this chapter has been designed to augment the more basic material of Chapters 2 through 4 by providing a variety of selective topics on the general subject of single variable, multiple predictor association.

Review Questions

1. Considering the television commercial data in Appendix A, choose variable C-2 (post exposure interest in the Alpha radial) as a criterion variable and A-1, A-2, A-3, and C-2 as predictors.
   a. Run an AID analysis of these data using a minimum group size of 40 and other stopping criteria set at the program's standard values. Impose monotonicity on predictors A-3 and C-2. Is there evidence of interaction in the results?
   b. Process the same data by means of stepwise regression (e.g., BMD-02R) and contrast the results.
2. Consider the following data obtained from a psychological experiment on risk taking:

Y	X_1	X_2	Y	X_1	X_2
2	4	4	32	17	23
7	9	6	28	32	50
3	13	3	36	41	48
9	7	2	30	37	47
14	19	9	31	45	59
18	26	16	43	58	47
22	31	19	47	46	43
17	22	21	48	47	46
24	28	42	50	53	41

a. Make up scatter plots of Y on X_1, Y on X_2 and X_1 on X_2. Is any particular functional form suggested by these plots? If so, what is it?

b. Regress Y on X_1 and X_2. Plot the residuals $Y_i - \hat{Y}_i$ versus X_1 and X_2 separately. Are there any clues to some other functional form?

c. Regress Y on X_1, X_2, and X_2^2. What does X_2^2 contribute to accounted-for variance beyond the contributions of X_1 and X_2?

d. Take the common logarithms of Y, X_1, and X_2 and perform a multiple regression. Compare your results with those of part b.

e. Regress Y on X_1, X_2, and X_1X_2. What does the cross product X_1X_2 contribute to accounted-for variance beyond the contributions of X_1 and X_2?

3. Referring to the same data in problem 2, define the dummy variable:

$$X_{i3} = \begin{matrix} 1 \\ 0 \end{matrix} \begin{matrix} \text{if } X_{i2} > 25 \\ \text{if } X_{i2} \leq 25 \end{matrix}$$

a. Regress Y on X_1, X_2, and X_3 via a piecewise linear regression procedure. Compare your results with part b of problem 2.

b. Define the dummy variable:

$$X_{i3} = \begin{matrix} 1 \\ 0 \end{matrix} \begin{matrix} \text{if } X_{i2} > 25 \\ \text{if } X_{i2} \leq 25 \end{matrix}$$

$$X_{i4} = \begin{matrix} 1 \\ 0 \end{matrix} \begin{matrix} \text{if } X_{i2} > 45 \\ \text{if } X_{i2} \leq 45 \end{matrix}$$

and regress Y on X_1, X_2, X_3, and X_4. Compare your results with part a above and part b of problem 2.

c. What other fitting procedures could be used in this problem?

4. Consider the following data, obtained from a 5 × 2 factorial design with two replications per cell:

				Factor B		
		1	2	3	4	5
Factor	1	12; 14	19; 21	17; 22	24; 29	19; 20
A	2	8; 7	11; 13	9; 12	17; 21	12; 14

a. Analyze these data via a two-factor ANOVA, using an alpha risk level of 0.05.

b. Compute the following confidence intervals:

$$\mu(B_3) - \frac{\mu(B_1) + \mu(B_2) + \mu(B_4) + \mu(B_5)}{4};$$

$$\mu(B_2) - \mu(B_1); \ \mu(B_2) - \mu(B_5); \ \mu(A_1) - \mu(A_2)$$

using, in each case, a confidence coefficient of $1 - \alpha = 0.95$.

c. What is the Bonferroni confidence for all four intervals?

d. Compute Hays' omega squared measure for A, B, and $A \times B$. Compare these results with eta squared for each effect.

e. Plot the sample means of the levels of B for levels 1 and 2 of A. How would you describe the interaction?

f. Transform the response variable to common logs and repeat the ANOVA. What is the effect of this transformation on the significance tests and omega squared?

5. Return to the data of problem 2 and compute the "importance" measures: (a) squared beta; (b) squared partial correlation; and (c) squared simple correlation, for X_1 and X_2 separately.

a. Which predictor is the more "important" under each measure?

b. Repeat the process for the regression (Part e of problem 2) of Y on X_1, X_2, and X_1X_2. How do the three predictors rank in importance under each of the importance measures?

c. In what order do the three predictors X_1, X_2, and X_1X_2 enter when stepwise regression is carried out?

Notes

1. The reader will recognize this as analogous to the correlation between treatment levels and a covariate in the context of ANCOVA.

2. While cross products involving three or more variables can be computed, the number of such combinations can become enormous in problems of realistic size.

3. Eta squared is also referred to as the correlation ratio by some writers (Freeman, 1965). Like its R^2 counterpart, η^2 varies between 0 and 1.

4. Notice that a two-step process is involved. First, each allowable split is examined for each predictor separately and the particular split that leads to the largest between-groups sum of squares is recorded. Then, AID scans all of these single-variable splits and selects the largest of the largest—that predictor with the highest between-groups sum of squares across all of the candidate predictors.

5. In practice the number of classes for X_j (whether ordered or not) is usually kept under 15. As illustrated here, however, all predictors—whether originally expressed as continuous variables or not—are discretized into a set of classes before the AID algorithm is applied. This recoding step can be accomplished by the program itself, if desired.

6. In the sample run of Figure 5.2, the tree diagram is only a partial representation of the output. In view of the small sample size to begin with, we only examined splitting up to the point where all predictor variables had an opportunity to enter the analysis.

7. The reason for this high generality is that each separate group constitutes a new round for examination of the predictors; as such, each set of effects can be idiosyncratic to each group.

8. In all of the cases of sections 5.3.1 and 5.3.2 we reiterate that the functions are *already* linear in the parameters; hence, substitution of new variables for the original ones is all that is required.

9. Of course, some functions such as:

$$Y = \beta_0 + \beta_1 e^{-\beta_2 X} \quad \text{or} \quad Y = [\beta_0 + \beta_1^{-\beta_2 X}]^{-1}$$

are inherently nonlinear in the parameters. Other procedures, such as nonlinear least squares, would have to be used for parameter estimation in these cases.

10. Carroll (in Green and Wind, 1973) extends the approach in still another way—in what he calls categorical conjoint measurement—by assuming a response variable that need only be nominally scaled.

11. If MONANOVA is later followed by application of a specific functional form, as suggested by the plot of Y_i versus $z_i\ (\beta)$, we note that its use is similar in spirit to that of AID followed by multiple regression.

12. The lack of association in Table 5.2 was substantiated by a four-group discriminant analysis in which the original ratings on the 25 profiles served as predictors and the Part D response category as the criterion. The discriminant analysis did not reveal significant differences in group centroids.

13. In this regard it is of interest to note that MONANOVA provides as part of its standard output a plot of the fitted values versus the original Y_i values. The nature of this scatter plot could suggest, in many instances, the specific functional form to be used in prespecification of a criterion-variable transformation prior to using standard ANOVA techniques. (In the illustrative problem, however, the plot was well approximated by a linear function.)

14. We shall be emphasizing an *estimation* approach to multiple comparisons, as opposed to a significance testing approach. However, the two approaches offer compatible and complementary ways to examine the same problems.

15. Other approaches are available for the detailed examination of factor level effects. For example, one could decompose the treatment sum of squares for B into separate components and make a series of separate F tests. Here, however, we emphasize the construction of confidence intervals and refer the reader to more detailed discussions of individual F tests in Mendenhall (1968) and Winer (1971).

16. The use of the t distribution is based on the fact that $[\overline{Y}_j - \mu_j]/s(\overline{Y}_j)$ is distributed as Student's t with degrees of freedom based on those associated with the residual mean square.

17. In particular, the topic of orthogonal contrasts that obey the condition:

$$\sum_{j=1}^{J} \frac{c_{1j}c_{2j}}{m_j} = 0$$

(where c_{1j} and c_{2j} are the weights associated with two different contrasts L_1 and L_2) represents a central topic in more advanced texts on experimental design.

18. Tukey, Newman and Keuls, Duncan, and Scheffé have all developed pocedures for making multiple comparisons. These are described in Winer (1971).

19. In keeping with the usual notation for elementary probability, we let:

$P(E_1 \cup E_2)$ denote the probability of either E_1, E_2, or both;

$P(E_1 \cap E_2)$ denote the joint probability of E_1 and E_2; and

$P(\overline{E}_1)$ denote the probability of not E_1

20. Looked at the other way around, it should be clear that if we want a family confidence coefficient of at least $1 - \alpha$, and if two intervals are involved we should compute separate confidence coefficients of $1 - \alpha/2$ each so that:

$$P(\overline{E}_1 \cap \overline{E}_2) \geqq 1 - \alpha/2 - \alpha/2 = 1 - \alpha$$

21. In particular, if the individual interval estimates are positively correlated, the Bonferroni confidence coefficient will tend to exceed $1 - \alpha$.

22. By the same token, each separate interval coefficient of $1 - \alpha/r$ will lead to the desired global confidence level of $1 - \alpha$.

23. Some multiple comparison procedures, such as Tukey's studentized range (Winer, 1971), can handle the problem of selective interval estimation after seeing the data (by being based on *all possible* confidence statements that could be made.) As such, any statements that might be made after the researcher sees the findings are covered in advance. The Bonferroni method does not exhibit this property.

24. It should be pointed out, however, that, in general, the ratio of two unbiased estimators is not an unbiased estimate of the ratio itself.

25. Computationally, $\hat{\omega}^2$ could be negative; if so, it would be set equal to zero. However, if $\hat{\omega}^2$ is computed only for *significant* effects—which will be the case in practice—it will always be positive.

26. Omega squared can also be employed in ANCOVA applications to measure the importance of factor effects *after* adjustment of responses for the effect of one or more covariates.

27. As before, s_y and s_j are the standard deviations of Y and X_j, respectively. R^2 is the squared correlation of Y with the full set of predictors, m denotes number of cases and n denotes number of predictors.

28. One of the nice features of controlled designs is the ability to make the design variables (i.e., the factors) uncorrelated. If so, measures, such as omega squared, can be interpreted without the ambiguity that surrounds their counterpart measures (such as beta weights and partial correlation coefficients) in situations involving correlated predictors.

29. This approach is discussed in more detail in Chapters 8 and 9.

30. However, BMD-03R also prints out Student t values and partial correlation coefficients for each predictor. Hence, one can find the contribution of each variable when *it* is considered the last to be entered.

31. In this case all of the common importance measures provide the same numerical results and, furthermore, individual contributions sum to R^2.

32. Relatively little research has been carried out as yet on the case illustrated in Panel V.

Multiple Criterion
Multiple Predictor Association

The Multivariate Normal Distribution and Canonical Correlation

6.1 Introduction

Chapter 6 marks the transition from analyses involving one criterion variable to those employing multiple criterion variables. Two principal topics are discussed in this chapter: the multivariate normal distribution and canonical correlation. The multivariate normal distribution is the basic model that underlies statistical inference in problems involving multiple criterion variables. We introduce this distribution in terms of a special case, the *bivariate* normal. Since the bivariate normal is a density function with only two variates, X_1 and X_2, we can show many of its aspects geometrically. Its basic characteristics are described and, in most cases, illustrated numerically as well.

We next turn to the multivariate normal involving n (> 2) variates and discuss some of its properties. The concluding part of this section discusses and illustrates numerically some rudimentary principles involving the sampling distribution of centroids as computed from samples drawn from a multivariate normal universe.

The next main section of the chapter deals with canonical correlation. In canonical correlation we seek pairs of linear composites—one composite computed over the criterion set and one computed over the predictor set—that are maximally correlated, subject to meeting certain side conditions. In general, more than a single pair of linear composites can be formed. If so, each successive pair will exhibit maximal correlation subject to its members being uncorrelated with previously computed pairs.

Formal techniques for solving canonical correlation problems are illustrated. We then turn to a description of structure correlations and Stewart and Love's redundancy measure—devices that can assist the researcher in the interpretation of canonical correlation results. Emphasis on the interpretation of canonical correlation continues as aspects of the TV commercial study are next analyzed. The chapter concludes with brief descriptions of various extensions of canonical correlation.

6.2 The Bivariate Normal Density Function

In univariate statistics one of the most common tests of significance is whether some population mean μ is equal to some numerical value μ_h hypothesized by the researcher. As recalled, one employs either a test based on the Z (unit normal) distribution or one based on Student's t distribution.[1]

In multivariate analysis, we consider analogous types of significance tests. However, rather than dealing with a single mean, we must cope with a set of means (i.e., a centroid) by which the sample is described and, in place of a single variance, with a covariance matrix of variance and covariance terms. While basic principles remain the same, the problem becomes considerably more complex.

Accordingly, we shall first examine some of the basic characteristics of the bivariate and multivariate normal distributions. We can then consider counterpart tests to those employed in univariate significance testing.

In univariate statistics the univariate normal density is probably the most often used frequency function.[2] Its counterpart in the case of two variables is the bivariate normal density function, an illustration of which appears in Panel I of Figure 6.1. As observed from Panel I, the bivariate normal density function

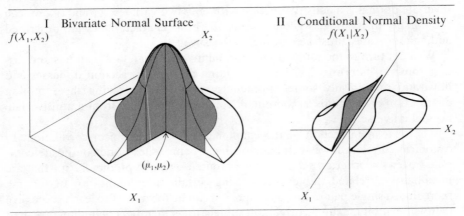

Figure 6.1 Bivariate Normal Surface and an Illustrative Conditional Density Function

displays a general bell-shaped appearance.[3] The surface is centered at the point (μ_1, μ_2), representing the centroid of the distribution. For each point on the X_1, X_2 plane, we have a point $f(X_1, X_2)$ lying on the surface of the bell-shaped mound.

Panel II of Figure 6.1 shows another bivariate normal (centered this time at the origin of the X_1, X_2 plane) that is sliced along the X_1 axis. As noted, the shaded area is also bell-shaped and represents a frequency function of X_1, conditioned upon X_2 being held at zero. If we hold X_2 at some other level we would still find that $f(X_1 | X_2)$ is normally distributed.[4] By the same token, if we hold X_1 constant at some level we would find that $f(X_2 | X_1)$ is normally distributed. In this case the slice would be parallel to the X_2 axis.

6.2.1 Mathematical Formulation

The bivariate normal density function is written as follows:

$$f_2(X_1, X_2) = \frac{1}{2\pi\sigma_1\sigma_2\sqrt{1-\rho^2}} e^{-\left[\frac{1}{2(1-\rho^2)}\left\{\frac{(X_1-\mu_1)^2}{\sigma_1^2} - \frac{2\rho(X_1-\mu_1)(X_2-\mu_2)}{\sigma_1\sigma_2} + \frac{(X_2-\mu_2)^2}{\sigma_2^2}\right\}\right]}$$

where the expression in brackets is the (negative) exponent of $e = 2.7183 \ldots$, the base of the natural logarithm system.

Alternatively, it is somewhat more convenient notationally to write the same expression as:

$$f_2(X_1, X_2) = \frac{1}{2\pi\sigma_1\sigma_2\sqrt{1-\rho^2}} \exp\left[-\frac{1}{2(1-\rho^2)}\left\{\frac{(X_1-\mu_1)^2}{\sigma_1^2}\right.\right.$$
$$\left.\left. -\frac{2\rho(X_1-\mu_1)(X_2-\mu_2)}{\sigma_1\sigma_2} + \frac{(X_2-\mu_2)^2}{\sigma_2^2}\right\}\right]$$

where $\exp(-y)$ is understood to mean $e^{(-y)}$. As we shall note subsequently, the quantity:

$$\frac{1}{2\pi\sigma_1\sigma_2\sqrt{1-\rho^2}}$$

is a positive constant and $\mu_1, \mu_2, \sigma_1, \sigma_2$, and ρ are also constants insofar as any specific case is concerned. Moreover, values of $\exp(-y)$ will always be positive for all functions $(-y)$ of interest.[5] Hence, $f_2(X_1, X_2)$ will always be positive. We note that $f_2(X_1, X_2)$ is a function of five parameters:

  μ_1, μ_2: the means of the variates X_1 and X_2, respectively

  σ_1, σ_2: the standard deviations of X_1 and X_2, respectively

  ρ: the product moment correlation between X_1 and X_2

The bell-shaped surface itself will reach a maximum height at $X_1 = \mu_1$ and $X_2 = \mu_2$, where the exponent of e is zero (since $e^0 = 1$) and will decrease in altitude as the expression in brackets gets larger on an absolute value basis.[6]

In most cases of interest, the bell-shaped surface will be stretched out in one direction, the specific direction and intensity of the stretch being dependent upon the values assumed by σ_1, σ_2 and ρ.

It is also relevant to point out that the marginal distributions for each variable taken singly are:

$$f_1(X_1) = \frac{1}{\sqrt{2\pi}\sigma_1} \exp\left[-\frac{\frac{1}{2}(X_1 - \mu_1)^2}{\sigma_1^2} \right]$$

$$f_1(X_2) = \frac{1}{\sqrt{2\pi}\sigma_2} \exp\left[-\frac{\frac{1}{2}(X_2 - \mu_2)^2}{\sigma_2^2} \right]$$

and each of these *univariate* density functions, if plotted, would be a bell-shaped curve, or univariate normal.

Next, let us concentrate on that part of the expression for $f_2(X_1, X_2)$ that is in brackets and represents the exponent of e:

$$\frac{-1}{2(1-\rho^2)}\left\{ \frac{(X_1 - \mu_1)^2}{\sigma_1^2} - \frac{2\rho(X_1 - \mu_1)(X_2 - \mu_2)}{\sigma_1\sigma_2} + \frac{(X_2 - \mu_2)^2}{\sigma_2^2} \right\}$$

What we would like to do now is find a more familiar representation for this rather complex expression.

First, let us consider the population covariance matrix for two variables X_1 and X_2, denoted by:

$$\Sigma = \begin{bmatrix} \sigma_1^2 & \rho\sigma_1\sigma_2 \\ \rho\sigma_2\sigma_1 & \sigma_2^2 \end{bmatrix}$$

where σ_1 and σ_2 are the standard deviations of X_1 and X_2, respectively, and ρ is their product moment correlation. Here, in line with common usage, we use the symbol Σ to denote a full-rank (universe) covariance matrix.[7]

If we wish to compute the inverse of a 2×2 matrix Σ, we can proceed quite simply by first reversing the main diagonal elements of Σ and then affixing a minus sign to each off diagonal element of Σ. The resulting matrix—called the adjoint of Σ—is then multiplied by the reciprocal of the determinant of Σ to obtain the inverse:

$$\Sigma^{-1} = \frac{1}{|\Sigma|}\begin{bmatrix} \sigma_2^2 & -\rho\sigma_1\sigma_2 \\ -\rho\sigma_2\sigma_1 & \sigma_1^2 \end{bmatrix}$$

And, in turn, we find the determinant of Σ to be:

$$\left|\sum\right| = \sigma_1^2 \sigma_2^2 (1 - \rho^2)$$

By simple algebra the expression for $\sum^{-1}$ can then be represented as:

$$\sum\nolimits^{-1} = \frac{1}{(1 - \rho^2)}\begin{bmatrix} 1/\sigma_1^2 & -\rho/\sigma_1\sigma_2 \\ -\rho/\sigma_2\sigma_1 & 1/\sigma_2^2 \end{bmatrix}$$

Next, by considering X_1, X_2 and μ_1, μ_2 as two-component vectors, we can set up the function:

$$F_2 = (\mathbf{x} - \boldsymbol{\mu})' \sum\nolimits^{-1} (\mathbf{x} - \boldsymbol{\mu})$$

This, in turn, can be written out as:

$$F_2 = \frac{1}{(1 - \rho^2)}\left\{ (X_1 - \mu_1, X_2 - \mu_2)\begin{bmatrix} 1/\sigma_1^2 & -\rho/\sigma_1\sigma_2 \\ -\rho/\sigma_2\sigma_1 & 1/\sigma_2^2 \end{bmatrix}\begin{bmatrix} X_1 - \mu_1 \\ X_2 - \mu_2 \end{bmatrix} \right\}$$

$$= \frac{1}{(1 - \rho^2)}\left\{ \frac{(X_1 - \mu_1)^2}{\sigma_1^2} - \frac{2\rho(X_1 - \mu_1)(X_2 - \mu_2)}{\sigma_1\sigma_2} + \frac{(X_2 - \mu_2)^2}{\sigma_2^2} \right\}$$

We see immediately that, aside from the absence of a constant of $-1/2$, F_2 is the *same* as the expression in brackets that follows the exponential in the bivariate normal density function.[8]

Next, let us examine the constant:

$$\frac{1}{2\pi\sigma_1\sigma_2\sqrt{1 - \rho^2}}$$

that represents the first factor in the bivariate density function. From the foregoing remarks in which we computed the determinant of $\sum$ as $\sigma_1^2\sigma_2^2(1 - \rho^2)$, this constant factor can now be written as the product:

$$(2\pi)^{-1}\left|\sum\right|^{-1/2}$$

Hence, we can write the *complete* bivariate normal density function in more compact notation as:

$$f_2(\mathbf{x}) = (2\pi)^{-1}\left|\sum\right|^{-1/2} \exp\left[-\frac{1}{2}(\mathbf{x} - \boldsymbol{\mu})' \sum\nolimits^{-1} (\mathbf{x} - \boldsymbol{\mu}) \right]$$

where, in this case, $\mathbf{x}$ is understood to be the two-component vector with entries X_1, X_2.

Note that this formula is analogous to the univariate normal density function:

$$f_1(X) = \frac{1}{\sqrt{2\pi}\sigma} \exp\left[-\frac{\frac{1}{2}(X - \mu)^2}{\sigma^2} \right]$$

where X and μ are scalars (rather than vectors) and the scalar σ^2 replaces the matrix Σ. Note also that the constant $(2\pi)^{-1/2}$ replaces the constant $(2\pi)^{-1}$.

If X_1 and X_2 are each standardized to zero mean and unit standard deviation, we have the counterpart formula for the bivariate normal density function:

$$f_2(\mathbf{x}_s) = f(X_{s1}, X_{s2}) = (2\pi)^{-1}|\mathbf{R}_p|^{-1/2} \exp\left[-\frac{1}{2}\mathbf{x}_s'\mathbf{R}_p^{-1}\mathbf{x}_s \right]$$

where $\mathbf{R}_p$ is the universe correlation matrix. Also, in this case $\mu_1 = \mu_2 = 0$ since X_1 and X_2 are each standardized.

6.2.2 Equal Density Contours

Returning to the function described above:

$$F_2 = (\mathbf{x} - \mu)' \sum\nolimits^{-1} (\mathbf{x} - \mu)$$

it is of interest to note that this expression, when set equal to some positive constant K_i denotes an ellipse, with center point given by μ and tilted at an angle θ with respect to the X_1 axis. By varying the size of K_i we can construct a series of concentric ellipses.

Panel I of Figure 6.2 shows the general idea. If we slice through the bivariate normal surface with a plane parallel to the X_1, X_2 plane, the value of the function $f_2(X_1, X_2)$ is constant and can be set equal to K_i. By raising or lowering the cutting plane (through smaller or larger values of K_i), we can trace out a series of concentric ellipses, as illustrated in Panel II of Figure 6.2. The figures will be ellipses rather than some other type of conic section because $(X_1 - \mu_1)^2$ and $(X_2 - \mu_2)^2$ will always have positive coefficients. Smaller values of K_i lead to larger altitudes since the height of $f_2(X_1, X_2)$ is proportional to the (negative) exponential.

The common "center" of the ellipses is the centroid of the points; as we know, this is also the maximum height of the surface. For any particular bivariate normal, all ellipses will have the same shape and same angle, relative to the X_1 axis. As the various equal density ellipses become larger, we see intuitively that they bound a larger proportion of the observations.

By methods that will be introduced shortly we can construct a series of equal density ellipses mathematically. As remembered from univariate statistics we can find areas under the standard unit normal curve if we know the value of the standard deviate. Similarly, we shall be able to do the same type of thing with

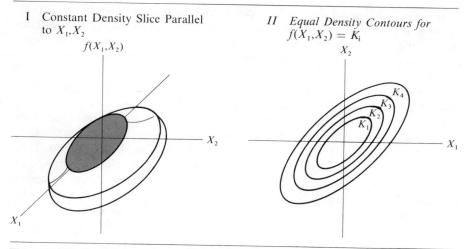

I   Constant Density Slice Parallel
    to X_1, X_2
       $f(X_1, X_2)$

II   *Equal Density Contours for*
     $f(X_1, X_2) = K_i$

Figure 6.2 Cutting Plane and Equal Density Contours

the bivariate normal density—namely, to find values that represent a certain proportion of the total *volume* enclosed by the bivariate normal surface. In so doing we can also make probability statements about observations drawn randomly from a bivariate normal with certain parameters.

A particular equal density ellipse can be specified by either the height of the implied ordinate at any point on the ellipse or by the length of (say) the major axis of the ellipse.[9] As we shall see in a moment, there is still another way of characterizing an equal density ellipse of a bivariate normal—namely as a *function* of the value of its ordinate.

The shape and orientation of the equal density contours are governed by the parameters, σ_1, σ_2, and ρ. If $\rho = 0$ and $\sigma_1 = \sigma_2$, then the equal density contours become circles. If $\rho = 0$ and $\sigma_1 \neq \sigma_2$ the equal density contours are ellipses whose major and minor axes are parallel with the original X_1, X_2 axes. If $\rho \neq 0$ the ellipses become tilted. For example, if $\rho \neq 0$ and $\sigma_1 \neq \sigma_2$, the angle θ that the major axis of the ellipse makes with the X_1 axis can be obtained from the relationship:

$$\tan 2\theta = \frac{2\sigma_1 \sigma_2 \rho}{\sigma_1^2 - \sigma_2^2}$$

The line with this angle relative to the X_1 axis coincides with the *major* axis of the ellipse if $0 < \rho < 1$ (positive correlation) and with the minor axis if $-1 < \rho < 0$ (negative correlation). However, if $\sigma_1 = \sigma_2$ and $\rho \neq 0$, the equal density contours will be oriented so that the major axis (if $\rho > 0$) or minor axis (if $\rho < 0$) makes a 45° angle with the X_1 and X_2 axes.

In other words, if the standard deviations of X_1 and X_2 are unequal, the orientation of the ellipse will be pulled toward the axis with the larger standard deviation. If $\sigma_1 = \sigma_2$, however, the angle of orientation is either 45° or 135° (with respect to the X_1 axis), depending upon whether the correlation is positive or negative.

Figure 6.3 shows a number of possible relations among σ_1, σ_2, and ρ. (For ease of presentation all equal density contours are drawn about a centroid-centered origin.) Notice that in Panels III and IV the absolute value of the angle that the major axis makes with the X_1 axis is less than 45° and reflects the facts that $\sigma_1 > \sigma_2$ and $\rho > 0$. On the other hand, if $\rho = 0$ (Panel II of Figure 6.3) the axes of the ellipse are superimposed on X_1 and X_2. While not illustrated in Figure 6.3, higher (absolute) values of the correlation between X_1 and X_2 would be represented by "thinner" ellipses while lower values would be represented by "fatter" ellipses that enclose a specified proportion of the observations.

6.2.3 A Numerical Illustration

It should be useful to take a numerical example and explore further the concept of equal density contours. Consider the following parameters of a bivariate normal density function:

$$\mu_1 = 2; \mu_2 = 1; \sigma_1 = \sigma_2 = 2; \rho = 0.75$$

That is, to make things particularly simple, we have set $\sigma_1 = \sigma_2 = 2$ so that the major axis of each equal density ellipse will make a 45° angle with X_1 (since $\rho > 0$).

We next inquire as to whether the equation of the bivariate normal surface itself can be simplified in any way. As has already been illustrated, we can find the determinant and adjoint of Σ, leading to the inverse:

$$\Sigma^{-1} = \frac{1}{1-\rho^2}\begin{bmatrix} 1/\sigma_1^2 & -\rho/\sigma_1\sigma_2 \\ -\rho/\sigma_2\sigma_1 & 1/\sigma_2^2 \end{bmatrix}$$

$$= \frac{1}{1-(0.75)^2}\begin{bmatrix} 1/4 & -0.75/4 \\ -0.75/4 & 1/4 \end{bmatrix}$$

$$= \begin{bmatrix} 0.57 & -0.43 \\ -0.43 & 0.57 \end{bmatrix}$$

Next, if we define:

$$\mathbf{x}_d = \begin{bmatrix} X_1 - \mu_1 \\ X_2 - \mu_2 \end{bmatrix}$$

as a vector of deviations from the centroid, we can write $F_2(\mathbf{x}_d)$ as:

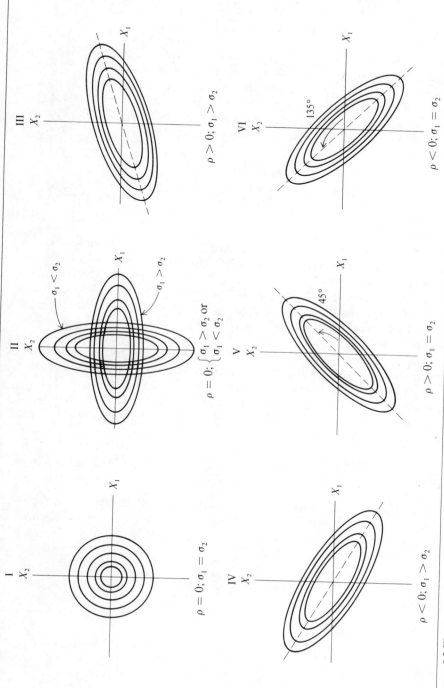

Figure 6.3 Illustrative Patterns of Equal Density Contours

$$F_2(\mathbf{x}_d) = \mathbf{x}_d' \sum{}^{-1} \mathbf{x}_d$$

The scalars that are given by $F_2(\mathbf{x}_d)$ for various values of $\mathbf{x}_d$ turn out to be distributed as chi square (χ^2) with 2 degrees of freedom (Graybill, 1961).

This is an important consequence, since percentage points of the chi square distribution are readily available in tabular form. We can examine this aspect of the problem numerically by choosing a set of values for X_1 and X_2. Illustratively, if we let $X_1 = 2.5$ and $X_2 = 2$, we have the deviation vector $\begin{bmatrix} 2.5 \\ 2 \end{bmatrix} - \begin{bmatrix} 2 \\ 1 \end{bmatrix}] = \begin{bmatrix} 0.5 \\ 1 \end{bmatrix}$ and, hence:

$$F_2(\mathbf{x}_d) = \chi_{[2]}^2 = (0.5, 1) \begin{bmatrix} 0.57 & -0.43 \\ -0.43 & 0.57 \end{bmatrix} \begin{bmatrix} 0.5 \\ 1 \end{bmatrix}$$

$$= 0.28$$

We can then look up the probability associated with $\chi^2 = 0.28$, with 2 degrees of freedom. Not surprisingly, the chance of obtaining an observation at least this close to the universe centroid of $\begin{bmatrix} 2 \\ 1 \end{bmatrix}$ is quite high ($P \cong 0.85$). The important point to remember is that *the chi square distribution is relevant for finding the desired probabilities associated with various vector-valued observations in the same way that the standard unit normal is used to find probabilities associated with various Z values in the univariate case.*

Returning to the numerical example, let us now turn the problem around by assuming that we wished to draw up an equal density ellipse that would enclose 95 percent of the vector-valued observations in our sample problem. The chi square value, with 2 degrees of freedom, associated with the 95 percent point is 5.99. If we set $\chi_{0.95[2]}^2 = K = 5.99$ we can then express the equal density ellipse by the formula:

$$\chi_{0.95[2]}^2 = \frac{1}{1-\rho^2} \left\{ \frac{(X_1 - \mu_1)^2}{\sigma_1^2} - \frac{2\rho(X_1 - \mu_1)(X_2 - \mu_2)}{\sigma_1 \sigma_2} + \frac{(X_2 - \mu_2)^2}{\sigma_2^2} \right\} = 5.99$$

Here, we omit the constant term $1/(2\pi\sigma_1\sigma_2 \sqrt{1-\rho^2})$ and also the multiplicative constant $-\frac{1}{2}$ in the brackets of $f_2(X_1, X_2)$ since these are not essential to constructing the equal density contours. By doing this we end up, as noted above, with the function:

$$\chi_{0.95[2]}^2 = \mathbf{x}_d' \sum{}^{-1} \mathbf{x}_d$$

Earlier, we found $\sum^{-1}$ to be:

$$\sum{}^{-1} = \frac{1}{1 - (0.75)^2} \begin{bmatrix} 1/4 & -0.75/4 \\ -0.75/4 & 1/4 \end{bmatrix}$$

Substituting these values in the function leads to:

$$\chi^2_{0.95[2]} = \frac{1}{1 - (0.75)^2}\left\{\frac{X^2_{d1}}{4} - \frac{2(0.75)(X_{d1})(X_{d2})}{4} + \frac{X^2_{d2}}{4}\right\} = 5.99$$

$$= \frac{X^2_{d1}}{1.75} - \frac{1.5X_{d1}X_{d2}}{1.75} + \frac{X^2_{d2}}{1.75} = 5.99$$

or, upon simplification:

$$X^2_{d1} - 1.5X_{d1}X_{d2} + X^2_{d2} = 10.5$$

Note that we can express X_{d1} as a function of X_{d2} (or vice versa). Then, if we assume various numerical values for X_{d2} we can solve for X_{d1} via the general quadratic formula. By taking enough values for X_{d2} we can eventually trace out the character of the equal density ellipse.

Figure 6.4 shows the 95 percent equal density ellipse that is consistent with the above chi square value.[10] Notice that the expression is given in terms of a (universe) centroid-centered origin since $\mathbf{x}_d$ denotes a deviation-from-mean vector. What the ellipse in Figure 6.4 now tells us is that if we were to draw single observations at random from a bivariate normal distribution with the specific parameters:

$$\mu_1 = 2;\ \mu_2 = 1;\ \sigma_1 = \sigma_2 = 2;\ \rho = 0.75$$

then 95 percent of the observations, expressed as deviations from the universe centroid, would fall within the elliptical region plotted in Figure 6.4. For example, if we have the two observations:

$$R = (5, 2);\ S = (7, 1)$$

we can express them in deviation form as:

$$R^* = (3, 1);\ S^* = (5, 0)$$

We note that R^* falls within the 95 percent equal density ellipse while S^* does not.

In summary, for the bivariate normal density we can construct equal density ellipses corresponding to various percentiles of interest, such as the 95 percentile. This is done by first finding the χ^2 value, with 2 degrees of freedom, associated with the percent ellipse of interest. Then, given the parameter values of the function, we can solve for (say) X_{d1}, given specified values of X_{d2} as illustrated by the preceding formula:

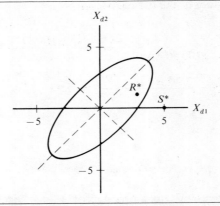

**Figure 6.4 95 Percent Ellipse for Individual Observations —
Sample Problem**

$$K = X_{d1}^2 - 1.5 X_{d1} X_{d2} + X_{d2}^2 = 10.5$$

in the sample problem. Similarly, we could find appropriate values of X_{d1} and X_{d2} for other percentile ellipses of interest by substituting the numerical values of χ^2 (with 2 degrees of freedom) associated with each of these percentiles.

In practice, of course, we would not go through the elaborate procedure of constructing the equal density ellipses. Rather, the value of the χ^2 statistic itself would be used to test whether the observation in question was drawn from the hypothesized universe. To illustrate for the above case of $S^* = (5, 0)$, we have a quantity that can be denoted U:

$$
U = \overset{\mathbf{x}'_d}{(5, 0)} \overset{\Sigma^{-1}}{\begin{bmatrix} 0.57 & -0.43 \\ -0.43 & 0.57 \end{bmatrix}} \overset{\mathbf{x}_d}{\begin{bmatrix} 5 \\ 0 \end{bmatrix}}
$$

$$= 14.25$$

which, of course, exceeds the tabular $\chi^2_{0.95[2]}$ of 5.99 that is associated with the 95 percentile. On the other hand, the U statistic for $R^* = (3, 1)$ turns out to be equal to 3.21, which is less than the tabular value of $\chi^2_{0.95[2]} = 5.99$.

In brief, then, the chi square distribution provides the analog to the (square of the) Z distribution for either: (a) testing to see whether a vector drawn from a bivariate normal with known covariance matrix is consistent with some hypothesized centroid or (b) finding equal density ellipses that enclose specified percentages of the sample observations. And, as we shall see, this feature of the chi square distribution extends to the multivariate normal distribution, as well.

6.3 The Multivariate Normal Density Function

Most of the characteristics described in the context of the bivariate normal carry over to the case of the n-variate ($n > 2$) normal density function. Here, however, graphical aids will be less relevant for picturing various relationships.

By way of introduction to the multivariate normal, let us first review the formulas for the univariate and bivariate normal density functions:

$$f_1(X) = (2\pi)^{-1/2}(\sigma^2)^{-1/2} \exp\left[-\frac{1}{2}(X - \mu)^2(\sigma^2)^{-1}\right]$$

$$f_2(X_1, X_2) = (2\pi)^{-1}\left|\sum_2\right|^{-1/2} \exp\left[-\frac{1}{2}(\mathbf{x} - \boldsymbol{\mu})'\sum_2^{-1}(\mathbf{x} - \boldsymbol{\mu})\right]$$

The multivariate normal generalization turns out to be a natural extension of the above equations to:

$$f_n(X_1, X_2, \ldots, X_n) = (2\pi)^{-n/2}\left|\sum_n\right|^{-1/2} \exp\left[-\frac{1}{2}(\mathbf{x} - \boldsymbol{\mu})'\sum_n^{-1}(\mathbf{x} - \boldsymbol{\mu})\right]$$

The expression in brackets after the exponential symbol "exp" is just $-\frac{1}{2}\chi^2_{[n]}$; that is, $-\frac{1}{2}$ times a variable that is distributed as chi square with n degrees of freedom.

The corresponding covariance matrix is:

$$\sum_n = \begin{bmatrix} \sigma_1^2 & \rho_{12}\sigma_1\sigma_2 & \cdots & \rho_{1n}\sigma_1\sigma_n \\ \rho_{21}\sigma_2\sigma_1 & \sigma_2^2 & \cdots & \rho_{2n}\sigma_2\sigma_n \\ \vdots & \vdots & & \vdots \\ \rho_{n1}\sigma_n\sigma_1 & \rho_{n2}\sigma_n\sigma_2 & \cdots & \sigma_n^2 \end{bmatrix}$$

where σ_i^2 is the variance of X_i and ρ_{ij} is the product moment correlation of X_i and X_j.

In the n-variate normal, the constant factor is:

$$(2\pi)^{-n/2}\left|\sum_n\right|^{-1/2}$$

That is, the exponent of 2π turns out to be $-\frac{1}{2}$ *times the number of variables* while the exponent of $|\Sigma_n|$ remains constant at $-\frac{1}{2}$.

In summary, we can write the general expression for the n-variate normal density as:

$$f_n(X_1, X_2, \ldots, X_n) = (2\pi)^{-n/2}\left|\sum_n\right|^{-1/2} \exp\left[-\frac{1}{2}\chi^2_{[n]}\right]$$

where, $\chi^2_{[n]}$ is, in turn, defined as the function:

$$\chi^2_{[n]} = \mathbf{x}'_d \sum_n^{-1} \mathbf{x}_d$$

Once we have found $\chi^2_{[n]}$, we can determine the appropriate equal density contours of interest, just as was done in the bivariate case.

Of course, there is no need actually to try to draw the type of ellipse illustrated in Figure 6.4 (and we could not do so even if we wanted to since, in general, hyperellipsoids would be involved). If we want to see whether a particular observation vector $\mathbf{x}_d$ has arisen from the n-variate normal distribution of interest, all we need to do is first compute the χ^2 value by substituting the specific value of $\mathbf{x}_d$ and the specific parameters of the multivariate normal in the quadratic form. This gives us a computed value of χ^2 that we can call U. We can then find the tabular value of χ^2 (with n degrees of freedom) associated with some percent point of interest, such as 95 percent. If the U statistic exceeds the tabular value, we assume that the observation $\mathbf{x}_d$ did not come from the hypothesized universe, with centroid μ and (assumed known) dispersion matrix $\sum$.

In brief, then, the n-variate normal density function can be used for the same purposes as its special case, the bivariate normal. The n-variate normal, of course, has more parameters and is more complex in this sense. However, the same potential uses exist and, as observed, the bivariate and univariate normals can be obtained as specializations of the n-variate normal.

6.4 Some Sampling Characteristics of the Multivariate Normal

Up to this point we have said relatively little about distributions of centroids computed from samples drawn from n-variate normal universes. Rather, our attention has been focused on the simpler case involving the distribution of just *single observations* drawn from the universe of interest. However, fully analogous to the univariate case, sampling distributions for centroids, computed from samples of size m, of the n-variate normal can be defined and illustrated numerically.

As mentioned earlier, in applying univariate sampling theory to inferences regarding the sample mean, two cases are distinguished. In situations where the universe variance is assumed to be known, the sampling distribution of the mean is distributed as the standard unit normal or Z distribution (with expectation μ and variance σ^2/m.) If the variance is not known and, hence, is estimated from the sample, the sampling distribution follows Student's t distribution (with expectation μ and variance s^2/m, where s^2 is an unbiased estimate of σ^2).

In the case of the multivariate normal we have two analogous situations where: (a) $\sum$ is assumed to be known and (b) $\sum$ must be estimated from the sample.

6.4.1 The Multivariate Analog of the Z Test

If random samples of size m are drawn from a univariate normal distribution with parameters μ and σ^2, the sample means are normally distributed with:

$$E(\overline{X}_i) = \mu$$

$$\text{Var}(\overline{X}_i) = \frac{\sigma^2}{m}$$

And, it turns out that the multivariate counterpart is fully analogous to this case. That is, if random samples of size m are drawn from a multivariate normal distribution of n variates, the sample centroids follow an n-variate normal distribution with:

$$E(\overline{\mathbf{x}}_i) = \mu$$

$$\text{Cov}(\overline{\mathbf{x}}_i) = \frac{\Sigma}{m}$$

To illustrate this idea in terms of the sample problem introduced earlier, let us assume that we are taking random samples of size $m = 25$ from a bivariate normal distribution with parameters:

$$\mu = \begin{bmatrix} 2 \\ 1 \end{bmatrix}$$

$$\Sigma = \begin{bmatrix} \sigma_1^2 & \rho\sigma_1\sigma_2 \\ \rho\sigma_2\sigma_1 & \sigma_2^2 \end{bmatrix} = \begin{bmatrix} 4 & 3 \\ 3 & 4 \end{bmatrix}$$

That is, we shall continue to work with the universe values assumed earlier:

$$\mu_1 = 2;\ \mu_2 = 1;\ \sigma_1 = 2;\ \sigma_2 = 2;\ \rho = 0.75$$

From section 6.2 we know that:

$$\Sigma^{-1} = \frac{1}{(1-\rho^2)} \begin{bmatrix} 1/\sigma_1^2 & -\rho/\sigma_1\sigma_2 \\ -\rho/\sigma_2\sigma_1 & 1/\sigma_2^2 \end{bmatrix} = \begin{bmatrix} 0.57 & -0.43 \\ -0.43 & 0.57 \end{bmatrix}$$

But, if the covariance matrix of the sampling is Σ/m, its inverse will be $m\Sigma^{-1}$. Hence, in this case the equal density ellipse for the 95 percent probability point is:

$$m\left[\overline{\mathbf{x}}_d' \sum{}^{-1} \overline{\mathbf{x}}_d \right] = 5.99$$

Substituting the appropriate numerical values from section 6.2.3 gives us:

$$25[\overline{X}_{d1}^2 - 1.5\overline{X}_{d1}\overline{X}_{d2} + \overline{X}_{d2}^2] = 1.75(5.99) = 10.5$$

or:

$$\overline{X}_{d1}^2 - 1.5\overline{X}_{d1}\overline{X}_{d2} + \overline{X}_{d2}^2 = 0.42$$

as the value of K_i specifying the 95 percent ellipse. Notice, then, that the new 95 percent bounding ellipse is considerably smaller—actually, the length of its major axis is $1/\sqrt{m}$ times the length of the original that was based on single observations.

One then proceeds in exactly the same way as before. That is, we could first compute a χ^2 statistic via the expression:

$$\chi^2 = 25\left[\overline{\mathbf{x}}_d' \sum{}^{-1} \overline{\mathbf{x}}_d\right]$$

where we have some sample centroid of interest, expressed as a vector of deviations:

$$\mathbf{x}_d = \begin{bmatrix} \overline{X}_{d1} \\ \overline{X}_{d2} \end{bmatrix} = \begin{bmatrix} \overline{X}_1 - \mu_1 \\ \overline{X}_2 - \mu_2 \end{bmatrix}$$

We could then see if this χ^2 statistic exceeds the tabular $\chi^2_{0.95[2]}$ value of 5.99. If so, we reject the null hypothesis that the universe centroid is as hypothesized, namely:

$$\boldsymbol{\mu} = \begin{bmatrix} 2 \\ 1 \end{bmatrix}$$

Alternatively, we can compute a 95 percent ellipse that is associated with the sample-problem parameters. This is done by substituting various values of (say) $\overline{X}_{d2}$ in the equation:

$$\overline{X}_{d1}^2 - 1.5\overline{X}_{d1}\overline{X}_{d2} + \overline{X}_{d2}^2 = 0.42$$

and then solving for $\overline{X}_{d1}$. Having plotted the ellipse, we can then plot any sample centroid that is actually found (expressed as a deviation vector and based on a sample size of $m = 25$) in order to see if it falls inside the ellipse. This graphical approach, while less precise, can be useful from an intuitive standpoint.

Figure 6.5 shows the 95 percent ellipse associated with the equation:

$$\overline{X}_{d1}^2 - 1.5\overline{X}_{d1}\overline{X}_{d2} + \overline{X}_{d2}^2 = 0.42$$

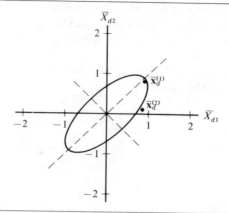

Figure 6.5 Bivariate Generalization of the Z-Test—Samples of Size 25

Suppose, now, that we obtain one random sample of $m_1 = 25$ cases in which the sample centroid is $\overline{X}_1 = 2.85$, $\overline{X}_2 = 1.80$. Hypothesizing that $\mu = \begin{bmatrix} 2 \\ 1 \end{bmatrix}$, we have the deviation vector $\overline{x}_d^{(1)} = \begin{bmatrix} 0.85 \\ 0.80 \end{bmatrix}$.

Let us also obtain another random sample of $m_2 = 25$ cases in which the centroid is $\overline{X}_1 = 2.75$, $\overline{X}_2 = 1.10$. The deviation vector of the second sample is $\overline{x}_d^{(2)} = \begin{bmatrix} 0.75 \\ 0.10 \end{bmatrix}$. Both sample deviation vectors are plotted in Figure 6.5. We see that $\overline{x}_d^{(1)}$ plots just inside the 95 percent ellipse while $\overline{x}_d^{(2)}$ falls outside. Hence, we would accept the null hypothesis:

$$\mu = \begin{bmatrix} 2 \\ 1 \end{bmatrix}$$

in the case of the first sample and reject it in the case of the second.

Similar results occur when we solve for the χ^2 statistic in each case;

$$\chi^{2(1)} = 25(0.85 \quad 0.8)\begin{bmatrix} 0.57 & -0.43 \\ -0.43 & 0.57 \end{bmatrix}\begin{bmatrix} 0.85 \\ 0.80 \end{bmatrix} = 4.80$$

$$\chi^{2(2)} = 25(0.75 \quad 0.1)\begin{bmatrix} 0.57 & -0.43 \\ -0.43 & 0.57 \end{bmatrix}\begin{bmatrix} 0.75 \\ 0.10 \end{bmatrix} = 6.54$$

In the case of the first sample $\chi^{2(1)} = 4.80$, which is less than tabular $\chi^2_{0.95[2]} = 5.99$. In the case of the second sample $\chi^{2(2)} = 6.54$ which exceeds the tabular chi square value of 5.99.

These conclusions may seem a bit strange when we note that $\overline{x}_d^{(2)}$ is actually closer geometrically to the universe centroid than is $\overline{x}_d^{(1)}$. However, the point to remember is that we are concerned here with "statistical" distance. Under this latter type of reasoning the population correlation $\rho = 0.75$ indicates that we should expect to find samples in which *high values on variable 1 tend to occur with high values on variable 2*. The first sample exhibits this expected correlation but the second sample shows a rather low mean on the second variate, coupled with a high mean on the first variate.

However, suppose we had decided to run two separate univariate Z tests on each sample by making the substitutions:

	Sample 1	Sample 2
Variate 1	$Z = \dfrac{0.85}{2/\sqrt{25}}$	$Z = \dfrac{0.75}{2/\sqrt{25}}$
	$= 2.13$	$= 1.88$
Variate 2	$Z = \dfrac{0.80}{2/\sqrt{25}}$	$Z = \dfrac{0.10}{2/\sqrt{25}}$
	$= 2.00$	$= 0.25$

The tabular Z value associated with the 95 percent probability level (two-tailed test) is 1.96. Under this criterion, had we made two separate univariate tests in sample 1, we would have *rejected* the null hypothesis each time. On the other hand, had we made two separate univariate tests of sample 2 we would have accepted the null hypothesis each time.

While this example is contrived, it nonetheless demonstrates the point that the bivariate test which, in this case, leads to conclusions that are *opposite* to those found from the two separate univariate tests, is the relevant test to make. This test considers the *total* information available on both variates considered simultaneously rather than the partial information obtained by running a test on one variate while ignoring information on the other.

The fact that the two variates are presumed to be positively correlated as is, in fact, evinced in the first sample but not so in the second, is important and relevant to the test based on the bivariate normal. On the other hand, this information is neglected (erroneously) under the two separate Z tests. Only if the variates were uncorrelated and possessed equal standard deviations would we find that Euclidean distance from each centroid to the origin agrees with "statistical" distance. If this situation prevailed, the equal density contours would be represented by a series of concentric circles. (Note that this distinction parallels Chapter 4's discussion of Mahalanobis' D^2 versus ordinary Euclidean squared distance.)

6.4.2 The Multivariate Analog of the *t* Test

It is fairly rare in empirical problems to encounter cases in which one knows the covariance matrix. Analogous to Student's univariate t test, which replaces the Z test in cases where the variance is unknown, Hotelling's T^2 statistic can be used for making tests about some hypothesized centroid in the case of more than one variate if Σ is unknown.

If we let **C** denote the (unbiased) sample-based estimate of Σ, we can define T^2 as follows:

$$T^2 = m[\bar{\mathbf{x}}_d' \mathbf{C}^{-1} \bar{\mathbf{x}}_d]$$

which is analogous to the χ^2 statistic, computed earlier. The only difference is that **C**, the estimate of Σ, replaces Σ, the universe covariance matrix.

However, T^2 does *not* follow the χ^2 distribution. Instead, we have the relation:

$$\frac{m - n}{n(m - 1)} T^2 \sim F_{[n, m-n]}$$

where $\sim$ denotes "is distributed as." That is, $[(m - n)/n(m - 1)]T^2$ follows the F distribution with n degrees of freedom in the numerator and $m - n$ degrees of freedom in the denominator.

In summary, in the one-sample problem if Σ is assumed known, we can compute the test statistic:

$$\chi^2 = m\left[\bar{\mathbf{x}}_d' \sum{}^{-1} \bar{\mathbf{x}}_d\right]$$

which follows the chi square distribution with n degrees of freedom, if n variates are involved.

Then we can choose some percentile of chi square, such as 0.95, and find the associated tabular value $\chi^2_{0.95[n]}$ for comparison with χ^2. If χ^2 exceeds this tabular value we reject the null hypothesis that $\mu = \mu_h$.

However, if Σ is estimated from the sample covariance matrix **C**, we have the statistic:

$$T^2 = m[\bar{\mathbf{x}}_d' \mathbf{C}^{-1} \bar{\mathbf{x}}_d]$$

with the following relationship between T^2 and F:

$$\frac{m - n}{n(m - 1)} T^2 \sim F_{[n, m-n]}$$

In this case we reject the null hypothesis that $\mu = \mu_h$ if $[(m - n)/n(m - 1)]T^2 > F_{[n, m-n]}$ for some specified value of α, such as $\alpha = 0.05$.[11]

So much for single-sample tests of the centroid. We now turn to a multivariate analog of the two-sample test.

6.4.3 The Case Involving Two Independent Samples

In univariate statistics situations are often encountered in which we have obtained two independent samples with respective means $\overline{X}_1$ and $\overline{X}_2$. As reviewed in Appendix B, a test of the null hypothesis that $\mu_1 - \mu_2 = 0$—that is, that the universe means are equal—is based on the t statistic.

Corresponding to this univariate test is the multivariate case in which Hotelling's T^2 plays the counterpart role of t. The test statistic is:

$$T_2^2 = \frac{m_1 m_2}{m_1 + m_2} \cdot \mathbf{d}'_{\overline{x}_1 - \overline{x}_2} \mathbf{C}_w^{-1} \mathbf{d}_{\overline{x}_1 - \overline{x}_2}$$

where $\mathbf{d}_{\overline{x}_1 - \overline{x}_2}$ is a difference vector between the sample centroids $\overline{\mathbf{x}}_1$ and $\overline{\mathbf{x}}_2$. The subscript 2 in T_2^2 is now used to denote the fact that two independent samples are involved. We assume that both covariance matrices, while unknown, are equal. Their (common) covariance matrix is estimated from the pooled samples.

In the two-sample case, we have the following function of T_2^2 that is distributed as F:

$$\frac{m - n - 1}{n(m - 2)} T_2^2 \sim F_{[n, m-n-1]}$$

where $m = m_1 + m_2$.[12]

We first encountered Hotelling's T_2^2 statistic in the context of two-group discriminant analysis in section 4.4 of Chapter 4. As recalled, T_2^2 was computed from a difference vector and a sample-based covariance matrix developed from the data of Table 4.1:

$$
T_2^2 = \frac{5(5)}{10} \overset{\mathbf{d}'_{\overline{x}_1 - \overline{x}_2}}{(-5 \quad -2)} \overset{\mathbf{C}_w^{-1}}{\begin{bmatrix} 0.776 & -0.471 \\ -0.471 & 0.588 \end{bmatrix}} \overset{\mathbf{d}_{\overline{x}_1 - \overline{x}_2}}{\begin{bmatrix} -5 \\ -2 \end{bmatrix}}
$$

$$= 30.9$$

$$F_{[2,7]} = \frac{m - n - 1}{n(m - 2)} T_2^2 = \frac{10 - 2 - 1}{2(10 - 2)}(30.9)$$

$$= 13.5$$

In this case $m_1 = m_2 = 5$ and $n = 2$. The associated tabular $F_{0.05[2,7]}$ was 4.8; hence, the null hypothesis of equality of group centroids was rejected.

It may seem surprising that we employed a multivariate test in two-group discriminant analysis since the criterion variable is "univariate" and codable as zero or one. However, in testing for the significance of the *difference between centroids*, the roles of criterion and predictors are *reversed*. In this case the two predictors become the criterion variables and we have a case of a two-sample test of the difference in centroids involving the variables:

X_1: attitude toward energy conservation

X_2: attitude toward state government intervention in the control of corporate practices

where the two samples are composed of those who support the bill versus those who oppose it (as shown in Table 4.1 of Chapter 4).

Thus, even in the case of two-group discrimination the hypothesis under test is *multivariate* since our interest is in whether the group *centroids* differ significantly on the "predictor" variables.

Since two-sample tests of centroids assume commonality of covariance matrices, Bartlett's test (described in Chapter 4) can be used to examine the equality of the covariance matrices prior to running the significance test for centroid differences. We see, then, that Hotelling's T^2 can be useful in both the single-sample and two-sample tests where covariance matrices are unknown.[13]

6.4.4 Concluding Comments

While we have discussed both the single-sample and two-sample tests for centroids (namely, the χ^2 and T^2 statistics), we have not as yet considered the most general case of all: the K-sample ($K > 2$) test of centroid equality in situations where the covariance matrices are assumed to be unknown but equal.

This test is a multivariate analog to the (univariate) one-way ANOVA test that was discussed in detail in Chapter 3. As described there, in the case of K samples we compute an F ratio, based on the among-groups mean square to the within-groups mean square:

$$F = \frac{\text{SSA}/(K - 1)}{\text{SSW}/(m - K)}$$

where SSA and SSW are, respectively, the among-groups and within-groups sums of squares, K is the number of samples and m is the total sample size, where $m_1 + m_2 + \cdots + m_K = m$.

As recalled, the null hypothesis is:

$$H_0: \mu_1 = \mu_2 = \cdots = \mu_K$$

and the alternative hypothesis is that at least one universe mean differs from the rest.

It turns out that a multivariate analog to the K-sample test also exists. This statistic is known as *Wilks' lambda* and is defined in its most general form as:

$$\Lambda = \frac{|\mathbf{S}_E|}{|\mathbf{S}_H + \mathbf{S}_E|}$$

where $\mathbf{S}_E$ is the SSCP matrix for error and $\mathbf{S}_H$ is the SSCP matrix for hypothesis.

In the case of the multivariate analog of the one-way ANOVA test, we have the following version of Λ:

$$\Lambda = \frac{|\mathbf{W}|}{|\mathbf{T}|}$$

where $\mathbf{W}$ is the pooled within-groups SSCP matrix and $\mathbf{T} = (\mathbf{A} + \mathbf{W})$ is the total-sample SSCP matrix.[14]

Note that the determinants of $\mathbf{W}$ and $\mathbf{T}$ play the role of generalized scalar measures of variance; hence Λ itself is a scalar. Indeed, if the problem involves only a single criterion variable, it turns out that Λ is inversely related to the familiar F ratio.[15]

Wilks' lambda plays so central a role in the analysis of multiple criterion, multiple predictor association that we devote a full section to it, and various functions of it, at the beginning of the next chapter. (Accordingly, we also defer discussion of hypothesis testing in canonical correlation to Chapter 7.)

6.5 Canonical Correlation

In canonical correlation the multivariate normal distribution plays a central role in describing the universes underlying both the criterion and predictor sets (when significance testing is involved). As an illustration of canonical correlation, consider the small illustration that was introduced in Table 2.1 in Chapter 2. There we had various subjects rate stimulus paragraphs (describing selected high school seniors on extra-curricular activity and scholastic achievement) in terms of overall liking. Now, however, suppose that each subject was also asked to rate each paragraph on the degree to which the person would be successful in his or her subsequent career.[16]

In this (hypothetical) situation, one subject's ratings of 12 such stimulus paragraphs are shown in Table 6.1. Also shown is the correlation matrix $\mathbf{R}$. We note from this matrix that all pairs of variables are positively associated. In particular, the two criterion variables display a positive correlation of 0.57.

6.5.1 An Intuitive Introduction

In canonical correlation our interest centers on the linear relationship between one battery of variables, illustrated by Y_1 and Y_2, and another battery of variables, illustrated by X_1 and X_2. The objective in canonical correlation is to find a linear composite of the Y variables and a (different) linear composite of the X variables so that when this pair of derived variables (i.e., pair of linear composites) is correlated, the resulting two-variable correlation is the highest attainable.

Having done this, it is generally possible to find a second pair of linear composites, chosen to be uncorrelated with the first pair, such that the correlation between this second pair of derived variables is (conditionally) maximal. In general, with p criteria and q predictors (where $p + q = n$) we can obtain min (p, q) different pairs of linear composites. The correlations between successive pairs will, in general, decline in size; that is, the first pair will exhibit the maximum correlation, the second pair the next highest, and so on.

To show what is involved in terms of the sample problem of Table 6.1 let us examine the scatter plots of Figure 6.6. These plots were made after standardizing each data column of Table 6.1 to zero mean and unit standard deviation. The dotted lines $\hat{Y}_{s2}$ and $\hat{X}_{s2}$ denote the regression of Y_{s2} on Y_{s1} and X_{s2} on X_{s1}, respectively. The slopes of the dotted lines in each panel of the figure represent the simple within-set correlations:

$$r_{y_2 y_1}; r_{x_2 x_1}$$

This is because both pairs of variables are expressed in standardized form. In this case the regression coefficients of Y_{s2} on Y_{s1} and X_{s2} on X_{s1} equal their respective correlation coefficients. So far, then, nothing new.

However, let us next examine the vectors $\mathbf{t}_1$ and $\mathbf{u}_1$ in Panels I and II of Figure 6.6. To be specific, suppose we were to project the points in Panel I onto $\mathbf{t}_1$, giving us 12 scores that represent linear composites of the standardized variates Y_{s1} and Y_{s2}. Let us do the same thing for the points in Panel II with respect to the vector $\mathbf{u}_1$.

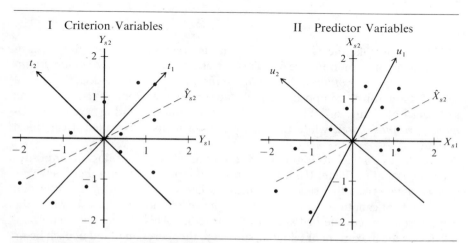

Figure 6.6 Scatter Plots of Standardized Data, Computed from Table 6.1

Table 6.1 Ratings of Paragraphs Describing Twelve High School Seniors

Paragraph Description of Stimulus Person	Overall Liking Y_1	Success in College Career Y_2	Degree of Extra-Curricular Activity X_1	Degree of Scholastic Achievement X_2
a	1	3	2	4
b	3	2	4	3
c	4	6	5	7
d	5	3	6	4
e	7	5	8	6
f	6	8	7	9
g	9	7	6	8
h	8	9	9	7
i	5	7	3	6
j	9	4	9	6
k	7	6	8	8
l	8	9	9	9
Mean	6.00	5.75	6.33	6.42
Standard Deviation	2.49	2.38	2.42	1.98

Correlation Matrix

	Y_1	Y_2	X_1	X_2
Y_1	1.000	0.569	0.844	0.648
Y_2		1.000	0.473	0.876
X_1			1.000	0.576
X_2				1.000

Next, let us take the 12 scores on t_1 and u_1, respectively, and plot them in scatter diagram form in Figure 6.7. This correlation $r_{t_1 u_1}$ is the *maximum attainable for a single pair of linear composites of the Y's and the X's* and, in the special case of standardized composites, equals the (regression) slope of the dotted line in Panel I of Figure 6.7. The various separate values, t_i and u_i, are called *canonical variate scores,* while the linear composite itself is called a canonical variate.

Now let us go back to Figure 6.6 and find another pair of linear composites t_2 (Panel I) and u_2 (Panel II). If the 12 points are projected onto these new vectors we obtain two other sets of canonical variate scores. We form a scatter plot of this pair of canonical variates and find the correlation $r_{t_2 u_2}$, as shown as the slope of the dotted line in Panel II of Figure 6.7. This second correlation coefficient is the maximum obtainable, subject to the scores on t_2 and u_2 each being uncorrelated with those on t_1 and u_1.[17]

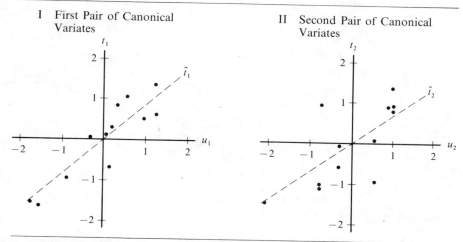

Figure 6.7 Plots of First and Second Pairs of Canonical Variates for the Sample Problem

This is essentially what canonical correlations is all about—namely finding successive pairs of linear composites that are maximally correlated, subject to being uncorrelated with previously found pairs of linear composites.

6.5.2 A Vector Representation

As recalled from Chapter 2, one can discuss various multivariate procedures in terms of either the point model, as just described, or the vector model, where each variable is represented as a unit length vector in "person" space. Figure 6.8 shows a vector representation for the sample problem.

Panel I of Figure 6.8 shows the unit-length criterion vectors y_1 and y_2, separated by the angle $\theta = 55°$. The cosine of this angle is 0.569 and denotes the correlation between y_2 and y_1. Similarly, in Panel II the cosine of $\theta = 54°$ is 0.576 and denotes the correlation between the predictor vectors x_1 and x_2.[18] In the vector model representation of canonical correlation, we seek new vectors, t_1 and u_1, represented as linear composites of their respective original vectors—y_1 and y_2 in Panel I and x_1 and x_2 in Panel II of Figure 6.8. The coefficients a_1 and a_2 and b_1 and b_2, respectively, denote the combining weights.

In the vector model interpretation of canonical correlation we attempt to find **a** and **b** so that the angle between t_1 and u_1 is minimized, *if we imagine that all four vectors are embedded in a four-dimensional space.* This, of course, is analogous to the multiple regression case in which we tried to minimize the angle between the single criterion variable **y** and the predicted criterion $\hat{y}$. In the present situation we allow linear transformations of *both* the criterion and predictor sets in order to maximize the correlation (cosine of the angle separating t_1 and u_1).

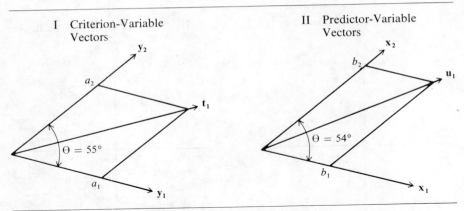

Figure 6.8 Vector Model Representation – Sample Problem

While not shown in Figure 6.8 we could then apply the same reasoning to obtain another pair of linear composites t_2 and u_2, whose correlation is maximal, subject to their being uncorrelated with the previously found canonical variates.

6.5.3 A Formal Representation

The point and vector models describe the essentials of canonical correlation in an intuitive way. We can also characterize the model in a more formal manner. In two-set canonical correlation we have $p + q = n$ variables. If we correlate every variable with every other, we can summarize the results in the form of a "super" matrix:

$$\begin{bmatrix} \mathbf{R}_{yy} & | & \mathbf{R}_{yx} \\ \hline \mathbf{R}_{xy} & | & \mathbf{R}_{xx} \end{bmatrix}$$

which has been partitioned into within-set criterion variable correlations $\mathbf{R}_{yy}$, within-set predictor variable correlations $\mathbf{R}_{xx}$, and the cross-set correlations $\mathbf{R}_{yx}$ and $\mathbf{R}_{xy} (= \mathbf{R}'_{yx})$.

Let us continue to assume that all n variables have been standardized across persons to mean zero and unit standard deviation.[19] With this understanding, any person's data can be represented by two row vectors:

$$\mathbf{y}' = (Y_1, Y_2, \ldots, Y_p)$$

$$\mathbf{x}' = (X_1, X_2, \ldots, X_q)$$

We wish to find sets of weights, $\mathbf{a}_1$ and $\mathbf{b}_1$, for criterion and predictor sets, respectively. Application of the weights results in two sets of canonical variate scores with general entry:

$$t_{i1} = \mathbf{y}_i' \mathbf{a}_1 \,; \; u_{i1} = \mathbf{x}_i' \mathbf{b}_1$$

We shall want to choose the vectors $\mathbf{a}_1$ and $\mathbf{b}_1$ so that the scores (t_{i1} and u_{i1}) are also standardized. However, in keeping with earlier discussion, we wish to find particular sets of weights $\mathbf{a}_1$ and $\mathbf{b}_1$ that maximize the scalar:

$$R_1 = \frac{1}{m-1} \sum_{i=1}^{m} t_{i1} u_{i1}$$

where R_1 denotes the canonical correlation between the sets of canonical variate scores.[20] That is, R_1 is simply an average cross product of standardized scores.

We can then seek new sets of weights $\mathbf{a}_2$ and $\mathbf{b}_2$ and their counterpart scores t_{i2} and u_{i2} (where t_{i2} and u_{i2} are constrained to be uncorrelated with t_{i1} and u_{i1}) so as to maximize the second canonical correlation:

$$R_2 = \frac{1}{m-1} \sum_{i=1}^{m} t_{i2} u_{i2}$$

and so on for min (p, q) sets of weights.

The solution to a canonical correlation problem is prototypical of techniques that appear in this and subsequent chapters, namely techniques whose implementation calls for solutions to various characteristic equations. (See section B.5 of Appendix B for details.) In the present case of canonical correlation we have to solve for the eigenstructure of a rather complicated expression:

$$(\mathbf{R}_{xx}^{-1} \mathbf{R}_{xy} \mathbf{R}_{yy}^{-1} \mathbf{R}_{yx} - \lambda_j \mathbf{I}) \mathbf{b}_j = 0$$

subject to the scaling condition:

$$\mathbf{b}_j' \mathbf{R}_{xx} \mathbf{b}_j = 1$$

This latter restriction guarantees that the canonical weights are scaled so that the resulting canonical variate scores will have unit standard deviations. Since, as it turns out, the $\mathbf{b}_j$ are eigenvectors, their components are unique only up to a scale transformation and we are free to apply this type of normalization. While eigenvectors are often normalized to unit sum of squares, here we adopt a different normalization in line with our wish to have the canonical variate scores exhibit unit standard deviation.

Solving for the eigenstructure of the matrix product $\mathbf{R}_{xx}^{-1}\mathbf{R}_{xy}\mathbf{R}_{yy}^{-1}\mathbf{R}_{yx}$ is almost a routine matter in the age of computers. While, for expository purposes, we go through the sample-problem calculations by hand, in practice eigenstructure routines would be employed. Of course, in addition to obtaining the eigenvectors $\mathbf{b}_j$, the routines also obtain their associated eigenvalues λ_j.[21]

Once we find the $\mathbf{b}_j$ and λ_j from the eigenstructure calculations, we can go on to solve for the second set of weights $\mathbf{a}_j$ (assuming columns of the data matrix have been arranged so that $q \leqq p$) for counterpart application to the matrix of standardized criterion variables. The second set of weights is found from the equation:

$$\mathbf{a}_j = \frac{(\mathbf{R}_{yy}^{-1}\mathbf{R}_{yx}\mathbf{b}_j)}{\sqrt{\lambda_j}}$$

In canonical correlation the scalar $\sqrt{\lambda_j}$ denotes the j-th canonical correlation R_j, where $0 \leqq \sqrt{\lambda_j} \leqq 1$.

As observed above, we shall have to find the inverses $\mathbf{R}_{xx}^{-1}$ and $\mathbf{R}_{yy}^{-1}$ and then the matrix product. One other difficulty arises from the fact that the matrix product:

$$\mathbf{R}_{xx}^{-1}\mathbf{R}_{xy}\mathbf{R}_{yy}^{-1}\mathbf{R}_{yx}$$

is nonsymmetric, even though each of the separate factors in the product is a symmetric matrix.

6.5.4 Sample Problem Calculations

In the case of a square, nonsymmetric matrix one has two options for obtaining eigenstructures. One can either solve the characteristic equation:

$$|\mathbf{R}_{xx}^{-1}\mathbf{R}_{xy}\mathbf{R}_{yy}^{-1}\mathbf{R}_{yx} - \lambda_j\mathbf{I}| = 0$$

for λ_j just as it stands or, alternatively, one can modify the matrix product in order to make it symmetric. We describe the first of these alternatives while section B.5 of Appendix B illustrates the second alternative.

Let us first examine the preliminary calculations of Table 6.2. We note that:

$$\mathbf{F} = \mathbf{R}_{xx}^{-1}\mathbf{R}_{xy}\mathbf{R}_{yy}^{-1}\mathbf{R}_{yx} = \begin{bmatrix} 0.599 & 0.122 \\ 0.196 & 0.729 \end{bmatrix}$$

does not have equal off diagonal entries.

Table 6.2 Canonical Correlation of Sample Data from Table 6.1

	Y_1	Y_2	X_1	X_2
Y_1	1.000	0.569	0.844	0.648
Y_2	0.569	1.000	0.473	0.875
X_1	0.844	0.473	1.000	0.576
X_2	0.648	0.875	0.576	1.000

Finding $\mathbf{R}_{yy}^{-1}$ and $\mathbf{R}_{xx}^{-1}$

$$\mathbf{R}_{yy}^{-1} = \frac{1}{0.676}\begin{bmatrix} 1 & -0.569 \\ -0.569 & 1 \end{bmatrix} \qquad \mathbf{R}_{xx}^{-1} = \frac{1}{0.668}\begin{bmatrix} 1 & -0.576 \\ -0.576 & 1 \end{bmatrix}$$

$$= \begin{bmatrix} 1.479 & -0.841 \\ -0.841 & 1.479 \end{bmatrix} \qquad\qquad = \begin{bmatrix} 1.497 & -0.862 \\ -0.862 & 1.497 \end{bmatrix}$$

Finding the Eigenstructure of $\mathbf{F}$

$$\mathbf{F} = \mathbf{R}_{xx}^{-1}\mathbf{R}_{xy}\mathbf{R}_{yy}^{-1}\mathbf{R}_{yx}$$

$$= \begin{bmatrix} 1.497 & -0.862 \\ -0.862 & 1.497 \end{bmatrix}\begin{bmatrix} 0.844 & 0.473 \\ 0.648 & 0.875 \end{bmatrix}\begin{bmatrix} 1.479 & -0.841 \\ -0.841 & 1.479 \end{bmatrix}\begin{bmatrix} 0.844 & 0.648 \\ 0.473 & 0.875 \end{bmatrix}$$

$$= \begin{bmatrix} 0.599 & 0.122 \\ 0.196 & 0.729 \end{bmatrix}$$

$$\mathbf{D} = \begin{bmatrix} 0.832 & 0 \\ 0 & 0.498 \end{bmatrix}; \qquad \mathbf{V} = \begin{bmatrix} 0.464 & 0.770 \\ 0.886 & -0.638 \end{bmatrix}$$

Finding the Canonical Weight Matrices

$$\mathbf{B} = \begin{bmatrix} 0.382 & 1.162 \\ 0.730 & -0.981 \end{bmatrix}; \qquad \mathbf{B}'\mathbf{R}_{xx}\mathbf{B} = \begin{bmatrix} 1 & 0 \\ 0 & 1 \end{bmatrix}$$

$$\mathbf{A} = \overset{\mathbf{R}_{yy}^{-1}}{\begin{bmatrix} 0.850 & 0.221 \\ -0.012 & 0.748 \end{bmatrix}}\overset{\mathbf{R}_{yx}}{}\overset{\mathbf{B}}{\begin{bmatrix} 0.382 & 1.162 \\ 0.730 & -0.981 \end{bmatrix}}\overset{\mathbf{D}^{-1/2}}{\begin{bmatrix} 1/0.912 & 0 \\ 0 & 1/0.706 \end{bmatrix}}$$

$$= \begin{bmatrix} 0.534 & 1.092 \\ 0.595 & -1.060 \end{bmatrix}; \qquad \mathbf{A}'\mathbf{R}_{yy}\mathbf{A} = \begin{bmatrix} 1 & 0 \\ 0 & 1 \end{bmatrix}$$

Finding the Canonical Correlations

$$\mathbf{M} = \mathbf{B}'\mathbf{R}_{xy}\mathbf{A} = \mathbf{D}^{1/2}$$

$$= \overset{\mathbf{B}'}{\begin{bmatrix} 0.382 & 0.730 \\ 1.162 & -0.981 \end{bmatrix}}\overset{\mathbf{R}_{xy}}{\begin{bmatrix} 0.844 & 0.473 \\ 0.648 & 0.875 \end{bmatrix}}\overset{\mathbf{A}}{\begin{bmatrix} 0.534 & 1.092 \\ 0.595 & -1.060 \end{bmatrix}}$$

$$\mathbf{M} = \begin{bmatrix} 0.912 & 0 \\ 0 & 0.706 \end{bmatrix} = \begin{bmatrix} \sqrt{\lambda_1} & 0 \\ 0 & \sqrt{\lambda_2} \end{bmatrix} = \begin{bmatrix} \sqrt{0.832} & 0 \\ 0 & \sqrt{0.498} \end{bmatrix}$$

In the case of a square *nonsymmetric* matrix we can introduce the following triple-product decomposition:

$$F = VDV^{-1}$$

where V is nonsingular (but not orthogonal). The matrix D is diagonal.[22]

In this case, V is referred to as a matrix of *right-hand* eigenvectors of F that satisfy the relationship:

$$FV = VD$$

(Of course, if V were orthogonal, then it would be the case that $V^{-1} = V'$.)

Furthermore, although we do not pursue the matter in detail here, we could also find a nonsingular matrix of eigenvectors W, called the *left-hand* eigenvectors of F, that satisfy the equation:

$$W'F = DW'$$

In any event, we can develop a decomposition of square, nonsymmetric matrices in a way that is analogous to the case of symmetric matrices by proceeding just as though F were symmetric and solving for D and V. While the columns of V can be normalized to unit sums of squares, *it is not the case that V is orthogonal.* However, V is nonsingular and, hence, V^{-1} exists.

Table 6.2 shows the solutions for D and V to be:

$$D = \begin{bmatrix} 0.832 & 0 \\ 0 & 0.498 \end{bmatrix} ; V = \begin{bmatrix} 0.464 & 0.770 \\ 0.886 & -0.638 \end{bmatrix}$$

The reader can then verify that:

$$F = VDV^{-1}$$

$$\overset{\textstyle V}{} \qquad \overset{\textstyle D}{} \qquad \overset{\textstyle V^{-1}}{}$$

$$= \begin{bmatrix} 0.464 & 0.770 \\ 0.886 & -0.638 \end{bmatrix} \begin{bmatrix} 0.832 & 0 \\ 0 & 0.498 \end{bmatrix} \begin{bmatrix} 0.652 & 0.787 \\ 0.906 & -0.474 \end{bmatrix}$$

$$= \begin{bmatrix} 0.599 & 0.122 \\ 0.196 & 0.729 \end{bmatrix}$$

as desired.

Finding the eigenstructure of F is not at all difficult in the special case of a 2×2 matrix. By way of review Table 6.3 shows the steps that are involved. As noted, we first expand the determinantal equation to obtain a quadratic equation

Table 6.3 Finding the Eigenstructure of F

Characteristic Equation	Expansion of Determinant
$\begin{vmatrix} 0.599 - \lambda_j & 0.122 \\ 0.196 & 0.729 - \lambda_j \end{vmatrix} = 0;$	$\lambda_j^2 - 1.328\lambda_j + 0.413 = 0$

Quadratic Formula

$$a = 1; \qquad b = -1.328; \qquad c = 0.413$$

$$\frac{-b \pm \sqrt{b^2 - 4ac}}{2a} = \frac{1.328 \pm \sqrt{1.764 - 1.652}}{2}$$

$$\lambda_1 = 0.832$$

$$\lambda_2 = 0.498$$

Solving for **V**

$$\begin{bmatrix} (0.599 - 0.832) & 0.122 \\ 0.196 & (0.729 - 0.832) \end{bmatrix} \begin{bmatrix} v_{11} \\ v_{21} \end{bmatrix} = \begin{bmatrix} 0 \\ 0 \end{bmatrix}$$

$$\mathbf{v}_1 = \begin{bmatrix} 0.122 \\ 0.233 \end{bmatrix}; \qquad \mathbf{v}_1 \text{ (norm)} = \begin{bmatrix} 0.464 \\ 0.886 \end{bmatrix}$$

$$\begin{bmatrix} (0.599 - 0.498) & 0.122 \\ 0.196 & (0.729 - 0.498) \end{bmatrix} \begin{bmatrix} v_{12} \\ v_{22} \end{bmatrix} = \begin{bmatrix} 0 \\ 0 \end{bmatrix}$$

$$\mathbf{v}_2 = \begin{bmatrix} 0.122 \\ -0.101 \end{bmatrix}; \qquad \mathbf{v}_2 \text{ (norm)} = \begin{bmatrix} 0.770 \\ -0.638 \end{bmatrix}$$

Eigenstructure of **F**

$$\mathbf{D} = \begin{bmatrix} 0.832 & 0 \\ 0 & 0.498 \end{bmatrix}; \qquad \mathbf{V} = \begin{bmatrix} 0.464 & 0.770 \\ 0.886 & -0.638 \end{bmatrix}$$

in λ_j. The general quadratic formula can then be used to solve for the eigenvalues $\lambda_1 = 0.832$ and $\lambda_2 = 0.498$.

These, in turn, are substituted in the matrix equation to solve for the eigenvectors $\mathbf{v}_1$ and $\mathbf{v}_2$. As is conventionally the case, we normalize $\mathbf{v}_1$ and $\mathbf{v}_2$ to unit sum of squares. The eigenvalues are assembled in the diagonal matrix **D** and their associated (normalized) eigenvectors are assembled in the matrix **V**, as shown in Table 6.3. (Appendix B provides further material on the nature and computation of eigenstructures.)

Let us return to Table 6.2. At this point we have the diagonal matrix **D** of eigenvalues and the nonsingular matrix of (right-hand) normalized eigenvectors, **V**. Recalling that we wish to develop weights that result in *standardized* *canonical variate scores*, we now have to consider a further rescaling of the columns of $\mathbf{V}$ to obtain the canonical weights matrix $\mathbf{B}$.

In this case we would like each set of canonical weights scaled so that:

$$\mathbf{b}_j'\mathbf{R}_{xx}\mathbf{b}_j = 1$$

As it stands now, the triple product associated with the first eigenvector $\mathbf{v}_1$ is $\mathbf{v}_1'\mathbf{R}_{xx}\mathbf{v}_1$. This form leads to the scalar:

$$\underset{\mathbf{v}_1'}{(0.464 \quad 0.886)} \underset{\mathbf{R}_{xx}}{\begin{bmatrix} 1 & 0.576 \\ 0.576 & 1 \end{bmatrix}} \underset{\mathbf{v}_1}{\begin{bmatrix} 0.464 \\ 0.886 \end{bmatrix}} = 1.474$$

The square root of this scalar is $\sqrt{1.474} = 1.214$. This value provides the divisor for obtaining the first *scaled* vector of canonical weights:

$$\mathbf{b}_1 = \frac{1}{1.214}\begin{bmatrix} 0.464 \\ 0.886 \end{bmatrix} = \begin{bmatrix} 0.382 \\ 0.730 \end{bmatrix}$$

The same procedure can then be applied to $\mathbf{v}_2' = (0.770 \quad -0.638)$ to find $\sqrt{0.439} = 0.663$ and the second scaled vector of canonical weights:

$$\mathbf{b}_2 = \frac{1}{0.663}\begin{bmatrix} 0.770 \\ -0.638 \end{bmatrix} = \begin{bmatrix} 1.162 \\ -0.981 \end{bmatrix}$$

The two scaled vectors appear in matrix $\mathbf{B}$ in Table 6.2. We see that they satisfy the matrix equation: $\mathbf{B}'\mathbf{R}_{xx}\mathbf{B} = \mathbf{I}$. This scaling assures us that the linear composites $\mathbf{u}_1$ and $\mathbf{u}_2$ are each standardized to unity. That is:

$$\frac{1}{m-1}\mathbf{u}_j'\mathbf{u}_j = \frac{1}{m-1}[\mathbf{b}_j'\mathbf{X}_s'\mathbf{X}_s\mathbf{b}_j] = \mathbf{b}_j'\mathbf{R}_{xx}\mathbf{b}_j = 1$$

or, in terms of the above matrix equation:

$$\mathbf{B}'\mathbf{R}_{xx}\mathbf{B} = \mathbf{I}$$

Next, we solve for $\mathbf{A}$, the weights matrix for the criterion variables, via the matrix equation:

$$\mathbf{A} = \mathbf{R}_{yy}^{-1}\mathbf{R}_{yx}\mathbf{B}\mathbf{D}^{-1/2}$$

$$\mathbf{A} = \begin{bmatrix} 0.534 & 1.092 \\ 0.595 & -1.060 \end{bmatrix}$$

as also shown in Table 6.2. Since **B** is already expressed in scaled form in the process of solving for **A**, the latter matrix of canonical weights is also scaled so that:

$$\frac{1}{m-1}\mathbf{t}'_j\mathbf{t}_j = \frac{1}{m-1}[\mathbf{a}'_j\mathbf{Y}'_s\mathbf{Y}_s\mathbf{a}_j] = \mathbf{a}'_j\mathbf{R}_{yy}\mathbf{a}_j = 1$$

And, we have the counterpart matrix equation:

$$\mathbf{A}'\mathbf{R}_{yy}\mathbf{A} = \mathbf{I}$$

Thus, at this point we have found the two sets of scaled canonical weights, represented by the matrices **B** and **A**, that lead to the statements:

1. The canonical weight vectors $\mathbf{a}_1$ and $\mathbf{b}_1$ when applied to the standardized data matrices $\mathbf{Y}_s$ and $\mathbf{X}_s$, respectively, yield the linear composites $\mathbf{t}_1$ and $\mathbf{u}_1$ that exhibit maximum (two-variable) correlation.

2. Subject to being uncorrelated with $\mathbf{t}_1$ and $\mathbf{u}_1$, the second pair of canonical weight vectors, $\mathbf{a}_2$ and $\mathbf{b}_2$, when applied to the standardized data matrices $\mathbf{Y}_s$ and $\mathbf{X}_s$, respectively, yield the linear composites $\mathbf{t}_2$ and $\mathbf{u}_2$ with maximum (two-variable) correlation.

The canonical correlations themselves can be expressed in matrix form by:

$$\mathbf{M} = \mathbf{B}'\mathbf{R}_{xy}\mathbf{A}$$

This matrix (see Table 6.2) will be of diagonal form:

$$\mathbf{M} = \begin{bmatrix} 0.912 & 0 \\ 0 & 0.706 \end{bmatrix}$$

That is, the first diagonal element (0.912) is the canonical correlation between the first pair of canonical variates and the second diagonal element (0.706) is the canonical correlation between the second pair of canonical variates. We note that the first pair of canonical variates does, indeed, exhibit higher correlation than the second pair.

6.5.5 Structure Correlations and Canonical Variate Scores

In addition to computing the canonical weights, represented by the matrices **B** and **A**, the analyst may wish to examine the linear association between the canonical variates and the original variables. Cooley and Lohnes (1971) call

272 The Multivariate Normal Distribution and Canonical Correlation

these *structure correlations* and they often provide help in interpreting the canonical variates: t_j and u_j.

The structure correlations for the first battery of (criterion) variables are simply:

$$G = R_{yy}A = \begin{matrix} \mathbf{R}_{yy} \\ \begin{bmatrix} 1 & 0.569 \\ 0.569 & 1 \end{bmatrix} \end{matrix} \begin{matrix} \mathbf{A} \\ \begin{bmatrix} 0.534 & 1.092 \\ 0.595 & -1.060 \end{bmatrix} \end{matrix} = \begin{matrix} & \mathbf{t_1} & \mathbf{t_2} \\ Y_1 & \begin{bmatrix} 0.872 & 0.489 \\ Y_2 & 0.898 & -0.439 \end{bmatrix} \end{matrix}$$

Similarly, the structure correlations for the second battery of (predictor) variables are:

$$H = R_{xx}B = \begin{matrix} \mathbf{R}_{xx} \\ \begin{bmatrix} 1 & 0.576 \\ 0.576 & 1 \end{bmatrix} \end{matrix} \begin{matrix} \mathbf{B} \\ \begin{bmatrix} 0.382 & 1.162 \\ 0.730 & -0.981 \end{bmatrix} \end{matrix} = \begin{matrix} & \mathbf{u_1} & \mathbf{u_2} \\ X_1 & \begin{bmatrix} 0.802 & 0.597 \\ X_2 & 0.950 & -0.312 \end{bmatrix} \end{matrix}$$

In the case of the criterion variables Y_1 and Y_2, each exhibits about the same degree of positive association with t_1. Moreover, this association is rather high (0.872 and 0.898 for Y_1 and Y_2, respectively). In the case of t_2, we note that Y_1 and Y_2 carry opposite signs, although their absolute values are approximately the same (and considerably lower than the correlations noted under t_1).

Insofar as the predictors, X_1 and X_2 are concerned, we see that X_2 has the higher correlation (0.950) with u_1 while X_1 has the higher correlation (0.597) with u_2. The differences between the contributions of X_1 and X_2 are not striking, however.

We should also note in passing that canonical correlation also produces a useful byproduct, namely the standardized partial regression coefficients (the betas) for predicting each standardized variable, $\hat{Y}_{s1}$, $\hat{Y}_{s2}$ from the battery of X-variables and, similarly so, those for predicting $\hat{X}_{s1}$ or $\hat{X}_{s2}$ from the battery of Y-variables.

To illustrate, let us return to Table 6.2 and record the following:

$$R_{xx}^{-1}R_{xy} = \begin{bmatrix} 1.497 & -0.862 \\ -0.862 & 1.497 \end{bmatrix} \begin{bmatrix} 0.844 & 0.473 \\ 0.648 & 0.875 \end{bmatrix} = \begin{bmatrix} 0.704 & -0.046 \\ 0.242 & 0.902 \end{bmatrix}$$

The first column vector:

$$\begin{bmatrix} 0.704 \\ 0.242 \end{bmatrix}$$

represents the beta coefficients to be applied to $\mathbf{X}_{s1}$ and $\mathbf{X}_{s2}$ to obtain $\hat{\mathbf{Y}}_{s1}$ while the second column vector:

$$\begin{bmatrix} -0.046 \\ 0.902 \end{bmatrix}$$

is the set of betas for finding $\hat{\mathbf{Y}}_{s2}$ from $\mathbf{X}_{s1}$ and $\mathbf{X}_{s2}$.

Although not illustrated here, one could use the matrix product $\mathbf{R}_{yy}^{-1}\mathbf{R}_{yx}$ to find the analogous betas for predicting $\hat{\mathbf{X}}_{s1}$ and $\hat{\mathbf{X}}_{s2}$ from $\mathbf{Y}_{s1}$ and $\mathbf{Y}_{s2}$. Canonical variate scores are obtained from the multiplications:

$$\mathbf{X}_s\mathbf{B} = \mathbf{U}$$

$$\mathbf{Y}_s\mathbf{A} = \mathbf{T}$$

where $\mathbf{X}_s$ and $\mathbf{Y}_s$ are each expressed in standardized form of mean zero and unit standard deviation, by columns.

Table 6.4 shows the four covariate scores for the sample problem. It could be easily verified that these sets of scores each have mean zero and unit standard deviation. These are the scores, incidentally, that appear in the scatter plots of Figure 6.4.

6.5.6 Recapitulation

At this point we have reported a rather large and diverse set of results. Accordingly, it seems appropriate to summarize the major outcomes of interest.

1. A typical canonical correlation problem involving p criteria and q predictors will provide min (p, q) canonical correlations, $\sqrt{\lambda_j}$.
2. Each of the min (p, q) canonical correlations has associated with it:
   a. A pair of canonical weight vectors $\mathbf{a}_j$ and $\mathbf{b}_j$ to be applied to the *standardized scores* of the criterion and predictor matrices, respectively, to derive a pair of canonical variates $\mathbf{t}_j$ and $\mathbf{u}_j$ with correlation $\sqrt{\lambda_j}$.

Table 6.4 Canonical Variate Scores for Sample Problem

Case	Scores for $\mathbf{t}_1$ and $\mathbf{t}_2$		Scores for $\mathbf{u}_1$ and $\mathbf{u}_2$	
	t_1	t_2	u_1	u_2
1	−1.76	−0.97	−1.57	−0.88
2	−1.58	0.35	−1.63	0.58
3	−0.37	−0.99	0.01	−0.93
4	−0.90	0.79	−0.95	1.04
5	0.03	0.77	0.11	1.01
6	0.56	−1.00	1.06	−0.96
7	0.96	0.76	0.53	−0.95
8	1.24	−0.57	0.64	0.99
9	0.10	−1.00	−0.68	−1.39
10	0.21	2.10	0.27	1.48
11	0.28	0.33	0.85	0.01
12	1.24	−0.57	1.37	−0.01

b. A pair of structure correlation vectors, $\mathbf{g}_j$ and $\mathbf{h}_j$ that, respectively, summarize the correlation of each criterion variable with $\mathbf{t}_j$ and each predictor variable with $\mathbf{u}_j$.

3. The correlations, $\sqrt{\lambda_j}$, are the square roots of a diagonal matrix of eigenvalues and the canonical weight vectors $\mathbf{a}_j$ and $\mathbf{b}_j$ are eigenvectors, scaled so that:

$$\mathbf{a}_j' \mathbf{R}_{yy} \mathbf{a}_j = 1$$

$$\mathbf{b}_j' \mathbf{R}_{xx} \mathbf{b}_j = 1$$

4. The eigenstructure itself can be found for the nonsymmetric matrix $\mathbf{R}_{xx}^{-1} \mathbf{R}_{xy} \mathbf{R}_{yy}^{-1} \mathbf{R}_{yx}$ by direct polynomial expansion.[23]

5. Other outcomes of interest, namely the beta coefficients of each criterion regressed on the set of predictors (and, conversely, each predictor regressed on the set of criteria) are easily computed in the course of finding the canonical correlations.

6. All pairs of linear composites $\mathbf{t}_j$ and $\mathbf{u}_j$ are uncorrelated with other pairs for $j \neq j^*$. Pairs of composites are "extracted" sequentially so as to maximize $\sqrt{\lambda_j}$, subject to their being uncorrelated with previously obtained pairs.[24]

While the above considerations represent the main points, a further topic of interest in canonical correlation is the concept of redundancy.

6.6 Stewart and Love's Redundancy Measure

The canonical correlation $\sqrt{\lambda_j}$ is a symmetric measure of association. Its square is a measure of *shared* variance between the two batteries of variables: criteria and predictors. However, occasions arise in which the researcher is interested in finding out how much variance in one set of variables (e.g., the criterion set) is accounted for by variation in the other set of the variables. That is, interest may center on a *nonsymmetric measure of accounted-for variance,* analogous to the problem of accounting for variance in a single criterion variable in the context of multiple regression.

Stewart and Love (1968) have proposed a measure that they call an *index of redundancy.* Their index expresses the proportion of variance accounted for in some set of variables by a two-step calculation:

1. The proportion of variance in the original battery that is accounted for by that battery's canonical variate, times

2. The proportion of variance that this variate shares with the correspondent canonical variate of the second set of variables.

Considering the first pair of canonical variates, we can denote the redundancy measure of the criterion set as $\mathrm{Rd}_{t_1|u_1}$ and that of the predictor set as $\mathrm{Rd}_{u_1|t_1}$. One could have a high $\sqrt{\lambda_j}$ and, consequently, a high shared variance between the two sets of variables and yet find that $\mathbf{t}_1$ (say) accounts for very little of the variance in the criterion set. If so, the redundancy measure $\mathrm{Rd}_{t_1|u_1}$ would be low.

On the other hand, $Rd_{u_1|t_1}$ may be quite high if u_1 accounts for an appreciable proportion of the variance in the predictor set. Accordingly, high redundancy requires both high λ_j (shared variance) and high variance accounted-for by that battery's canonical variate.

6.6.1 A Numerical Example

An example should help clarify the meaning of the redundancy measure. In terms of the sample problem we have already computed the structure correlations:

$$\mathbf{G} = \begin{array}{c} \\ Y_1 \\ Y_2 \end{array} \begin{bmatrix} \overset{\mathbf{t_1}}{0.872} & \overset{\mathbf{t_2}}{0.489} \\ 0.898 & -0.439 \end{bmatrix} ; \mathbf{H} = \begin{array}{c} \\ X_1 \\ X_2 \end{array} \begin{bmatrix} \overset{\mathbf{u_1}}{0.802} & \overset{\mathbf{u_2}}{0.597} \\ 0.950 & -0.312 \end{bmatrix}$$

The average variance in Y_1 and Y_2 that is extracted by $\mathbf{t_1}$ is simply:

$$\tfrac{1}{2}\mathbf{g}_1'\mathbf{g}_1 = \tfrac{1}{2}(0.872 \quad 0.898)\begin{bmatrix} 0.872 \\ 0.898 \end{bmatrix}$$

$$= 0.783$$

The squared canonical correlation between $\mathbf{t_1}$ and $\mathbf{u_1}$ is, as we know already, 0.832. Hence, the redundancy is 0.783 (0.832) = 0.651.
   That is, we can write:

$$Rd_{t_1|u_1} = \frac{\mathbf{g}_1'\mathbf{g}_1}{p}\lambda_1 = 0.783(0.832) = 0.651$$

where p denotes the number of criterion variables. The average redundancy of $t_2|u_2$ is found analogously as:

$$Rd_{t_2|u_2} = \frac{\mathbf{g}_2'\mathbf{g}_2}{p}\lambda_2$$

$$= 0.216(0.498)$$

$$= 0.108$$

Corresponding redundancy measures for u_1 and u_2 of the predictor set are:

$$Rd_{u_1|t_1} = \frac{\mathbf{h}_1'\mathbf{h}_1}{q}\lambda_1 = 0.773(0.832) = 0.643$$

$$Rd_{u_2|t_2} = \frac{\mathbf{h}_2'\mathbf{h}_2}{q}\lambda_2 = 0.227(0.498) = 0.113$$

where q denotes the number of predictor variables. In this particular problem the counterpart redundancy measures for criteria and predictors, respectively, are fairly similar:

0.651 versus 0.643 (first pair of variates)

0.108 versus 0.113 (second pair of variates)

In other problems, however, the differences could be striking, depending upon how much variance t_1 accounts for in the criterion set versus how much variance u_1 accounts for in the predictor set.[25]

Finally, an overall redundancy measure (across all retained canonical variates) can be found by merely summing the separate measures:

$$Rd_{y|x} = 0.651 + 0.108 = 0.759$$

$$Rd_{x|y} = 0.643 + 0.113 = 0.756$$

In practice, individual redundancy measures would be obtained only for those pairs of canonical variates that pass significance levels, a topic to be taken up in the next chapter.

6.6.2 The Relationship of Redundancy to Squared Multiple Correlations

The redundancy measure bears a simple relationship to the set of squared multiple correlations between each *original* variable of one set and the full set of variables in the other set. More specifically, if we were to compute the squared multiple correlation R_k^2 for each variable in the criterion set, as regressed on the full set of predictor variables, we would find that their average equals the redundancy of the Y-battery, given the X-battery:

$$Rd_{y|x} = \sum_{k=1}^{p} R_k^2/p$$

where p is the number of criterion variables. Thus, the total redundancy of the criterion set, given the predictor set, is nothing more than the average squared multiple correlation of each variable in the Y set with the full set of X variables. Similar conclusions pertain to the relationship of $Rd_{x|y}$ to the averaged squared multiple correlation of each predictor variable on the full set of criterion variables.

If the number of criterion variables p is not equal to the number of predictors q one finds separate redundancy measures for (at most):

$$\min (p, q)$$

In the sample problem, of course, we had the case in which $p = q = 2$.

Stewart and Love's redundancy measure is receiving increased attention in applied studies using canonical correlation. Gleason (1976) shows how the index can be interpreted as a measure of the extent to which one matrix (either the Y-battery or the X-battery) can be reconstructed from knowledge about the other matrix. This view has both practical and theoretical usefulness, similar in spirit to the objectives underlying multiple regression. Moreover, Gleason also shows how the measure generalizes easily to other types of cross-products, such as covariance matrices.

6.7 A Large-Scale Application of Canonical Correlation

At this point we can return to the television commercial data and examine the application of canonical correlation to the criterion-variable set:

C-1: Believability of Alpha's television commercial.

C-2: Post-exposure interest in Alpha's brand of radial tires.

In keeping with the discussion of Chapter 2, two sets of predictor variables come to mind:

A-1: Pre-exposure interest in the product class of radial tires

A-2: Whether Alpha was the brand selected in the respondent's last purchase of replacement tires

A-3: Pre-exposure interest in Alpha's brand of radial tires

and the demographic variables: E-1 through E-7 (age, family size, education, marital status, occupation—professional or white collar, blue collar, other—and income).

In the multiple regression analyses of Chapter 2 we recall that the demographics accounted for virtually no variance at all in the single criterion variable C-2. Given this to be the case, we should not be surprised if the demographics continue to play no predictive role in the canonical correlation that now involves both C-2 and C-1 as criterion variables.

Accordingly, two canonical correlations were run—one with and one without the demographics. The program used here was the BMD-P6M Canonical Correlation Analysis Program (Dixon, 1975). In each case C-1 and C-2 served as the criterion variables and A-1, A-2, and A-3 served as predictors. Table 6.5 shows the results of the two canonical correlations, denoted as run I, without demographics, and Run II, with demographics. Also shown is the input correlation matrix for run I, the main analysis of interest.

6.7.1 Results of Computer Run

Only the results for the first pair of variates are shown since the second pair of canonical variates in each run was clearly not significant, even at the (lenient)

0.2 alpha risk level. The demographic variables' lack of predictive efficacy—first illustrated in Chapter 2—continues to be evident here. When A-1, A-2, and A-3, alone serve as predictors, the canonical correlation is 0.582. Introducing the seven demographic variables as additional predictors results in a canonical correlation of 0.591—certainly not worth the effort.[26]

Accordingly, we confine our attention from this point on to the canonical correlation involving the first pair of variates of run 1 only. The canonical weight vector:

$$\mathbf{a}_1 = \begin{bmatrix} -0.110 \\ 1.069 \end{bmatrix}$$

associated with the criterion set indicates that C-2 (post-exposure interest) carries the higher importance. In the case of the predictor set, the canonical weight vector:

$$\mathbf{b}_1 = \begin{bmatrix} 0.346 \\ 0.141 \\ 0.817 \end{bmatrix}$$

indicates that A-3 (pre-exposure interest) appears most important.

6.7.2 Structure Correlations and Redundancy

Table 6.5 also shows the structure correlations for the first (significant) variate of run I only. In the case of the criterion set these are obtained as:

$$\mathbf{g}_1 = \begin{bmatrix} 1 & 0.659 \\ 0.659 & 1 \end{bmatrix} \begin{bmatrix} -0.110 \\ 1.069 \end{bmatrix} = \begin{bmatrix} 0.594 \\ 0.997 \end{bmatrix}$$

The accompanying structure correlations for the predictor set are:

$$\mathbf{h}_1 = \begin{bmatrix} 1 & 0.086 & 0.222 \\ 0.086 & 1 & 0.258 \\ 0.222 & 0.258 & 1 \end{bmatrix} \begin{bmatrix} 0.346 \\ 0.141 \\ 0.817 \end{bmatrix} = \begin{bmatrix} 0.539 \\ 0.382 \\ 0.930 \end{bmatrix}$$

While the vectors $\mathbf{g}_1$ and $\mathbf{h}_1$ are not proportional to $\mathbf{a}_1$ and $\mathbf{b}_1$, respectively, in this application we note that they do continue to rank the variables in the same way. In particular, C-2 continues to be the most highly related variable to $\mathbf{t}_1$, the canonical variate made up of the criterion variables, while A-3 is the most highly related variable to $\mathbf{u}_1$, the counterpart variate of the predictor set.[27]

Stewart and Love's redundancy measure for the criterion set is simply:

$$Rd_{t_1|u_1} = \tfrac{1}{2}\mathbf{g}_1'\mathbf{g}_1\lambda_1 = 0.674(0.582)^2 = 0.228$$

Table 6.5 Results of Two Canonical Correlations Involving C-1 and C-2 as Criterion Variables — First Variate Only

Criterion Set	Canonical Weights Run I	Run II	Structure Correlations—Run I
C-1	−0.110	−0.093	0.594
C-2	1.069	1.059	0.997
Predictor Set			
A-1	0.346	0.335	0.539
A-2	0.141	0.154	0.382
A-3	0.817	0.797	0.930
E-1		−0.140	
E-2		−0.060	
E-3		−0.030	
E-4 Demographics		0.014	
E-5		0.053	
E-6		−0.010	
E-7		0.138	
Canonical Correlation	0.582	0.591	

Input Correlations for Run I

	C-1	C-2	A-1	A-2	A-3
C-1	1.000				
C-2	0.659	1.000			
A-1	0.202	0.315	1.000		
A-2	0.097	0.218	0.086	1.000	
A-3	0.321	0.539	0.226	0.258	1.000

while that for the predictor set is:

$$Rd_{u_1|t_1} = \tfrac{1}{3}\mathbf{h}_1'\mathbf{h}_1\lambda_1 = 0.434(0.582)^2 = 0.147$$

As observed, neither redundancy value is at all high.[28] Even the "higher" value of 0.228 indicates that only 23 percent of the variance of the criterion set is accounted for by the predictor battery.

In summary, variation in the criterion set, composed of C-1 and C-2, is not highly correlated with either pre-exposure responses or demographics. Variable A-3, pre-exposure interest in Alpha radial tires, is the most important of the predictor variables, while C-2 is the most important criterion variable. Interestingly enough, the simple correlation of A-3 with C-2 is 0.539. This correlation is not appreciably lower than the canonical correlation of 0.582 that is associated with the full batteries.

6.8 Other Aspects of Canonical Correlation

Up until the last decade or so, relatively few applications had been made of canonical correlation. However, more recently a variety of canonical correlation studies have appeared in both the behavioral and administrative fields. In addition, research has been conducted on various extensions of canonical correlation. Here we briefly discuss some newer developments in canonical analysis that could serve to increase its flexibility in the future.

6.8.1 Canonical Analysis of Covariance Matrices

So far, our discussion of canonical analysis has emphasized the *correlation* of two batteries of variables. However, as we know, other types of cross products involving raw sums of squares and cross products, the mean-corrected SSCP matrix, or the covariance matrix can be computed.

In particular one might wish to solve for a set of **a** and **b** canonical weight vectors such that the resulting canonical variates exhibit the property of maximizing the *covariance*, rather than the correlation, between **Y** and **X**. As shown by Van de Geer (1971)—also see Cliff (1966) and Kettenring (1972)—this approach leads to sets of canonical weights, denoted by the matrices **A** and **B**, that are direction cosines with the further property:

$$\mathbf{A'A} = \mathbf{AA'} = \mathbf{I}$$

$$\mathbf{B'B} = \mathbf{BB'} = \mathbf{I}$$

The matrices **A** and **B** are each orthogonal; hence simple rotations, rather than linear transformations involving dimension shrinking or stretching of each space, are involved in the case of covariance matrices.

In this formulation of the problem we let $\mathbf{Y}_d$ and $\mathbf{X}_d$ denote deviation-from-mean matrices (rather than standardized data) and **V** denotes their covariance matrix $\mathbf{Y}_d'\mathbf{X}_d/(m-1)$. Maximizing $\mathbf{a'Vb}$, subject to $\mathbf{a'a} = 1$ and $\mathbf{b'b} = 1$, involves solving the matrix equations:

$$\mathbf{V'Vb} = \mu_j\mathbf{b}$$

$$\mathbf{VV'a} = \mu_j\mathbf{a}$$

where μ_j denotes the j-th common eigenvalue of the *symmetric* matrices $\mathbf{V'V}$ and $\mathbf{VV'}$. The eigenvectors of $\mathbf{V'V}$, and $\mathbf{VV'}$, respectively, are the required matrices of canonical weights **A** and **B** such that:

$$\mathbf{V'V} = \mathbf{BD^*B'} \text{ and } \mathbf{VV'} = \mathbf{AD^*A'}$$

with $\mathbf{D^*}$ denoting the diagonal matrix of μ_j's.

Since $\mathbf{V'V}$ and $\mathbf{VV'}$ are each symmetric, this type of canonical analysis is actually easier to carry out than canonical correlation. Geometrically, the analysis can be viewed as rotating $\mathbf{Y}_d$(via $\mathbf{A}$) and $\mathbf{X}_d$ (via $\mathbf{B}$) so that the projections of $\mathbf{T} = \mathbf{Y}_d\mathbf{A}$ and $\mathbf{U} = \mathbf{X}_d\mathbf{B}$ in each separate subspace have maximum covariance.

While we do not delve into the topic deeply, Table 6.6 shows the covariance matrix $\mathbf{V}$ for the sample problem of Table 6.1, as found from:

$$\mathbf{V} = \mathbf{Y}_d'\mathbf{X}_d/(m - 1)$$

Table 6.6 also shows the various matrices obtained by computing the eigenstructures of $\mathbf{VV'}$ and $\mathbf{V'V}$, namely $\mathbf{A}$, $\mathbf{B}$, and $\mathbf{D}^*$. Finally, the derived canonical variate matrices $\mathbf{T}$ and $\mathbf{U}$ are shown as well.

If we then compute the covariance matrix of $\mathbf{T}$ and $\mathbf{U}$ we get:

$$\mathbf{N} = \frac{1}{m - 1}\mathbf{T'U} = \frac{1}{11}\begin{bmatrix} 76.76 & 12.29 \\ 0.99 & 11.07 \end{bmatrix}$$

$$= \begin{array}{c} \\ t_1 \\ t_2 \end{array}\begin{array}{cc} u_1 & u_2 \\ \begin{bmatrix} 6.98 & 1.12 \\ 0.09 & 1.01 \end{bmatrix} \end{array}$$

The first diagonal entry of $\mathbf{N}$ is the covariance of t_1 and u_1:

$$\text{cov}(t_1, u_1) = 76.76/11$$
$$= 6.98$$

which is the maximum that can be achieved subject to $\mathbf{a'a} = \mathbf{b'b} = 1$.

The second diagonal entry of $\mathbf{N}$ is the covariance of t_2 and u_2; their covariance is 1.01. Since $\mathbf{A}$ and $\mathbf{B}$ are each orthogonal, t_2 and u_2 are determined once t_1 and u_1 are found.

Unlike the case of canonical correlation we note that the off-diagonal entries of $\mathbf{N}$ are *not* zero. That is, in this case the canonical variates t_1 and u_2 are correlated (as are t_2 and u_1). This can be a distinct disadvantage in many kinds of applications.

Van de Geer also discusses canonical rotation towards a fixed "target" matrix, a topic that we reconsider in later chapters dealing with factor analysis. However, for the moment assume that we wished to transform $\mathbf{X}_d$ to best match $\mathbf{Y}_d$. However, if we also were to find $\mathbf{Y}_d\mathbf{A}$, in terms of the above canonical analysis problem, we could express the transformation of $\mathbf{X}_d$ that attempts to best match $\mathbf{Y}_d$ as utilizing the product of $\mathbf{B}$ and $\mathbf{A}^{-1}$:

$$\boxed{\mathbf{P} = \mathbf{X}_d\mathbf{B}\mathbf{A}^{-1}}$$

Table 6.6 Canonical Analysis of the Covariance Matrix—Sample Problem of Table 6.1

Original Covariance Matrix

$$\mathbf{V} = \frac{\mathbf{Y'X}}{(m - 1)}$$

$$= \begin{array}{c} \\ Y_1 \\ Y_2 \end{array} \begin{array}{cc} X_1 & X_2 \\ \begin{bmatrix} 5.09 & 3.18 \\ 2.73 & 4.11 \end{bmatrix} \end{array}$$

Canonical Weight Matrices and Eigenvalues

$$\mathbf{A} = \begin{bmatrix} 0.747 & -0.665 \\ 0.665 & 0.747 \end{bmatrix}; \quad \mathbf{B} = \begin{bmatrix} 0.778 & -0.628 \\ 0.628 & 0.778 \end{bmatrix}; \quad \mathbf{D}* = \begin{bmatrix} 5.777 & 0 \\ 0 & 2.593 \end{bmatrix}$$

Canonical Variates

	t_1	t_2			u_1	u_2
	-5.56	1.27			-4.89	0.84
	-4.73	-0.81			-3.96	-1.20
	-1.33	1.52			-0.68	1.29
	-2.58	-1.39			-1.78	-1.68
	0.25	-1.23			1.04	-1.38
$\mathbf{T} = \mathbf{Y}_d\mathbf{A} =$	2.24	1.02		$\mathbf{U} = \mathbf{X}_d\mathbf{B} =$	2.15	1.59
	0.83	0.93			0.74	1.44
	4.40	0.43			2.43	-1.22
	2.33	-0.40			-2.85	1.76
	-1.91	-0.64			1.81	-2.00
	2.41	-1.81			2.29	0.19
	3.66	1.10			3.70	0.34

That is, the compromise solution can readily be transformed into the fixed-target solution.

The product of **B** and $\mathbf{A}^{-1}$ is:

$$\mathbf{BA}^{-1} = \begin{bmatrix} 0.999 & 0.048 \\ -0.048 & 0.999 \end{bmatrix}$$

indicating that $\mathbf{Y}_d$ and $\mathbf{X}_d$ are *already* lined up quite closely (since $\mathbf{BA}^{-1}$ is very close to an identity matrix). This also helps explain why the covariance (6.98) of t_1 and u_1 is not much larger than that (5.09) of the original case, $\mathbf{y}_{d1}$ and $\mathbf{x}_{d1}$. Not surprisingly, then, the rotation matrix **A** has direction cosines that are almost equal to their counterparts in the rotation matrix **B**, as can be noted in Table 6.6.

In summary, the canonical analysis of covariance matrices displays some simple algebraic properties in that both **A** and **B** are orthogonal; hence *rotations* of $\mathbf{Y}_d$ and $\mathbf{X}_d$ to maximal covariance are entailed. Since $\mathbf{V}'\mathbf{V}$ and $\mathbf{VV}'$ are each symmetric, the eigenstructure problem is particularly simple, leading to the aforementioned orthogonal eigenvectors and common eigenvalues.

However, in this case we assume that the variances of $\mathbf{Y}_d$ and $\mathbf{X}_d$ are meaningful, which can sometimes be questioned in the case of behavioral data. Moreover, there is no guarantee that the off-diagonal pairs of covariates, such as $\mathbf{t}_1$ and $\mathbf{u}_2$, will be uncorrelated. (As a matter of fact, they usually will be correlated, as was the case in this example.)

6.8.2 Generalized Canonical Correlation

Cases can arise in applied multivariate research where the analyst encounters three or more matrices of data for the same individuals or objects. For example, one might be interested in the correlation among three batteries of variables for a specific set of individuals: (a) their brand-choice behaviors; (b) demographics; and (c) life styles. While one could perform three pairs of two-set canonical correlations, we might also be interested in what all three data sets exhibit in common.

Generalized canonical correlation is the name given to procedures used to study association among three or more data sets. One of the first procedures for doing this was proposed by Horst (1961). In Horst's procedure one tries to find a linear composite of each separate data set whose average correlation, over all pairs of linear composites, is maximized.

Most recently, Carroll (1968) has developed a procedure that finds a $K + 1$ canonical variate such that the sum of the (possibly differentially weighted) squared correlations between linear composites of the K original data matrices and the new $K + 1$ variate is maximized.

We let $\mathbf{X}_{d1}, \mathbf{X}_{d2}, \ldots, \mathbf{X}_{dk}, \ldots, \mathbf{X}_{dK}$ denote K data matrices, each of order $m \times n_k$, where all matrices are expressed as deviations about their respective column means. In Carroll's procedure—implemented in a computer program by Carroll and Chang, called CANCOR, standing for CANonical CORrelation—the problem is stated as maximizing:

$$R^2 = \sum_{k=1}^{K} w_k [r(\mathbf{z}, \mathbf{X}_{dk}\mathbf{a}_k)]^2$$

where $r(\mathbf{z}, \mathbf{X}_{dk}\mathbf{a}_k)$ denotes the product moment correlation between $\mathbf{z}$, the $K + 1$ variate, and each linear composite $\mathbf{X}_{dk}\mathbf{a}_k$. The w_k are user-supplied weights and may, of course, be each set to unity if no differential weighting of data sets is desired.

We assume here, for each of illustration, that all w_k *are* equal and, hence, can be ignored in what follows. Carroll shows that $\mathbf{z}$, the $K + 1$ variate (of order $m \times 1$), can be obtained by first forming the symmetric $m \times m$ matrix:

$$Q = \sum_{k=1}^{K} X_{dk}(X'_{dk}X_{dk})^{-1}X'_{dk}$$

which will be either positive definite or positive semidefinite. The variate z is the eigenvector associated with the largest eigenvalue of Q. Additional z-variates can also be obtained, all mutually orthogonal, with successively smaller eigenvalues, up to the rank of $Q(\leqq \sum n_k)$.

Having found a z-variate, one can then find a transformation vector a_k:

$$a_k = (X'_{dk}X_{dk})^{-1}X'_{dk}z$$

that "converts" each X_{dk} to its linear composite that is most highly correlated with the $K + 1$ vector z. Note that a_k is simply a regression vector of the sort described in Chapter 2 and elsewhere. This can be easily seen by observing the structure of the expression on the right, involving the inverse of $(X'_{dk}X_{dk})$. One can, of course, find an a_k-vector for each subsequent z that is extracted. In general, the a_k vectors will not be orthogonal.

We can illustrate the application of CANCOR by considering the illustrative data of Table 6.7. We show three 12×2 matrices, X_{d1}, X_{d2}, and X_{d3}, respectively, where each appears in column-centered form.

CANCOR was then applied to these input data. Two z-type vectors were found and are shown as the matrix Z in Table 6.7. The first column z_1 represents that $K + 1$ vector whose sum of r^2's with the separate linear composites of each input matrix is maximal. The second column z_2 denotes the $K + 2$ vector, orthogonal to z_1, whose sum of r^2's with separate linear composites of each input matrix is maximal (and so on for all z-vectors that might be extracted).

Also shown are the separate matrices of regression vectors A_1, A_2, and A_3. When these matrices postmultiply their respective X_d matrices, one finds linear composites (not shown) that are most highly correlated with their respective z vectors. As noted, each column of the transformation matrix is normalized to unit sum of squares but neither the A_k nor the resulting matrices of linear composites are orthogonal.[29]

The rank of the input matrix Q is 6. The first two eigenvalues of A, $\lambda_1 = 2.68$ and $\lambda_2 = 1.79$, collectively account for 75 percent of the variation in Q, suggesting that further z-variates may not be required.

The correlations of each linear composite with each z-variate are also shown in Table 6.7. In the case of z_1, all correlations exceed 0.9 while in the case of z_2 only the composite associated with the X_{d2} input matrix exceeds 0.9. However, it is clear that all three data sets are highly intercorrelated.[30]

CANCOR and other types of generalized canonical correlation are still relatively little applied, but there are indications that this set of techniques will receive increased attention in the future.

Table 6.7 Applying CANCOR to Three Sets of Column-Centered Data

Case	X_{d1}		X_{d2}		X_{d3}		z_1	z_2
			Column-Centered Input Data				Z-Matrix of Eigenvectors of Q	
1	−5	−2.75	−4.33	−2.42	−4.17	−3.83	0.56	−0.16
2	−3	−3.75	−2.33	−3.42	−1.17	−3.83	0.38	0.32
3	−2	0.25	−1.33	0.58	−3.17	−4.83	0.23	−0.26
4	−1	−2.75	−0.33	−2.42	−0.17	−1.83	0.06	0.38
5	1	−0.75	1.67	−0.42	−2.17	1.67	−0.01	0.18
6	0	2.25	0.67	2.58	1.83	−2.83	−0.15	−0.26
7	3	1.25	−0.33	1.58	−0.17	3.17	−0.17	−0.17
8	2	3.25	2.67	0.58	2.83	−1.83	−0.28	0.09
9	−1	1.25	−3.33	−0.42	−2.17	2.17	0.18	−0.42
10	3	−1.75	2.67	−0.42	1.83	0.17	−0.21	0.54
11	1	0.25	1.67	1.58	2.83	4.17	−0.25	−0.01
12	2	3.25	2.67	2.58	3.83	8.17	−0.45	−0.25
							$\lambda_1 = 2.68$	$\lambda_2 = 1.79$

Transformation Matrices

$$\mathbf{A}_1 = \begin{bmatrix} -0.92 & 0.57 \\ -0.39 & -0.82 \end{bmatrix} \begin{matrix} \mathbf{a}_1 & \mathbf{a}_2 \end{matrix}$$

$$\mathbf{A}_2 = \begin{bmatrix} -0.78 & 0.59 \\ -0.63 & -0.81 \end{bmatrix} \begin{matrix} \mathbf{a}_1 & \mathbf{a}_2 \end{matrix}$$

$$\mathbf{A}_3 = \begin{bmatrix} -0.97 & 0.81 \\ -0.26 & -0.59 \end{bmatrix} \begin{matrix} \mathbf{a}_1 & \mathbf{a}_2 \end{matrix}$$

Correlations Between Each Canonical Variate and the z-Vectors

Data Set	z_1	z_2
1	0.95	0.86
2	0.96	0.95
3	0.92	0.39

6.9 Summary

This chapter marks the introduction to multivariate criterion variables per se. The prototype of techniques for dealing with multiple criteria and multiple predictors is canonical correlation. However, before describing the essentials of canonical correlation, attention was given to the multivariate normal distribution and its relevance for statistical inference in multivariate analysis.

We first described the essentials of the bivariate normal density function and then generalized this function to the multivariate normal. Hypothesis testing, under conditions of known and unknown dispersion matrices, was described illustratively for the case of a single-sample centroid and the case of two independent samples as well.

With this as background, our attention then turned to canonical correlation. The canonical model was described both intuitively and formally. A small problem involving artificial data was solved and various ancillary measures, such as structure correlations and Stewart and Love's redundancy index, were computed. Canonical correlation was then applied to a more realistic data set drawn from the television commercial study.

More recent developments in the field were then described briefly. In particular, some comments were made on the canonical analysis of covariance, rather than correlation, matrices and generalized canonical correlation. Generalized (involving three or more data sets) canonical correlation shows particular promise for coping with various applied problems in multivariate analysis, including the matching of different solutions in factor analysis and finding composite configurations in multidimensional scaling, topics that are considered in Chapters 8 and 9.

Review Questions

1. Assume a bivariate normal density function with:

$$\mathbf{u} = \begin{bmatrix} 15 \\ 10 \end{bmatrix} ; \sum = \begin{bmatrix} 16 & 8 \\ 8 & 9 \end{bmatrix}$$

   a. Compute $\sum^{-1}$ and find the quadratic form.
   b. Draw the 95 percent ellipse for individual observations.
   c. What is the probability that a sample observation would be as discrepant as:

$$\mathbf{x} = \begin{bmatrix} 25 \\ 20 \end{bmatrix} ?$$

2. Considering the same bivariate normal density function as in question 1, assume that a sample of 50 observations was drawn with a centroid of

$$\mathbf{x} = \begin{bmatrix} 16 \\ 11 \end{bmatrix}$$

   a. How likely would a centroid as discrepant as this occur under the null hypothesis?
   b. What is the probability of a centroid as discrepant as this, if the sample size was 100?

3. Referring to question 2, perform two separate Z tests on the centroid $\begin{bmatrix} 16 \\ 11 \end{bmatrix}$ given: (a) a sample size of 50 and (b) a sample size of 100.

a. If the alpha risk were set at 0.05, what is the conclusion under each sample size?

b. How do these results compare with those of question 2?

4. Consider the following data:

Case	Y_1	Y_2	X_1	X_2	Z_1	Z_2
1	2	1	9	2	12	2
2	4	5	11	7	13	7
3	3	3	15	3	7	4
4	9	4	7	5	9	9
5	7	8	8	9	5	6
6	5	9	4	4	4	14
7	8	11	7	8	6	3
8	6	14	3	2	2	11
9	4	7	2	7	3	9
10	9	13	4	13	3	10

Canonically correlate the 10×2 matrix $\mathbf{Y}$ with the 10×2 matrix $\mathbf{X}$ according to the procedure illustrated in Table 6.2. Next, compute and interpret the set of structure correlations and Stewart and Love's redundancy index.

5. Split the television commercial data into first and second halves of 126 respondents each, using C-1 and C-2 versus A-1, A-2, and A-3.

a. Perform split-half canonical correlations and check the stability of canonical weights across the separate halves.

b. How do the split-half results compare with the total-sample results shown in Table 6.5?

6. Refer to the data shown in question 4. Use Carroll and Chang's CANCOR procedure to find the first variate $\mathbf{z}_1$ only for the three column-centered matrices.

a. What are the product moment correlations between $\mathbf{z}_1$ and each of the three linear composites?

b. Use Carroll and Chang's CANCOR procedure to canonically correlate $\mathbf{Y}_d$ and $\mathbf{X}_d$ (only) in column centered form. How do these results compare with the two-set canonical correlation results from question 4?

Notes

1. The Z (standard unit normal) is used if the variance is assumed known; the t distribution is used when the variance is estimated from the sample.

2. Our discussion from this point on is in terms of universe (rather than sample) characteristics.

3. It should be mentioned that the bivariate normal is only a probability density function per se. Thus, one interprets the ordinate as a kind of average probability per unit

area in the small region, $X_1 \pm \Delta X_1, X_2 \pm \Delta X_2$. It is the *volume* under the surface of the density function that is interpretable as a probability.

4. However, the centroid and variance of X_1 will, in general, differ as X_2 assumes different values, assuming X_1 and X_2 are correlated.

5. For example, if the expression in brackets were -3, then $\exp(-3) = e^{-3}$ would still be positive; that is, $e^{-3} = 0.05$.

6. When $y = 0$, that is $\exp(y) = 1$, the ordinate of the surface is above the centroid (μ_1, μ_2). However, the farther the deviation $(X_1 - \mu_1, X_2 - \mu_2)$ from the centroid, the lower the ordinate, as can be observed from Panel I of Figure 6.1.

7. We shall assume throughout that the covariance matrix Σ is nonsingular and, hence, of full rank. While this is not a necessary assumption it is appropiate for an introductory discussion.

8. In matrix algebra, F_2 is often called a *quadratic form*. Quadratic forms provide a way to express a polynomial function that is homogeneous and of the second degree as the triple product of a vector by a matrix by a vector (where the first vector is the transpose of the second). For example, $f(x_1, x_2) = 2x_1^2 + 8x_1x_2 + 6x_2^2$ can be written in matrix format as the quadratic form:

$$f(\mathbf{x}) = (X_1, X_2) \begin{bmatrix} 2 & 4 \\ 4 & 6 \end{bmatrix} \begin{bmatrix} X_1 \\ X_2 \end{bmatrix}$$

Additional discussion of quadratic forms can be found in Chapter 5 of Green (1976).

9. The maximum height of the surface is when $X_1 = \mu_1$; $X_2 = \mu_2$. At this point (i.e., the centroid itself), the exponential equals zero and the height of the surface is given by the positive constant $1/(2\pi\sigma_1\sigma_2\sqrt{1-\rho^2})$. Otherwise the expression in brackets, after simplification and setting it equal to some constant K_i, is the equation of an ellipse or, by taking different values of K_i, a family of ellipses.

10. The contours in Figure 6.4 are found easily by setting X_{d1} equal to some positive constant and solving for X_{d2}. The general quadratic formula can be used for this purpose.

11. The reader should note that in the chi square example we simply used $1 - \alpha = 0.95$ for the percentile of interest. We could just as easily have used $\alpha = 0.05$ itself.

12. Notice that if the test is univariate (i.e., $n = 1$), we have the well-known special case in which $t^2 = F$ with 1 degree of freedom in the numerator and $m - 2$ degrees of freedom in the denominator.

13. The T^2 statistic can also be used in testing for the significance of the difference in centroids between two *matched* samples. If we let $\mathbf{d}_i$ denote the difference vector between members of the i-th matched pair, then:

$$T^2 \text{ (matched samples)} = m[\bar{\mathbf{d}}'\mathbf{C}_w^{-1}\bar{\mathbf{d}}]$$

where $\bar{\mathbf{d}}$ denotes the mean difference vector across all pairs of matched samples and $\mathbf{C}_w$ is the covariance matrix of the differences $\mathbf{d}_i$. Note that T^2 for matched samples is of the same form as T^2 for the one-sample test.

14. In this particular version $\mathbf{W}$ assumes the role of the error matrix $\mathbf{S}_E$ and $\mathbf{A}$ assumes the role of the hypothesis matrix $\mathbf{S}_H$. In this case $\mathbf{T}$ denotes the sum of $\mathbf{S}_H$ and $\mathbf{S}_E$.

15. More specifically, if we have only a single variate ($n = 1$), Wilks' lambda assumes the form:

$$\Lambda_{(n=1)} = \frac{1}{1 + [(K-1)/(m-K)F]}$$

16. We continue to assume a nine-point rating scale where higher numbers indicate higher degrees of the attribute being rated.

17. This does not mean that $\mathbf{t}_1$ and $\mathbf{t}_2$ (or $\mathbf{u}_1$ and $\mathbf{u}_2$) in Figure 6.6 are orthogonal, but only that the sets of *scores* across pairs of canonical variates are uncorrelated. We discuss this distinction later.

18. Both of these correlations have already been shown in Table 6.1.

19. For greater simplicity of notation we also assume that $\mathbf{y}$ and $\mathbf{x}$ represent standardized variables (as usually denoted by $\mathbf{y}_s$ and $\mathbf{x}_s$).

20. We use $m - 1$ as the divisor because the original standardized scores employed $m - 1$ in the computation of the standard deviations of the Y and X variables.

21. In this class of problems all of the λ_j's and $\mathbf{b}_j$'s will be real-valued. In addition, the λ_j's will be non-negative.

22. As described in section B.5 of Appendix B, if some matrix $\mathbf{A}$ is *symmetric*, it can be decomposed into the triple product:

$$\mathbf{A} = \mathbf{TDT}'$$

where $\mathbf{T}$ is *orthogonal* ($\mathbf{T}'\mathbf{T} = \mathbf{TT}' = \mathbf{I}$) and $\mathbf{D}$ is diagonal.

23. Still other (computer-based) methods are available (Stewart, 1973) for finding eigen-structures of nonsymmetric matrices. Also, see section B.5 for an alternative procedure that symmetrizes $\mathbf{F}$.

24. By "sequentially" is meant that the eigenvalues obtained from the decomposition of $\mathbf{F}$ are ordered in decreasing size. Their associated eigenvectors are correspondingly ordered as well.

25. It should be noted that if all canonical variates are extracted, one accounts for all of the variance in the battery. In our example:

$$\tfrac{1}{2}\mathbf{g}_1'\mathbf{g}_1 + \tfrac{1}{2}\mathbf{g}_2'\mathbf{g}_2 = 0.78 + 0.22 = 1.0$$

$$\tfrac{1}{2}\mathbf{h}_1'\mathbf{h}_1 + \tfrac{1}{2}\mathbf{h}_2'\mathbf{h}_2 = 0.77 + 0.23 = 1.0$$

26. It is of interest to note, however, that the canonical weights associated with C-1, C-2, A-1, A-2, and A-3 change very little in the presence of the seven demographic variables.

27. To maintain continuity with earlier sections we continue to use $\mathbf{t}$ to denote the ca-nonical variate with respect to criteria and $\mathbf{u}$ the canonical variate with respect to pre-dictors, even though $p < q$ in this application.

28. It should be mentioned that BMD-P6M computes structure correlations but not the Stewart-Love redundancy measures.

29. However, Carroll shows how one could orthogonalize each data matrix to its own linear composite, prior to computing the next regression vector; if this procedure is followed each time, the successive linear composites will be orthogonal.

30. Interset correlations of linear composites associated with the first ($\mathbf{z}_1$) vector are 0.90 for set 1 versus set 2, 0.79 for set 1 versus set 3, and 0.83 for set 2 versus set 3.

CHAPTER 7

Multivariate Discrimination and Analysis of Variance and Covariance

7.1 Introduction

In Chapters 3 and 4, ANOVA, ANCOVA, and two-group discriminant analysis were discussed as illustrations of techniques that entailed either a dummy-coded criterion (two-group discriminant analysis) or dummy-coded predictors (ANOVA and ANCOVA).

In this chapter we discuss the analogous techniques:

1. Multiple discriminant analysis (MDA)

2. Multivariate analysis of variance (MANOVA) and multivariate analysis of covariance (MANCOVA)

that are appropriate for the study of multiple criterion, multiple predictor association.

In multiple discriminant analysis the criterion variables are expressed as dummies while in MANOVA it is the predictor set that is dummy-coded. In MANCOVA there is a mixture of dummy-coded and interval-scaled predictors.

By means of a small numerical problem, we start off the chapter with a discussion of the principal statistic, Wilks' lambda, that is used in significance tests involving these techniques. Wilks' lambda—or, more usually, the Bartlett or Rao functions of it—is used to test for centroid equality in the case of multiple variates and several groups.

We then consider the technique of multiple discriminant analysis (MDA). The MDA model is described both intuitively and formally. Hypothesis testing is also examined within the MDA framework. By reversing the role of predictors and criteria we see how a test of the significance of multiple discrimination is akin to a test of centroid equality in the predictor space.

The same sample data are then analyzed via canonical correlation (with dummy-variable criteria) and the two solutions are compared. In this way we can see the commonality between tests of significance in MDA and counterpart tests in canonical correlation.

Multivariate analysis of variance and covariance is then introduced in the context of a sample problem involving a two-way factorial design. We also consider Wilks' lambda test in this new context.

MDA, MANOVA, and MANCOVA are then applied to various aspects of the TV commercial data so as to provide problems of realistic size and complexity. This section also gives us an opportunity to describe some of the better known computer programs that are available for carrying out multiple criterion, multiple predictor association.

7.2 Multivariate Tests of Centroid Equality for Several Groups

The reader will recall that Chapter 6 was introduced by a discussion of the multivariate normal distribution and multivariate generalizations of the Z test and t test, namely, χ^2 and Hotelling's T^2. These statistics are used in testing whether the centroid of a single sample is equal to some specified vector μ or whether the centroids of two sampled universes are equal to each other. If the dispersion matrix Σ is assumed to be known, chi square is the relevant statistic; if Σ has to be estimated from the sample(s), T^2 is used.

Wilks' lambda Λ is appropriate for testing centroid equality in the case of *three or more* groups where the covariance matrices are assumed to be unknown but equal.[1] It is also applicable to testing the significance of canonical correlations.

For illustrative purposes we can start out with the simplest case, namely a test of centroid equality involving two or more variates measured on three or more groups. We can later extend this basic case to more complex situations, such as canonical correlation and factorial designs in the context of MANOVA and MANCOVA.

7.2.1 Wilks' Lambda Statistic

To motivate the discussion of Wilks' lambda, consider the response data in Table 7.1. We shall assume that three test groups are exposed to three different treatments and two response measures, Y_1 and Y_2, are obtained. As shown, the sample centroids of the three groups are:

Table 7.1 Three Experimental Groups' Responses on Two Criterion Variables

Group	Y_1	Y_2	SSCP Matrices*
1 $\begin{cases} 1 \\ 2 \\ 3 \\ 4 \end{cases}$	1 2 2 3	1 1 2 2	$T = \begin{bmatrix} 170.25 & 128.25 \\ 128.25 & 106.92 \end{bmatrix}$
2 $\begin{cases} 1 \\ 2 \\ 3 \\ 4 \end{cases}$	5 5 6 7	4 6 5 4	$W = \begin{bmatrix} 6.75 & 1.75 \\ 1.75 & 8.75 \end{bmatrix}$
3 $\begin{cases} 1 \\ 2 \\ 3 \\ 4 \end{cases}$	10 11 11 12	8 7 9 10	$A = \begin{bmatrix} 163.50 & 126.50 \\ 126.50 & 98.17 \end{bmatrix}$
	$\overline{Y}_j$ 6.25	4.92	

Group Centroids

$$\overline{y}_1' = (2, 1.5); \quad \overline{y}_2' = (5.75, 4.75); \quad \overline{y}_3' = (11, 8.5)$$

Computing the Determinants of **W** and **T**

$$|W| = (6.75)(8.75) - (1.75)^2$$
$$= 56$$

$$|T| = (170.25)(106.92) - (128.25)^2$$
$$= 1755.07$$

*See Table 7.3 for detailed calculations of the SSCP matrices.

$$\overline{y}_1' = (2, \quad 1.5); \quad \overline{y}_2' = (5.75, \quad 4.75); \quad \overline{y}_3' = (11, \quad 8.5)$$

Are the universe centroids of the three groups significantly different?

In the preceding chapter Wilks' lambda statistic was defined as the ratio of two determinants:

$$\boxed{\Lambda = \frac{|W|}{|T|}}$$

where **W** is the pooled within-groups SSCP matrix and **T** is the total-sample SSCP matrix.

In testing for the significance of centroid differences involving G groups, the hypotheses are:

$$H_1: \boldsymbol{\mu}_1 = \boldsymbol{\mu}_2 = \cdots = \boldsymbol{\mu}_g = \cdots = \boldsymbol{\mu}_G$$

$H_1:$ Not all $\boldsymbol{\mu}_g$ are equal.

where each $\boldsymbol{\mu}_g$ is a centroid (i.e., a vector of means).[2]

Wilks' lambda is a *general* statistic for handling this problem. Indeed, although we do not pursue the matter here, such familiar statistics as t^2, F, Hotelling's T^2, and Mahalanobis' D^2 can all be viewed as special cases of Λ. This means that in all of our discussions of Part II one *could* have computed Λ and used appropriate functions of it in place of t^2, F, T^2, or D^2, as the case may be.

However, the relatively simple functions that convert Λ to these more familiar statistics disappear when we proceed to cases involving *several criteria and several groups*. In more general situations we find that the distribution of Λ under the null hypothesis is quite complicated, necessitating the employment of various approximations to it.

The exact sampling distribution of Λ under the null hypothesis has been derived (Schatzoff, 1966) only recently. Even at that, exact values are possible to obtain only when n, the number of variates, is an even number and/or G, the number of groups, is an odd number. Otherwise, interpolation between tabled values is required. For this reason and also because Schatzoff's tables do not include many combinations of parameters that may be of interest, statisticians have tended to work with *approximations* to the distribution of various functions of Λ.

However, there are certain special cases where (a function of) Λ is distributed exactly as the familiar F distribution. In the cases where:

1. $G = 2$ or $G = 3$ groups, for any number of variates, or

2. $n = 1$ or $n = 2$ variates for any number of groups

it turns out that a simple function of Λ is *exactly* distributed as F. These cases are shown in Table 7.2.

Since the sample problem of Table 7.1 involves only $n = 2$ variates and $G = 3$ groups, we can find an exact F-value equivalent to (a function of) Λ. First, we calculate Λ from Table 7.1 as follows:

$$\Lambda = \frac{|\mathbf{W}|}{|\mathbf{T}|} = \frac{56}{1755.07} = 0.0319$$

As should be noted, the *smaller* Λ is, the more significant are the centroid differences.[3]

Next, using the special case of $n = 2$ criterion variables, we can find an exact F value, by reference to Table 7.2. Since $n = 2$:

Table 7.2 Special Cases in Which Functions of Λ Are Exactly Distributed as F

$$
\text{Any } n \text{ variates,}
\begin{cases}
G = 2: & \left[\dfrac{1 - \Lambda}{\Lambda}\right] \dfrac{m - n - 1}{n} = F_{[n,\, m - n - 1]} \\[2em]
G = 3: & \left[\dfrac{1 - \Lambda^{1/2}}{\Lambda^{1/2}}\right] \dfrac{m - n - 2}{n} = F_{[2n,\, 2(m - n - 2)]}
\end{cases}
$$

$$
\text{Any } G \text{ groups,}
\begin{cases}
n = 1: & \left[\dfrac{1 - \Lambda}{\Lambda}\right] \dfrac{m - G}{G - 1} = F_{[G - 1,\, m - G]} \\[2em]
n = 2: & \left[\dfrac{1 - \Lambda^{1/2}}{\Lambda^{1/2}}\right] \dfrac{m - G - 1}{G - 1} = F_{[2(G - 1),\, 2(m - G - 1)]}
\end{cases}
$$

$$
\left[\frac{1 - \Lambda^{1/2}}{\Lambda^{1/2}}\right] \cdot \frac{m - G - 1}{G - 1} = F_{[2(G - 1),\, 2(m - G - 1)]}
$$

Substituting the appropriate numerical values, we have:

$$
F = \left[\frac{1 - (0.0319)^{1/2}}{(0.0319)^{1/2}}\right] \frac{(12 - 3 - 1)}{(3 - 1)} = 18.395
$$

with 4 and 16 degrees of freedom. This (exact) F value is highly significant beyond the 0.01 alpha level.[4]

Of course, cases arise which do not meet the special conditions of Table 7.2 and in these more general situations, various approximations to the exact distribution of Λ are needed. These approximations—one based on the χ^2 distribution and one based on the F distribution—apply to *functions of* Λ rather than Λ itself. The two functions of Λ are, respectively, (a) Bartlett's V statistic and (b) Rao's Ra statistic.

7.2.2 Bartlett's Chi Square Approximation

Bartlett (1947) proposed a logarithmic function of Λ that can be used as a test statistic:

$$
V = -[m - 1 - (n + G)/2]\, ln\, \Lambda
$$

where m denotes the total sample size, n denotes the number of variates, G denotes the number of groups, and ln is the natural logarithm of Λ. In this case V is distributed approximately as chi square with $n(G - 1)$ degrees of freedom.

Although we already have an exact F value for the special case of the sample data of Table 7.1, for comparison purposes let us still compute Bartlett's V statistic and the associated chi square approximation to the distribution of V.

$$V = -[m - 1 - (n + G)/2]\ ln\ \Lambda$$
$$= -[12 - 1 - (2 + 3)/2]\ ln\ (0.0319)$$
$$= -8.5(-3.4450)$$
$$= 29.282$$

Next, we consider V to be approximately distributed as a χ^2 variate with $n(G - 1) = 4$ degrees of freedom. The tabular χ^2 value (Table C.2 in Appendix C) at the 0.01 alpha level is only 13.28, suggesting that V is highly significant for the sample-problem data. In short, the chi square approximation to V leads to the same conclusion as the (exact) F test for the particular data of Table 7.1.

7.2.3 Rao's F Ratio Approximation

Rao (1952) has proposed a different function of Λ whose distribution is approximated by the F distribution. Rao's statistic is defined as:

$$Ra = \frac{1 - \Lambda^{1/s}}{\Lambda^{1/s}} \cdot \frac{1 + ts - n(G - 1)/2}{n(G - 1)}$$

where:

$$t = m - 1 - (n + G)/2$$

$$s = \sqrt{\frac{n^2(G - 1)^2 - 4}{n^2 + (G - 1)^2 - 5}}$$

$$s = 1 \text{ if } n^2 + (G - 1)^2 = 5$$

(As can be easily noted, this statistic is more tedious to compute than Bartlett's V.) The Ra statistic is approximately distributed as F with $n(G - 1)$ degrees of freedom in the numerator and $1 + ts - n(G - 1)/2$ degrees of freedom in the denominator.[5] However, Ra is *exactly* distributed as F in the cases of $n = 1$ or 2 or $G = 2$ or 3, as already shown in Table 7.2.

Although we have computed F for the special case represented by the sample problem, it is still useful to show the computations involved in the more general Ra formulation. We first make the necessary substitutions to find t and s:

$$t = m - 1 - \frac{(n + G)}{2}; \quad s = \sqrt{\frac{n^2(G - 1)^2 - 4}{n^2 + (G - 1)^2 - 5}}$$

$$= 12 - 1 - \frac{(2 + 3)}{2}; \quad = \sqrt{\frac{4(3 - 1)^2 - 4}{4 + (3 - 1)^2 - 5}}$$

$$= 8.5 \qquad = 2$$

Having found these quantities, we compute Ra as follows:

$$Ra = \frac{1 - \Lambda^{1/2}}{\Lambda^{1/2}} \cdot \frac{1 + ts - [n(G - 1)/2]}{n(G - 1)}$$

$$= \frac{1 - (0.0319)^{1/2}}{(0.0319)^{1/2}} \cdot \frac{1 + 8.5(2) - [2(3 - 1)/2]}{2(3 - 1)}$$

$$= 18.395 = F_{[n(G-1),\, 1 + ts - n(G-1)/2]}$$

The appropriate degrees of freedom for F are

Numerator: $n(G - 1) = 2(3 - 1) = 4$

Denominator: $1 + ts - \dfrac{n(G - 1)}{2} = 1 + 8.5(2) - \dfrac{2(3 - 1)}{2} = 16$

As expected, we obtain the same results from the general formula for Ra as found in the special (exact) case for the sample problem involving $n = 2$ and $G = 3$.

In practice, both Bartlett's V (approximated by χ^2) and Rao's Ra (approximated by F) are used in various computer programs where Wilks' Λ is computed. It should be reiterated that each of these tests of centroid equality assumes that the covariance matrices are equal across groups:

$$\sum_1 = \sum_2 = \cdots = \sum_G$$

As illustrated in Chapter 4, tests are available for first checking on the equality of covariance matrices before conducting the test of centroids.

7.3 Multiple Discriminant Analysis

Chapter 4's discussion of two-group discriminant analysis provides most of the essentials for our present extension to three or more groups. As recalled, the pooled within-groups covariance matrix C_w played a major role in the computation of discriminant weights. We also illustrated how two-group discriminant weights were related (up to a scale transform) to multiple regression weights obtained by employing a binary-valued criterion variable.

In multiple discriminant analysis (MDA) similar kinds of relationships hold; for example, in MDA the pooled within-groups covariance (actually the pooled within-groups SSCP) matrix plays a central role. Furthermore, a canonical correlation, in which the set of criterion variables is expressed as dummies and the predictors as interval-scaled variables, provides canonical weights that are the same, up to a scale transform, as the discriminant weights computed from MDA.

First, however, let us examine the multiple discriminant problem on its own terms. Here we shall be interested in how the problem is formulated and solved as the eigenstructure of a certain type of nonsymmetric matrix. Throughout the ensuing discussion the reader should bear in mind that the objectives of MDA are generalizations of those of the two-group case:

1. To find linear composites of the predictor variables with the property of maximizing among-groups to within-groups variability, subject to each linear composite being uncorrelated with previously obtained composites. The composites are computed so that the accounted-for variation appears in decreasing order of magnitude.

2. To test whether the group centroids are, indeed, different and, if so, the number of discriminant axes (i.e., the dimensionality of the discriminant space) for which this is the case.

3. To find which predictors contribute most to discriminating among the groups.

4. To assign new objects to groups on the basis of the objects' predictor-variable profiles.[6]

Moreover, as in the case with two-group discrimination, we assume in MDA that the within-groups SSCP matrices, while unknown, are equal. In the case of classification via multiple discriminant functions we further assume that the among-groups prior probabilities and misclassification costs are equal. (Otherwise, Bayesian-type procedures, of the sort described in Chapter 4, are appropriate here, as well.)

7.3.1 Informal Introduction

To provide background for formulating the MDA model, let us continue to consider the sample problem outlined in Table 7.1. In this case we have $G = 3$ groups. Now, however, let us give a different interpretation to the "criterion" variables Y_1 and Y_2. Let us assume that these are *predictor* variables—denoted X_1 and X_2—and our job is to find a linear composite of X_1 and X_2 with the property of maximizing among-groups to within-groups variation.[7]

In the two-group discriminant case of Chapter 4 we wished to find a *single axis* (linear composite) along which the groups were maximally separated. Similarly, in MDA we wish to find an axis with the property of maximizing among-groups to within-groups variability of projections onto this axis. A complicating feature in the case of three or more groups is that, in general, one axis does not exhaust the discrimination potential. Generally speaking, with G groups and n predictors, we can find:

$$\min (G - 1, n)$$

discriminant axes. In the data of Table 7.1 $G - 1 = n = 2$. In the usual type of applied problem, however, the number of predictor variables will greatly exceed the number of groups; hence, a good deal of parsimony can be achieved via the computation of, at most, $G - 1$ discriminant axes. We say "at most" since not all of the $G - 1$ axes may show statistically significant variation among the groups.

Before describing procedures for finding the discriminant axes of interest, it is helpful to plot the data of Table 7.1. Actually, we shall plot the data in mean-centered form so that the original centroid of the total sample:

$$\mathbf{y} = \begin{bmatrix} 6.25 \\ 4.92 \end{bmatrix}$$

is at the origin and each observation is expressed as a deviation from the total-sample centroid.

Figure 7.1 shows the scatter plot involving translation of the data to a centroid-centered origin. Notice also that two new axes, $\mathbf{z}_1$ and $\mathbf{z}_2$, have been plotted in the space.

Next, suppose we imagine projecting the points first onto $\mathbf{z}_1$ and then onto $\mathbf{z}_2$. In so doing we could also find the mean projection of each group on each axis. The objective of MDA is to find those axes ($\mathbf{z}_1$ and $\mathbf{z}_2$ in this illustration) with the property of maximizing among-groups to within-groups variability, subject to the scores on each new discriminant axis being uncorrelated with the scores on previously obtained discriminant axes. That is, $\mathbf{z}_1$ is the single axis that maximizes among-groups to within-groups variability. The second axis,

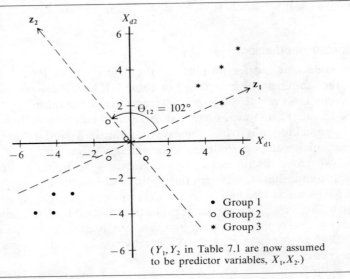

Figure 7.1 A Plot of the Mean-Centered Data of Table 7.1

z_2, maximizes residual among-groups to within-groups variability, subject to its point projections being uncorrelated with the point projections on z_1.

Notice that z_1 and z_2 are *not* orthogonal (their angle is 102°) even though the projections on them are uncorrelated.[8] The reader may also recall from Chapter 6 that the sets of axes t_j and u_j, called canonical variates in the context of canonical correlation, were also uncorrelated but not orthogonal within their respective spaces. The present problem reduces to an eigenstructure formulation involving a nonsymmetric matrix and is again analogous to the canonical correlation problem described in the preceding chapter.

7.3.2 Formal Representation

We recall from Chapter 4 (Table 4.2) that given a matrix of predictor variables, partitioned into groups (three groups in the present case), we can obtain W, the SSCP matrix representing the pooled within-groups variation and A, the SSCP matrix representing the among-groups variation. Table 7.3 shows the appropriate calculations for the present problem using the sample data of Table 7.1.

We also know from Chapter 4 that if a linear composite $z = X_d k$ of the mean-corrected matrix X_d is found, the pooled within-groups sum of squares will be:

$$\boxed{SSW(z) = k'Wk}$$

Similarly, the among-groups sum of squares SSA(z) is given by:

$$\boxed{SSA(z) = k'Ak}$$

The ratio of the among-groups to within-groups sums of squares of the linear composite z is:

$$\boxed{\frac{SSA(z)}{SSW(z)} = \frac{k'Ak}{k'Wk} = \lambda}$$

where λ, representing the ratio of two scalars, is itself a scalar.[9]

The problem is to maximize λ with respect to k. Appendix B shows that this involves finding the partial derivative of λ with respect to k and setting it equal to 0. After simplification the resulting matrix equation is:

$$\boxed{(A - \lambda W)k = 0}$$

We assume that W is nonsingular and, thus, that W^{-1} exists. Hence, we can premultiply both sides of the above equation by W^{-1} and get:

Table 7.3 Calculation of SSCP Matrices from Sample Data of Table 7.1

<table>
<tr><td align="center">Define</td><td align="center">Data Matrix</td></tr>
</table>

$$\mathbf{H} = \begin{bmatrix} 1 & 0 & 0 \\ 1 & 0 & 0 \\ 1 & 0 & 0 \\ 1 & 0 & 0 \\ \hline 0 & 1 & 0 \\ 0 & 1 & 0 \\ 0 & 1 & 0 \\ 0 & 1 & 0 \\ \hline 0 & 0 & 1 \\ 0 & 0 & 1 \\ 0 & 0 & 1 \\ 0 & 0 & 1 \end{bmatrix} \qquad \mathbf{X} = \begin{bmatrix} 1 & 1 \\ 2 & 1 \\ 2 & 2 \\ 3 & 2 \\ \hline 5 & 4 \\ 5 & 6 \\ 6 & 5 \\ 7 & 4 \\ \hline 10 & 8 \\ 11 & 7 \\ 11 & 9 \\ 12 & 10 \end{bmatrix}$$

Then, find

$$\overline{\mathbf{X}}_g = (\mathbf{H}'\mathbf{H})^{-1}\mathbf{H}'\mathbf{X} = \begin{bmatrix} 2.00 & 1.50 \\ \hline 5.75 & 4.75 \\ \hline 11.00 & 8.50 \end{bmatrix} \quad \begin{array}{l} \text{Group 1 centroid} \\ \\ \text{Group 2 centroid} \\ \\ \text{Group 3 centroid} \end{array}$$

$$\overline{\mathbf{x}}' = \frac{1}{12}\mathbf{1}'\mathbf{X} = (6.25, 4.92) \qquad \text{Total-sample centroid}$$

where $\mathbf{1}'$ is a 1×12 unit vector.

Matrix of Within-Groups Deviations	Matrix of Among-Groups Deviations	Matrix of Total-Sample Deviations
$\mathbf{P} = \mathbf{X} - \mathbf{H}\overline{\mathbf{X}}_g$	$\mathbf{Q} = \mathbf{H}\overline{\mathbf{X}}_g - \mathbf{1}\overline{\mathbf{x}}'$	$\mathbf{R} = \mathbf{X} - \mathbf{1}\overline{\mathbf{x}}'$

$$\mathbf{P} = \begin{bmatrix} -1 & -0.5 \\ 0 & -0.5 \\ 0 & 0.5 \\ 1 & 0.5 \\ \hline -0.75 & -0.75 \\ -0.75 & 1.25 \\ 0.25 & 0.25 \\ 1.25 & -0.75 \\ \hline -1 & -0.5 \\ 0 & -1.5 \\ 0 & 0.5 \\ 1 & 1.5 \end{bmatrix} \quad \mathbf{Q} = \begin{bmatrix} -4.25 & -3.42 \\ -4.25 & -3.42 \\ -4.25 & -3.42 \\ -4.25 & -3.42 \\ \hline -0.5 & -0.17 \\ -0.5 & -0.17 \\ -0.5 & -0.17 \\ -0.5 & -0.17 \\ \hline 4.75 & 3.58 \\ 4.75 & 3.58 \\ 4.75 & 3.58 \\ 4.75 & 3.58 \end{bmatrix} \quad \mathbf{R} = \begin{bmatrix} -5.25 & -3.92 \\ -4.25 & -3.92 \\ -4.25 & -2.92 \\ -3.25 & -2.92 \\ \hline -1.25 & -0.92 \\ -1.25 & 1.08 \\ -0.25 & 0.08 \\ 0.75 & -0.92 \\ \hline 3.75 & 3.08 \\ 4.75 & 2.08 \\ 4.75 & 4.08 \\ 5.75 & 5.08 \end{bmatrix}$$

Table 7.3 (continued)

where **1** is a 12×1 unit vector.

Within-Groups SSCP Matrix	Among-Groups SSCP Matrix	Total-Sample SSCP Matrix
$\mathbf{W} = \mathbf{P'P}$	$\mathbf{A} = \mathbf{Q'Q}$	$\mathbf{T} = \mathbf{R'R}$
$\mathbf{W} = \begin{bmatrix} 6.75 & 1.75 \\ 1.75 & 8.75 \end{bmatrix}$	$\mathbf{A} = \begin{bmatrix} 163.50 & 126.50 \\ 126.50 & 98.17 \end{bmatrix}$	$\mathbf{T} = \begin{bmatrix} 170.25 & 128.25 \\ 128.25 & 106.92 \end{bmatrix}$

$$\boxed{\mathbf{T} = \mathbf{W} + \mathbf{A}}$$

$$\boxed{(\mathbf{W}^{-1}\mathbf{A} - \lambda\mathbf{I})\mathbf{k} = \mathbf{0}}$$

We see that $\mathbf{W}^{-1}\mathbf{A}$ plays the role that $\mathbf{R}_{xx}^{-1}\mathbf{R}_{xy}\mathbf{R}_{yy}^{-1}\mathbf{R}_{yx}$ played in the canonical correlation formulation of Chapter 6. Analogously, the problem now is to solve for the eigenstructure of $\mathbf{W}^{-1}\mathbf{A}$.

Again, we find that $\mathbf{W}^{-1}\mathbf{A}$ is not symmetric and, under these conditions, the matrix of eigenvectors will not be orthogonal. Table 7.4 shows the eigenvalues and eigenvectors for the sample problem. These are obtained in the same manner as followed in Chapter 6's discussion of canonical correlation; hence we omit computational details.

From Table 7.4 we first note that $\lambda_1 = 29.444$ while the second eigenvalue, $\lambda_2 = 0.0295$, is virtually zero. Referring to Figure 7.1 we see that the points lie

Table 7.4 Eigenstructure of $\mathbf{W}^{-1}\mathbf{A}$
(See Table 7.3 for Preliminary Computations)

$$\mathbf{W}^{-1} = \begin{bmatrix} 0.156 & -0.031 \\ -0.031 & 0.121 \end{bmatrix}; \qquad \mathbf{A} = \begin{bmatrix} 163.50 & 126.50 \\ 126.50 & 98.17 \end{bmatrix}$$

$$\mathbf{W}^{-1}\mathbf{A} = \begin{bmatrix} 21.594 & 16.698 \\ 10.138 & 7.880 \end{bmatrix}$$

Matrix of Eigenvalues	Matrix of Eigenvectors
	$\quad\quad \mathbf{k}_1 \quad\quad \mathbf{k}_2$
$\mathbf{D} = \begin{bmatrix} 29.444 & 0 \\ 0 & 0.0295 \end{bmatrix};$	$\mathbf{K} = \begin{bmatrix} 0.905 & -0.612 \\ 0.425 & 0.791 \end{bmatrix}$

very close to the first discriminant axis z_1. Not surprisingly, then, there is little variability left over to be attributable to z_2. Thus, in this case the group centroids fall almost on the straight line z_1 in the original test space. Also, in general, the *centroids* of G groups will be in (at most) a $G - 1$ space, since we need only $G - 1$ axes to account for all of the variability in the group centroids, assuming that $n \geq G - 1$.

The matrix **K** in Table 7.4 contains, as columns, the two sets of discriminant weights, expressed here as direction cosines (i.e., each column is normalized to unit sum of squares). Thus, the discriminant scores of the first observation in Table 7.1, when it is expressed as a deviation vector about the total-sample centroid, are:

$$z_{11} = 0.905(-5.25) + 0.425(-3.92)$$

$$= -6.42$$

$$z_{12} = -0.612(-5.25) + 0.791(-3.92)$$

$$= 0.11$$

Similarly, we could compute discriminant scores for the remaining 11 observations. These scores would be represented by the projections of the points onto the two discriminant axes in Figure 7.1.

Note that z_1 and z_2 are not orthogonal in the X_{d1}, X_{d2} space, even though their scores are uncorrelated. From Table 7.4 we can compute the cosine between the weight vectors as follows:

$$\cos \theta_{12} = \overset{\mathbf{k}_1'}{(0.905 \quad 0.425)} \overset{\mathbf{k}_2}{\begin{bmatrix} -0.612 \\ 0.791 \end{bmatrix}} = -0.21$$

The angle θ_{12} separating z_1 and z_2 is $90° + 12° = 102°$, as is shown in Figure 7.1.

7.3.3 Significance Testing

As we know from the preceding steps, the eigenvalues of $\mathbf{W}^{-1}\mathbf{A}$ are:

$$\lambda_1 = 29.444$$

$$\lambda_2 = 0.0295$$

We might next inquire as to whether either or both of these discriminant values are statistically significant. Testing the statistical significance of individual discriminant axes involves a *decomposition approach to Bartlett's V statistic,* described in section 7.2.2.[10]

To recapitulate, V was defined as:

$$V = -\left[m - 1 - \frac{(n + G)}{2}\right] \ln \Lambda$$

We recall that V is distributed approximately as chi square with $n(G - 1)$ degrees of freedom.

As background to the decomposition idea, consider the following relationship:[11]

$$\frac{1}{\Lambda} = |\mathbf{W}^{-1}\mathbf{T}| = |\mathbf{W}^{-1}(\mathbf{W} + \mathbf{A})|$$

$$= |\mathbf{I} + \mathbf{W}^{-1}\mathbf{A}|$$

However, as shown in section B.5 of Appendix B, the determinant of a matrix is equal to the product of its eigenvalues. Hence, we can also write:

$$\frac{1}{\Lambda} = (1 + \lambda_1)(1 + \lambda_2) \cdots (1 + \lambda_r)$$

for the r (nonzero) eigenvalues of $\mathbf{I} + \mathbf{W}^{-1}\mathbf{A}$.[12] And, since $-\ln a = \ln 1/a$ we see that Bartlett's V can be modified to:

$$V = \left[m - 1 - \frac{(n + G)}{2}\right] \sum_{j=1}^{r} \ln (1 + \lambda_j)$$

The decomposition, then, concerns the manner in which each eigenvalue of Λ enters the statistic V. Retention of *all* eigenvalues of $\mathbf{W}^{-1}\mathbf{A}$ will provide exactly the same overall test as already illustrated in section 7.2.2.

What differs in the present case is the possibility of removing one or more of the $(1 + \lambda_j)$ terms and carrying out *further* tests on sequentially computed discriminant axes, such as $\mathbf{z}_2$ in the present problem.

In practice this means that we first test the null hypothesis that the group centroids are equal in the *full* discriminant space. If this null hypothesis is rejected, we can then test separate axes of the space to see if they provide incremental discriminatory power that is statistically significant beyond the earlier-considered discriminant axes. The approach assumes that rejection of the overall null hypothesis implies that at least the first (highest variance accounted-for) axis is statistically significant.

In terms of our numerical example, we have:

$$\lambda_1 = 29.444; \lambda_2 = 0.0295$$

Substituting and solving in the new formulation of V leads to:

$$V = \left[12 - 1 - \frac{5}{2} \right] [ln \, 30.444 + ln \, 1.0295]$$

$$= (8.5)(3.4159 + 0.0291)$$

$$= 29.035 + 0.247$$

$$= 29.282$$

We recall that $V = 29.282$ is approximately distributed as chi square with $n(G - 1) = 4$ degrees of freedom.[13] In this case V is significant at better than the 0.01 alpha level.

Let us next examine the two additive components, namely $V_1 = 29.035$ and $V_2 = 0.247$. These correspond, respectively, to the first and second discriminant axes. The value V_1, corresponding to the first discriminant axis, has already been tested in the context of the total discriminant space. Thus, our attention now focuses on $V_2 = 0.247$. If V_1 is, in a sense, partialed out of the relationship, is the second discriminant axis also statistically significant?

In the case of the second discriminant axis, $\chi^2 = 0.247$ has degrees of freedom given by:

$$\text{d.f.} = (n - 1)(G - 2)$$

$$= 1$$

Clearly, $\chi^2 = 0.247$ with one degree of freedom is not significant. Table C.2 in Appendix C shows a tabular value of 3.84 at the 0.05 level. In general, with $r \leq \min(n, G - 1)$ discriminants, the degrees of freedom for the j-th discriminant axis ($j = 1, 2, \ldots, r$) are given by:[14]

$$\text{d.f.} (j) = (n - j + 1)(G - j)$$

Assuming that the λ_j's are arranged in decreasing order of magnitude, as soon as we obtain a function whose V_j is less than the appropriate tabular χ^2 value, we need test no further. That is, all smaller λ_j's can be discarded as nonsignificant at the prescribed alpha level.

This decomposition property turns out to be quite useful and represents a special feature of the V statistic. Either V or Ra could be used to test the *primary* null hypothesis that the centroids in the total space of original variables are equal. However, in addition, the V statistic permits us to partition the variation into incremental tests for each successive discriminant axis. Although one would obtain, at most:

$$\min (G - 1, n)$$

discriminant axes, it could be the case that even further parsimony is achieved if one or more of the "lesser" axes turn out to be nonsignificant statistically.

For purposes of additional interpretation, the researcher may also be interested in a *descriptive* index of the relative importance of the discriminant axes:

$$I(\lambda_j) = \frac{\lambda_j}{\sum\limits_{j=1}^{r} \lambda_j}$$

In the sample problem we have:

$$I(\lambda_1) = \frac{29.444}{29.444 + 0.0295} = 0.999$$

$$I(\lambda_2) = \frac{0.029}{29.444 + 0.0295} = 0.001$$

Clearly, the first discriminant axis represents the major basis for separating the three groups.

While the decomposition property of Bartlett's V statistic is utilized by several popular computer programs and is advocated in many textbooks on the subject, some authors (Harris, 1975, 1976) present strong arguments *against* the sequential approach, illustrated above. The gist of their argument is that the first discriminant axis is never tested directly but only in the context of all the others. As Harris indicates, this is analogous to claiming that an overall significant F value in a one-way ANOVA implies that at least one pairwise difference (i.e., the largest mean minus the smallest) is significant.[15]

7.3.4 Scaling the Discriminant Weights

As we recall from discussion of two-group discrimination in Chapter 4, the researcher has a number of options available to him for scaling the discriminant weights. So far, each set of weights has been scaled to unit sum of squares. For ease of reference Table 7.5 reproduces the matrix of discriminant weights **K** (normalized eigenvectors) as obtained by solving the eigenstructure problem in Table 7.4.

Other scalings are possible. In Chapter 4 we considered the possibility of scaling the weights in a way that led to "spherizing" the average within-groups variation (i.e., providing an average within-groups variance of unity). In MDA

Table 7.5 Alternative Scalings of Discriminant Weights

Eigenvectors of $\mathbf{W}^{-1}\mathbf{A}$—Normalized to Unit Sum of Squares

Discriminant
Axis

$$
\begin{array}{cc}
1 & 2
\end{array}
$$

$$
\mathbf{K} = \begin{bmatrix} 0.905 & -0.612 \\ 0.425 & 0.791 \end{bmatrix}
$$

Scaled to Result in an Average Within-Groups Variance of Unity

$$
\underset{\mathbf{I}}{\begin{bmatrix} 1 & 0 \\ 0 & 1 \end{bmatrix}} = \underset{\mathbf{K}'_w}{\begin{bmatrix} 0.933 & 0.439 \\ -0.731 & 0.944 \end{bmatrix}} \underset{\mathbf{C}_w}{\begin{bmatrix} 0.750 & 0.194 \\ 0.194 & 0.972 \end{bmatrix}} \underset{\mathbf{K}_w}{\begin{bmatrix} 0.933 & -0.731 \\ 0.439 & 0.944 \end{bmatrix}}
$$

Scaled to Result in a Total-Sample Variance of Unity

$$
\underset{\mathbf{I}}{\begin{bmatrix} 1 & 0 \\ 0 & 1 \end{bmatrix}} = \underset{\mathbf{K}'_t}{\begin{bmatrix} 0.187 & 0.088 \\ -0.796 & 1.029 \end{bmatrix}} \underset{\mathbf{C}_t}{\begin{bmatrix} 15.477 & 11.659 \\ 11.659 & 9.720 \end{bmatrix}} \underset{\mathbf{K}_t}{\begin{bmatrix} 0.187 & -0.796 \\ 0.088 & 1.029 \end{bmatrix}}
$$

this can be done with the additional provision that successive discriminant axes are uncorrelated. Although we have found that only the first discriminant value is statistically significant, let us proceed to describe rescalings of the *complete* $\mathbf{K}$ matrix of weights.

The first rescaled matrix of weights, denoted $\mathbf{K}_w$, as shown in Table 7.5, exhibits the spherized property:

$$
\boxed{\mathbf{K}'_w \mathbf{C}_w \mathbf{K}_w = \mathbf{I}}
$$

The entries of $\mathbf{C}_w$ are obtained from the pooled within-groups SSCP matrix $\mathbf{W}$ (in Table 7.3) by the relation:

$$
\mathbf{C}_w = \frac{\mathbf{W}}{(m - G)}
$$

$$
= \frac{1}{9} \begin{bmatrix} 6.75 & 1.75 \\ 1.75 & 8.75 \end{bmatrix} = \begin{bmatrix} 0.750 & 0.194 \\ 0.194 & 0.972 \end{bmatrix}
$$

The matrix $\mathbf{K}_w$ is then found by first computing the covariance matrix of the linear composites $\mathbf{z}_1$ and $\mathbf{z}_2$ as:

$$
\mathbf{C}_w(\mathbf{Z}) = \mathbf{K}'\mathbf{C}_w\mathbf{K}
$$

$$
\mathbf{C}_w(\mathbf{Z}) = \begin{bmatrix} 0.938 & 0 \\ 0 & 0.701 \end{bmatrix} = \underset{\mathbf{K}'}{\begin{bmatrix} 0.905 & 0.425 \\ -0.612 & 0.791 \end{bmatrix}} \underset{\mathbf{C}_w}{\begin{bmatrix} 0.750 & 0.194 \\ 0.194 & 0.972 \end{bmatrix}} \underset{\mathbf{K}}{\begin{bmatrix} 0.905 & -0.612 \\ 0.425 & 0.791 \end{bmatrix}}
$$

We see that this derived covariance matrix shows (as it should) the two sets of discriminant scores to be uncorrelated.

The desired spherizing transformation is then obtained from:[16]

$$\mathbf{K}_w = \mathbf{K}[\mathbf{C}_w(\mathbf{Z})]^{-1/2}$$

$$\mathbf{K}_w = \begin{bmatrix} 0.933 & -0.731 \\ 0.439 & 0.944 \end{bmatrix} = \begin{bmatrix} 0.905 & -0.612 \\ 0.425 & 0.791 \end{bmatrix} \begin{bmatrix} 1/\sqrt{0.938} & 0 \\ 0 & 1/\sqrt{0.701} \end{bmatrix}$$

Hence, if the mean-corrected scores are postmultiplied by $\mathbf{K}_w$ we obtain two sets of discriminant scores, each of whose average within-groups variance is unity.

Still another way of scaling the weights is to consider the linear composites in the context of the total-sample covariance matrix $\mathbf{C}_t = \mathbf{T}/(m-1)$ in which:

$$\boxed{\mathbf{C}_t(\mathbf{Z}) = \mathbf{K}'\mathbf{C}_t\mathbf{K}}$$

$\mathbf{T}$, the total-sample SSCP matrix, is also found from Table 7.3. We then compute:

$$\mathbf{C}_t(\mathbf{Z}) = \overset{\mathbf{K}'}{\begin{bmatrix} 0.905 & 0.425 \\ -0.612 & 0.791 \end{bmatrix}} \overset{\mathbf{T}/(m-1)}{\begin{bmatrix} 15.477 & 11.659 \\ 11.659 & 9.720 \end{bmatrix}} \overset{\mathbf{K}}{\begin{bmatrix} 0.905 & -0.612 \\ 0.425 & 0.791 \end{bmatrix}}$$

$$\mathbf{C}_t(\mathbf{Z}) = \begin{bmatrix} 23.401 & 0 \\ 0 & 0.591 \end{bmatrix}$$

Notice that $\mathbf{C}_t(\mathbf{Z})$ reflects the fact that the discriminant scores are uncorrelated, irrespective of whether $\mathbf{W}$ or $\mathbf{T}$ is used in computing the covariance matrix of interest.

Proceeding analogously, the reciprocals of the square roots of $\mathbf{C}_t(\mathbf{Z})$ are found as:

$$[\mathbf{C}_t(\mathbf{Z})]^{-1/2} = \begin{bmatrix} 1/\sqrt{23.401} & 0 \\ 0 & 1/\sqrt{0.591} \end{bmatrix} = \begin{bmatrix} 0.207 & 0 \\ 0 & 1.301 \end{bmatrix}$$

Similar to the transformation $\mathbf{K}_w$ above, we can then compute the matrix:

$$\mathbf{K}_t = \mathbf{K}[\mathbf{C}_t(\mathbf{Z})]^{-1/2}$$

which results in two sets of discriminant scores whose variance across the total sample is each equal to unity.

In this case we have:

$$\mathbf{K}_t = \begin{bmatrix} 0.905 & -0.612 \\ 0.425 & 0.791 \end{bmatrix} \begin{bmatrix} 0.207 & 0 \\ 0 & 1.301 \end{bmatrix}$$

$$= \begin{bmatrix} 0.187 & -0.796 \\ 0.088 & 1.029 \end{bmatrix}$$

This scaling alternative also appears in Table 7.5.

We note that the ratios of all three scalings in Table 7.5:

	$\mathbf{K}$	$\mathbf{K}_w$	$\mathbf{K}_t$
Axis 1:	$\frac{0.905}{0.425}$;	$\frac{0.933}{0.439}$;	$\frac{0.187}{0.088}$
Axis 2:	$\frac{-0.612}{0.791}$	$\frac{-0.731}{0.944}$	$\frac{-0.796}{1.029}$

are the same (within rounding error); the *relative size of the coefficients is unaffected by the scaling.* In practice all three scalings are employed by various packaged MDA programs.

7.3.5 Predictor Variable Importance

Earlier we indicated that one of the questions examined in MDA is the relative importance of predictor variables in separating group means. A number of caveats apply in the (usual) case of correlated predictors: the ambiguity of "relative importance" was pointed out in some detail in Chapter 5. Here we discuss two approaches: (a) standardized weights and (b) structure correlations, for summarizing variable importance.

Despite the serious limitations of any measure of predictor-variable "importance" in the (usual) case of correlated predictors, researchers often make some attempt to answer the question. Assuming, contrary to the test results, that both discriminant axes in the sample problem are statistically significant, how would one proceed to ascertain the relative importance of the predictors in the sample problem of Table 7.1?

If we consider the discriminant weights matrix $\mathbf{K}$ we know that its components:

$$\mathbf{K} = \begin{bmatrix} 0.905 & -0.612 \\ 0.425 & 0.791 \end{bmatrix}$$

are affected by the units in which the original data are expressed. Hence, we need some form of standardization involving weights that would be applied to the predictors to express them in standardized form.[17]

Two possibilities are available for finding this rescaling. We could use either the square roots of the diagonal elements of $\mathbf{C}_w$ (here denoted $\mathbf{Y}_w$) or those of $\mathbf{C}_t$ (here denoted $\mathbf{Y}_t$), as found from Table 7.5.

$$\mathbf{K}_{st}(\mathbf{C}_w) = \overset{\mathbf{Y}_w}{\begin{bmatrix} \sqrt{0.750} & 0 \\ 0 & \sqrt{0.972} \end{bmatrix}} \overset{\mathbf{K}}{\begin{bmatrix} 0.905 & -0.612 \\ 0.425 & 0.791 \end{bmatrix}} = \begin{matrix} \\ X_1 \\ X_2 \end{matrix} \overset{\text{Discriminant Axis}}{\begin{matrix} 1 & 2 \\ \end{matrix}} \begin{bmatrix} 0.783 & -0.530 \\ 0.419 & 0.780 \end{bmatrix}$$

$$\mathbf{K}_{st}(\mathbf{C}_t) = \begin{bmatrix} \sqrt{15.477} & 0 \\ 0 & \sqrt{9.720} \end{bmatrix} \begin{bmatrix} 0.905 & -0.612 \\ 0.425 & 0.791 \end{bmatrix} = \begin{array}{c} \\ X_1 \\ X_2 \end{array} \begin{array}{c} \text{Discriminant Axis} \\ \begin{array}{cc} 1 & 2 \end{array} \\ \begin{bmatrix} 3.563 & -2.409 \\ 1.325 & 2.466 \end{bmatrix} \end{array}$$

As shown above, in the illustrative problem both standardization procedures lead to the same ranking of the predictors. In the case of the first discriminant axis, predictor X_1 is the most important, while in the case of the second discriminant axis, predictor X_2 has the larger absolute value.

However, the ratios differ rather considerably between the two procedures. For example, in examining the results of the standardization based on the $\mathbf{C}_w$ matrix, we see in the case of the first discriminant axis that $0.783/0.419 = 1.9$. The corresponding ratio, based on the $\mathbf{C}_t$ matrix is $3.563/1.325 = 2.7$. (That based on the original eigenvector solution is $0.905/0.425 = 2.1$.)

Both standardization procedures are used in practice. It should be noted that standardization based on $\mathbf{C}_t$ gives results that would be obtained had the original variables been standardized to mean zero and unit standard deviation *across the whole sample* prior to conducting the discriminant analysis. However, the procedure utilizing the $\mathbf{C}_w$ matrix is not influenced by among-groups differences in the standard deviations of the predictors.

In Chapter 6's discussion of canonical correlation we had occasion to compute *structure correlations* between each original variable of a battery of variables and various canonical variates computed for that battery. As has been suggested by Cooley and Lohnes (1971) a similar idea is applicable to MDA. Usually, structure correlations are computed across the *total* sample. In this case, the matrix $\mathbf{K}_t$ of Table 7.5 is modified as follows:

$$\begin{aligned} \mathbf{K}_t^* &= \mathbf{Y}_t \mathbf{K}_t \\ &= \begin{bmatrix} \sqrt{15.477} & 0 \\ 0 & \sqrt{9.720} \end{bmatrix} \begin{bmatrix} 0.187 & -0.796 \\ 0.088 & 1.029 \end{bmatrix} \\ &= \begin{bmatrix} 0.737 & -3.131 \\ 0.274 & 3.207 \end{bmatrix} \end{aligned}$$

This has the effect of standardizing the discriminant weights to apply to a data matrix that has previously been standardized by columns to mean zero and unit standard deviation.

Next, we find the total-sample *correlation matrix* $\mathbf{R}_t$ from:

$$\begin{aligned} \mathbf{R}_t &= \begin{bmatrix} 1/\sqrt{15.477} & 0 \\ 0 & 1/\sqrt{9.720} \end{bmatrix} \begin{bmatrix} 15.477 & 11.659 \\ 11.659 & 9.720 \end{bmatrix} \begin{bmatrix} 1/\sqrt{15.477} & 0 \\ 0 & 1/\sqrt{9.720} \end{bmatrix} \\ &= \begin{bmatrix} 1.000 & 0.952 \\ 0.952 & 1.000 \end{bmatrix} \end{aligned}$$

by the procedure that was first described in Chapter 1.

We then find the matrix of structure correlations as:

$$\mathbf{H}_t = \overset{\mathbf{R}_t}{\begin{bmatrix} 1.000 & 0.952 \\ 0.952 & 1.000 \end{bmatrix}} \overset{\mathbf{K}_t^*}{\begin{bmatrix} 0.737 & -3.131 \\ 0.274 & 3.207 \end{bmatrix}}$$

$$
\begin{array}{cc}
& \text{Discriminant Axis} \\
& \begin{array}{cc} 1 & 2 \end{array} \\
= \begin{array}{c} X_1 \\ X_2 \end{array} & \begin{bmatrix} 0.998 & -0.078 \\ 0.976 & 0.226 \end{bmatrix}
\end{array}
$$

The resulting structure correlations tend to support what we know already: both predictors are highly correlated with the first discriminant axis and virtually uncorrelated with the second (nonsignificant) discriminant axis.

Ordinarily, structure correlations would be computed only for statistically significant discriminant functions; as such, both predictors display high correlations with the first function. This is not surprising since the two predictors are highly correlated with each other.

A further interpretive device that is mentioned by Cooley and Lohnes involves the calculation of the sum of squares of each *row* of the matrix **H** of structure correlations. These sums of squares are called *communalities,* a term borrowed from the literature of factor analysis. They could be of interest in the more usual cases where $G - 1$ is less than n, the number of predictors; otherwise each sum of squares equals unity, as noted below for the sample problem:

$$X_1: [(0.998)^2 + (-0.078)^2] = 1$$

$$X_2: [(0.976)^2 + (0.226)^2] = 1$$

In cases in which $G - 1$ is less than n a high communality for a particular predictor variable indicates that it provides relatively high information for discriminating among the groups.[18] A low communality suggests that the predictor is not informative with respect to group separation, in general.

Predictor variable standardization measures, structure correlations, and communalities are useful devices for interpreting the results of a discriminant analysis. Their utilization in applied problems in the behavioral and administrative sciences is on the increase.

7.4 A Canonical Correlation Approach to MDA

Earlier it was asserted that if canonical correlation were modified to incorporate a set of dummy criterion variables, we could obtain canonical weights for the predictor set that were the same, up to a possible scale transform, as those obtained by MDA. (This is analogous to multiple regression involving a binary-valued criterion versus two-group linear discrimination, as described in Chapter 4.) We now demonstrate this equivalence.

In the present sample problem involving three groups and two predictors, we can code the three groups on two dummies so that: (a) all group 1 members receive the code (1, 0); (b) all group 2 members receive the code (0, 1); and (c) all group 3 members receive the code (0, 0).

From here on we can treat the problem as a form of canonical correlation in which there are two (dummy-valued) criteria and two predictors. Table 7.6 summarizes the results. The first canonical correlation of 0.983 is the only one that turns out to be statistically significant at the 0.05 level; hence the canonical weight vectors **a** and **b** are shown only for this variate.

$$\mathbf{a} = \begin{bmatrix} 1.151 \\ 0.657 \end{bmatrix}; \mathbf{b} = \begin{bmatrix} 0.737 \\ 0.274 \end{bmatrix}$$

The weight vector for predictors **b** is the vector of interest in this case. Its components should be equal (up to a possible scale transformation) to the first column vector of $\mathbf{K}_t^*$. The reason why $\mathbf{K}_t^*$ is used is because this is the scaling that is based on a prior standardization of the predictors to mean zero and unit standard deviation. And, as noted in the preceding section:

$$\mathbf{K}_t^* = \begin{bmatrix} 0.737 & -3.131 \\ 0.274 & 3.207 \end{bmatrix}$$

Thus, the first (column) vector of $\mathbf{K}_t^*$ is, indeed, precisely equal to the predictor-variable weights obtained from the canonical correlation in which group membership was coded by means of dummy criterion variables.

Table 7.6 Results of MDA via Canonical Correlation of Dummy-Valued Criterion Variables

Correlation Matrix

	Y_1	Y_2	X_1	X_2
Y_1	1.00	-0.50	-0.80	-0.81
Y_2		1.00	-0.09	-0.04
X_1			1.00	0.95
X_2				1.00

Correlation of First Pair of Canonical Variates:

$$\sqrt{\lambda_{1(can)}} = 0.983$$

Canonical Weights (Criteria) $\begin{cases} a_1 = 1.151 \\ a_2 = 0.657 \end{cases}$ Canonical Weights (Predictors) $\begin{cases} b_1 = 0.737 \\ b_2 = 0.274 \end{cases}$

Correlation of Second Pair of Canonical Variates:

$$\sqrt{\lambda_{2(can)}} = 0.169$$

7.4.1 Statistical Significance of Canonical Correlations

As might also be surmised, the canonical correlations of 0.983 and 0.169, respectively, for the first and second pairs of variates can also be tested for significance via Bartlett's V or Rao's Ra statistics, described earlier.

The λ_j's, denoting the *squares* of the canonical correlations are the eigenvalues of $\mathbf{R}_{xx}^{-1}\mathbf{R}_{xy}\mathbf{R}_{yy}^{-1}\mathbf{R}_{yx}$. But, we have also used the symbol λ_j to denote the eigenvalues of $\mathbf{W}^{-1}\mathbf{A}$; therefore, let us now call the former eigenvalue: $\lambda_{j(can)}$. The following relationship between the two sets of eigenvalues can be shown to hold:

$$\lambda_j = \frac{\lambda_{j(can)}}{1 - \lambda_{j(can)}}$$

In terms of our sample problem, we can see that this is so:

$$\lambda_1 = (0.983)^2/[1 - (0.983)^2]$$

$$\cong 29.444$$

$$\lambda_2 = (0.169)^2/[1 - (0.169)^2]$$

$$\cong 0.0295$$

Thus, by means of the rather simple transformation above, we can relate the eigenvalues of $\mathbf{W}^{-1}\mathbf{A}$, in the context of MDA, to squared canonical correlations (i.e., the eigenvalues of $\mathbf{R}_{xx}^{-1}\mathbf{R}_{xy}\mathbf{R}_{yy}^{-1}\mathbf{R}_{yx}$) in the case where the criterion variables are expressed as dummy variables.

In section 7.3.3 we defined the reciprocal of Wilks' Λ as:

$$\frac{1}{\Lambda} = (1 + \lambda_1)(1 + \lambda_2) \cdots (1 + \lambda_r)$$

$$= \prod_{j=1}^{r} (1 + \lambda_j)$$

Noting that:

$$\lambda_j = \frac{\lambda_{j(can)}}{[1 - \lambda_{j(can)}]}$$

by simple algebra, we obtain an expression for Λ in terms of $\lambda_{j(can)}$ as:

$$\Lambda = \prod_{j=1}^{r} [1 - \lambda_{j(can)}]$$

In the sample problem, we have:

$$\Lambda = [1 - (0.983)^2] \cdot [1 - (0.169)^2]$$
$$\cong 0.0319$$

Earlier, we defined Wilks' Λ as the ratio of two determinants:

$$\Lambda = \frac{|\mathbf{W}|}{|\mathbf{T}|}$$

in the context of MDA. In canonical correlation, however, the notion of a within-groups SSCP matrix is generalized to the idea of an SSCP matrix for error. That is, each $[1 - \lambda_{j(can)}]$ is a measure of error variance. The product of these error variances equals Λ. To the extent that the error variances are small, Λ will be small and, hence, the two batteries of variables in the canonical correlation will be highly related.

Bartlett's V statistic can also be employed in canonical correlation. In each case it is the (squared) canonical correlations $\lambda_{j(can)}$ that enter the test.

$$V = -\left[m - 1 - \frac{(p + 1 + q)}{2}\right] \sum_{j=1}^{r} ln\,[1 - \lambda_{j(can)}]$$

Note that this formula is consistent with that used in MDA:

$$V = \left[m - 1 - \frac{(n + G)}{2}\right] \sum_{j=1}^{r} ln\,(1 + \lambda_{j})$$

if one bears in mind that (a) the q predictors correspond to n, the number of predictors in MDA, and (b) the number of groups G in MDA is one more than the number of dummy variables in the p set of criterion variables, assuming that canonical correlation is used as an MDA procedure.

In the *general* case of canonical correlation the degrees of freedom are pq. If successive pairs of canonical variates are to be tested, the procedure that is followed is the same as that shown earlier. For example, the degrees of freedom associated with the second pair of canonical variates are $(p - 1)(q - 1)$, and so on.

In terms of the sample problem, we first note that $\lambda_{1(can)} = (0.983)^2 = 0.9672$ and $\lambda_{2(can)} = (0.169)^2 = 0.0287$. We then have:

$$V = -\left[m - 1 - \frac{(p + 1 + q)}{2}\right] \{ln\,[1 - \lambda_{1(can)}] + ln\,[1 - \lambda_{2(can)}]\}$$

$$= -\left[12 - 1 - \frac{5}{2}\right][ln\ 0.0328 + ln\ 0.9713]$$

$$= -8.5(-3.4159 - 0.0291)$$

$$= 29.035 + 0.247$$

$$= 29.282,\ \text{the value for}\ V\ \text{found in section 7.3.3.}$$

with $pq = 4$ degrees of freedom. The quantity $V_2 = 0.247$ with $(p - 1)(q - 1)$ = 1 degree of freedom agrees with that obtained earlier and is, of course, not statistically significant.

7.4.2 Testing Canonical Correlations in General

In Chapter 6 the topic of significance testing in canonical correlation was purposely deferred until the present chapter in which the Λ, V, and Ra statistics could first be developed in the simpler one-way MANOVA context.

However, for the sake of completeness let us return to the sample problem introduced in Table 6.1 of the preceding chapter. As can be recalled from Table 6.2, the squared canonical correlations in this earlier problem are:

$$\lambda_{1(can)} = (0.912)^2 = 0.8317$$

$$\lambda_{2(can)} = (0.706)^2 = 0.4984$$

and the other relevant parameters are: $m = 12; p = q = 2$.
 We first find Λ as follows:

$$\Lambda = \prod_{j=1}^{r} [1 - \lambda_{j(can)}]$$

$$= (1 - 0.8317)(1 - 0.4984)$$

$$= 0.0844$$

Using Bartlett's V statistic we then obtain:

$$V = -\left[m - 1 - \frac{(p + q + 1)}{2}\right] \sum_{j=1}^{r} ln\ [1 - \lambda_{j(can)}]$$

$$= -\left[12 - 1 - \frac{5}{2}\right][ln\ 0.1683 + ln\ 0.5016]$$

$$= -8.5(-1.7820 - 0.6900)$$

$$= 15.147 + 5.865$$

$$= 21.012$$

Applying the chi square approximation to the distribution of V, the overall relationship, $V = 21.012$, is clearly significant for $p \cdot q = 4$ degrees of freedom. The second variate $V_2 = 5.865$, with $(p - 1)(q - 1) = 1$ degree of freedom, also results in a χ^2 that is significant at the 0.05 level.

Rao's Ra statistic can also be used to ascertain the significance of Λ, although, as recalled, this statistic is not appropriate for tests of separate canonical variates. As recalled, the distribution of Ra is approximated by the F distribution. In the present case we have:

$$Ra = \frac{1 - \Lambda^{1/s}}{\Lambda^{1/s}} \cdot \frac{1 + ts - \dfrac{pq}{2}}{pq}$$

$$t = m - 1 - \frac{(p + q + 1)}{2}; \qquad s = \sqrt{\frac{p^2q^2 - 4}{p^2 + q^2 - 5}}$$

$$= 12 - 1 - \frac{5}{2} \qquad\qquad = \sqrt{\frac{4(4) - 4}{4 + 4 - 5}}$$

$$= 8.5; \qquad\qquad\qquad = 2$$

We have already calculated $\Lambda = 0.0844$. Next we find:

$$\Lambda^{1/s} = (0.0844)^{1/2} = 0.2905$$

Ra is then found as follows:

$$Ra = \frac{1 - 0.2905}{0.2905} \cdot \frac{1 + 8.5(2) - \dfrac{4}{2}}{4}$$

$$= \frac{0.7095}{0.2905} \cdot 4$$

$$= 9.769$$

In the special case of only two variables (either p or q), Ra is exactly distributed as $F = 9.769$ with $p \cdot q = 4$ degrees of freedom in the numerator and $1 + ts - pq/2 = 16$ degrees of freedom in the denominator. In this case the F value is highly significant at the 0.05 alpha level.

In summary, with the straightforward modifications described here, the Λ statistic, and the V and Ra statistics as well, are just as appropriate for testing the significance of canonical correlations as they are for testing the significance of discriminant functions. Indeed, once the relevant eigenvalues are obtained and one recalls that G groups are codable into $G - 1$ dummy variates, the procedures are identical.

7.4.3 Recapitulation

At this point it may be useful to pull together some of the material of the two preceding sections. In MDA the following summary remarks are appropriate:

1. The objective of MDA is to find linear composites of the original variables that maximize the ratio of among-groups to within-groups variability, subject to being uncorrelated with previously obtained composites.

2. In general, the composites are computed so that successive λ_j's, the discriminant values, decrease in magnitude.

3. Tests of the ability of the discriminant axes z_1, z_2, etc., to separate the groups are provided by either V or Ra at the total-sample level. Moreover, the V statistic provides the researcher with a way to test successive discriminants beyond the first axis (albeit with the caveat discussed by Harris, 1975, 1976).

4. The discriminant weight vectors can be scaled in various ways (Table 7.5) and further standardized for interpretive purposes. Similar to canonical correlation, structure correlations of each predictor with each discriminant axis can be computed as a measure of the predictor's relative importance.

We then discussed canonical correlation (with a dummy-coded set of $p = G - 1$ criterion variables) as a way to do multiple discriminant analysis. In particular we noted that Λ, the discriminant criterion in MDA, can be expressed in several equivalent ways. First, we have the formulation:

$$\Lambda = \frac{|\mathbf{W}|}{|\mathbf{T}|}$$

where $\mathbf{W}$ is the pooled within-groups SSCP matrix and $\mathbf{T}$ is the total-sample SSCP matrix. However, Λ can also be written as:

$$\Lambda = \frac{1}{\prod\limits_{j=1}^{r} (1 + \lambda_j)}$$

where $1 + \lambda_j$ is the j-th eigenvalue of $\mathbf{I} + \mathbf{W}^{-1}\mathbf{A}$. Finally, Λ can be formulated as:

$$\Lambda = \prod\limits_{j=1}^{r} [1 - \lambda_{j(\text{can})}]$$

where $\lambda_{j(\text{can})}$ is the j-th eigenvalue of $\mathbf{R}_{xx}^{-1}\mathbf{R}_{xy}\mathbf{R}_{yy}^{-1}\mathbf{R}_{yx}$ in the context of canonical correlation (with dummy criterion variables).

We illustrated how Bartlett's V statistic could be adapted to handle any of the three formulations of Λ. Furthermore, in the more usual case of canonical correlation applied to two sets of *interval-scaled* variables, V appears as:

$$V = -\left[m - 1 - \frac{(p + q + 1)}{2}\right] \sum_{j=1}^{r} ln\,[1 - \lambda_{j(\text{can})}]$$

where p denotes the number of criterion variables and q denotes the number of predictors.

7.5 Multivariate Analysis of Variance and Covariance

Multivariate analysis of variance and covariance embodies characteristics of both discriminant analysis and canonical correlation. In MANOVA one has a set of dummy design variables and a set of interval-scaled criterion variables. In MANCOVA one has a mixture of dummies and interval-scaled variables representing the set of predictors.

In the case of MDA the test involving Wilks' Λ is a test of centroid equality across three or more groups. This is akin to a one-way MANOVA test. Similarly, in canonical correlation, the Λ test concerns whether the two batteries exhibit any linear association.

In each of these cases it is important to note that a *single* null hypothesis: (a) equality of G population centroids or (b) zero canonical correlation between two batteries of variables, is being tested. However, in Chapter 3's discussion of (univariate) ANOVA and ANCOVA, we discussed two-way (and higher-way) experimental designs in which *several* tests might be carried out on the same response variable.

Designs analogous to (univariate) factorials and other types of higher-way experimental layouts are available for MANOVA and MANCOVA as well. Naturally, the computations become much more complex inasmuch as SSCP matrices replace the various sums of squares computed in the univariate cases. Accordingly, we confine our first illustration to a two-factor MANOVA design, with replications, in which only two criterion variables are involved. A few summary comments are then made about extensions to higher-way designs as well as the incorporation of one or more covariates in the experiment.

The objectives of MANOVA and MANCOVA are the same as their univariate counterparts. However, in the multivariate situation we have more than a single response variable and a design matrix that can consist of two or more treatment variables. In this case we are interested in the following questions:

1. Do response-variable centroids differ significantly across levels of each of the various treatment variables?

2. If so, which treatment levels are contributing most to significant variation across the response-variable centroids?

3. What are the significant dimensions of group separation within separate treatments?

4. If covariates are present (in the case of MANCOVA) do response-variable centroids differ across treatments after covariate adjustment?

MANOVA and MANCOVA are typically, but not necessarily, employed in experimental design contexts where an equal number of responses can be obtained for each combination of treatment levels. In the example to follow we focus on a relatively simple two-way factorial design, with an equal number of replications per cell.

We shall see that MANOVA and MANCOVA also entail the computation of linear composites, not unlike the illustration of section 7.4 where canonical correlation was applied to a problem in which one battery of (criterion) variables consisted of dummies and the other consisted of interval-scaled (predictor) variables. Now, however, it is the predictor set that consists of dummies while the criterion set is interval-scaled. ·

7.5.1 Informal Introduction to MANOVA

For purposes of illustration let us assume that a researcher is interested in the effect on employee productivity and morale of two factors:

1. Factor A—three levels of lighting intensity.
2. Factor B—two levels of liveliness of background music.

We assume that six random samples of two workers each have been chosen (workers being treated as replications) and assigned at random to the six combinations of work room conditions. Worker productivity is represented by the response Y_1 (average number of assemblies produced daily) and morale by Y_2 (a summary score made up of worker responses to a series of on-the-job attitude statements).

Table 7.7 shows the original (artificial) data from the experiment. Also shown are the cell means (based on two replications each) for the six combinations of factors A and B for Y_1 and Y_2, respectively. Marginal means also appear for factors A and B.

One of the problems in applying MANOVA is the complexity of the notation and the sheer labor involved in computing the needed SSCP matrices. (Fortunately, this is now handled quite easily by computer programs.) For expository purposes, however, let us assume the following notation for the general entry of **Y**, the response matrix of Table 7.7:

$$Y_{ijk\ell}$$

where:

$i = 1, 2$	denotes the i-th criterion variable
$j = 1, 2, 3$	denotes the j-th level of factor A
$k = 1, 2$	denotes the k-th level of factor B
$\ell = 1, 2$	denotes the ℓ-th replicate.

Table 7.7 Sample Data for Two-Way MANOVA with Two Replications per Treatment

| | | Factor B | | | |
| | | Level 1 | | Level 2 | |
		Y_1	Y_2	Y_1	Y_2
	1	1	1	2	2
		2	1	3	2
Factor A	2	5	4	6	5
		5	6	7	4
	3	10	8	11	9
		11	7	12	10

Cell Means on Response Y_1

		B 1	B 2	Marginal Means
	1	1.5	2.5	2.00
A	2	5.0	6.5	5.75
	3	10.5	11.5	11.00
Marginal Means		5.67	6.83	6.25

Cell Means on Response Y_2

		B 1	B 2	Marginal Means
	1	1	2	1.50
A	2	5	4.5	4.75
	3	7.5	9.5	8.50
Marginal Means		4.5	5.33	4.92

Our problem now is to compute the appropriate SSCP matrices:

$\mathbf{S}_t$:   the SSCP matrix for the total sample

$\mathbf{S}_A$:   the SSCP matrix for factor A

$\mathbf{S}_B$:   the SSCP matrix for factor B

$\mathbf{S}_{A \times B}$:   the SSCP matrix for the $A \times B$ interaction

$\mathbf{S}_E$:   the SSCP matrix for error.

Computation of the various SSCP matrices can be shown most simply in terms of summational notation. To illustrate, let us take the case of $\mathbf{S}_t$, the SSCP matrix for the total sample and assume that we wish to find the cross-products entry of this matrix. This is given by:

$$\mathbf{S}_{t(1,2)} = \sum_{j=1}^{3} \sum_{k=1}^{2} \sum_{\ell=1}^{2} (Y_{1jk\ell} - \overline{Y}_1 \ldots)(Y_{2jk\ell} - \overline{Y}_2 \ldots)$$

Table 7.8 shows the counterpart calculations of the cross-product term for the other SSCP matrices of interest. By similar means we could also find the diagonal entries of each SSCP matrix. As noted from Table 7.8, in the case of factor A each level is based on four observations of Y_1 and Y_2 each. Hence, the cross

Table 7.8 Formulas for Computing the Cross-Product Term of the SSCP Matrices in Two-Way MANOVA

Cross Product for Factor A

$$A_{(1,2)} = 4 \sum_{j=1}^{3} (\bar{Y}_{1j}.. - \bar{Y}_1...)(\bar{Y}_{2j}.. - \bar{Y}_2...)$$

Cross Product for Factor B

$$B_{(1,2)} = 6 \sum_{k=1}^{2} (\bar{Y}_{1 \cdot k}. - \bar{Y}_1...)(\bar{Y}_{2 \cdot k}. - \bar{Y}_2...)$$

Cross Product for $A \times B$ Interaction

$$A_{(1,2)} \times B_{(1,2)} = 2 \sum_{j=1}^{3} \sum_{k=1}^{2} [(\bar{Y}_{1jk}. - \bar{Y}_{1j}.. - \bar{Y}_{1 \cdot k}. + \bar{Y}_1...)$$
$$\times (\bar{Y}_{2jk}. - \bar{Y}_{2j}.. - \bar{Y}_{2 \cdot k}. + \bar{Y}_2...)]$$

Cross Product for Error

$$E_{(1,2)} = \sum_{j=1}^{3} \sum_{k=1}^{2} \sum_{\ell=1}^{2} (\bar{Y}_{1jk\ell} - \bar{Y}_{1jk}.)(\bar{Y}_{2jk\ell} - \bar{Y}_{2jk}.)$$

product is multiplied by 4; similar remarks pertain to the remaining cross-product terms.

Fully analogous to the univariate case, in MANOVA we obtain the following partitioning of the total-sample $\mathbf{S}_t$ matrix:

$$\mathbf{S}_t = \mathbf{S}_A + \mathbf{S}_B + \mathbf{S}_{A \times B} + \mathbf{S}_E$$

where the various SSCP matrices take the place of the sums of squares that are routinely computed in the univariate ANOVA case.

These computations have been worked out for the sample data of Table 7.7 and are shown in Table 7.9. Also shown are the accompanying degrees of freedom for each effect. We observe that the same rules for finding degrees of freedom in the univariate case apply to the multivariate case as well. (For example, since factor A has three design levels its number of degrees of freedom is $3 - 1 = 2$.)

The task at hand is to see if the multivariate response centroids differ across treatment variables and, in particular, whether significant effects exist for factors A, B, and the $A \times B$ interaction.

Table 7.9 SSCP Matrices and Degrees of Freedom for the Sample Data of Table 7.7

Source of Variation	SSCP Matrix	d.f.
Factor A	$\mathbf{S}_A = \begin{bmatrix} 163.50 & 126.50 \\ 126.50 & 98.17 \end{bmatrix}$	2
Factor B	$\mathbf{S}_B = \begin{bmatrix} 4.08 & 2.92 \\ 2.92 & 2.08 \end{bmatrix}$	1
$A \times B$ Interaction	$\mathbf{S}_{A \times B} = \begin{bmatrix} 0.17 & -0.67 \\ -0.67 & 3.17 \end{bmatrix}$	2
Error	$\mathbf{S}_E = \begin{bmatrix} 2.50 & -0.50 \\ -0.50 & 3.50 \end{bmatrix}$	6
Total	$\mathbf{S}_t = \begin{bmatrix} 170.25 & 128.24 \\ 128.25 & 106.92 \end{bmatrix}$	11

7.5.2 Significance Testing

Throughout the chapter we have emphasized Wilks' lambda:

$$\Lambda = \frac{|\mathbf{W}|}{|\mathbf{T}|}$$

as the primary statistic for testing whether the centroids of G population groups are equal. The lambda statistic also figured prominently in the test for the significance of canonical correlation coefficients.

In the context of (one-way) MANOVA we could formulate Λ in the following way:

$$\Lambda = \frac{|\mathbf{W}|}{|\mathbf{W} + \mathbf{A}|}$$

where we consider $\mathbf{W}$ as the "error" matrix and $\mathbf{A}$ as the "hypothesis" matrix. If so, the total SSCP matrix $\mathbf{T}$ denotes the *sum* of error and hypothesized effect.

In two-way (and higher-way) MANOVA designs, more than a single hypothesis is under test. For example, in Table 7.9 we observe that *three* hypothesis matrices:

$$\mathbf{S}_A; \mathbf{S}_B; \mathbf{S}_{A \times B}$$

are under test. However, as can be easily checked, any single one of these, if added to the error matrix $\mathbf{S}_E$, will not equal $\mathbf{S}_t$ (i.e., $\mathbf{T}$).

Table 7.10 Computation of Wilks' Lambda Measures for Each Separate Effect
(See Table 7.9 for SSCP Matrices)

Source of Variation	Value of Determinant $\|S_H + S_E\|$		Wilks' Lambda Λ_H
Factor A: $S_A + S_E$	$\begin{vmatrix} 166 & 126 \\ 126 & 101.67 \end{vmatrix}$	$= 1{,}001.22$	$\Lambda_A = 0.0085$ $F_{[4,10]} = 26.24$
Factor B: $S_B + S_E$	$\begin{vmatrix} 6.58 & 2.42 \\ 2.42 & 5.58 \end{vmatrix}$	$= 30.86$	$\Lambda_B = 0.275$ $F_{[2,5]} = 6.59$
$A \times B$ Interaction: $S_{A \times B} + S_E$	$\begin{vmatrix} 2.67 & -1.17 \\ -1.17 & 6.67 \end{vmatrix}$	$= 16.44$	$\Lambda_{A \times B} = 0.517$ $F_{[4,10]} = 0.98$
Error: S_E	$\begin{vmatrix} 2.50 & -0.50 \\ -0.50 & 3.50 \end{vmatrix}$	$= 8.50$	

Hence, in the present situation we shall want to replace the denominator $|T|$ by a determinant of the more general form:

$$|S_H + S_E|$$

where S_H stands for the *particular* SSCP matrix whose hypothesized effect is being tested and S_E is the appropriate error matrix.

Under this view, we can now define a more general Wilks' Λ as:

$$\Lambda_H = \frac{|S_E|}{|S_H + S_E|}$$

Note that the determinant of the error SSCP matrix is in the numerator while the determinant of the sum of hypothesis and error SSCP matrices appears in the denominator.

Table 7.10 summarizes these computations with values of Λ_H of 0.0085, 0.275, and 0.517, respectively, for the factor A, factor B and $A \times B$ interaction effects. Once the various Λ_H values have been computed, either Bartlett's V or Rao's Ra functions of Λ can be used, just as before. Some changes are made in the formulas to reflect the particular degrees of freedom that are being used. To illustrate, Bartlett's V statistic is now:

$$V = -\left[d_E + d_H - \frac{(n + d_H + 1)}{2} \right] \ln \Lambda_H$$

The degrees of freedom for error d_E are 6 (see Table 7.9). Similarly, the degrees of freedom for the various hypotheses d_H are:

$$d_A = 2; \; d_B = 1; \; d_{A \times B} = 2$$

In the present application Bartlett's V is still approximately distributed as χ^2 with nd_H degrees of freedom.

To illustrate testing the factor A effect, we note from Table 7.10 that $\Lambda_H = 0.0085$. We then find:

$$V = -\left[d_E + d_H - \frac{(n + d_H + 1)}{2}\right] \ln \Lambda_H$$

$$= -\left[6 + 2 - \frac{(2 + 2 + 1)}{2}\right] \ln 0.0085$$

$$= -[8 - 2.5](-4.768)$$

$$= 26.224$$

The value 26.224 with $n(d_H) = 2(2) = 4$ degrees of freedom is approximately distributed as chi square. The associated tabular χ^2 at the 0.05 alpha level (Table C.2 in Appendix C) is only 9.49. We conclude that the factor A effect is highly significant.

Similarly, the Ra statistic can be employed in the present example. In this case the statistic is modified as follows:

$$Ra = \frac{1 - \Lambda_H^{1/s}}{\Lambda_H^{1/s}} \cdot \frac{1 + ts - \dfrac{nd_H}{2}}{nd_H}$$

where:

$$t = d_E + d_H - \frac{(n + d_H + 1)}{2}$$

$$s = \sqrt{\frac{(nd_H)^2 - 4}{n^2 + d_H^2 - 5}}; \; s = 1 \text{ if } n^2 + d_H^2 = 5$$

As before, Ra, in general, is approximately distributed as F with nd_H degrees of freedom in the numerator and $1 + ts - (nd_H)/2$ degrees of freedom in the denominator. Also as before, when $n = 1$ or 2 or $d_H = 1$ or 2 (the latter corresponding to either $G = 2$ or $G = 3$), the Ra statistic is exactly distributed as F.

In the example of Table 7.10 we can use the (general) Ra statistic to test the significance of the A effect. Noting that Λ_A (Table 7.10) is 0.0085 we find:[19]

$$Ra = \frac{1 - \Lambda_H^{1/s}}{\Lambda_H^{1/s}} \cdot \frac{1 + ts - \dfrac{nd_H}{2}}{nd_H}$$

$$= \frac{1 - (0.0085)^{1/2}}{(0.0085)^{1/2}} \cdot \frac{1 + 5.5(2) - \dfrac{4}{2}}{2(2)}$$

$$= 24.62$$

Since $n = 2$, $Ra = 24.62$ is exactly distributed as F with degrees of freedom:

$$2d_H = 4 \text{ for the numerator}$$

$$2(d_E - 1) = 10 \text{ for the denominator}$$

We can repeat the process for the B effect and the $A \times B$ interaction. As noted from Table 7.10, the F ratio for the B effect is:

$$F_{[2, 5]} = 6.59$$

and that for the $A \times B$ interaction is:

$$F_{[4, 10]} = 0.98$$

Clearly the latter F ratio is not significant. However, if we adopt an 0.05 alpha level, the B effect is significant since 6.59 exceeds the tabular F value (Table C.4 of Appendix C) of 5.79.

In conclusion, we see that Wilks' Λ, in the form modified for MANOVA, still constitutes the principal test statistic. Moreover, Bartlett's V statistic and Rao's Ra statistic can still be used to make various significance tests using the more familiar χ^2 and F approximations to V and Ra, respectively.

7.5.3 Interpretative Aspects of MANOVA

Once the researcher has established that one or more effects are significant, a variety of questions arise in the general *interpretation* of findings. Illustrative of these questions are the following:

1. How do the various groups making up the levels of a significant factor differ in terms of the criterion variables?

2. What are the underlying "dimensions" of group separation on the significant factors?

3. What is the relative importance of the criterion variables in indicating differences across groups in terms of the significant factors?

A wide variety of techniques—MDA, the computation of structure correlations, stepdown F ratios, univariate F ratios, simultaneous confidence intervals—have been used. Even a cursory examination of these specialized MANOVA approaches would far exceed the scope of this chapter. Accordingly, the interested reader is referred to discussions by Roy (1958), Bock and Haggard (1968, 1975), Stevens (1972, 1973), Finn (1974), and Morrison (1976).

What can be said here is that MANOVA becomes exceedingly complex, once the significance tests have been carried out. This is particularly true if the designs are not balanced in the sense of having an equal number of replications per cell. While a number of computer programs provide the flexibility for dealing with nonorthogonal MANOVA designs, the problems of interpreting one's findings become increasingly difficult in the case of both correlated predictors and correlated criterion variables.

7.6 Applications Involving the TV Commercial Study

Having applied MDA and MANOVA to small-scale data sets, it is of interest to examine their suitability for analyzing real-world data of the type represented by the TV commercial study. Accordingly, we apply MDA and MANOVA/MANCOVA to some of the data drawn from this study.

7.6.1 MDA Analysis of Brand Choice

In section 5.4 of Chapter 5 we examined the trade-off data of Part B of the questionnaire via monotonic ANOVA. Three clusters of respondents who displayed similar trade-off responses were also compared regarding their brand choice in Part D of the questionnaire. No significant association between the clusters formed from the utility functions of Part B and those based on the brand choices of Part D was found.

Still, one could inquire as to whether respondents who chose Beta, Gamma, Delta, or Alpha in Part D (excluding the 12 respondents who chose none of the four brands) might differ systematically in terms of the *original* ratings data in Part B. In this way of looking at things we first partition the group of 240 respondents into one of four brand classes based on their Part D responses. We then examine their 25-component profiles of original ratings on tire purchase combinations involving different levels of:

1. Advertised tread mileage
2. Brand
3. Tire price
4. Driving time to tire store

of Part B in order to see if intergroup differences are found.

Given the earlier findings it would not be surprising to find no significant differences among the four groups. However, the present approach does differ in its emphasis on the partitioning variable—brand choice in Part D in the current instance—and the results could differ.

In this MDA application the original ratings data are used as 25 predictor variables in an attempt to predict respondent membership in one of the four brand groups: Beta, Gamma, Delta or Alpha. The sample size is 240 since we exclude the 12 respondents who chose "none" in Part D. The program used to carry out the MDA is BMD-07M, a stepwise discriminant program (Dixon, 1973). We describe the essentials of this program first and then comment on the empirical findings.

The BMD-07M stepwise discriminant program is a large-capacity program capable of performing MDA with 2 to 80 groups and up to 80 predictor variables. The program has a number of features that are worth describing in some detail:

1. Stepwise selection of predictor variables and computation of a classification function for each group.
2. Significance tests, both pairwise and an overall test across all groups.
3. Posterior probability calculations and classification table.
4. Eigenstructure decomposition and discriminant axis plots.

Stepwise Selection of Variables   Prior to the selection of predictor variables BMD-07M computes and prints out the pooled within-groups covariance and correlation matrices. It then proceeds to compute single predictor-variable F ratios, as though each was the only predictor variable in the problem. Hence, a one-way ANOVA on each "predictor" (viewed as a criterion variate) is performed.

The predictor variable with the largest F value is then chosen to enter the discriminant functions. Successive steps add (or delete) new predictors based on their F values conditioned on predictors already in the discriminant functions.[20] In short, this part of the process proceeds just like BMD-02R, the stepwise regression program described in Chapter 2. (And, as a matter of fact, this phase of BMD-07M is identical to BMD-02R in the special case of two groups.)

The classification functions obtained in this phase of the program are based on an approach that resembles the one followed in the two-group case described in Chapter 4. To be specific, discriminant weights are obtained for *each* of G groups by the formula:

$$\mathbf{k}_g = \mathbf{C}_w^{-1}\overline{\mathbf{x}}_g$$

where $\mathbf{C}_w^{-1}$ is the inverse of the pooled within-groups covariance matrix and $\overline{\mathbf{x}}_g$ is the centroid of the g-th group ($g = 1, 2, \ldots, G$). The program also computes a constant for each group by the formula:

$$k_{0g} = -\tfrac{1}{2}\mathbf{k}_g'\overline{\mathbf{x}}_g$$

In practice, a new profile's predictor-variable values would be substituted in each of the G discriminant functions and a *score computed on each*. The observation would then be assigned to that group for which the discriminant score was highest across the G groups.

It is appropriate to point out that this method of computing discriminant functions is based on a procedure proposed by Rao and Slater (1949) *before* the current emphasis on approaches emphasizing eigenstructures. As such, not all G discriminants are independent and a more parsimonious representation could be obtained from the eigenstructure formulation described earlier in this chapter. (This step is carried out later by BMD-07M.)

Significance Tests   In BMD-07M a series of significance tests (in addition to the F tests that are run for predictor variable inclusion and retention) are carried out. First, at each stage in the addition of predictor variables, Mahalanobis' D^2 is computed between each pair of groups. Each (transformed) D^2 can be tested as an F statistic to see which pairs of groups are most widely separated. The program also prints the appropriate number of degrees of freedom to use for each F test.

Second, the program computes Wilks' Λ as a statistic for summarizing *overall* group separation. Rao's Ra function of Λ is then used to relate Λ to the (approximate) F distribution. Again, appropriate degrees of freedom for both Λ and F are printed.

Posterior Probabilities and Classification   At the conclusion of the stepwise computation the program computes the Mahalanobis D^2 of each observation from the centroid of each group. These D^2's are printed out and can be used descriptively in order to see how far each case is from each centroid in a space that has been "spherized" on a pooled within-groups covariance basis.

In addition, the program uses the discriminant scores to compute a posterior probability that the observation in question comes from each group under study. This is done in two steps. First, one finds the discriminant value of the case in question on each function.

$$t_g = k_{0g} + \mathbf{k}'_g \mathbf{x}$$

for the $g = 1, 2, \ldots, G$ groups.

Next, given user-supplied prior probabilities q_g for $g = 1, 2, \ldots, G$ groups, the following substitution is made:

$$P(g \mid \mathbf{x}) = \frac{q_g \exp [t_g]}{\displaystyle\sum_{g=1}^{G} q_g \exp [t_g]}$$

The program then assigns the observation $\mathbf{x}$ to that group with the highest posterior probability and a classification matrix (actual group versus group assigned by the function) is prepared.

Eigenstructure Decomposition and Function Plots   BMD-07M also performs a type of eigenstructure decomposition in which only $G - 1$ discriminant axes are found. The $G - 1$ discriminants found via this approach are standardized so that:

$$\mathbf{K}'\mathbf{C}_w\mathbf{K} = \mathbf{I}$$

where $\mathbf{K}$ denotes the normalized (to unit sum of squares) eigenvectors from $\mathbf{W}^{-1}\mathbf{A}$ and $\mathbf{C}_w$ denotes the pooled within-groups covariance matrix. Canonical correlations are also computed from the relation:

$$\sqrt{\lambda_{j(\text{can})}} = \sqrt{\frac{\lambda_j}{(1 + \lambda_j)}}$$

where λ_j is an eigenvalue of $\mathbf{W}^{-1}\mathbf{A}$.

Finally, the program plots discriminant axis scores on a scatter diagram for visual inspection. Mean scores for each of the groups are also computed and plotted.

Applying the BMD-07M Program   The BMD-07M program was applied to the television commercial data with the Part B responses serving as predictors of group membership based on brand selected in Part D of the questionnaire. As mentioned earlier, the four groups were: Beta, Gamma, Delta, and Alpha brands, while the predictors consisted of the Part B ratings on 25 different combinations of advertised tread mileage, brand, price, and nearness to retail store.[21]

The results of the MDA were not very promising. Table 7.11 shows the classification matrix obtained from BMD-07M. As inferred from the sum of the main diagonal entries, 133/240 cases are correctly assigned by the function — a "hit" ratio of approximately 55 per cent. Insofar as Alpha's misclassified cases are concerned, they tend to be misassigned about equally across Beta, Gamma, and Delta.

The overall F approximation to Ra shows statistical significance, but the canonical correlations are not impressive. (Since 25 predictor variables are involved, no attempt is made to list complete results here.)

In summary, the results of Table 7.11 support those already described in section 5.4 of Chapter 5. The upshot of all of this is that brand choice in Part D of the questionnaire is *not* highly related to the utility functions computed from the Part B trade-off data. Some indication of why this occurs can be inferred from Figure 5.5 in Chapter 5. As recalled, the two most important factors in the trade-off analysis were tread mileage and price. As might be expected, all four brand groups' responses were quite similar with regard to these factors. While utility for brand name did follow the expected pattern of being related to brand chosen in Part D, this factor, in itself, was not important enough to lead to high association between overall responses to Parts B and D.

Table 7.11 Selected Output from BMD-07M Analysis of Part D Brand Choice and Part B Ratings Data

Classification Matrix

Actual	Predicted by Function				Total
	Beta	Gamma	Delta	Alpha	
Beta	21	6	3	6	36
Gamma	2	15	4	3	24
Delta	13	21	48	20	102
Alpha	11	9	9	49	78
Total	47	51	64	78	240

Eigenvalues of $\mathbf{W}^{-1}\mathbf{A}$ Canonical Correlations

$$\lambda_1 = 0.40 \qquad \sqrt{\lambda_{1(can)}} = \sqrt{\frac{0.40}{1.40}} = 0.54$$

$$\lambda_2 = 0.24 \qquad \sqrt{\lambda_{2(can)}} = \sqrt{\frac{0.24}{1.24}} = 0.44$$

$$\lambda_3 = 0.10 \qquad \sqrt{\lambda_{3(can)}} = \sqrt{\frac{0.10}{1.10}} = 0.30$$

F Value Approximation to Ra

$$F_{[75,635]} = 2.03$$

7.6.2 The Essentials of MANCOVA

Our discussion of MANCOVA is quite brief, largely because the basic ideas of covariance adjustment have already been covered in the univariate case of Chapter 3. In the multivariate case we partial out the effects of one or more covariates *from each criterion variable separately* before applying MANOVA to the adjusted criterion variables.

To illustrate, consider three sets of variables: (a) a set of n_1 criterion variables, denoted by the matrix of cases by variables, $\mathbf{Y}$; (b) a set of dummy-coded design variables $\mathbf{X}$; and (c) a set of n_2 covariates $\mathbf{Z}$.

The logic of MANCOVA is simple. We wish to see if the groups (constructed from the design matrix $\mathbf{X}$) show centroid differences on $\mathbf{Y}$, the criterion variables, once the effect of the covariates is removed from $\mathbf{Y}$.

First, we compute the $\mathbf{T}$ and $\mathbf{W}$ matrices across the *full* set of $n_1 + n_2$ criteria and covariates. We then partition $\mathbf{T}$ (the total-sample SSCP matrix) and $\mathbf{W}$ (the within-groups SSCP matrix) as follows:

$$\mathbf{T} = \begin{bmatrix} \mathbf{T}_{yy} & \mathbf{T}_{yz} \\ \mathbf{T}_{zy} & \mathbf{T}_{zz} \end{bmatrix}; \mathbf{W} = \begin{bmatrix} \mathbf{W}_{yy} & \mathbf{W}_{yz} \\ \mathbf{W}_{zy} & \mathbf{W}_{zz} \end{bmatrix}$$

In MANCOVA the next step is to regress each variable of the criterion set **Y** separately on the full covariate set **Z** leading to two matrices of residuals:

$$\mathbf{T}_{y \cdot z} = \mathbf{T}_{yy} - \mathbf{T}_{yz}\mathbf{T}_{zz}^{-1}\mathbf{T}_{zy}$$

$$\mathbf{W}_{y \cdot z} = \mathbf{W}_{yy} - \mathbf{W}_{yz}\mathbf{W}_{zz}^{-1}\mathbf{W}_{zy}$$

These two matrices of residuals—net of the linear association of each criterion variable in **Y** on the *full set of covariates* **Z**—are then analyzed via MANOVA. (Adjustments must be made in degrees of freedom, however, for the subsequent MANOVA.)

The same basic idea pertains to two-way (and higher-way) design matrices once the design variables are used to find the appropriate groups. As might be expected, the computational burden becomes quite heavy and computer programs—like the one illustrated next—are used in virtually all MANCOVA applications.

7.6.3 MANOVA and MANCOVA Applications

In illustrating larger-scale applications of MANOVA and MANCOVA we continue to use the real data of the TV commercial study. In Chapter 3 it may be recalled that two design variables were selected for the univariate ANOVA and ANCOVA applications:

1. Whether or not Alpha was the last brand purchased (A-2) and
2. High versus low pre-exposure interest in the Alpha brand of radial tires (A-3)

to examine post-exposure interest in Alpha radials C-2. One analysis was done without covariates and one was done in which the demographic variables served as covariates. These applications employed the univariate portion of BMD-12V (Dixon, 1973) and appeared in section 3.6.[22] We use this same program for the multivariate analyses to follow.

To make the analysis "multivariate" we now introduce a second criterion variable—believability in the Alpha commercial claims (C-1). All other conditions of the analysis remain as before.

BMD-12V is primarily designed for MANOVA and MANCOVA, although, as we know, univariate ANOVA and ANCOVA can be performed as well. The algorithm is restricted to a fixed-effects analysis involving fully crossed or nested designs with *equal* cell sizes. Up to 10 design variables can be included with criteria and covariates limited only by constraints on total storage. By the use of subproblem cards a wide variety of univariate and multivariate analyses can be carried out with the same input data in a single run.

Table 7.12 shows the cell means on the two criterion variables and the MANOVA results while Table 7.13 shows the summary output of the full MANCOVA.

As noted from Table 7.12, average post-exposure believability (C-1) and interest (C-2) are higher, on average, for those with high pre-exposure interest

Table 7.12 Criterion-Variable Means on the Two Design Variables and MANOVA Results
(17 Respondents Per Cell)

| | Pre-Exposure Interest in Alpha Radials (A-3) | | | | | |
| | High | | Low | | Marginal Means | |
Was Alpha the Last Brand Purchased (A-2)?	C-1	C-2	C-1	C-2	C-1	C-2
Yes	7.82	8.00	5.76	5.59	6.79	6.79
No	7.94	7.35	6.65	5.06	7.29	6.21
Marginal Means	7.88	7.68	6.21	5.32	7.04	6.50

MANOVA Results

Source	Logarithm of Generalized Variance	Wilks' Lambda	F Ratio	d.f.
A-2	11.396	0.936	2.16	2; 63
A-3	11.545	0.806	7.57	2; 63 ($p < 0.01$)
Interaction	11.340	0.989	0.36	2; 63
Full Model	11.329			

in Alpha radials. The mean post-exposure interest is also higher if Alpha is the last brand purchased, but mean post-exposure believability in the commercial claims is lower if Alpha is last brand purchased.

Table 7.13 shows, however, that only the former effect is significant at the 0.01 level. That is, after adjustment for the set of demographic covariates, only the effect of pre-exposure interest in Alpha radials is significant with respect to the set of two criterion variables. (As can also be noted from the MANOVA results of Table 7.12, the same conclusion is also reached without covariate adjustment.)

Table 7.13 shows the typical kind of output associated with BMD-12V. The logarithm of the generalized variance is the *natural logarithm of the determinant of each SSCP (hypothesis plus error) matrix of interest.* Also shown is Wilks' lambda for the general component:

$$\Lambda_H = \frac{|\mathbf{S}_E|}{|\mathbf{S}_E + \mathbf{S}_H|}$$

as described earlier.

Rao's *Ra* function of Λ (distributed exactly as *F* in this illustration) is also computed and the *F* ratio and degrees of freedom printed. We see, of course, that design variable A-3 leads to a highly significant ($p < 0.01$) *F* ratio.[23]

Also shown is the *F* ratio for each covariate, none of which is significant at the 0.01 level used here. As can be seen by comparing the MANOVA and MANCOVA results in Tables 7.12 and 7.13, the covariates contributed little in the way of accounted-for variation in the criterion variables.

Table 7.13 Summary of MANCOVA with Post-Exposure Believability (C-1) and Interest (C-2) as Criteria

Source		Logarithm of Generalized Variance	Wilks' Lambda	F Ratio	d.f.
Alpha Last Brand Purchased (A-2)		11.175	0.887	3.58	2; 56
Pre-Exposure Interest (A-3)		11.250	0.823	6.04	2; 56 ($p < 0.01$)
Interaction		11.060	0.994	0.15	2; 56
Covariate	E-1	11.527	0.954	1.20	2; 56
	E-2	11.125	0.993	0.18	2; 56
	E-3	12.291	0.899	3.13	2; 56
	E-4	11.471	0.963	1.06	2; 56
(Occupation)	$\{ \begin{matrix} E\text{-5} \\ E\text{-5} \end{matrix}$	11.506 / 11.907	0.961 / 0.928	1.14 / 2.16	2; 56 / 2; 56
	E-6	11.169	0.990	0.29	2; 56
Full Model		11.055			

An increasing number of computer programs are becoming available for MANOVA and MANCOVA (Finn, 1974; Clyde, Cramer and Sherin, 1966). Most of these programs are rather complex but do provide a high degree of flexibility for performing various types of analysis of variance and covariance, both univariate and multivariate.[24]

7.7 Descriptive Measures in MANOVA and MANCOVA

In our discussion of regression analysis in Chapter 2 considerable emphasis was given to descriptive indexes of association, such as the squared multiple correlation and measures of part and partial correlation. Similarly, in section 5.5 of Chapter 5, eta squared and Hays' omega squared were described and illustrated numerically as counterpart measures for ANOVA and ANCOVA.

Analogous measures have also been proposed for MANOVA and MANCOVA. Our discussion is brief and limited to generalizations of eta squared (η^2).[25] Perhaps the most straightforward generalization of η^2 is due to Wilks. Wilks (1932) defined a generalization of η^2 in the one-way multivariate case as:

$$\eta^2_{(mult.)} = 1 - \frac{|\mathbf{S}_E|}{|\mathbf{S}_t|}$$

$$= 1 - \Lambda$$

It can be noted that as the determinant of the error matrix increases relative to the determinant of the total SSCP matrix, multivariate eta squared decreases. Moreover, $\eta^2_{(\text{mult.})}$ will range between zero (no treatment effect) and unity (no error).

Hsu (1940) also proposed a generalization of η^2 to deal with the multivariate case. Hsu's generalization appears particularly well suited for higher-way designs (e.g., factorials) and appears as follows:

$$\eta^{2(H)}_{(\text{mult.})} = \frac{H}{1 + H}$$

and H, in turn, is defined as:

$$H = tr[\mathbf{S}_E^{-1}\mathbf{S}_H]$$

where tr denotes the trace (i.e., sum of the main diagonal elements) of $[\mathbf{S}_E^{-1}\mathbf{S}_H]$. Notice, here, that the definition of H is general enough to include tests of effects for two-way (or higher-way) designs.

Indeed, if we desire, we can also modify Wilks' generalization to:

$$\eta^2_{(\text{mult.})} = 1 - \Lambda_H$$

where Λ_H is the ratio $|\mathbf{S}_E|/|\mathbf{S}_H + \mathbf{S}_E|$, in accord with our earlier discussion. (Again, we see that $\eta^2_{(\text{mult.})}$ varies between 0 and 1.) In this way we can extend the scope of the Wilks' generalization to handle designs other than the one-way case.

We can illustrate the two η^2 generalizations by reference to the original sample problem with the relevant statistics appearing in Table 7.14. First, we compute:

$$H = tr(\mathbf{S}_E^{-1}\mathbf{S}_H)$$

for each hypothesis—factor A and factor B.[26]

Table 7.14 Generalized Eta Squared and Omega Squared for Multivariate Case

Source	Λ_H	d_H	$H = tr(\mathbf{S}_E^{-1}\mathbf{S}_H)$	$\eta^2_{(\text{mult.})}$	$\eta^{2(H)}_{(\text{mult.})}$
Factor A	0.0085	2	111.1	0.99	0.99
Factor B	0.275	1	2.6	0.73	0.72
Interaction	0.517	2			

Wilks' generalization provides the measures:

$$\text{Factor } A: \eta^2_{(\text{mult.})} = 1 - \Lambda_A = \frac{|\mathbf{S}_E|}{|\mathbf{S}_A + \mathbf{S}_E|} = 1 - 0.0085$$

$$= 0.99$$

$$\text{Factor } B: \eta^2_{(\text{mult.})} = 1 - \Lambda_B = 1 - 0.275$$

$$= 0.73$$

Hsu's generalization provides the measures:

$$\text{Factor } A: \eta^{2(H)}_{(\text{mult.})} = \frac{111.1}{112.1} = 0.99$$

$$\text{Factor } B: \eta^{2(H)}_{(\text{mult.})} = \frac{2.6}{3.6} = 0.72$$

Both generalizations yield quite similar results in the illustrative problem. While neither measure has been used extensively in applied studies as yet, we expect that this situation will change as information about descriptive measures of association in MANOVA and MANCOVA undergoes greater diffusion in the applied research area.

7.8 Summary

Multivariate discriminant analysis and multivariate analysis of variance and covariance were the principal topics of this chapter. The primary criterion for conducting tests of significance in multiple criterion, multiple predictor association is Wilks' lambda. Accordingly, this statistic was described first, along with the Bartlett and Rao functions of it. The chi square distribution approximates the distribution of Bartlett's V statistic while the F distribution approximates the distribution of Rao's Ra statistic.

Attention then turned to MDA. By means of a small problem involving three groups and two predictors we illustrated the application of MDA, both intuitively and formally. Related topics dealing with significance tests and structure correlations for interpreting results were also presented. Canonical correlation with dummy criterion variables was shown to yield the same discriminant weights (up to a possible scale transformation) as MDA.

MANOVA and MANCOVA were introduced next. A two-way MANOVA was described and analyzed. MDA and MANOVA/MANCOVA were then applied to some of the data of the TV study, along with a brief discussion of descriptive measures of association in MANOVA and MANCOVA.

In concluding this chapter, it should be reiterated that MDA, MANOVA, and MANCOVA are all *special cases of the canonical analysis model* in which dummy variables appear as either criterion or predictor variables. While statistical assumptions may differ, the fact remains that in all of these cases we are seeking linear composites of each set of variables that exhibit maximum association, subject to satisfying various restrictions. Since the multiple regression model is also a special case of the canonical model, it follows that the canonical analysis model is the most general of all models for the analysis of dependence structures.

Review Questions

1. Consider the sample problem of Tables 7.1 and 7.4.
   a. Compute the matrix of discriminant scores on both discriminant axes:

$$\mathbf{Z} = \mathbf{X}_d\mathbf{K}$$

   b. Find the three group-centroid scores on each of the two discriminant axes. Then find their respective among-groups and pooled within-groups sums of squares.
   c. Show numerically that the eigenvalues $\lambda_1 = 29.444$ and $\lambda_2 = 0.0295$ in Table 7.4 are the values obtained by computing the ratio of among-groups to within-groups sums of squares in part b, above.

2. Consider the following data involving three groups and two predictors:

	Predictors	
Group	X_1	X_2
1	2	8
1	5	7
1	6	4
1	4	3
2	6	12
2	7	9
2	10	11
2	12	8
3	8	15
3	14	13
3	10	11
3	12	12

   a. Run a three-group discriminant analysis by finding the eigenstructure of $\mathbf{W}^{-1}\mathbf{A}$.

    b. Check the statistical significance (alpha $= 0.05$) for λ_1 and λ_2 by means of Bartlett's V and Rao's Ra transformation of Wilks' lambda criterion.

    c. Find the two types of standardized discriminant weights and the structure correlations. How do these sets of importance measures compare from an interpretive standpoint?

3. Using the same dummy-variable coding as employed in section 7.4, apply the canonical correlation approach to the sample-problem data of Table 7.1. This time, however, find the eigenstructure of the matrix product:

$$\mathbf{S}_{xx}^{-1}\mathbf{S}_{xy}\mathbf{S}_{yy}^{-1}\mathbf{S}_{yx}$$

where SSCP (rather than correlation) matrices are involved.

    a. How do your results compare with those of Table 7.6 insofar as the canonical correlations are concerned?

    b. How do the results compare with the discriminant weight vectors in Table 7.5?

    c. How would these results be affected if covariance (rather than SSCP) matrices were used?

4. Return to the data of question 2 but now assume that: (a) X_1 and X_2 are criterion variables: Y_1 and Y_2, respectively, and (b) the experimental design is as follows:

		\multicolumn{4}{c}{Factor B}			
		\multicolumn{2}{c}{1}	\multicolumn{2}{c}{2}
		Y_1	Y_2	Y_1	Y_2
Factor A	1	2	8	6	4
		5	7	4	3
	2	6	12	10	11
		7	9	12	8
	3	8	15	10	11
		14	13	12	12

    a. Run a two-way MANOVA (with two replications per cell) and test for the A, B, and $A \times B$ effects using Bartlett's V statistic and an alpha level of 0.01.

    b. Compute multivariate η^2 (by the Wilks' procedure) for A, B, and $A \times B$.

    c. Compute multivariate η^2 (by the Hsu procedure) for the same effects and compare the results.

5. Select from the TV commercial data the 138 Alpha, Beta, and Gamma brand choosers in Part D.

    a. Run a three-group MDA analysis using the first 10 variables of the Part B trade-off data as predictors.

    b. Compare the results of this analysis with those obtained by running three separate two-group discriminant analyses.

    c. In what ways (if any) do the conclusions differ across the approaches?

Notes

1. In some texts and computer programs Wilks' Λ is referred to as the U statistic.

2. As discussed in Chapter 4, if one suspects heterogeneity among the sample covariance matrices, this test should be carried out, prior to testing for centroid differences.

3. Since $|\mathbf{W}|$ is a scalar measure of within-groups variability and the denominator is $|\mathbf{T}| = |\mathbf{W}| + |\mathbf{A}|$, the smaller $|\mathbf{W}|$ is relative to $|\mathbf{A}|$, the greater the among-groups variation is relative to the within-groups variation.

4. The tabular F value (Table C.4 in Appendix C) at the 0.01 alpha level is only 4.77.

5. If $1 + ts - n(G - 1)/2$ turns out not to be an integer, one can use the integer in the F table that is closest to it.

6. The fourth goal, in actuality, is a separate problem since many techniques other than MDA can be used to classify profiles. Moreover, one can study the structure of group differences for its own sake. Still, MDA is often used to classify multivariate profiles, a topic that is discussed in section 7.6.

7. In the context of Chapter 4, we could view the three groups of Table 7.1 as supporting, neutral, and opposing some ecology-oriented legislation. The two predictors X_1 and X_2 could be ratings on two attitude scales, assumed to be related to one's views about ecology.

8. This statement may seem confusing. What should be pointed out are two things: (a) the sets of weights (to be described) in computing $\mathbf{z}_1$ and $\mathbf{z}_2$ are *not* orthogonal and (b) relative to the X_{d1}, X_{d2} coordinate system, $\mathbf{z}_1$ and $\mathbf{z}_2$ are not orthogonal; that is, $\mathbf{z}_1$ and $\mathbf{z}_2$ are *not* the results of a simple rotation of the original X_{d1}, X_{d2} axes, even though $\mathbf{z}_1$ and $\mathbf{z}_2$ are uncorrelated themselves.

9. Section 4.3 of Chapter 4 illustrated this line of development in the context of Fisher's two-group discriminant function.

10. It should be noted the Rao's Ra statistic (described in section 7.2.3) does not exhibit this decomposition property.

11. This relationship follows from the fact that the determinant of the product of two matrices is equal to the product of their determinants; that is, $|\mathbf{W}^{-1}\mathbf{T}| = |\mathbf{W}^{-1}||\mathbf{T}| = |\mathbf{T}|/|\mathbf{W}|$.

12. Recalling (Table 7.4) that the eigenvalues of $\mathbf{W}^{-1}\mathbf{A}$ are $\lambda_1 = 29.444$ and $\lambda_2 = 0.0295$, we note that in the present example:

$$\Lambda = \frac{|\mathbf{W}|}{|\mathbf{T}|} = \frac{1}{1 + \lambda_1} \cdot \frac{1}{1 + \lambda_2}$$

$$= \frac{1}{1 + 29.444} \cdot \frac{1}{1 + 0.0295}$$

$$= 0.0319$$

which is the correct value for the sample problem.

13. Note that we get the same overall result via the original formulation in section 7.2.2:

$$V = -\left[m - 1 - \frac{(n + G)}{2}\right] ln \Lambda$$

$$= -\left[12 - 1 - \frac{(2 + 3)}{2}\right] ln\, 0.0319$$

$$= 29.282$$

14. Hence, with (say) $n = 5$ and $G = 4$, we would have:

$$\text{d.f. (1)} = (5)(3) = 15$$

$$\text{d.f. (2)} = (4)(2) = 8$$

$$\text{d.f. (3)} = (3)(1) = 3$$

for the first, second and third discriminant axes, respectively.

15. In practice, researchers are more often guided by interpretability and practical importance than by statistical significance. With large samples it is not unusual to find several significant discriminant functions. However, only the first one or two may be important from an interpretative viewpoint. Moreover, the first one or two functions may account for a relatively large amount of the variability. For example, in the sample problem one would ordinarily not bother with the second function even if it had turned out to be significant, given its tiny share of accounted-for variability.

16. Note that the diagonal entries of $[\mathbf{C}_w(\mathbf{Z})]^{-1/2}$ are the reciprocals of the square roots of $\mathbf{C}_w(\mathbf{Z})$.

17. This problem is analogous to finding beta weights in the context of multiple regression.

18. Looked at the other way around, we see that the first discriminant function accounts for a high proportion of variation in *both* X_1 and X_2, namely $(0.988)^2 = 0.996$ and $(0.976)^2 = 0.953$, respectively.

19. This follows from the numerical substitutions:

$$t = d_E + d_H - \frac{(n + d_H + 1)}{2}; \quad s = \sqrt{\frac{(nd_H)^2 - 4}{n^2 + d_H^2 - 5}}$$

$$= 6 + 2 - \frac{(2 + 2 + 1)}{2} \qquad\qquad = \sqrt{\frac{(2 \cdot 2)^2 - 4}{2^2 + 2^2 - 5}}$$

$$= 5.5 \qquad\qquad\qquad\qquad = 2$$

20. One way of describing these "partial" F ratios is to consider the predictor variables already in the analysis as covariates. Their linear association with each candidate predictor to enter is removed prior to computing the F ratio for each. In each case we are interested in how highly separated the group means are on each candidate predictor (as conditioned by the predictors already selected at that stage).

21. In this computer run equal prior probabilities were assigned to each of the four groups.

22. We found that factor A-3 was significant, both before and after covariate adjustment, while factor A-2 was not.

23. While not shown in Table 7.13, it is also of interest to point out that the A-3 effect is significant insofar as each *separate* criterion variable (C-1 or C-2) is concerned.

24. In particular, Finn's program computes useful ancillary statistics such as step-down F ratios and discriminant functions for each effect. However, it is more complex to apply than BMD-12V.

25. While multivariate generalizations of omega squared have also been proposed, eta squared has appeared to receive the most attention in the multivariate case.

26. Tables 7.9 and 7.10 provide relevant input for calculation of the H statistic in Table 7.14. For example, $tr(\mathbf{S}_E^{-1}\mathbf{S}_A) = 111.1$, as can be verified by first finding $\mathbf{S}_E^{-1}$ from $\mathbf{S}_E$ in Table 7.9. $\mathbf{S}_E^{-1}$ is then postmultiplied by $\mathbf{S}_A$, in Table 7.9, to find the product, $\mathbf{S}_E^{-1}\mathbf{S}_A$.

PART IV

The Analysis of Interdependence

CHAPTER **8**

Introductory Aspects
of Factor Analysis

8.1 Introduction

Up to this point our discussion of multivariate data analysis has involved procedures for the analysis of dependence structures. In Chapters 2 through 7 the focus has always been on some aspect of criterion variable, predictor variable association. This chapter, and the succeeding one as well, depart from this path by emphasizing the analysis of *interdependence* among entries of an intact (i.e., nonpartitioned) data matrix.

This topic, in turn, can be subdivided into procedures that focus on the variables versus those that concentrate on the objects of the data matrix. Reduced space procedures are representative of the first category and their purpose is to simplify the original data by representing the (same) objects in fewer than the original number of variables (dimensions). Cluster analysis, to be considered in the latter portion of Chapter 9, is concerned with the obverse problem of simplifying the data by reducing the number of objects. Both approaches can be applied to the same data matrix with a consequent reduction in both dimensionality and the number of objects.

Reduced space analysis involves a variety of techniques—factor analysis, multidimensional scaling, and others. If we let each original variable denote a dimension of the data space, then reduced space techniques, as the name suggests, attempt to find a smaller number of dimensions that retain most of the information in the original space.[1]

Probably the most important and well known set of techniques for accomplishing this objective are factor analysis methods. Over the years factor analysis has become something of a generic term that includes a wide variety of separate procedures for effecting the desired dimensional reduction. One of the most popular approaches to factor analysis is *principal components* analysis.

Accordingly, the emphasis of this chapter is on the principal components model.[2] In principal components analysis we seek linear composites of the original variables that display certain desirable properties, namely, scores that exhibit maximal variance, subject to being uncorrelated with previously computed composites.[3] A small-scale numerical example is used to illustrate the basic concepts; component scores and component loadings are some of the topics that are defined and illustrated at this point.

We next discuss the question of *how many* principal components to retain in reducing the initial space. Various techniques (such as the sphericity test, the Scree diagram, and lambda criterion) for assisting the researcher in making his decision, are described and illusrated in the context of the sample problem.

A companion problem to any type of factoring procedure is how to rotate the original solution to some more interpretable orientation. Accordingly, at this point we introduce data from the TV commercial study and compare two popular procedures—Varimax and Biquartimin—for transforming initial principal components solutions to more interpretable configurations. Factor transformations for external or target-matrix fitting are also described briefly and related to previous discussion in Chapter 6.

8.2 Fundamentals of Principal Components Analysis

In the analysis of behavioral data—particularly, data obtained from personality and aptitude testing—it is not unusual to have a large number of test scores available for each of a set of subjects. While each test may purport to measure some specific part of one's personality, it would be most unusual if various groups of tests did not show some tendency to covary, suggesting that they might represent somewhat different ways of expressing a relatively few central characteristics of the person. These central characteristics are often called *latent* traits in the sense that they are not observed directly but are only inferred from the associations noted among a set of manifest variables, such as scores on specific test questions.

In the analysis of business-type data the researcher also must deal with cases in which he has data on more variables than he can effectively assimilate. For example, in marketing research it is not unusual to collect responses to 100 or more attitudinal questions dealing with consumer interests, activities and opinions. In addition, one may have collected information on 20 or 30 demographic variables from the same respondents. Similar surfeits of data may appear in credit screening, personnel record analysis, comparative cost studies of distribution outlets, profitability studies of corporate subsidiaries and the like. However, the analyst is usually interested in much more limited descriptions of the data, particularly when he suspects that individual variables may be either unreliable on a test-retest basis or highly correlated with other variables and, thus, redundant.

As pointed out by Rummel (1970), factor analysis has been used for a variety of objectives, including:

1. Untangling complex patterns of intervariable association in multivariate data.

2. Exploratory research in the identification of latent characteristics for future experimentation.

3. Developing empirical typologies of variables.

4. Reducing the dimensionality of a set of multivariate data.

5. Developing a data-based unidimensional index that maximally separates individuals.

6. Testing hypothesized relationships between certain variables in specific content areas.

7. Transforming the matrix of predictor variables prior to applying some other technique like multiple regression or canonical correlation.

8. Scaling and the spatial representation of perception and preference data.

Still other uses of factor analysis could be mentioned but the preceding ones should be sufficient to convey the versatility of factoring procedures.

The success of a factoring technique depends upon the existence of correlation across at least some of the original variables. Factor analysis—in particular, the principal components method—proceeds in a sequence of steps:

1. Rotation of the initial configuration of points (objects) to a new orientation, of the same dimensionality, that exhibits the characteristic of mutually orthogonal dimensions with sequentially maximal variance. That is, the first dimension displays the largest variance of point projections. The second dimension displays the next largest variance, subject to being orthogonal to the first, and so on.

2. Reducing the dimensionality of this transformed space, usually by discarding those higher dimensions that exhibit the smallest variance of point projections.

3. Finding still a new orientation of the reduced space that makes the retained dimensions more interpretable from a content point of view.

4. Substantive interpretation of the reoriented dimensions in terms of the variables that show high association with each dimension.

We shall discuss each of these steps in due course in the context of principal components analysis. Principal components analysis represents only one technique for factoring data, albeit one that is highly useful and popular in both behavioral and business research. In principal components analysis the derived dimensions—called components in this case—are actual linear composites of the original variables.

8.2.1 A Numerical Illustration

Certain aspects of principal components analysis are best conveyed by numerical example. Consider the (artificial) data shown in Table 8.1. The three variables X_1, X_2, and X_3 could represent respondent ratings on three attributes of a particular brand in which we suspect that a halo effect may be prominent.[4] R, the correlation matrix, also shown in the table, suggests that the variables are highly correlated positively.

The data in Table 8.1 are shown in three ways, namely as: (a) original ratings X_j; (b) deviations from their respective column means X_{dj}; and (c) standardized variates X_{sj}. As recalled from Chapter 1, various cross products matrices can be obtained:

1. Raw sums of squares and cross products (**B**).
2. Mean corrected sums of squares and cross products (**S**).
3. Variances and covariances (**C**).
4. Correlations (**R**).

Table 8.1 Illustrative Data and Derived Association Matrices

Observation	X_1	X_{d1}	X_{s1}	X_2	X_{d2}	X_{s2}	X_3	X_{d3}	X_{s3}
1	2	-5.25	-1.92	1	-6.67	-1.85	2	-4.92	-1.29
2	4	-3.25	-1.18	3	-4.67	-1.30	1	-5.92	-1.55
3	6	-1.25	-0.46	7	-0.67	-0.19	5	-1.92	-0.50
4	7	-0.25	-0.09	6	-1.67	-0.46	3	-3.92	-1.02
5	5	-2.25	-0.82	4	-3.67	-1.02	7	0.08	0.02
6	8	0.75	0.27	9	1.33	0.37	6	-0.92	-0.24
7	9	1.75	0.64	8	0.33	0.09	7	0.08	0.02
8	7	-0.25	-0.09	10	2.33	0.65	9	2.08	0.55
9	10	2.75	1.01	11	3.33	0.93	11	4.08	1.07
10	8	0.75	0.27	9	1.33	0.37	8	1.08	0.28
11	12	4.75	1.74	11	3.33	0.93	10	3.08	0.81
12	9	1.75	0.64	13	5.33	1.48	14	7.08	1.85
$\bar{X}_j$	7.25	0	0	7.67	0	0	6.92	0	0
s_j	2.73	2.73	1	3.60	3.60	1	3.83	3.83	1

B: Raw Sums of Squares and Cross Products				**S**: Mean-Corrected Sums of Squares and Cross Products			
	X_1	X_2	X_3		X_1	X_2	X_3
X_1	713	763	688	X_1	82.25	96.00	86.25
X_2	763	848	771	X_2	96.00	142.67	134.67
X_3	688	771	735	X_3	86.25	134.67	160.92

C: Covariance Matrix				**R**: Correlation Matrix			
X_1	7.477	8.727	7.841	X_1	1.000	0.886	0.750
X_2	8.727	12.970	12.242	X_2	0.886	1.000	0.889
X_3	7.841	12.242	14.629	X_3	0.750	0.889	1.000

Each of these matrices appears in Table 8.1 and is a type of sums of squares and cross products matrix. The matrices differ in terms of corrections made for differences in means and/or standard deviations among the variables of interest.

In most behavioral and administrative research studies, principal components analysis is applied at the standardized data matrix level X_s or, in terms of the associated cross products, the correlation matrix R. This is because much of the data of the applied researcher involves variables in which the means and scale units are arbitrary, such as consumer ratings of brands on various desired attributes. Accordingly, we first discuss a principal components analysis of the standardized data matrix X_s and its associated cross products matrix R. We then describe related procedures for analyzing the other cross products matrices that are shown in Table 8.1.

As background to the analysis, let us plot the standardized data of Table 8.1, X_{s1}, X_{s2} and X_{s3}, in two-variable scatter diagram form. Figure 8.1 shows the

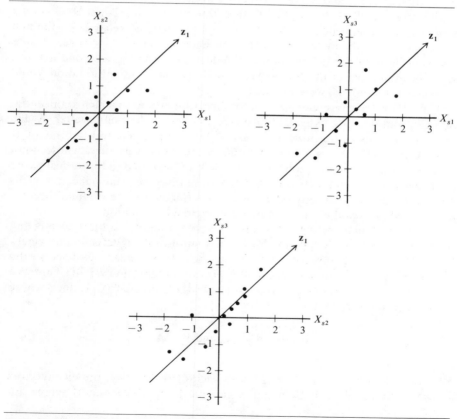

Figure 8.1 Scatter Plots of Standardized Data of Table 8.1 and the First Principal Component z_1

results. As noted from the figure, all three pairs of variables exhibit high positive correlation, as already observed from the correlation matrix of Table 8.1. This being the case, we might wonder if a *single* axis or linear composite could be computed whose derived scores (found by projecting the points onto it) would maximally separate the 12 individual respondents.[5]

This, in essence, is what principal components analysis is all about. In brief, we seek a linear composite of the matrix $\mathbf{X}_s$ that displays maximum variance if the points are projected onto it. However, once this has been done we can seek a second linear composite or axis, orthogonal to the first, that maximizes residual variance, and so on, until we extract as many linear composites—all mutually orthogonal—as are needed to exhaust the original variance in the data. Having done this, we can then drop off the higher-dimensional axes whose accounted-for variance is "small."

Figure 8.2 shows some of the geometric aspects of principal components analysis. In Panel I we assume that only two variables are involved and, in this illustrative case, X_{s1} and X_{s2} are perfectly correlated. Z_1, the first principal axis, would preserve all of the original information in the data. Panel II shows the more usual case in which X_{s1} and X_{s2} are not perfectly correlated. Z_1, the first principal axis, is the major axis of the concentration ellipse. However, it does not exhaust all of the information in the data, necessitating a second axis Z_2. Z_2 is the minor axis of the ellipse. We note that *most* of the information would be preserved if only Z_1 were retained.

Panel III shows the analogous case for three dimensions in which a concentration ellipsoid (looking like a flattened cigar) describes the point scatter. Again, the principal axis Z_1 (or possibly Z_1 *and* Z_2) would preserve most of the information. Panel IV shows a case of three uncorrelated variables, whose point concentration appears as a sphere (in which X_{s3} is assumed to be vertical to the plane of the paper and, hence, is not shown). In this case principal components analysis is not relevant; any set of three axes (including the original three) is just as good as any other in preserving the original information.

The idea of maximizing some variance-type quantity, subject to meeting certain constraints (orthogonality of linear composites) suggests that an eigenstructure problem is involved, a topic that has already been discussed in the context of canonical correlation and multiple discriminant analysis in Chapters 6 and 7. However, in canonical correlation and MDA, the matrix products whose eigenstructures were found:

$$\mathbf{R}_{xx}^{-1}\mathbf{R}_{xy}\mathbf{R}_{yy}^{-1}\mathbf{R}_{yx} \text{ and } \mathbf{W}^{-1}\mathbf{A}$$

were each *nonsymmetric*. As recalled, in each of these cases, the eigenvectors are *not* orthogonal even though the canonical (or discriminant) scores are uncorrelated.

In the present case, however, $\mathbf{R}$, the correlation matrix, *is* symmetric. Hence, the (normalized) eigenvectors of $\mathbf{R}$, denoting direction cosines, will be *orthogonal* and the overall effect will consist of a *rotation* of the original data space to

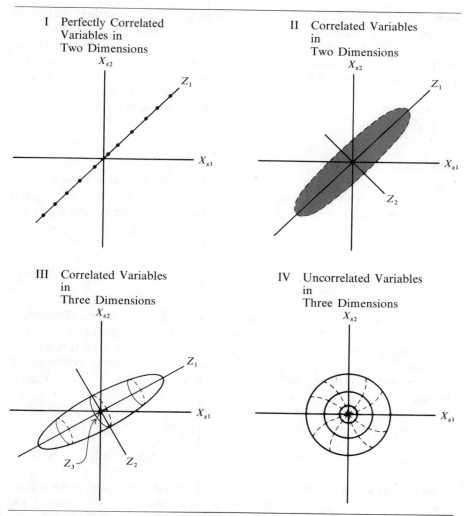

Figure 8.2 Geometric Aspects of Principal Components Analysis

the new orientation whose axes z_1, z_2, and so on, are the principal components axes.

Section B.5 of Appendix B shows the algebraic arguments that lead to finding the eigenstructure of R:

$$\boxed{R = UDU'}$$

where U is an orthogonal matrix ($U'U = UU' = I$), whose columns are the eigenvectors of R, and D is a diagonal matrix whose entries are the eigenvalues of $\mathbf{R}$.[6] The columns of $\mathbf{U}$ can be viewed as sets of direction cosines that rotate $\mathbf{X}_s$, the matrix of standardized data, to principal components orientation.[7]

The entries of $\mathbf{D}$ have an interpretation as well: namely, as the variances of the point projections along the new axes, z_1, z_2, and so on, that represent the principal components of $\mathbf{X}_s$. However, before working out the specific details, let us examine another, and related, way of approaching the problem—one that operates on $\mathbf{X}_s$ *directly* (rather than through the correlation matrix $\mathbf{R}$).

8.2.2 Singular Value Decomposition of $\mathbf{X}_s$

One of the most important decompositions in matrix algebra is known as singular value decomposition. It can be applied to any type of matrix—rectangular or square, singular or nonsingular. Singular value decomposition expresses any matrix, such as an $m \times n$ matrix $\mathbf{A}$, as the product of three other matrices:

$$\begin{array}{cccc} \mathbf{A} & = & \mathbf{P} & \boldsymbol{\Delta} & \mathbf{Q}' \\ m \times n & & m \times r & r \times r & r \times n \end{array}$$

where $\mathbf{P}$ and $\mathbf{Q}$ are each orthonormal sections and $\boldsymbol{\Delta}_{r \times r}$ is a diagonal matrix of ordered positive values.[8] The rank of $\mathbf{A}$ is given by r, the number of such positive values (which are called singular values). Furthermore, $r(\mathbf{A}) \leq \min(m, n)$.

One of the interesting things about singular value decomposition is that $\mathbf{P}$ and $\mathbf{Q}$ are matrices whose columns are eigenvectors of product moment matrices that are *derived* from $\mathbf{A}$. Specifically,

$\mathbf{P}$ is the matrix of eigenvectors of (symmetric) $\mathbf{AA}'$, of order $m \times m$

$\mathbf{Q}$ is the matrix of eigenvectors of (symmetric) $\mathbf{A}'\mathbf{A}$, of order $n \times n$

Of additional interest is the fact that $\boldsymbol{\Delta}$ is a diagonal matrix whose main diagonal entries are the square roots of $\boldsymbol{\Delta}^2$, the *common* matrix of eigenvalues of $\mathbf{AA}'$ and $\mathbf{A}'\mathbf{A}$. The diagonal matrix $\boldsymbol{\Delta}^2$ consists of r *ordered* positive eigenvalues of either $\mathbf{AA}'$ or $\mathbf{A}'\mathbf{A}$.[9] (Section B.6 of Appendix B provides additional details on this method.)

In terms of the sample problem, the singular value decomposition of $\mathbf{X}_s$ is shown in Table 8.2. We note from the table that $\mathbf{P}$ is a 12×3 orthonormal section ($\mathbf{P}'\mathbf{P} = \mathbf{I}$) whose columns are all mutually orthogonal and of unit sums of squares. In this specific case $\mathbf{Q}$ is square since $\mathbf{X}_s$ is of full column rank where $r = n = 3$.[10] Hence, $\mathbf{Q}'\mathbf{Q} = \mathbf{QQ}' = \mathbf{I}$ and we see that $\mathbf{Q}$ is *orthogonal*.

If we next postmultiply both sides of:

$$\mathbf{X}_s = \mathbf{P}\boldsymbol{\Delta}\mathbf{Q}'$$

by $\mathbf{Q}$, the rotation matrix, we get:

**Table 8.2 Singular Value Decomposition of the
Standardized Data Matrix X_s of Table 8.1**

X_s				P		
-1.92	-1.85	-1.24		-0.538	-0.270	0.163
-1.18	-1.30	-1.55		-0.428	0.152	-0.136
-0.46	-0.19	-0.50		-0.120	0.017	-0.302
-0.09	-0.46	-1.02		-0.167	0.397	-0.123
-0.82	-1.02	0.02		-0.196	-0.358	0.573
0.27	0.37	-0.24		0.045	0.218	-0.338
0.64	0.09	0.02	$=$	0.079	0.264	0.238
-0.09	0.65	0.55		0.119	-0.272	-0.385
1.01	0.93	1.07		0.318	-0.025	0.156
0.27	0.37	0.28		0.099	-0.004	-0.073
1.74	0.93	0.81		0.366	0.399	0.383
0.64	1.48	1.85		0.423	-0.517	-0.156

Δ			Q'		
5.434	0	0	0.567	0.597	0.567
0	1.659	0	0.711	-0.006	-0.704
0	0	0.854	0.417	-0.802	0.428

$$X_s Q = Z = P\Delta$$

as the rotation of X_s (with direction cosines given by the columns of Q) that produces the three principal component axes: $Z = P\Delta$. Based on the properties of P and Δ, the three sets of component scores, denoted by Z, are uncorrelated. Futhermore, their positions, as defined by Q, are all mutually orthogonal, and the variances of the columns of $Z = P\Delta$ successively decline. These variances are obtained by finding the covariance matrix of the new linear composite matrix Z:

$$C(Z) = C(P\Delta) = \frac{1}{m-1}(P\Delta)'(P\Delta)$$

$$= \frac{1}{11}[\Delta P' P\Delta]$$

$$= \frac{1}{11}\Delta^2$$

$$= \frac{1}{11}\begin{bmatrix} (5.434)^2 & 0 & 0 \\ 0 & (1.659)^2 & 0 \\ 0 & 0 & (0.845)^2 \end{bmatrix}$$

$$\mathbf{C(Z)} = \begin{bmatrix} 2.685 & 0 & 0 \\ 0 & 0.250 & 0 \\ 0 & 0 & 0.065 \end{bmatrix}$$

Note that all covariances of $\mathbf{Z}$ are zero and the variances of the three columns of $\mathbf{Z}$ successively decline; furthermore, their sum is:

$$2.685 + 0.250 + 0.065 = 3$$

which is also the sum of the main diagonal entries (i.e., the trace) of the original product moment matrix:

$$\mathbf{R} = \frac{1}{m-1}\mathbf{X}_s'\mathbf{X}_s$$

since each of the three diagonal entries of $\mathbf{R}$ is 1.0.

In summary, a *direct way to find the principal components* $\mathbf{Z}$ *of* $\mathbf{X}_s$ *is to find its singular value decomposition:*

$$\mathbf{X}_s = \mathbf{P\Delta Q'}$$

The rotation of $\mathbf{X}_s$ that leads to the principal components orientation is defined by $\mathbf{Q}$ and the principal component axes themselves are given by $\mathbf{Z} = \mathbf{P\Delta}$. These axes are all mutually orthogonal with successive variances computed from the diagonal matrix $\mathbf{\Delta}$ as:

$$\frac{\text{Variances of the}}{\text{columns of } \mathbf{Z}} = \frac{\mathbf{\Delta}^2}{(m-1)}$$

In short, it would seem that this is all that we need. Each $\mathbf{z}_j$ is a linear composite of the original variables given by $\mathbf{X}_s$ with weights defined by $\mathbf{q}_j$, the successive columns of $\mathbf{Q}$. All of the $\mathbf{z}_j$ are mutually orthogonal and their variances decrease in the order: 2.685, 0.250 and 0.065. Moreover, the sum of these variances equals 3.0, the sum of the main diagonal entries (individual-variable variances) of $\mathbf{R}$. Our reason for taking this "side excursion" is that singular value decomposition provides a direct and intuitively appealing way to show how principal components are computed.[11]

One further point of interest concerns the decomposition itself in Table 8.2. If we were to retain only the first principal component $\mathbf{z}_1$, it is of interest to note that we could approximate $\mathbf{X}_s$ by means of the matrix product:

$$\boxed{\hat{\mathbf{X}}_s^{(1)} = \delta_1\mathbf{p}_1\mathbf{q}_1'}$$

where $\delta_1 = 5.434$ is the first entry of $\mathbf{\Delta}$ and $\mathbf{p}_1$ and $\mathbf{q}_1'$ are, respectively, the first column of $\mathbf{P}$ and the first row of $\mathbf{Q}'$. The rank of $\hat{\mathbf{X}}_s^{(1)}$ is, of course, 1 (even though its order is 12×3). It is a property of singular value decomposition that $\hat{\mathbf{X}}_s^{(1)}$ is the *closest rank-1 approximation to* $\mathbf{X}_s$ *that can be found*, in the sense of minimizing the sum of the squared deviations between corresponding elements of $\mathbf{X}_s$ and $\hat{\mathbf{X}}_s^{(1)}$.

Similarly, we could find another approximation of $\mathbf{X}_s$ via:

$$\hat{\mathbf{X}}_s^{(2)} = \delta_2 \mathbf{p}_2 \mathbf{q}_2'$$

in just the same manner, and also an approximation via $\hat{\mathbf{X}}_s^{(3)}$. A further property of singular value decomposition is that these three "slices" are additively composable so that we could reproduce $\mathbf{X}_s$ perfectly by their addition:

$$\mathbf{X}_s = \hat{\mathbf{X}}_s^{(1)} + \hat{\mathbf{X}}_s^{(2)} + \hat{\mathbf{X}}_s^{(3)}$$

As we shall see, this idea provides a key for retaining component dimensions since, clearly, $\hat{\mathbf{X}}_s^{(1)}$ accounts for more of $\mathbf{X}_s$ than $\hat{\mathbf{X}}_s^{(2)}$, and $\hat{\mathbf{X}}_s^{(2)}$ accounts for more of $\mathbf{X}_s$ than $\hat{\mathbf{X}}_s^{(3)}$. (One could say that $\hat{\mathbf{X}}_s^{(2)}$ is the best rank-1 approximation to the *residual* matrix $\mathbf{X}_s - \hat{\mathbf{X}}_s^{(1)}$, and so on.)

However, in most expositions of principal components analysis, emphasis is placed on finding the decomposition of $\mathbf{R}$, the correlation matrix. As we shall see, this more traditional approach bears a strong resemblance to what has just been covered.

8.2.3 Principal Components Obtained from the Correlation Matrix

As shown above, we define the correlation matrix, as:

$$\mathbf{R} = \frac{1}{m-1} \mathbf{X}_s' \mathbf{X}_s$$

and note that $\mathbf{R}$ is simply a matrix of averaged cross products of the standardized data matrix $\mathbf{X}_s$.

In the case of a symmetric cross products matrix, such as $\mathbf{R}$, its singular value decomposition takes the special form:

$$\mathbf{R} = \mathbf{U}\mathbf{D}\mathbf{U}'$$

where, as noted earlier, $\mathbf{U}$ is an orthogonal matrix of eigenvectors of $\mathbf{R}$ and $\mathbf{D}$ is a diagonal matrix of $\mathbf{R}$'s eigenvalues.[12]

As we might suppose, there is a close connection between the singular value decomposition of **R** and the decomposition of $\mathbf{X}_s$ on which **R** is based. This connection is shown in Table 8.3 by first writing out the singular value decomposition of **R**. In the special case in which **R** is symmetric and represents a product moment matrix, namely, $\mathbf{R} = \mathbf{X}_s'\mathbf{X}_s/(m - 1)$, we note that:

1. The matrix of (ordered) eigenvalues **D** in the decomposition of **R** is equal to $\boldsymbol{\Delta}^2/(m - 1)$, as obtained from the decomposition of $\mathbf{X}_s$.[13]

2. The matrix of eigenvectors **U** in the decomposition of **R** is exactly the same as **Q** in the decomposition of $\mathbf{X}_s$.

In other words, the rotation matrix **U** (= **Q**) can be found from the decomposition of either $\mathbf{X}_s$ or **R**.[14] Finding the decomposition of $\mathbf{X}_s$ has the advantage of directly producing the principal components axes, given by the matrix $\mathbf{Z} = \mathbf{P}\boldsymbol{\Delta}$. The variances of the projections on these axes are given by the diagonal entries of $\boldsymbol{\Delta}^2/(m - 1) = \mathbf{D}$.

If we return to Table 8.3, we note that the direction cosines for locating the first principal component axis are found from the first column of **U**:

$$\mathbf{u}_1 = \begin{bmatrix} 0.567 \\ 0.597 \\ 0.567 \end{bmatrix}$$

The resulting axis $\mathbf{z}_1$ has been plotted in the scatter diagrams of Figure 8.1. If the points are projected onto this axis, the variance of their scores is maximal and equals 2.685, the first element in the main diagonal of **D** in the singular value decomposition of **R**.

However, if all three principal axes are employed, it is clear that no spatial reduction at all has been carried out. Rather, what has been done is to consider the set of 12 points in the three original dimensions as roughly resembling a flattened football, or ellipsoid. We then found the three axes of this ellipsoid; the first axis ($\mathbf{z}_1$) corresponded to its length, the second ($\mathbf{z}_2$) to the longer axis of its cross section and the third ($\mathbf{z}_3$) to the shorter axis.

Table 8.3 Singular Value Decomposition of the Correlation Matrix, R

$$\underset{\mathbf{R}}{\begin{bmatrix} 1.000 & 0.886 & 0.750 \\ 0.886 & 1.000 & 0.889 \\ 0.750 & 0.889 & 1.000 \end{bmatrix}} = \underset{\mathbf{U}}{\begin{bmatrix} 0.567 & 0.711 & 0.417 \\ 0.597 & -0.006 & -0.802 \\ 0.567 & -0.704 & 0.428 \end{bmatrix}}$$

$$\underset{\mathbf{D}}{\begin{bmatrix} 2.685 & 0 & 0 \\ 0 & 0.25 & 0 \\ 0 & 0 & 0.065 \end{bmatrix}} \underset{\mathbf{U}'}{\begin{bmatrix} 0.567 & 0.597 & 0.567 \\ 0.711 & -0.006 & -0.704 \\ 0.417 & -0.802 & 0.428 \end{bmatrix}}$$

Somewhat more formally, we could ask: how much information about $\mathbf{R}$ would be lost if we portrayed the configurations of Figure 8.1 in only one dimension, albeit, the dimension represented by the first principal component $\mathbf{z}_1$?

Since $\mathbf{R}$ is symmetric and is composed of real-valued entries, it can be represented in terms of the specific decomposition $\mathbf{R} = \mathbf{UDU}'$ outlined above. Then, if we wish to find a matrix $\hat{\mathbf{R}}^{(1)}$ of rank 1, with a minimum sum of squared deviations between $\mathbf{R}$ and $\hat{\mathbf{R}}^{(1)}$, it turns out that singular value decomposition of $\mathbf{R}$ provides $\hat{\mathbf{R}}^{(1)}$ for us:

$$\hat{\mathbf{R}}^{(1)} = \lambda_1 \mathbf{u}_1 \mathbf{u}_1'$$

where λ_1 denotes the first eigenvalue of $\mathbf{R}$ and $\mathbf{u}_1$ denotes the associated eigenvector (the first column of $\mathbf{U}$).

In general, any $n \times n$ correlation matrix can be decomposed into a set of additive matrices:[15]

$$\mathbf{R} = \lambda_1 \mathbf{u}_1 \mathbf{u}_1' + \lambda_2 \mathbf{u}_2 \mathbf{u}_2' + \cdots + \lambda_n \mathbf{u}_n \mathbf{u}_n'$$

which, in turn, can be written as:

$$\mathbf{R} = \mathbf{UDU}'$$

Moreover, approximations to $\mathbf{R}$ can be portrayed by a matrix $\hat{\mathbf{R}}^{(r)}$ of rank r as:

$$\hat{\mathbf{R}}^{(r)} = \mathbf{U}_r \mathbf{D}_r \mathbf{U}_r'$$

by keeping the first r ($\leq n$) eigenvalues and eigenvectors of $\mathbf{R}$.

In summary, decomposition of $\mathbf{X}_s$ directly, via its singular value decomposition, or decomposition of $\mathbf{R}$ produces fully comparable results by providing the matrix $\mathbf{U}$ ($= \mathbf{Q}$) that rotates $\mathbf{X}_s$ to principal components. The variances of the point projections are given by $\mathbf{D} = \mathbf{\Delta}^2/(m - 1)$. Moreover, we can "build up" $\mathbf{X}_s$ or $\mathbf{R}$ by means of various rank-r approximations $\hat{\mathbf{X}}_s^{(r)}$, or $\hat{\mathbf{R}}^{(r)}$, depending upon the number of dimensions retained.[16]

If $\mathbf{X}_s$ is of full rank initially (and it generally will be if empirical data are involved), then $\mathbf{U}'\mathbf{U} = \mathbf{UU}' = \mathbf{I}$, assuming that all eigenvalues are retained. However, if only r ($r < n$) eigenvalues are retained, we still have $\mathbf{U}'\mathbf{U} = \mathbf{I}$ but now $\mathbf{UU}' \neq \mathbf{I}$ since, in this case, $\mathbf{U}$ is not square but of order $n \times r$. (As such, it is an orthonormal section.)

This particular type of decomposition—applied either to $\mathbf{X}_s$ or $\mathbf{R}$—serves us in good stead if we wish to reduce the dimensionality of the original set of points to some smaller number of dimensions, with the least loss of information.

8.2.4 Component Scores and Loadings

Virtually all computer programs for performing principal components analysis calculate *two* sets of numbers of interest to the applied researcher: (a) component scores and (b) component loadings. Each of these sets of numbers is obtained quite easily from the singular value decompositions described in the preceding section.

We have already computed the matrix of principal component scores, obtained from the decomposition of $\mathbf{X}_s$, as:

$$\mathbf{Z} = \mathbf{P}\mathbf{\Delta} = \mathbf{X}_s\mathbf{U}$$

in Table 8.2 (where $\mathbf{Q} = \mathbf{U}$). As noted, the product of $\mathbf{Z}$ and $\mathbf{U}'$ equals the original data matrix $\mathbf{X}_s$.

However, we may be interested in other scalings of the principal components matrix $\mathbf{Z}$. One scaling that is appropriate for many kinds of applications adjusts the principal components so that their variances are each equal to unity. This is easily accomplished by means of the transformation:

$$\mathbf{Z}_s = \sqrt{m - 1}\,\mathbf{X}_s\mathbf{U}\mathbf{\Delta}^{-1}$$

where $\mathbf{Z}_s$ denotes a matrix of unit-variance component scores. Noting that $\mathbf{X}_s = \mathbf{P}\mathbf{\Delta}\mathbf{U}'$, we can find the covariance matrix of $\mathbf{Z}_s$ as follows:

$$\begin{aligned}
\mathbf{C}(\mathbf{Z}_s) &= \frac{\mathbf{Z}_s'\mathbf{Z}_s}{m - 1}\\[2mm]
&= \frac{[\sqrt{m - 1}\,\mathbf{P}\mathbf{\Delta}\mathbf{U}'\mathbf{U}\mathbf{\Delta}^{-1}]'[\sqrt{m - 1}\,\mathbf{P}\mathbf{\Delta}\mathbf{U}'\mathbf{U}\mathbf{\Delta}^{-1}]}{m - 1}\\[2mm]
&= \mathbf{\Delta}^{-1}\mathbf{U}'\mathbf{U}\mathbf{\Delta}\mathbf{P}'\mathbf{P}\mathbf{\Delta}\mathbf{U}'\mathbf{U}\mathbf{\Delta}^{-1}\\[2mm]
&= \mathbf{\Delta}^{-1}\mathbf{I}\mathbf{\Delta}\mathbf{I}\mathbf{\Delta}\mathbf{I}\mathbf{\Delta}^{-1}\\[2mm]
&= \mathbf{I}
\end{aligned}$$

We see that the columns of $\mathbf{Z}_s$ must each have unit variance since $\mathbf{C}(\mathbf{Z}_s) = \mathbf{I}$ is an identity matrix.[17] Alternatively, we could apply the (equivalent) transformation:

$$\mathbf{Z}_s = \mathbf{X}_s\mathbf{U}\mathbf{D}^{-1/2}$$

since $\mathbf{D} = \mathbf{\Delta}^2/(m - 1)$, as observed in the preceding section.

Standardizing the components to unit variance has the effect of first rotating X_s to principal components orientation, via U. Then the diagonal matrix $D^{-1/2}$ stretches the shorter axes and shrinks the longer ones until they are all of the same length. Thus, after the orientation of the ellipsoid of points has been found (or hyperellipsoid for four or more variables), the ellipsoid is changed into a sphere (or hypersphere), via $D^{-1/2}$, along the axes of that orientation.

Component loadings are just as easily found from singular value decomposition. By "loading" is meant the correlation of some original variable X_{s1}, X_{s2} or X_{s3} with a principal component.[18] Since a correlation is simply an average cross product between two standardized variables, we could find the loadings on (say) the first standardized component z_{s1} from:

$$f_1 = \frac{X'_s z_{s1}}{m - 1}$$

By means of relatively simple algebra,[19] the whole matrix of factor loadings can be found from the matrix product:

$$F = UD^{1/2}$$

Again, we note that the component loadings matrix utilizes the basic decomposition of R (or X_s, if desired).

Finally, we can reproduce X_s, the original matrix of standardized scores, by postmultiplying the matrix of unit-variance component scores Z_s by the transpose of the matrix of component loadings:

$$X_s = Z_s F'$$
$$= X_s UD^{-1/2}(UD^{1/2})'$$
$$= X_s UD^{-1/2}D^{1/2}U'$$
$$= X_s$$

Note that to do this we assume that all components are retained and, hence, $UU' = I$ (as well as $U'U = I$). If not all components are retained, U is no longer orthogonal (but will still be an orthonormal section satisfying $U'U = I$). In this case X_s will be matched as closely as possible by the best rank-r approximation to it obtained from $U^{(r)}$ and $D^{(r)}$, as described earlier.

Table 8.4 shows numerical values of F and Z_s (for a selected set of observations) based on the sample data of Table 8.1. Also shown for comparison purposes are a selected set of unstandardized principal component scores $Z = P\Delta$, as obtained from Table 8.2.

Table 8.4 Component Loadings Matrix and Selected Values of the Z (Unstandardized) and Z_s (Standardized) Component Score Matrices

Component Loadings

$\mathbf{F} = \mathbf{UD}^{1/2}$

Variables	Components I	II	III
1	0.930	0.356	0.108
2	0.980	-0.003	-0.203
3	0.930	-0.352	0.110
λ_j	2.685	0.250	0.065
Cumulative Variance Accounted-For	0.89	0.98	1.00

Unstandardized Component Scores

$\mathbf{Z} = \mathbf{P\Delta}$

Case	Components I	II	III
1	-2.923	-0.448	0.138
2	-2.326	0.252	-0.115
3	-0.652	0.028	-0.255
.	.	.	.
11	1.989	0.662	0.324
12	2.299	-0.858	-0.132

Standardized Component Scores

$\mathbf{Z}_s = \mathbf{X}_s \mathbf{UD}^{-1/2}$

Case	Components I	II	III
1	-1.784	-0.899	0.420
2	-1.419	0.502	-0.548
3	-0.399	0.057	-1.030
.	.	.	.
11	1.218	1.324	1.355
12	1.402	-1.712	-0.420

If we compare the two sets of component scores first, we note that (within rounding error) they are in the ratios:

First column: $\dfrac{\mathbf{z}_{i1}}{\mathbf{z}_{is1}} = \sqrt{\lambda_1} = \sqrt{2.865} = 1.639:1$

Second column: $\dfrac{\mathbf{z}_{i2}}{\mathbf{z}_{is2}} = \sqrt{\lambda_2} = \sqrt{0.250} = 0.5:1$

Third column: $\dfrac{\mathbf{z}_{i3}}{\mathbf{z}_{is3}} = \sqrt{\lambda_3} = \sqrt{0.065} = 0.255:1$

as desired.

The matrix of component loadings shows that all three variables exhibit high positive correlations (0.93, 0.98 and 0.93, respectively) with principal component I. The correlations drop off markedly, however, in the case of components II and III.

It is also a property of the principal components solution that the sum of the *squared* loadings in each column of the component loadings matrix equals its respective eigenvalue λ_j (obtained from the diagonal matrix **D**). For example, in the case of component I, we have:

$$\lambda_1 = (0.93)^2 + (0.98)^2 + (0.93)^2 = 2.685$$

Viewed in this way, we can see how much each variable contributes to each component's variance. (In the case of λ_1 the contributions of the three variables are almost equal.)

The row labeled Cumulative Variance Accounted-For indicates the cumulative proportion of total variance in **R** that is accounted for by each component. Not surprisingly, the first principal component accounts for 89 percent of the total variance, since:

$$\frac{\lambda_1}{\sum\limits_{j=1}^{3} \lambda_j} = \frac{2.685}{3} = 0.89$$

while the third component accounts for only 2 percent of the total variance.

By the same token we could compute the sum of the squared loadings for each row and examine the relative size of the entries in this sum. For example, in the case of variable X_{s1}, we have:

$$(0.93)^2 + (0.356)^2 + (0.108)^2 = 1$$

Since all components have been computed, all row sums of squares will equal unity. However, insofar as X_{s1} is concerned, the first component accounts for

$(0.93)^2 = 0.86$ of its total variation. The sum of the squared row entries is called the *communality* of the variable with the set of components of interest.[20]

8.2.5 Principal Components Analysis of Other Cross-Product Matrices

While principal components analysis is typically applied to $\mathbf{X}_s$, the standardized matrix or, equivalently, to $\mathbf{R}$, the correlation matrix, occasions arise where other cross product matrices are used. The main candidates are:

1. **B**, the raw sums of squares and cross products matrix.
2. **S**, the mean-corrected or SSCP matrix.
3. **C**, the covariance matrix.

Each of these appears in Table 8.1, as computed in conjunction with the sample-problem data.

The decomposition of **B**, **S** and **C** into the triple product of **U**, an orthogonal matrix, **D** a diagonal matrix, and **U′** can be undertaken in just the same manner as described in the case of **R**, the correlation matrix. Moreover, one can find the singular value decomposition of **X** or $\mathbf{X}_d$ in just the same way that the singular value decomposition of $\mathbf{X}_s$ was found.

Table 8.5 shows the **D** (diagonal) and **U** (rotation) matrices that are found by computing the eigenstructure of **B**, **S** and **C**. These can be compared to each other and to their counterpart matrices in Table 8.3. First, we see that the eigenstructure of **B**, the raw sums of squares and cross products matrix, differs from **S**, the SSCP matrix, because of the differences in means across X_1, X_2 and X_3. The

Table 8.5 Eigenstructures of Other Cross Product Matrices

Raw Sums of Squares and Cross Products Matrix **B**

$$\mathbf{D} = \begin{bmatrix} 2250.83 & 0 & 0 \\ 0 & 36.12 & 0 \\ 0 & 0 & 9.05 \end{bmatrix}; \quad \mathbf{U} = \begin{bmatrix} 0.555 & 0.672 & 0.490 \\ 0.612 & 0.069 & -0.788 \\ 0.563 & -0.738 & 0.372 \end{bmatrix}$$

Mean-Corrected SSCP Matrix **S**

$$\mathbf{D} = \begin{bmatrix} 348.85 & 0 & 0 \\ 0 & 28.77 & 0 \\ 0 & 0 & 8.21 \end{bmatrix}; \quad \mathbf{U} = \begin{bmatrix} 0.435 & 0.685 & 0.585 \\ 0.626 & 0.238 & -0.743 \\ 0.648 & -0.689 & 0.325 \end{bmatrix}$$

Covariance Matrix **C**

$$\mathbf{D} = \begin{bmatrix} 31.71 & 0 & 0 \\ 0 & 2.62 & 0 \\ 0 & 0 & 0.75 \end{bmatrix}; \quad \mathbf{U} = \begin{bmatrix} 0.435 & 0.685 & 0.585 \\ 0.626 & 0.238 & -0.743 \\ 0.648 & -0.689 & 0.325 \end{bmatrix}$$

mean of X_2 is the largest of the three variables' means and this influences the positions of the principal components, as well as the eigenvalues in **D**.

The eigenvalues of **S** versus **C**, however, are in the ratio of $(m - 1): 1$ and their eigenvectors are exactly the same. This is true since **S** and **C** differ initially only by the ratio $(m - 1): 1$. On the other hand, the eigenstructure of **C** differs from that of **R** since the variances of the original variables differ and these differences affect the eigenstructure of **C** but not that of **R**.

In short, *the results of a principal components analysis of a set of data are dependent upon the particular type of cross products matrix that is used for decomposition.* Differences in means and variances affect the results and there is no simple way to compare results over solutions based on different cross products matrices. (This lack of invariance of results over scaling differences is not true of all factoring methods, as will be pointed out in Chapter 9.)

8.2.6 Recapitulation

At this point we have introduced a variety of concepts which should be summarized prior to moving on to the next main topic.

1. The (unstandardized) principal components of a standardized data matrix $\mathbf{X}_s$ consist of a matrix **Z** of linear composites. The linear composites are orthogonal and their variances successively decrease. That is, the first composite has the largest variance, the second, the next largest variance, and so on.

2. The (unstandardized) principal components **Z** are given by a rotation of $\mathbf{X}_s$ that is defined by:

$$\mathbf{Z} = \mathbf{X}_s\mathbf{U}$$

where **U** is orthogonal (if $\mathbf{X}_s$ is of full rank). **U** can be found from the singular value decomposition of either $\mathbf{X}_s$ or **R**, the correlation matrix.

3. If $\mathbf{X}_s$ is used, the singular value decomposition is: $\mathbf{X}_s = \mathbf{P\Delta U'}$.[21] If **R** is used, the singular value decomposition is $\mathbf{R} = \mathbf{UDU'}$.

4. The diagonal matrix **D** (above) contains the variances of the linear composites appearing in **Z**.

5. Alternatively, one can decompose $\mathbf{X}_s$ via the product of a matrix of standardized (to unit variance) component scores $\mathbf{Z}_s$ and the transpose of a matrix of component loadings **F**. In this case the decomposition is:

$$\mathbf{X}_s = \mathbf{Z}_s\mathbf{F'}$$

6. The standardized (to unit variance) component scores matrix $\mathbf{Z}_s$ is computed by:

$$\mathbf{Z}_s = \mathbf{X}_s\mathbf{U}\mathbf{D}^{-1/2}$$

7. The component loadings matrix is computed by:

$$\mathbf{F} = \mathbf{U}\mathbf{D}^{1/2}$$

In the case of a correlation matrix $\mathbf{R}$, component loadings are correlations of each variable with each principal component.[22]

8. The sum of squares of each column of component loadings equals that component's variance λ_j (see Table 8.4).

9. The sum of squares of each row of component loadings equals that variable's communality (variance accounted for by the components).

10. In general, finding the principal components of other cross products matrices, such as the raw sums of squares and cross products matrix or the covariance matrix leads to *different results* that are not easily compared across solutions.

11. A symmetry exists in components analysis in the sense that a standardized variable $\mathbf{x}_{sj}$ can be represented as a linear composite of the (standardized) components by means of:

$$\mathbf{x}_{sj} = f_{j1}\mathbf{z}_{s1} + f_{j2}\mathbf{z}_{s2} + \cdots + f_{jn}\mathbf{z}_{sn}$$

where the $\mathbf{z}_s$'s are columns of $\mathbf{Z}_s$, the standardized component scores, and the f_j's are component loadings for the j-th variable.

Similarly, an (unstandardized) component $\mathbf{z}_j$ can be represented as a linear composite of the standardized variables by means of:

$$\mathbf{z}_j = u_{j1}\mathbf{x}_{s1} + u_{j2}\mathbf{x}_{s2} + \cdots + u_{jn}\mathbf{x}_{sn}$$

where the $\mathbf{x}_s$'s are the columns of $\mathbf{X}_s$, the standardized data, and the u_j's are the entries in the j-th column of $\mathbf{U}$, the rotation matrix.

At this point, then, we have a way to factor either the data matrix $\mathbf{X}_s$ or the derived average cross products matrix $\mathbf{R}$ into a set of orthogonal components with sequentially maximal variance.

The problem now is to decide *how many* of these components to retain—that is, how much the original space should be reduced in dimensionality.

8.3 How Many Components to Retain?

So far, our discussion of principal components analysis has concentrated on finding an orientation of the data space in which the axes are mutually orthogonal with projections that exhibit sequentially maximal variance. However, as indicated at the beginning of the chapter, one of the primary purposes of factor analysis is *spatial reduction*. Data-based matrices are generally of full rank and, hence, after applying principal components analysis we usually end up with as many components as original dimensions. Moreover, we have been assuming all along that the data were highly enough correlated to justify the factoring in the first place.

Accordingly, this section describes two questions dealing with spatial reduction via principal components analysis:

1. Is it worthwhile reducing the original space at all?

2. If so, how many components (dimensions) should be retained?

The first question can be answered relatively straightforwardly by means of Bartlett's sphericity test. The second question is usually approached more judgmentally, as we shall see in due course.

8.3.1 Bartlett's Sphericity Test

The sphericity test, developed by Bartlett (1950), is directly addressed to the question of whether the correlation matrix should be factored in the first place.[23] The test makes use of the determinant of the correlation matrix as a generalized measure of variance.

The rationale for the test is simple enough to describe. Assume that the universe from which a sample is drawn is made up of n mutually *uncorrelated* variables. If so, the universe correlation matrix would be an identity matrix since all off-diagonal correlations would be zero and all diagonal (self) correlations would be unity. As recalled from matrix algebra the determinant of an identity matrix, denoted $|\mathbf{I}|$, equals unity. However, suppose through sampling variability that the sample-based correlation matrix exhibits some (or all) non-zero off-diagonal elements. We wish to find out if these departures from an identity matrix could have arisen from sampling fluctuations alone.

Somewhat more formally, let us assume that a sample of size m has been taken from a multivariate normally distributed universe of n random variables. Let $\mathbf{R}_p$ denote the universe correlation matrix and $\mathbf{R}$ denote the sample estimate of it. The null hypothesis is that $|\mathbf{R}_p| = 1$; that is, that the matrix of universe correlations is an $n \times n$ identity matrix. Bartlett sets up a chi square approximate test with the following test statistic:

$$\chi^2_{[0.5(n^2 - n)]} = -[m - 1 - \tfrac{1}{6}(2n + 5)]\, ln\,|\mathbf{R}|$$

were n denotes the number of variates in $\mathbf{R}$; m denotes the sample size; and $ln\,|\mathbf{R}|$ denotes the natural logarithm of the determinant of the sample correlation matrix $\mathbf{R}$.

The computed χ^2 value is compared to the tabular χ^2 for a selected α risk with 0.5 $(n^2 - n)$ degrees of freedom. In the sample problem of Table 8.1 the determinant of $\mathbf{R}$, the sample correlation matrix, is 0.0436, $n = 3$, and $m = 12$. Hence, we have:

$$\chi^2_{[0.5(9 - 3)]} = -[12 - 1 - \tfrac{1}{6}\{2(3) + 5\}]\, ln\, 0.0436$$
$$= (-9.17)(-3.1327)$$
$$= 28.73$$

The tabular χ^2 value for $\alpha = 0.01$ and $0.5(3^2 - 3) = 3$ degrees of freedom is only 11.34; hence, we reject the null hypothesis.

Bartlett's sphericity test is particularly useful when one is dealing with relatively small samples of data (e.g., $m < 100$) with a relatively large number of variates (e.g., $n \geq 10$). The determinant of $\mathbf{R}$ is readily computed from the relationship:

$$|\mathbf{R}| = \prod_{j=1}^{n} \lambda_j$$

for n eigenvalues of $\mathbf{R}$ (since we assume $\mathbf{R}$ to be of full rank).[24]

From Table 8.4, we note that the eigenvalues of $\mathbf{R}$ are:

$$\lambda_1 = 2.685; \lambda_2 = 0.250; \lambda_3 = 0.065$$

Hence, the determinant is:

$$|\mathbf{R}| = 2.685(0.250)(0.065)$$
$$= 0.0436$$

as indicated above.

As simple as Bartlett's test statistic is to compute, relatively few packaged computer programs for principal components analysis include the test.[25] An exception is Cooley and Lohnes' PRINCO (principal components) program (1971) which does compute Bartlett's (approximate) χ^2 value.

Bartlett's test statistic results in an approximate test but its power appears to be quite high. Cooley and Lohnes report the results of a study involving a sample size of only $m = 20$. Even with the number of variates $n = 10$, the test is almost sure to reject the null hypothesis that $\mathbf{R}_p = \mathbf{I}$ if an alpha level of 0.05 is used and the variates are correlated 0.36 (absolute value) or more. If $m = 200$, the same test is almost certain to reject H_0 if the universe variates are correlated 0.09 or more.

8.3.2 Eliminating Higher-Dimensional Components via Statistical Tests

With rather large sample sizes (e.g., $m \geq 200$), it will not be unusual to find that many components are statistically significant. However, from a practical standpoint, some of the components may account for a very small proportion of the total variance. While various significance tests can be conducted, more often than not the researcher is primarily interested in *operational significance and interpretability;* as such, he may retain fewer components than turn out to be statistically significant.

A number of procedures for determining how many principal components to retain have been suggested:

1. Run significance tests on the separate eigenvalues of $\mathbf{R}$.

2. Retain all eigenvalues of $\mathbf{R}$ that are greater than unity.

3. Prepare a plot of residuals from the r-th ($r < n$) rank approximation matrix to $\mathbf{R}$ and see if these residuals are small and approximately normally distributed.

4. Prepare a plot of variance accounted-for versus the rank number of the extracted components and look for an "elbow" in the graph.

5. Split the sample in half and retain only those components with high correspondence of factor loadings across the two subsamples.

We describe these procedures in more detail. However, it should be stated at the outset that there is a large degree of judgment required in applying any of these "rules."

Bartlett's test can be extended rather straightforwardly to deal with testing the significance of residual matrices following the extraction of each principal component, in turn.[26] To illustrate, suppose that r components have already been extracted and one wishes to test whether the determinant of the universe *residual* matrix is zero. If so, the formula shown above is modified as follows:

$$\chi^2_{[0.5(n-r)(n-r-1)]} = -[(m + 1) - \tfrac{1}{6}(2n + 5)] \, ln \, W_{[n-r]}$$

where $W_{[n-r]}$ can be considered as a type of a residual value that is computed after the r largest eigenvalues have been removed. The formula for $W_{[n-r]}$ is:

$$W_{[n-r]} = \frac{|\mathbf{R}|}{\lambda_1 \lambda_2 \cdots \lambda_r \left[\dfrac{n - \lambda_1 - \lambda_2 \cdots - \lambda_r}{n - r} \right]^{n-r}}$$

on the assumption that r ordered eigenvalues have been removed.[27]

To illustrate, in the sample problem let us test whether one or more of the remaining eigenvalues are significant *after* removing the first ($\lambda_1 = 2.685$) of the ordered (large to small) eigenvalues. Since $n = 3$ and $r = 1$ we wish to find $W_{[3-1]}$:

$$W_{[3-1]} = \frac{0.0436}{2.685 \left[\dfrac{3 - 2.685}{3 - 1} \right]^{3-1}}$$

$$= \frac{0.0436}{2.685(0.1575)^2}$$

$$= 0.654$$

We then substitute $W_{[3-1]} = 0.654$ into the (modified) general formula:

$$\chi^2_{[0.5(2)(1)]} = -\left[(12 - 1) - \tfrac{1}{6}[2(3) + 5]\right] \ln 0.654$$
$$= (-9.17)(-0.425)$$
$$= 3.9$$

The computed $\chi^2 = 3.9$ with $0.5(3 - 1)(3 - 1 - 1) = 1$ degree of freedom is not significant at the $\alpha = 0.01$ level. Hence only the first principal component is retained, according to this test.

As first indicated in Chapter 7, it should be noted that this sequential approach to principal component retention assumes that a significant overall result implies that at least the largest eigenvalue is significant and that one can remove this effect to test for the significance of the remaining eigenvalues in the stepwise manner indicated above. The critical remarks by Harris (1975, 1976) in the context of Bartlett's V statistic are also relevant to the present context.

Other tests have been developed (Lawley and Maxwell, 1963) for examining the significance of eigenvalues obtained from a principal components analysis. The fact remains, however, that the researcher is usually interested in more than statistical significance, even if all of the statistical assumptions are satisfied by his data.

Another approach to the problem of deciding how many components to retain has been proposed by Kaiser (1959). Kaiser has recommended that only principal components of $\mathbf{R}$ with eigenvalues exceeding unity be retained. The argument is based, in part, on Guttman's earlier work dealing with minimum-rank matrices and, in part, on the common sense rationale that any principal component, being a measure of *common* variance, should account for more variance than any single variable in the standardized score space. A number of computer program developers have incorporated this simple rule in their computation routines. Under this rule only the first component (with $\lambda_1 = 2.685$) in the sample problem of Table 8.1 would be retained.

8.3.3 Graphical and Related Criteria

The preceding "statistical-like" criteria do not exhaust the kinds of procedures that have been proposed for determining how many principal components to retain. Other criteria based on various pictorial representations have been suggested as rule-of-thumb guides.

One rather flexible approach is to examine residual correlation matrices after each successive component has been extracted. The motivation here is to look for "small" and normally distributed residuals as a graphical aid to when to stop component extraction. Cooley and Lohnes (1971), among others, discuss this type of examination.

As recalled, by taking advantage of the relationship:

$$\mathbf{R} = \hat{\mathbf{R}}^{(1)} + \hat{\mathbf{R}}^{(2)} + \cdots + \hat{\mathbf{R}}^{(n)}$$
$$= \lambda_1 \mathbf{u}_1 \mathbf{u}_1' + \lambda_2 \mathbf{u}_2 \mathbf{u}_2' + \cdots + \lambda_n \mathbf{u}_n \mathbf{u}_n'$$

one can decompose **R** into additive slices and examine the residual matrices at any desired level in the component extraction procedure.

Using the eigenstructure decomposition of Table 8.3, Table 8.6 shows the computations leading to the three additive slices for the sample problem. As can be noted from the table, after the first slice $\hat{\mathbf{R}}^{(1)}$ has been found, the next two slices $\hat{\mathbf{R}}^{(2)}$ and $\hat{\mathbf{R}}^{(3)}$ exhibit relatively small residuals. While these residuals have not been plotted, it seems fairly obvious from the size of the values themselves that little additional information is found in the second and third slices.

Another technique, called the *scree test,* has been proposed by Cattell (1966). This procedure entails plotting the variance accounted-for by each principal component in the order extracted (see Table 8.4) and then looking for an elbow in the curve. Panel I of Figure 8.3 shows a hypothetical case in which the elbow appears with the extraction of the fourth eigenvalue. Presumably, one would retain only the first three eigenvalues.[28]

Panel II of Figure 8.3 shows a plot of the actual eigenvalues obtained from the sample problem. Here we have only three eigenvalues and it is not clear at what point we should keep components, since there are too few to show just where the elbow might be. Should only the first component be retained? This

Table 8.6 Computing Additive Slices of the Sample-Problem Correlation Matrix

From Table 8.3

$$\mathbf{R} = \begin{bmatrix} 1.000 & 0.886 & 0.750 \\ 0.886 & 1.000 & 0.889 \\ 0.750 & 0.889 & 1.000 \end{bmatrix};$$

Eigenvectors

$$\mathbf{U} = \begin{bmatrix} 0.567 & 0.711 & 0.417 \\ 0.597 & -0.006 & -0.802 \\ 0.567 & -0.704 & 0.428 \end{bmatrix};$$

Eigenvalues

$$\mathbf{D} = \begin{bmatrix} 2.685 & 0 & 0 \\ 0 & 0.25 & 0 \\ 0 & 0 & 0.065 \end{bmatrix}$$

Computing the Additive Slices

$$\hat{\mathbf{R}}^{(1)} = 2.685 \begin{bmatrix} 0.567 \\ 0.597 \\ 0.567 \end{bmatrix} [0.567 \quad 0.597 \quad 0.567] = \begin{bmatrix} 0.863 & 0.909 & 0.863 \\ 0.909 & 0.957 & 0.909 \\ 0.863 & 0.909 & 0.863 \end{bmatrix}$$

$$\hat{\mathbf{R}}^{(2)} = 0.25 \begin{bmatrix} 0.711 \\ -0.006 \\ -0.704 \end{bmatrix} [0.711 \quad -0.006 \quad -0.704] = \begin{bmatrix} 0.126 & -0.001 & -0.125 \\ -0.001 & 0 & 0.001 \\ -0.125 & 0.001 & 0.123 \end{bmatrix}$$

$$\hat{\mathbf{R}}^{(3)} = 0.065 \begin{bmatrix} 0.417 \\ -0.802 \\ 0.428 \end{bmatrix} [0.417 \quad -0.802 \quad 0.428] = \begin{bmatrix} 0.011 & -0.022 & 0.012 \\ -0.022 & 0.043 & -0.022 \\ 0.012 & -0.022 & 0.014 \end{bmatrix}$$

$$\mathbf{R} = \hat{\mathbf{R}}^{(1)} + \hat{\mathbf{R}}^{(2)} + \hat{\mathbf{R}}^{(3)}$$

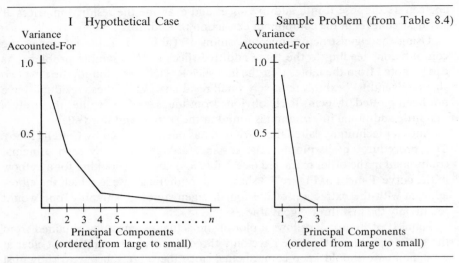

Figure 8.3 Application of Cattell's Scree Test

example illustrates the rather high degree of subjectivity in using the scree diagram for principal components retention. Real data often fail to show a clear elbow effect even if a relatively large number of components are present. One might find that the first couple of eigenvalues account for over 80 per cent of the variance (and appear sufficient for representing the data) even though no sharp elbow appears after the second eigenvalue.

A less ambiguous variation on the scree test theme has been provided by Horn (1965). Horn has proposed that the eigenvalues of **R** be plotted in the same diagram with a set generated from an appropriate null-hypothesis distribution, involving independently (and normally) distributed standardized variables. The idea is to draw one or more random samples, each of size m, and see what the pattern of eigenvalues of **R** is in the case where the universe correlation matrix is known to be the identity matrix. One then plots the eigenvalues from the random data together with the eigenvalues obtained from the real data. Components are retained up to the point where the two lines cross.

Figure 8.4 illustrates the basic concept. Suppose that we have $n = 28$ variables and a sample size of $m = 300$ in some problem of interest. According to Horn's proposal one would generate K matrices (where K is decided by the researcher) of random variables, each of order 300×28. Each data set would be correlated and its principal components extracted. One would then average the first, the second, the third, and so on, eigenvalues of **R** over the K randomly generated matrices and plot the averages on the same graph.

If the universe correlation matrix is actually an identity matrix, we know that its determinant is unity. Furthermore, the eigenvalues of an identity matrix are each equal to unity. However, based on sampling variation, a procedure that

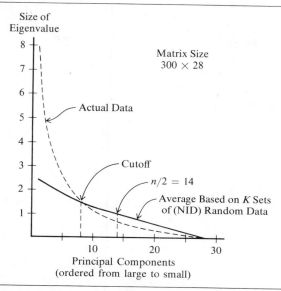

Size of
Eigenvalue

Matrix Size
300 × 28

Figure 8.4 A Hypothetical Application of Horn's Test for Principal Component Retention

extracts components for a *sample*-based correlation matrix $\mathbf{R}$ drawn from $\mathbf{R}_p = \mathbf{I}$ will show departures from the value of unity for various eigenvalues. Over many such samples we would expect the average curve of ordered (large to small) eigenvalues to cross the ordinate scale at the value of 1.0 for component number $n/2$ or, in this case, $28/2 = 14$, as noted in Figure 8.4.

However, the first $n/2$ eigenvalues would be expected to exceed 1.0 and the second $n/2$ eigenvalues would be expected to be less than 1.0.[29] This type of behavior is also shown in Figure 8.4. Moreover, the smaller the sample size the more we would expect the reference curve (based on random data) to depart from a horizontal line plotted at an eigenvalue of 1.0. For example, in the hypothetical case of Figure 8.4 the first (largest) eigenvalue for the reference curve is 2.5 (strictly as a function of sampling error).

Under the Horn criterion, one chooses the number of eigenvalues to be retained on the basis of where the actual curve (based on the specific data) crosses the reference curve. This point will occur *before* an eigenvalue of 1.0 is reached, as is illustrated in Figure 8.4. One chooses the number of eigenvalues from the number on the abcissa—in this case 8 components—nearest to where the two lines cross. The rationale is that the reference line, aside from sampling error, really represents the case in which the *eigenvalues under the null hypothesis are unity*.

Horn advocates the incorporation of this test in standard principal components routines but, so far, relatively few researchers have used the test. Crawford and Koopman (1973) have found some difficulties with the accuracy of Horn's procedure in factoring methods other than principal components. In the case of components analysis, however, Horn's test appeared to work well in the examples that they considered.

With today's computer capability, it is relatively easy to generate a number of random matrices of desired size for application of Horn's test. Perhaps this approach will see increasing application in the future as information about it becomes more widely disseminated.

8.3.4 Concluding Comments

Despite the availability of various statistical tests, such as Bartlett's sphericity test, and graphical approaches, such as Horn's, the fact remains that most applications researchers employ fairly mechanical procedures for deciding how many components to retain.

Probably the most prevalent of these more simplistic criteria is to retain all components whose eigenvalues exceed unity. (In this sample problem only λ_1 has a value exceeding unity.) As noted above, however, this approach, in the presence of sampling error, could retain *more eigenvalues than are appropriate* according to Horn's approach.

A second approach is to retain only those eigenvalues that account, on a cumulative basis, for some rather high proportion of the total variance in **R**, such as 75 or 80 per cent. This rule is followed in many kinds of behavioral and business applications on the pragmatic rationale that lesser components—even if statistically significant and reliable over subsequent studies—may just not be important enough to consider.[30]

At this stage in the application of principal components analysis, the researcher might consider retaining varying numbers of principal components for later study. In particular, one may examine the interpretability of the subsequently rotated factors (to be discussed in the next section) as a function of how many components are retained.

Another useful approach to the problem of determining how many principal components to keep is based on split-half sample cross-validation. The idea here is simply to split the sample observations randomly into halves and perform a separate principal components analysis on each half. Then by application of various factor matching procedures, such as those described in Chapter 6, the component loading matrices are orthogonally rotated to best agreement and correlations are computed for each pair of first-half versus second-half component loadings. Those components displaying high split-half correlations are retained for subsequent analysis.

In summary, there is no fool-proof way to decide on how many principal components to retain. Data-based matrices are typically full-rank matrices and all components need to be extracted for full reproduction of the matrix. Thus, the researcher, in desiring parsimony, must eliminate some of the information by dropping lesser components. One hopes that it is the "noisy" or unreliable part of the data that is eliminated. However, whatever method or combination of methods is employed, one has to trade off the possible loss of real information for a reduction in the dimensionality of the space. *Statistical significance, in the case of large samples particularly, does not provide much help in the decision.* The researcher will probably want to retain fewer components than turn out to be statistically significant.

8.4 Factor Rotation

Factor analysis—whatever the variety—is usually concerned with two major problems:

1. Reducing the dimensionality of the original data space, whether by principal components or some other factoring procedure.

2. Rotation of the factor loading solution in the reduced space to some more interpretable orientation and recomputation of factor scores in the new orientation.[31]

It is now time to consider the second problem—one involving interpretation of the analysis from a content viewpoint.

Principal components analysis, in attempting to find a set of orthogonal dimensions that result in successive variance maximization, may not result in a configuration of points that is substantively interpretable. Indeed, the tendency in principal components analysis is to produce a first component that is a general factor on which almost all variables are highly correlated. This is accompanied by a series of bipolar components on which some variables are correlated positively and others correlated negatively.

In applied problems we are usually interested in orientations of the factor axes in which each dimension (including the first) has only a *few variables* defining it. Accordingly, we first need a rationale for permitting further transformations of the initial principal components solution. Then we can consider various criteria for achieving a more interpretable representation of the factor loadings matrix, to be followed by a companion transformation of the points (observations) themselves.

8.4.1 Rationale for Subsequent Rotation of Principal Components

The rationale for subsequent rotation of the component loadings and scores matrices derives from a basic indeterminacy in representing one matrix by the product of only two other matrices.

In section 8.3, it will be recalled that the following matrices were computed as a fundamental to the principal components problem:

Component scores (standardized): $\mathbf{Z}_s = \mathbf{X}_s \mathbf{U} \mathbf{D}^{-1/2}$

Component loadings: $\mathbf{F} = \mathbf{U} \mathbf{D}^{1/2}$

The basic equation of principal components analysis then consists of the product of these two matrices (actually the product of $\mathbf{Z}_s \mathbf{F}'$) as representing the original matrix of standardized scores:

$$\boxed{\mathbf{X}_s = \mathbf{Z}_s \mathbf{F}'}$$

assuming that all components are retained. What is now asserted is that $\mathbf{X}_s$ can still be reproduced if certain transformations are made of $\mathbf{Z}_s$ and $\mathbf{F}$. This is called the indeterminacy problem regarding $\mathbf{X}_s$ and provides the researcher with the opportunity to examine more interpretable orientations of $\mathbf{Z}_s$ and $\mathbf{F}$ that will reproduce $\mathbf{X}_s$.

To be more specific, suppose we had some nonsingular matrix $\mathbf{T}$ that was otherwise arbitrary. Further, suppose that we decided to postmultiply $\mathbf{Z}_s$ by $\mathbf{T}$:

$$\mathbf{Z}_s \mathbf{T}$$

We now wonder if we can "undo" this effect by making some companion transformation of $\mathbf{F}$, so that $\mathbf{X}_s$ is not affected by all of this.

It turns out that we can. First, let us define another nonsingular matrix by the transformation $(\mathbf{T}')^{-1}$. Next, let us postmultiply $\mathbf{F}$ by this matrix:

$$\mathbf{F}(\mathbf{T}')^{-1}$$

Then, if the product of $\mathbf{Z}_s \mathbf{T}$ and the transpose of $\mathbf{F}(\mathbf{T}')^{-1}$ is computed, we obtain:

$$\mathbf{X}_s = (\mathbf{Z}_s \mathbf{T})[\mathbf{F}(\mathbf{T}')^{-1}]'$$

$$= \mathbf{Z}_s \mathbf{F}'$$

and we see that $\mathbf{X}_s$ is not disturbed by the two transformations.[32]

Next, consider indeterminacy in the major product moment of the component loadings matrix $\mathbf{F}$, defined as:

$$\mathbf{R} = \mathbf{F}\mathbf{F}'$$

Here we note that $\mathbf{R}$, the correlation matrix, can be represented as the major product moment of $\mathbf{F}$.

Let $\mathbf{J}$ be an $r \times r$ *orthogonal* matrix (but otherwise arbitrary) by which we postmultiply $\mathbf{F}$:

$$\mathbf{F}\mathbf{J}$$

Next, since $\mathbf{J}'\mathbf{J} = \mathbf{J}\mathbf{J}' = \mathbf{I}$, we find the major product moment of $\mathbf{F}\mathbf{J}$ to be:

$$(\mathbf{FJ})(\mathbf{FJ})' = \mathbf{FJ'JF'}$$

$$= \mathbf{FF'}$$

Hence, we can reproduce $\mathbf{R}$ with either the major product moment of $\mathbf{F}$ or that of $\mathbf{FJ}$ where $\mathbf{J}$, while orthogonal, is otherwise arbitrary. This property also holds true for various slices of $\mathbf{R}$: $\hat{\mathbf{R}}^{(1)}$, $\hat{\mathbf{R}}^{(2)}$, and so on, that were described in section 8.3.

What this all boils down to is that we can choose some *additional* rotation $\mathbf{J}$ of the component loadings matrix without destroying the property by which the major product moment (of $\mathbf{FJ}$) is able to reconstitute $\mathbf{R}$, or any lower rank approximation to $\mathbf{R}$ that one is working with.

The factor loading rotation problem, thus, consists of attempting to find an orthogonal matrix $\mathbf{J}$ such that the matrix of rotated loadings $\mathbf{FJ}$ is more interpretable in some sense.[33] The most popular approach to finding $\mathbf{J}$ and, hence, the new loading matrix $\mathbf{FJ}$ is based on principles of simple structure.

8.4.2 Simple Structure

Simple structure refers to a particular type of pattern in the factor loadings matrix *after* transformation by $\mathbf{J}$. The desiderata underlying this pattern are somewhat vague and the pattern itself will only tend to approximate the objectives of simple structure.

Researchers who use factor analysis in the behavioral and business fields are usually interested in finding factor loadings matrices that exhibit approximations to Thurstone's (1947) idea of simple structure. Thurstone believed that most content domains would probably involve several latent factors. He also assumed that any single observed variable would be correlated with only one or a few of the several factors and any single factor would be associated with only a few variables. The general idea, then, was to find clusters of variables, each of which defined only one factor.

If the reasoning underlying simple structure is valid, the factor loadings matrix should tend to exhibit a particular kind of pattern (Comrey, 1973):

1. Most of the loadings on any specific factor (column) should be small (as close to zero as possible) and only a few loadings should be large in absolute value.[34]

2. A specific row of the loadings matrix, containing the loadings of a given variable with each of the factors, should display nonzero loadings on only one or, at most, a few factors.

3. Any pair of factors (columns) should exhibit different patterns of loadings. Otherwise, one could not distinguish the two factors represented by the columns.

In empirical problems one may fail to find simple structure, even after applying rotations designed to approximate it. However, in any case the motivation is to have each variable associated with only one or, at most, a few factors and, similarly, to have each factor identified, where possible, with a distinct cluster of variables.

Figure 8.5 shows the basic ideas involved. Panel I of the figure portrays a matrix of factor loadings as might be obtained initially from a principal components analysis. An "X" indicates a "high" loading (e.g., a variable-factor correlation of 0.5 or higher in absolute value) while a blank indicates a relatively low loading. We note from Panel I that in this illustrative case all nine variables load highly on the first principal component. Moreover, each variable continues to load highly on at least one other component.

Panel II, in contrast, indicates a different pattern, following a rotation of $\mathbf{F}$ to approximate Thurstone's simple structure. In this case each variable loads highly on only a single factor. Moreover, each factor (column) can be described in terms of only a few variables.

Most factor analyses entail a sequence of steps. First, principal components analysis, or some other factoring procedure, is used to reduce the dimensionality of the original space. Then, the factor loadings in the *reduced space* are again rotated in an attempt to improve on the interpretability of the solution. Finally, a companion rotation is applied to the component scores so that the rank-*r* approximation to $\mathbf{X}_s$ can still be reproduced.

It is important to note that in the second and third steps the variance-maximizing characteristics of principal components are lost.[35] That is, while the retained components *in total* account for just as much variance in the original data set as they did before, this variance is now split up differently across the new dimensions of the *rotated* factor loadings configuration. Hence, it is no longer necessarily the case that the first (rotated) dimension accounts for the highest variance, the second dimension accounts for the next highest, and so on.

	I High Loadings Before Rotation Components						II High Loadings After Rotation Factors				
Variables	I	II	III	IV	V	Variables	I	II	III	IV	V
1	X		X		X	1	X				
2	X	X		X		2		X			
3	X		X		X	3		X			
4	X	X		X		4					X
5	X	X		X	X	5			X		
6	X		X		X	6	X				
7	X	X			X	7				X	
8	X		X			8			X		
9	X	X		X		9				X	

Figure 8.5 Principal Component Loadings Versus Rotated Matrix of Factor Loadings (Hypothetical)

Indeed, rotation schemes designed to approximate simple structure will generally lead to a set of axes in which the variance of point projections does not vary appreciably across the new dimensions.

The idea of simple structure is brought out geometrically in the hypothetical example of Figure 8.6 for the case of two factors and several variables. Panel I shows a case in which all variables load relatively highly (absolute-value basis) on factor I, the horizontal axis, but differ markedly on factor II, the vertical axis. In this case the factor II scale is bipolar, with both large positive and large negative loadings.

As a case in point, the data could represent consumers' ratings of new model cars on various attributes, such as roominess, style, prestige, and so on, as obtained from an initial components-type factoring. The first dimension might represent general desirability while the second dimension distinguishes between desirability as a family car and desirability as a personal (sports-type) car.

Panel II, showing the same picture *after* rotation, neatly separates the one general and one bipolar factor into two separate factors with a relatively high loading of each attribute on one and only one of the two factors—either desirability as a family car or desirability as a personal car.

8.4.3 An Illustrative Problem

Having described the conceptual rationale for subsequent rotation of the component loadings matrix and the characteristics of simple structure, an empirical illustration would seem to be helpful. At this point an example based on real data is more appropriate; we return to the TV commercial data for this purpose.

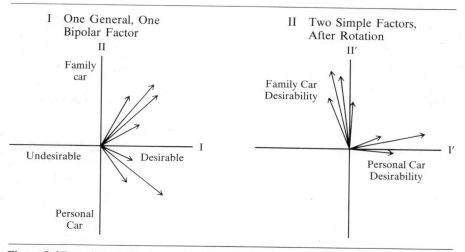

Figure 8.6 Rotation of Initial Factor Loadings to a Better Approximation of Simple Structure

As can be noted in Appendix A, data for the 252 male respondents of the study were available on several demographics: (a) age; (b) family size; (c) education; (d) marital status; (e) occupation; and (f) income. Mean values appear below:

Variable	Sample Means
1. Age	33.2 years
2. Family size	3.5 persons
3. Education	13.8 years
4. Married	57 per cent
5. White collar ⎫ Occupation	44 per cent
6. Blue collar ⎭	39 per cent
7. Annual income	$14,946

These seven variables (involving three dummies) were first factored via principal components analysis, using the BMD-08M program (Dixon, 1973). BMD-08M is a highly versatile computer program for factoring either covariance or correlation matrices involving up to 198 variables for any number of observations. Either principal components analysis, with unities in the diagonal of the correlation matrix, can be employed or, alternatively, one can factor analyze the input matrix with diagonal entries that are less than unity (to be discussed in Chapter 9). Factor loadings and factor scores are computed and a wide variety of rotational options are available.

Let us first consider the initial principal components analysis of the 7×7 correlation matrix, computed across the 252 respondents. For illustrative purposes we elected to retain all components whose associated eigenvalues exceeded unity (Kaiser, 1959).

Table 8.7 shows the initial component loadings matrix as obtained from BMD-08M, as well as **R**, the correlation matrix being factored. We note that the first three eigenvalues (those with λ_j's greater than unity) collectively account for 73 per cent of the variance in the 7×7 correlation matrix. The first component alone accounts for 35 per cent. Furthermore, with the exception of family size, all variables load highly on this general factor (as suggested in the preceding section).[36]

Of additional interest is the column headed h_i^2. This column lists the *communalities* of each variable with each component. A variable's communality is an R^2 measure based on a regression of a row variable on the components; it is equal to the sum of the squared loadings (with each component) in that variable's row. For example, the communality for the first variable, age, is:

$$(0.57)^2 + (0.68)^2 + (0.14)^2 = 0.80$$

as shown in the table.

Table 8.7 Initial Loadings Matrix from Principal Components Analysis

Variable	Retained Components			
	I	II	III	h_i^2
1. Age	0.57	0.68	0.14	0.80
2. Family Size	−0.07	−0.03	−0.86	0.75
3. Education	0.68	−0.29	−0.01	0.54
4. Marital Status	0.52	0.73	−0.07	0.81
5. White Collar	0.84	−0.25	0.14	0.79
6. Blue Collar	−0.74	0.47	−0.09	0.79
7. Income	0.40	−0.01	−0.70	0.65
λ_j	2.47	1.38	1.28	5.13
Cumulative VAF	35%	55%	73%	

Variable	Original Correlation Matrix **R**						
	1	2	3	4	5	6	7
1	1.00						
2	−0.14	1.00					
3	0.15	−0.03	1.00				
4	0.61	0.03	0.17	1.00			
5	0.29	−0.12	0.47	0.25	1.00		
6	−0.15	0.05	−0.46	−0.04	−0.71	1.00	
7	0.13	0.29	0.21	0.16	0.21	−0.19	1.00

Each component's contribution to a variable's communality is provided by the squared loading for that column in the row of interest. Hence, component II, $(0.68)^2 = 0.46$, provides the largest contribution to the communality for age. Communalities, being R^2 measures, range from 0 to 1. If the communality is 0, none of the variance in the (standardized) variable is accounted for by the components. If the communality is 1, all of the variance is accounted for by the components.[37]

Thus, it is a rather neat feature of component loadings that the loadings do double duty. Looked at on a column basis their squares indicate the contribution of each variable to each component's variance. Looked at on a row basis their squares indicate the contribution of each component to each variable's total R^2. In the present case at least half of the variance in each row variable is accounted for by the first three components.

However, the component loadings matrix of Table 8.7 does *not* come even close to the conditions of simple structure. In particular, none of the seven rows displays a single large loading and zero loadings elsewhere. Accordingly, it is time to examine ways in which the loadings of Table 8.7 can be rotated to a more interpretable orientation.

8.4.4 Kaiser's Varimax Rotation Procedure

Probably the most popular of rotation schemes for approximating simple structure is Kaiser's Varimax procedure, one of the options available in BMD-08M. If we continue to let $\mathbf{F}$ denote the initial loadings matrix (Table 8.7) obtained by principal components analysis, Varimax rotation entails finding an orthogonal matrix $\mathbf{J}$ in which:

$$\boxed{\mathbf{G} = \mathbf{FJ}}$$

Following the principles of simple structure we try to find a new loadings matrix $\mathbf{G}$ with columns containing as many near-zero loadings as possible. If so, the few variables remaining in each column that do *not* exhibit low loadings would serve (by their high loadings) to identify the factor. A simple geometric representation should facilitate understanding of this idea. In Figure 8.7 we plot the loadings of a single hypothetical variable, denoted by the vector $\mathbf{x}_1$, in a plane of two principal components $\mathbf{f}_1$ and $\mathbf{f}_2$.

Next, suppose we rotate the initial axis $\mathbf{f}_1$ to the position $\mathbf{g}_1$. If so, variable $\mathbf{x}_1$, with length $l(\mathbf{x}_1)$, makes a small angle θ with $\mathbf{g}_1$. Its projection on $\mathbf{g}_1$ is:

$$\boxed{g_1(\mathbf{x}_1) = l(\mathbf{x}_1)\cos\theta}$$

and this projection will be relatively large, as can be noted in the figure.[38] On the other hand, the projection of $\mathbf{x}_1$ on $\mathbf{g}_2$, chosen to be orthogonal to $\mathbf{g}_1$, is:

$$\boxed{g_2(\mathbf{x}_1) = l(\mathbf{x}_1)\sin\theta}$$

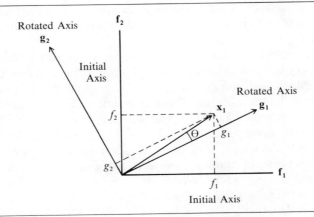

Figure 8.7 Rotation of Initial Loadings Matrix to New Axes g_1 and g_2

and this projection will be relatively small. The problem, in general, is to find new orthogonal axes $g_1, g_2, \ldots, g_r$ so that for all variables $x_1, x_2, \ldots, x_n$, considered simultaneously, we want *either* large projections *or* small projections.

However, if we desire either large or small projections of variables on factors, this can be interpreted as saying that we wish to maximize some type of column loading *variance* measure. As it turns out the Varimax criterion is quite close to this intuitive idea except that in Varimax we operate on the *squared* loadings rather than on the loadings themselves. Use of the squared loadings possesses certain advantages since all quantities will be positive and, in any given problem, the total sum of the squared loadings equals:

1. The sum of the eigenvalues $\lambda_1, \lambda_2, \ldots, \lambda_r$ of the *r* retained components.
2. The sum of the communalities $h_1^2, h_2^2, \ldots, h_n^2$ of the *n* variables.

Thus, with *n* variables and *r* retained components we seek a rotation of **F** (given by **J**) that leads to a new loadings matrix **G** in which we try to maximize each column's variance.

However, if the rows of **F**, the initial loadings matrix, have different communalities (as they usually will), then the variables with the largest communalities will contribute more to the solution than the others. Accordingly, *Kaiser normalization* (an additional option in BMD-08M) enables the user to normalize each row of **F** by dividing each entry in the *i*-th row by the square root h_i of that row's communality. Having done this, the sum of squares for each row will then equal unity and initial differences in the variables' communalities will not affect the Varimax solution.

The Kaiser Varimax procedure, then, attempts to simplify the *columns* of the initial loadings matrix **F** by finding a rotation **J** that results in a matrix **G** in which each column consists of either high or low loadings. Varimax proceeds through a series of iterations, whose total number is specifiable by the researcher, until the criterion being maximized does not change more than some allowable numerical value between successive iterations or until the maximum number of iterations is reached.[39]

8.4.5 The Varimax Solution

The Kaiser Varimax option in BMD-08M was applied to the original loadings shown in Table 8.7. The new loadings matrix **G** (after Varimax) appears in Table 8.8.

The first thing to be noticed about the Varimax-rotated matrix is that the *communalities* h_1^2 and the *total* variance accounted-for are *unchanged* from the original. What we do notice, however, is that the total variance accounted-for becomes split up differently *across factors,* with factor 2 accounting for more, and factor 1 for less variance than each did originally.[40] The major change is in the pattern of factor loadings *inside* the table.

Factor I now shows high loadings on only three variables. Of these, education and white collar workers evince the same sign while blue collar workers display

Table 8.8 Varimax-Rotated Factor Loadings of Table 8.7

Variable	Factor			h_i^2
	I	II	III	
1. Age	0.14	0.88*	−0.09	0.80
2. Family Size	−0.14	−0.09	0.85*	0.75
3. Education	0.72*	0.10	0.09	0.54
4. Marital Status	0.05	0.89*	0.11	0.81
5. White Collar	0.86*	0.23	−0.03	0.79
6. Blue Collar	−0.89*	0.02	−0.01	0.79
7. Income	0.26	0.17	0.74*	0.65
λ_j	2.16	1.67	1.30	5.13
Cumulative VAF	31%	55%	73%	

*Highest row loading.

the opposite sign. Factor I would appear to be a type of occupational status scale with educated, white collar workers defining one end and blue collar workers the other.

Factor II displays high loadings on age and marital status. Not surprisingly, both loadings carry the same sign. Factor III shows high loadings on family size and income; moreover, the loadings carry the same sign, suggesting that family size is positively associated with income.

Note further that each row of Table 8.8 now shows only a single high loading. That is, each variable has been assigned to only one factor. Such clear-cut results may not occur in other applications, however, and the reader should bear in mind that some variables may fail to load highly on only a single factor.

In summary, Kaiser Varimax "cleaned up" the loadings matrix originally obtained from the principal components analysis quite well. The variance-maximizing property of principal components no longer holds after Varimax rotation, although the factors, in total, continue to account for the same proportion of variance in each variable and in the total set of variables.

A few other features of orthogonal transformations, such as Varimax, are useful to note:

1. After Varimax rotation the new loadings matrix **G** maintains the property that **GG′** = **R**, if all factors are retained.

2. If only r ($r < n$) factors are retained (where $\mathbf{F}_r \mathbf{F}'_r = \hat{\mathbf{R}}^{(r)}$), then $\mathbf{G}_r \mathbf{G}'_r = \hat{\mathbf{R}}^{(r)}$.

3. Factor scores, after their companion Varimax rotation, will continue to be mutually uncorrelated.

Finding the new (standardized) factor scores, denoted $\mathbf{Z}_{sv}$, after Varimax rotation, makes use of the new loadings matrix **G** in the following way:

$$\mathbf{Z}_{sv} = \mathbf{X}_s \mathbf{G}(\mathbf{G}'\mathbf{G})^{-1}$$

In other words, having found the loadings matrix **G**, it is a relatively simple matter to go on to solve for $\mathbf{Z}_{sv}$, the standardized factor scores on the new (rotated) factors.

The BMD-08M program computes and prints out the new (rotated) standardized scores as part of its regular output. The program also prints out:

$$\mathbf{N} = \mathbf{G}(\mathbf{G}'\mathbf{G})^{-1}$$

the transformation that produces the rotated factor scores. Table 8.9 shows this transformation.

However, BMD-08M does *not* print out **J**, the rotation matrix that transforms the initial loadings matrix **F** into **G** the new loading matrix. Fortunately, **J** can be obtained from previously computed matrices by the equation:

$$\mathbf{J} = \mathbf{D}^{-1/2}\mathbf{U}'\mathbf{G}$$

This matrix appears in Table 8.9.

Table 8.9 also shows some algebraic demonstrations of interest. First, it is shown that the matrix **N** used to find the rotated factor scores $\mathbf{Z}_{sv}$ is equal to $\mathbf{Z}_s\mathbf{J}$. The rotation matrix, in turn, is shown to equal $\mathbf{D}^{-1/2}\mathbf{U}'\mathbf{G}$, as asserted above.

In summary, Varimax rotation simply transforms the retained axes from the principal components solution to a new orthogonal space of the same dimensionality. The new dimensions (found by **J**) are chosen to approximate a criterion based on simple structure objectives and the whole procedure is one that is designed to improve the interpretability of the reference dimensions. The new factors are no longer principal components. While the new axes are still orthogonal they no longer display sequentially maximal variances. This requirement is sacrificed in the interest of obtaining better interpretability of the factor loadings matrix and, hence, the reoriented dimensions themselves.

8.5 Other Types of Factor Transformations

Most applied work in factor analysis tends to favor orthogonal transformations, such as Varimax, to approximate simple structure. On occasion, however, *oblique* or nonorthogonal transformations may be applied.[41] Oblique transformations result in new factor loadings in which the factors are correlated rather than being mutually orthogonal.

Furthermore, as described earlier in section 6.8 of Chapter 6, one might want to transform a (factor loadings) matrix to a prespecified *target* matrix, rather than going through a procedure that involves "blind" transformation to maximize some criterion that is related to simple structure objectives.

Both oblique and target-matrix transformations are rather specialized topics in factor analysis. Our discussion here is necessarily brief and introductory.

Table 8.9 Additional Output Related to Varimax Rotation

Rotated Factor Score Coefficients Matrix	Rotation Matrix

$$
\mathbf{N} = \begin{bmatrix}
 & \text{I} & \text{II} & \text{III} \\
 & -0.05 & 0.54 & -0.10 \\
 & -0.09 & -0.06 & 0.66 \\
 & 0.34 & -0.03 & 0.04 \\
 & -0.11 & 0.56 & 0.07 \\
 & 0.40 & 0.02 & -0.06 \\
 & -0.44 & 0.14 & 0.02 \\
 & 0.08 & 0.06 & 0.56
\end{bmatrix}
\qquad
\mathbf{J} = \begin{bmatrix}
\text{I} & \text{II} & \text{III} \\
0.84 & 0.52 & 0.12 \\
-0.51 & 0.85 & -0.03 \\
0.11 & 0.04 & -0.99
\end{bmatrix}
$$

Relationship of $\mathbf{Z}_{sv}$ to Original Factor Scores $\mathbf{Z}_s$	Solving for the Rotation Matrix $\mathbf{J}$

$$\mathbf{Z}_{sv} = \mathbf{X}_s\mathbf{N}$$

$$\qquad = \mathbf{X}_s\mathbf{G}(\mathbf{G}'\mathbf{G})^{-1}; \ \mathbf{G} = \mathbf{FJ} = \mathbf{UD}^{1/2}\mathbf{J}$$

$$\qquad = \mathbf{X}_s\mathbf{UD}^{1/2}\mathbf{J}\{\mathbf{J}'\mathbf{D}^{1/2}\mathbf{U}'\mathbf{UD}^{1/2}\mathbf{J}\}^{-1}$$

$$\qquad = \mathbf{X}_s\mathbf{UD}^{1/2}\mathbf{JJ}'\mathbf{D}^{-1/2}\mathbf{D}^{-1/2}\mathbf{J}$$

$$\qquad = \mathbf{X}_s\mathbf{UD}^{-1/2}\mathbf{J}$$

$$\qquad = \mathbf{Z}_s\mathbf{J}$$

since $\mathbf{J}'\mathbf{J} = \mathbf{JJ}' = \mathbf{U}'\mathbf{U} = \mathbf{I}$ and

$$\mathbf{Z}_s = \mathbf{X}_s\mathbf{UD}^{-1/2}$$

$$\mathbf{G} = \mathbf{FJ}$$

$$\mathbf{G} = \mathbf{UD}^{1/2}\mathbf{J}$$

$$(\mathbf{UD}^{1/2})'\mathbf{G} = (\mathbf{UD}^{1/2})'\mathbf{UD}^{1/2}\mathbf{J}$$

$$\mathbf{D}^{1/2}\mathbf{U}'\mathbf{G} = \mathbf{DJ}$$

$$\mathbf{D}^{-1}[\mathbf{D}^{1/2}\mathbf{U}'\mathbf{G}] = \mathbf{D}^{-1}[\mathbf{DJ}]$$

$$\mathbf{J} = \mathbf{D}^{-1/2}\mathbf{U}'\mathbf{G}$$

8.5.1 Oblique Transformations

In contrast to Varimax and other types of orthogonal transformations of principal component loadings matrices, oblique transformations drop the requirement that the new factor axes be orthogonal. The rationale underlying their application is the belief that clusters of variables may not exist in orthogonal directions. The intent of oblique transformations is to put the factor axes through clusters of variables, whether or not the resulting orientation of axes is orthogonal.[42]

In oblique transformations there are three primary matrices of interest. First, there is the $r \times r$ matrix, denoted $\mathbf{J}^*$, that transforms the initial loadings matrix $\mathbf{F}$ (of order $n \times r$) to the new loadings matrix. In general, $\mathbf{J}^*$ will not have orthogonal columns even though each of its columns *will* be of unit length, since direction cosines are involved in the transformation. Second, there is the new $n \times r$ loadings matrix itself, denoted $\mathbf{G}^*$. Third, there is an $r \times r$ matrix of cosines of angles separating the pairs of factors in the transformed space; this matrix is denoted $\boldsymbol{\Psi}$. The entries of $\boldsymbol{\Psi}$ represent the correlations between factors in the transformed space.

Oblique transformations display a type of ambiguity that is related to whether the *structure* or the *pattern* version of the new factor loadings matrix is being examined.[43] As it turns out, what has just been called $\mathbf{G}^*$ is really the matrix of structure loadings.

Figure 8.8 illustrates the problem. Suppose we have some point P in the oblique system that is obtained after transforming $\mathbf{F}$, the original factor loadings matrix to the new loadings matrix $\mathbf{G}^*$. The angle Ψ shows the oblique character of the two axes of the transformed loadings matrix. If we elect to interpret the new loadings in terms of (perpendicular) projections onto $\mathbf{g}_1^*$ and $\mathbf{g}_2^*$, then the projections x_1 and x_2 are the values that are reported.

On the other hand, if we elect to report coordinates in the oblique reference system of $\mathbf{g}_1^*$ and $\mathbf{g}_2^*$, then w_1 and w_2 are the values reported. The relationship between the two sets of values is given by:[44]

$$x_1 = w_1 + w_2 \cos \Psi$$

$$x_2 = w_2 + w_1 \cos \Psi$$

In factor analysis parlance, the perpendicular projections x_1 and x_2 are called *structure* loadings while the coordinates w_1 and w_2 on the oblique axes themselves are called *pattern* loadings.

Structure loadings are correlations between variables and factors (and hence vary between -1 and $+1$) while pattern loadings are regression coefficients of each variable on each of the factors. These latter coefficients are not constrained to vary within -1 and $+1$. It is the pattern loadings that are usually of most interest, since these are the ones that result from attempts to pass factor axes

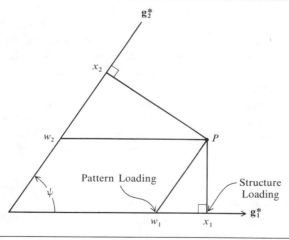

Figure 8.8 Projection Versus Coordinate Interpretation of Oblique Transformation

through clusters of variables (in the quest for some approximation to simple structure).

For comparison purposes it should be useful to take one oblique transformation procedure—in this case Carroll's Biquartimin method—and apply it to the same loadings matrix of Table 8.7 that was used earlier in the Varimax rotation analysis of Table 8.8.

8.5.2 Carroll's Biquartimin Procedure

The Biquartimin procedure was developed by J. B. Carroll (1957). Virtually all oblique transformation procedures (and there are several of them) attempt to minimize a sum of cross products of squared factor loadings in the new loadings matrix. As we know from earlier discussion, if a row variable increases its loading on a particular column (factor) it tends to decrease its loading on one or more of the remaining factors.

Accordingly, if we were to find the cross products of each row of transformed squared loadings with itself, we would have $r(r - 1)/2$ such cross products, assuming that the transformed loadings matrix is of order n variables by r factors. The cross products are summed within each row. We then find the total sum of squared cross products by summing the row subtotals across the n rows.

What we would like to do is *minimize* this total sum of cross products of (squared) loadings. If so, then each row of the transformed loadings matrix would tend to have only one large squared loading and the rest of the loadings would be close to zero. Nontechnically, this is what Carroll's Biquartimin procedure tries to do. BMD-08M has an option for performing various kinds of oblique transformations, including Biquartimin. Accordingly, the principal components loadings matrix of Table 8.7 was obliquely transformed by the Biquartimin technique.

Table 8.10 shows the results of the Biquartimin transformation for both the structure (projections) and the pattern (oblique coordinates) loadings matrices. Interestingly enough, in this specific problem the solutions are quite close to each other and also to the Varimax solution in Table 8.8. This result is also implied by the fact that the matrix of factor correlations, Ψ, also shown in Table 8.10, has off-diagonal entries that are practically zero. In effect, then, the Biquartimin transformation in this specific problem behaved similarly to the orthogonal Varimax procedure. It need hardly be added that in other problems such correspondence between Varimax and Biquartimin may not hold.

In general, oblique transformations exhibit a number of complexities of interpretation:

1. The communality (h_i^2) of a variable cannot be computed as the sum of the squared loadings across factors in that variable's row.

2. The variance (λ_j) of a factor cannot be computed as the sum of the squared loadings across variables in that factor's column.

3. A distinction between factor structure and factor pattern needs to be maintained. In particular, the pattern matrix is the one that tends to reveal clusters of variables,

Table 8.10 Biquartimin Transformed Loadings of Table 8.7

	Factor					
	I		II		III	
Variable	Structure	Pattern	Structure	Pattern	Structure	Pattern
1. Age	0.20	0.09	0.89*	0.87*	−0.09	−0.10
2. Family Size	−0.15	−0.13	−0.09	−0.08	0.84*	0.84*
3. Education	0.73*	0.72*	0.14	0.05	0.09	0.09
4. Marital Status	0.11	0.00	0.89*	0.89*	0.11	0.10
5. White Collar	0.87*	0.85*	0.28	0.17	−0.04	−0.03
6. Blue Collar	−0.88*	−0.89*	−0.03	−0.08	−0.01	−0.01
7. Income	0.28	0.26	0.19	0.15	0.74*	0.74*

Correlation Matrix of
Transformed Factors Ψ

	I	II	III
I	1.00	0.13	−0.01
II	0.13	1.00	0.01
III	−0.01	0.01	1.00

*Highest row loading, by method.

while the structure matrix is a matrix of correlations of the variables with the oblique factors.

4. The pattern matrix appears most appropriate for interpreting the oblique factors while the structure matrix measures the variance (squared loading) of each variable as *jointly* accounted for by the factor itself and the interrelationships of that factor with the others. (Entries in the pattern matrix may, on occasion, exceed unity since they are not product moment correlations; the absolute value of entries in the structure matrix, however, will not exceed unity.)

5. Factor scores are no longer uncorrelated. Their product moment correlations are given by Ψ, the matrix of factor correlations (see Table 8.10).

BMD-08M provides the capability for performing various types of oblique transformations. It will print out both the pattern and structure loadings matrices as illustrated in Table 8.10 and, in addition, will print out the matrix of factor scores whose correlations are also shown in Table 8.10. However, these factor scores are no longer simply related to the transformed factor loadings matrix, which was the case in Varimax. (Rather, they must be estimated by regression techniques.)

Given these difficulties of interpretation, the applied researcher is well advised to carry out a Varimax or some other type of orthogonal transformation for comparison purposes, any time he wishes to apply an oblique transformation, such as Biquartimin.

8.5.3 Target-Matrix Fitting

Both types of factor loading transformations (orthogonal and oblique) discussed so far have been based on some internal criterion involving various operationalizations of Thurstone's simple structure desiderata. In some studies, however, the researcher may have an *external* matrix as a target to which he wishes to transform his currently obtained factor loadings matrix to achieve some best fit. The target may be constructed on the basis of a theory about how the variables should load on the factors or, alternatively, represent an actual (and interpreted) factor loadings matrix from some previous study.[45]

In other cases the researcher may have two factor loadings matrices that display equal status from a hypothesis viewpoint. This situation might occur when split halves of a sample are separately factor analyzed for cross-validation purposes or when two studies are done concurrently on two different samples. In these instances the researcher may wish to transform both matrices simultaneously to some best compromise fit.

Accordingly, target-matrix procedures vary according to:

1. Whether one matrix is left unaltered (serves as a fixed target) or whether all matrices are transformed to some mutual best fit.

2. Whether two versus more than two matrices are involved.

3. The nature of the transformation by which the matrices are changed, the requirements imposed on the solution, and the measure of the closeness of fit.

In section 6.8 of Chapter 6 we discussed, in the context of canonical correlation, various procedures for transforming matrices to some "best" agreement. Accordingly, our present discussion is brief and complementary to that of Chapter 6.

One of the earliest contributions to target-matrix fitting was the procedure known as Procrustes, suggested by Hurley and Cattell (1962). We continue to let $\mathbf{F}$ denote the loadings matrix. We let $\mathbf{H}$ denote a fixed target matrix of the same order as $\mathbf{F}$. The idea behind Procrustes transformation is to find a matrix $\mathbf{S}$ that will transform $\mathbf{F}$ to best match the target matrix $\mathbf{H}$ in a least squares sense.

To be more specific, if $\mathbf{F}$ and $\mathbf{H}$ are of order $n \times r$ we seek an $r \times r$ matrix $\mathbf{S}$ such that:

$$\hat{\mathbf{H}} = \mathbf{FS}$$

best matches $\mathbf{H}$ in the sense of displaying the smallest sum of squared residuals from its counterpart entries in $\mathbf{H}$.

The transformation $\mathbf{S}$ that represents the best least squares fit to $\mathbf{H}$ is given by:

$$\mathbf{S} = (\mathbf{F'F})^{-1}\mathbf{F'H}$$

which, when normalized by columns, in order to represent direction cosines, will give:

$$\hat{H} = FS$$

This procedure is nothing more elaborate than regressing each column of **H**, in turn, on **F**. The resulting vectors of regression coefficients are each normalized to represent direction cosines and placed in the matrix **S**. Notice that the columns of **S** will not, in general, be orthogonal, although they will each be of unit length.

The Procrustes transformation, then, is a case of factor matching in which a matrix **H** is selected as a fixed target and the transformation **S** that best matches **F** to the target is permitted to be oblique. Not surprisingly, if we were to allow both **H** and **F** to undergo separate oblique transformations so that **H** is no longer a fixed target, we could use canonical analysis for this type of problem. Canonical analysis could be applied (see section 6.8 of Chapter 6) to mean-corrected-only matrices as well as to the more common case of standardized-variable matrices.

Other researchers, such as Cliff (1966), have placed restrictions on **S** so as to make it orthogonal. Cliff's procedure enables the researcher either to rotate a data matrix to best fit a fixed target or to rotate both of them to a best compromise position: one that maximizes the sum of their cross products of projections on the axes of the transformed space. Schönemann (1966) has also developed a solution for orthogonal Procrustes while Browne (1967) has made contributions to the oblique Procrustes case. Schönemann and R. M. Carroll (1970) have developed a fitting procedure that permits translation of origin and uniform expansion or contraction (as well as rotation) of the data matrix.

Extensions of factor matching procedures to three or more matrices were illustrated in section 6.8 of Chapter 6. Horst's generalized canonical correlation procedure was mentioned briefly and J. D. Carroll's approach was described and illustrated in some detail.

More recently, Kristof and Wingersky (1971) have developed an orthogonal Procrustes procedure that simultaneously rotates $t \geq 2$ matrices to best mutual fit where "best" is described as minimizing the sum of squares between each of the n points—involving t replicates of point 1, point 2, and so on—and the cluster's centroid, summed over all n clusters. Gower (1975) has proposed a method that generalizes Schönemann and R. M. Carroll's two-matrix approach to three or more matrices. Gower also uses the Kristof and Wingerksy criterion but permits translation and uniform expansion or contraction of the configurations (as well as rotation and reflection of the dimensions).

In short, factor matching procedures are now available for dealing with a variety of situations. It is to be expected that they will receive increased attention as more applications researchers learn about their availability and utility.

8.6 Summary

This chapter has been concerned with the study of interdependent data matrices via factor analysis. Principal components analysis, although only one of many techniques for factoring data, was selected as the primary model for discussion, given its particular relevance for applications in the behavioral and administrative sciences. By way of a small numerical example we described the essential characteristics of principal components analysis, its relationship to singular value decomposition, and the concepts of component loadings and component scores matrices.

We next discussed the difficult problem of determining how many components to extract. A number of procedures, some statistical and some graphical, were presented and illustrated numerically.

We then turned to another important part of factor analysis, namely transforming initial solutions to more interpretable orientations. Both orthogonal and oblique transformations were described and illustrated with data taken from the TV commercial study. A brief description of transformations to target matrices concluded the chapter.

Review Questions

1. Find the eigenstructure of the following symmetric matrices:

a. $A = \begin{bmatrix} 5 & 7 \\ 7 & 8 \end{bmatrix}$

b. $A = \begin{bmatrix} -3 & 4 \\ 4 & -2 \end{bmatrix}$

c. $A = \begin{bmatrix} 2 & 6 \\ 6 & 7 \end{bmatrix}$

d. $A = \begin{bmatrix} 0 & 2 & 1 \\ 2 & 3 & 0 \\ 1 & 0 & 9 \end{bmatrix}$

2. Calculate the trace and determinant of each of the above matrices and verify that:

$$tr(A) = \sum_{j=1}^{n} \lambda_j; \quad |A| = \prod_{j=1}^{n} \lambda_j$$

where the λ_j's are the eigenvalues of A.

3. Find the singular value decomposition and rank of:

a. $A = \begin{bmatrix} 0 & 1 \\ 1 & 0 \end{bmatrix}$

b. $A = \begin{bmatrix} 0.707 & 0.707 \\ -0.707 & 0.707 \end{bmatrix}$

c. $A = \begin{bmatrix} 2 & 2 \\ -1 & 1 \\ 3 & 4 \end{bmatrix}$

d. $A = \begin{bmatrix} 1 & 2 \\ 3 & 0 \\ 4 & 2 \\ 6 & -1 \\ 7 & 1 \end{bmatrix}$

4. Return to the data of Table 8.1. Run a principal components analysis of the covariance matrix:
   a. How does the rotation matrix $\mathbf{U}$ compare to that found from a principal components analysis of $\mathbf{R}$?
   b. Compute the standardized component scores and component loadings matrices and compare these to their counterparts in Table 8.7.
   c. Show empirically that $\mathbf{FF'} = \mathbf{C}$ and $\mathbf{F'F} = \mathbf{D}$, a diagonal matrix.

5. Consider the same demographic data used in Table 8.7. Split the data into first and second halves of 126 respondents each.
   a. Run separate principal components analyses and Varimax rotations on each half of the data; retain all components where λ_j exceeds unity.
   b. Which factors are stable across split halves in the sense of showing similar loading patterns after Varimax rotation?

Notes

1. In the case of multidimensional scaling (to be discussed in Chapter 9), the approach is modified to account for the fact that the original input data are not profiles of m objects in n dimensions but distance-like numbers between the $m(m - 1)/2$ pairs of objects.

2. Chapter 9 discusses other approaches to factor analysis, including the common factor model.

3. Some researchers prefer to exclude principal components as a subcategory of factor analysis. In this book, however, we use the term factor analysis as a generic label that includes principal components as a separate case.

4. By halo effect is meant that if a consumer likes some brand on an overall basis there is a tendency to rate the brand high on any generally desirable specific attribute (i.e., overall liking is transferred to particularized attribute ratings).

5. This axis, labeled z_1, is plotted in Figure 8.1; its computation is described in subsequent sections of the chapter.

6. Conventionally, the main diagonal entries in $\mathbf{D}$ are ordered from large to small and the associated columns of $\mathbf{U}$ are similarly ordered to agree with the order of their eigenvalues.

7. Since $\mathbf{R}$ is symmetric, all distinct eigenvectors are real and mutually orthogonal. Since $\mathbf{R}$ is a product moment matrix (see section B.5 of Appendix B) all diagonal entries of $\mathbf{D}$ are nonnegative.

8. An orthonormal section is defined as a matrix $\mathbf{P}$ for which $\mathbf{P'P} = \mathbf{I}$ (but $\mathbf{PP'} \neq \mathbf{I}$). That is, orthogonal sections are rectangular matrices that are columnwise orthonormal; each column is orthogonal to every other column and each column displays unit sum of squares.

9. In practice one computes the eigenstructures of the *smaller* of $\mathbf{AA'}$ and $\mathbf{A'A}$ to find the matrix of eigenvalues $\mathbf{\Delta}^2$. However, in either case only $r \leq \min(m, n)$ positive entries will appear along the main diagonal of $\mathbf{\Delta}^2$.

10. In this case we found the eigenstructure of $\mathbf{A'A} = \mathbf{Q\Delta^2Q'}$ and then made the substitutions in:

$$\mathbf{P} = \mathbf{AQ\Delta}^{-1}$$

to find $\mathbf{P}$ (see section B.6 of Appendix B).

11. In data-based applications, the entries along the main diagonal of Δ in the singular value decomposition of $\mathbf{X}_s$ will all be distinct. If so, the full decomposition is unique except for the possibility of column reflections in $\mathbf{P}$ or $\mathbf{Q}$.

12. If a matrix $\mathbf{A}$ is symmetric, then $\mathbf{AA}' = \mathbf{A}'\mathbf{A} = \mathbf{A}^2$. However, the eigenvectors of $\mathbf{A}^2$ are the same as those of $\mathbf{A}$ and the eigenvalues of $\mathbf{A}^2$ are $\mathbf{D}^2$, assuming that those of $\mathbf{A}$ are $\mathbf{D}$. Hence, if $\mathbf{A}$ is symmetric and of product moment form its *singular value decomposition* is equal to its eigenstructure:

$$\mathbf{A} = \mathbf{UDU}'$$

where $\mathbf{U}$ is orthogonal. All values of $\mathbf{D}$ are nonnegative and can be arranged in decreasing order (large to small).

13. In general the eigenvalues of a matrix given by $1/k$ (where k is a scalar) times $\mathbf{A}$ are equal to $1/k$ times the matrix of eigenvalues of $\mathbf{A}$.

14. Since $\mathbf{U} = \mathbf{Q}$ in the context of interest, we shall use $\mathbf{U}$ from this point on to denote the matrix of direction cosines that rotates $\mathbf{X}_s$ to principal components.

15. Note that this follows the same idea in which $\mathbf{X}_s$ was built up as the sum of a set of additive slices $\hat{\mathbf{X}}_s^{(1)}$, $\hat{\mathbf{X}}_s^{(2)}$, etc.

16. Moreover, we shall always order the eigenvalues of either $\mathbf{X}_s'\mathbf{X}_s$ or $\mathbf{R}$ from large to small and order the associated eigenvectors as well.

17. Notice that $\mathbf{Z}_s$ is related to $\mathbf{Z}$ as:

$$\mathbf{Z}_s = \sqrt{m-1}\,\mathbf{Z}\Delta^{-1}$$

and, furthermore:

$$\mathbf{Z}_s = \sqrt{m-1}\,\mathbf{P}\Delta\Delta^{-1}$$
$$= \sqrt{m-1}\,\mathbf{P}$$

18. More generally, the loading of a variable on a factor is a weight that measures the contribution to variance in a variable (factor) that is made by a factor (variable).

19. This follows from:

$$\mathbf{F} = \frac{\mathbf{X}_s'\mathbf{Z}_s}{m-1}$$
$$= \frac{(\mathbf{P}\Delta\mathbf{U})'(\mathbf{X}_s\mathbf{UD}^{-1/2})}{m-1}$$
$$= \frac{\mathbf{U}\Delta\mathbf{P}'\mathbf{P}\Delta\mathbf{U}'\mathbf{UD}^{-1/2}}{m-1}$$

However, since $\mathbf{D} = \Delta^2/(m-1)$, and $\mathbf{P}$ and $\mathbf{U}$ are orthonormal sections, we can simplify the preceding expression to:

$$\mathbf{F} = \frac{(m-1)\mathbf{UDD}^{-1/2}}{m-1}$$
$$= \mathbf{UD}^{1/2}$$

as desired.

20. Further discussion of communalities is deferred until later in the chapter.

21. In this case $\mathbf{P}$ is an orthonormal section ($\mathbf{P'P} = \mathbf{I}$) and $\mathbf{\Delta}$ is diagonal with r ordered (from large to small) positive entries. The rank of $\mathbf{X}_s$ is r. If $\mathbf{X}_s$ is of full rank, then $\mathbf{U}$ is orthogonal with $\mathbf{UU'} = \mathbf{U'U} = \mathbf{I}$; otherwise $\mathbf{U}$ is also an orthonormal section.

22. Although not shown earlier, the following relationships also hold in the case of $\mathbf{F}$, the component loadings matrix.

$$\mathbf{RF} = \mathbf{FD}$$

$$\mathbf{F'F} = [\mathbf{UD}^{1/2}]'[\mathbf{UD}^{1/2}] = \mathbf{D}$$

$$\mathbf{FF'} = [\mathbf{UD}^{1/2}][\mathbf{UD}^{1/2}]' = \mathbf{UDU'} = \mathbf{R}$$

Thus, the original correlation matrix $\mathbf{R}$ can be reproduced as the major product moment of $\mathbf{F}$, the component loadings matrix (assuming that the full eigenstructure of $\mathbf{R}$ has been retained).

23. The sphericity test derives its name from the fact that a scatter plot of observations in a three-dimensional space (whose axes are variables) would resemble a sphere if the variables were each standardized, and were mutually uncorrelated. (In higher dimensions, of course, hyperspheres would be involved.)

24. See section B.5 of Appendix B for an illustration of this property.

25. One possible reason for this is that with relatively large samples, one will almost always reject the null hypothesis and, hence, conclude that $\mathbf{R}_p \neq \mathbf{I}$. Moreover, when Bartlett's test is applied to individual components (to be described below), it is likely that too many "significant" components will be found, relative to the researcher's ability to interpret them.

26. This extension is carried out by Cooley and Lohnes' PRINCO program, cited above.

27. In general, the test can be applied sequentially for $r = 1$ up to $r = n - 2$ eigenvalues.

28. The fourth eigenvalue is part of the progression associated with the error components; we choose the last component *before* the start of the set of (assumed) error components.

29. We would expect the first half to exceed 1.0 because sampling fluctuations would produce some off-diagonal correlations greater than zero and, hence, some eigenvalues greater than the expected value. Since the sum of the eigenvalues must equal the order ($n \times n$) of the correlation matrix, the second half of the eigenvalues would show values less than 1.0.

30. This rule can cause problems. With highly noisy data it may be the case that only 30 or 40 percent of the variance in the data is capable of being accounted for systematically.

31. It is not necessary that transformation of the (reduced) components space involve a rotation. However, there are a number of advantages associated with orthogonal transformations and we shall emphasize this type of transformation in the chapter.

32. This follows from the facts that $(\mathbf{T'})^{-1} = (\mathbf{T}^{-1})'$ and $[\mathbf{F}(\mathbf{T}^{-1})']' = \mathbf{T}^{-1}\mathbf{F'}$. Furthermore, $\mathbf{TT}^{-1} = \mathbf{I}$.

33. Having done this one can then find a suitable transformation of the (standardized) component scores matrix to maintain the decomposition of $\mathbf{X}_s$ into the product of a matrix of scores and a matrix of loadings.

34. The reader should remember that any column of loadings in the factor loadings matrix can be reflected (multiplied by -1) without changing the reconstruction of the correlation matrix by $\mathbf{FF'} = \mathbf{R}$, or any slice of it, as the case may be.

35. It is also of interest to note that the decomposition of some matrix $\mathbf{A}$ into the product of three matrices via singular value decomposition does *not* entail the indeterminacy problem. That is:

$$\mathbf{A} = \mathbf{P} \Delta \mathbf{Q}'$$

(where $\mathbf{P}'\mathbf{P} = \mathbf{I}$, $\mathbf{Q}'\mathbf{Q} = \mathbf{I}$, and Δ is diagonal with ordered, large to small, positive entries) is unique up to reflections in the columns of $\mathbf{P}$ and $\mathbf{Q}$ as long as there are no tied entries in Δ. Data matrices are typically of full rank with no tied values in the diagonal matrix Δ and, hence, meet the uniqueness condition.

36. The high negative loading for variable 6 (blue collar occupation) is due to the use of a dummy-coded classification for occupation in which negative correlations between classes necessarily occur.

37. If *all* components are retained all communalities will equal 1; more interesting is the case in which only some of the components are retained, as illustrated here.

38. For purposes of simplifying the demonstration we assume that $\mathbf{f}_1$ and $\mathbf{f}_2$ are each of unit length.

39. The actual Varimax algorithm is rather complex computationally and is not discussed here. Interested readers should see more specialized books on factor analysis, such as those by Mulaik (1972) and Comrey (1973).

40. The more generic term "factors" is used now because, after rotation, the principal components property of successively decreasing variance no longer applies.

41. By oblique is meant that the angles separating factor axes are not right angles.

42. If the reader returns to Panel II of Figure 8.6, in the case of oblique transformations the axes $\mathbf{I}'$ and $\mathbf{II}'$ are not required to be orthogonal if, by being oblique, they match their respective clusters of vectors more closely on the average.

43. There is further confusion in oblique transformations about whether primary or reference axes are to be employed. Here we assume that the transformed axes are primary; these are the ones that go through clusters of variables. Reference axes (see Rummel, 1970) are orthogonal to primary axes and are employed in some types of oblique transformations.

44. If we let $\boldsymbol{\Psi}$ denote a 2×2 matrix whose entries are direction cosines between, $\mathbf{g}_1^*$ and $\mathbf{g}_2^*$, then $\mathbf{w}'$, a row vector of pattern loadings, is given by:

$$\mathbf{w}' = \mathbf{x}'\boldsymbol{\Psi}$$

where $\mathbf{x}'$ is a row vector of structure loadings.

45. The field of *confirmatory* factor analysis deals with the objective of testing substantive theory via factor analysis. This topic is described briefly in Chapter 9.

Selected Topics in Reduced Space And Cluster Analysis

9.1 Introduction

While principal components factor analysis is a very popular dimension reduction model, it nonetheless has many competitors. In actuality, an almost bewildering array of factor models and methods have been proposed. Furthermore, other classes of techniques, such as multidimensional scaling and cluster analysis have been employed for the purpose of revealing patterns of one sort or another in interdependent data structures. The present chapter considers some of these alternatives.

We first discuss some of the models and methods that have been developed as alternatives to the principal components approach. In particular, the common factor model is discussed in some detail. This model exhibits some special features that could be of interest to the applied researcher. Other factor models and methods are described much more briefly and appropriate references to more detailed discussions are given.

We then turn to a description of Q-technique factor analysis, a procedure in which one factors cross products matrices based on the rows (people or objects), rather than the variables of the data matrix. Q-technique is contrasted with the usual (R-technique) type of factor analysis and applications of Q-technique to person or object grouping are discussed.

Other topics of related interest to factor analysis—preliminary orthogonalization of predictor or criterion variables in the context of analyzing dependence structures, higher order factoring, three-mode factor analysis—are described,

if only briefly. We also comment on some of the problems of solution interpretation and inference associated with factor analysis.

Our attention then turns to other techniques for dimensional reduction or
object grouping. In particular, metric and nonmetric methods of multidimensional scaling are described and illustrated numerically. The chapter concludes
with some introductory material on cluster analysis, including the description
of various proximity measures and algorithms for clustering objects.

Like Chapter 5, this chapter consists of a comparatively large miscellany of
topics. Factor analysis and other techniques for the study of interdependent
data structures represent an already vast (and continually growing) field. Accordingly, references are made to specialized literature on various methods that,
by necessity, can be described only briefly here.

9.2 Communalities and the Common Factor Model

Up to this point only one factoring procedure, principal components analysis,
has been discussed (in Chapter 8). While principal components analysis can be
applied to any type of cross products matrix, typically it is the correlation matrix
that is factored. An important aspect of the principal components procedure is
that *unities appear along the main diagonal* of the correlation matrix, since each
variable is assumed to be perfectly correlated with itself.

In principal components analysis we are interested in common variance
across *all* variables in the data matrix; each component is a linear composite
of the full set of variables. And this is true whether we retain some or all of the
components for further analysis.

This is not the only point of view that can be taken. Factor analysts, particularly those in psychology, have applied another type of model quite extensively.
This is called the *common factor model* and, operationally, is distinguished by
the fact that *numbers less than unity appear in the main diagonal* of the correlation matrix to be factored.

The numbers that appear in the main diagonal are called *communalities*. In
Chapter 8 we used the same term to refer to the sum of squared loadings for
each row variable across the r columns of the retained principal components.
In that context, a variable's communality was an R^2 value that measured how
much of its variance was accounted-for by the set of components retained in the
loadings matrix. While our present interpretation of communality is similar,
the task now is to estimate the communality numbers (in some way) for placement in the main diagonal of the correlation matrix *before the matrix is factored*.

Most data matrices will be of full rank; hence if **R** is $n \times n$ we will need all
n components to account for all of the variance in $\mathbf{X}_s$. However, suppose the
diagonal entries in **R** are changed to values that are less than unity but greater
than zero. In all probability the "effective" rank of this matrix can be reduced
substantially even though its actual rank is still n. As we shall see, the basic idea

behind the common factor model is to find a set of diagonal entries for $\mathbf{R}$ that will closely match this reduced form of $\mathbf{R}$, which we shall call $\mathbf{R}_c$, by a low-rank product $\mathbf{F}_c\mathbf{F}_c'$ of the loadings matrix, even though $\mathbf{R}_c$ is de facto of rank n. (We shall use the subscript c to distinguish the *common* factor model from the principal components model.)

The key to how well a low-rank product $\mathbf{F}_c\mathbf{F}_c'$ approximates $\mathbf{R}_c$ lies in the off-diagonal entries of $\mathbf{R}_c - \mathbf{F}_c\mathbf{F}_c'$. If we can find a low-rank $\mathbf{F}_c$ and a set of communalities that *jointly* produce *small off-diagonal residuals,* then we achieve the objectives underlying application of the common factor model. In general, by being able to manipulate the communalities (in the common factor model) we shall be able to obtain *smaller residuals on the average than can be found by the principal components model,* for a given rank of $\mathbf{F}_c$, the loadings matrix. (These loadings matrices will, of course, be different numerically for the common factor model versus the components model.)

This represents the operational problem. However, the *philosophical* basis of the common factor model differs markedly from that of principal components. In the common factor model we assume two sets of entities: (a) a set of common factors and (b) a set of unique factors that represent the variance contributed by specific variables (plus error).

9.2.1 Characteristics of the Common Factor Model

The common factor model is most easily described by contrasting it with the principal components model. In the components model, each of the n column variables $\mathbf{x}_{sj}$ of the $m \times n$ standardized data matrix $\mathbf{X}_s$ can be represented as a linear composite:

$$\mathbf{x}_{sj} = f_{j1}\mathbf{z}_{s1} + f_{j2}\mathbf{z}_{s2} + \cdots + f_{jn}\mathbf{z}_{sn}$$

where, for convenience, we assume that $\mathbf{X}_s$ is of full rank. The $\mathbf{z}_s$'s are columns of $\mathbf{Z}_s$, the $m \times n$ matrix of standardized component scores. The coefficients f_j are factor loadings for the j-th row of the $n \times n$ factor loadings matrix $\mathbf{F}$. As recalled, the full data matrix is represented by:

$$\mathbf{X}_s = \mathbf{Z}_s\mathbf{F}'$$

where $\mathbf{F}$ is an $n \times n$ matrix of variables-by-components loadings.

In the common factor model the situation changes. In this case $\mathbf{x}_{sj}$ is represented as the linear composite:

$$\mathbf{x}_{sj} = f_{cj1}\mathbf{z}_{cs1} + f_{cj2}\mathbf{z}_{cs2} + \cdots + f_{cjn}\mathbf{z}_{csn} + \mathbf{t}_j$$

Here, t_j is a vector that represents a unique contribution (plus error) to x_{sj}. Since each of the n variables has its own unique contribution, the common factor model, in matrix form, becomes:

$$\mathbf{X}_s = \mathbf{Z}_{cs}\mathbf{F}'_c + \mathbf{T}$$

where $\mathbf{T}$ is an $m \times n$ matrix of specific factor contributions, one for each variable.

In the common factor model, there are two assumptions made about $\mathbf{T}$. First, we assume that the specific factors are uncorrelated with each other so that their covariance matrix is:

$$\frac{\mathbf{T}'\mathbf{T}}{m - 1} = \mathbf{V}$$

where $\mathbf{V}$ is a *diagonal* matrix of order $n \times n$.

Second, we assume that the specific factors, represented by $\mathbf{T}$, are also uncorrelated with the standardized factor scores $\mathbf{Z}_{cs}$, so that we can write:

$$\mathbf{T}'\mathbf{Z}_{cs} = \mathbf{\Phi}$$

where $\mathbf{\Phi}$ denotes the null matrix of order $n \times n$.

The variance v_i^2 of each specific factor contribution represents the complement of that variable's communality:

$$v_i^2 = 1 - h_i^2$$

since, as noted above, each specific factor is assumed to be uncorrelated with the common factors. The v_i^2's are usually called *uniquenesses* and are the diagonal entries of $\mathbf{V}$.

The theoretical[1] correlation matrix for the common factor model can now be written as:

$$\mathbf{R} = \frac{\mathbf{X}'_s\mathbf{X}_s}{m - 1} = \frac{(\mathbf{Z}_{cs}\mathbf{F}'_c + \mathbf{T})'(\mathbf{Z}_{cs}\mathbf{F}'_c + \mathbf{T})}{m - 1}$$

However, since $\mathbf{T}'\mathbf{Z}_s = \mathbf{\Phi}$, the above expression reduces to:

$$\mathbf{R} = \frac{\mathbf{F}_c\mathbf{Z}'_{cs}\mathbf{Z}_{cs}\mathbf{F}'_c}{m - 1} + \mathbf{V}$$

Since it can be assumed that the common factor scores are (theoretically) un-correlated,[2] we can set $\mathbf{Z}'_{cs}\mathbf{Z}_{cs}/(m-1) = \mathbf{I}$. Given this assumption we obtain:

$$\mathbf{R} = \mathbf{F}_c\mathbf{F}'_c + \mathbf{V}$$

At this point we see that the data based correlation matrix $\mathbf{R}$ is the sum of two matrices $\mathbf{F}_c\mathbf{F}'_c$ and $\mathbf{V}$. The matrix $\mathbf{F}_c$ contains the factor loadings of the common factor part while $\mathbf{V}$ is a diagonal matrix with diagonal entries that are between 0 and 1. This is the point where we assume that $\mathbf{F}_c\mathbf{F}'_c$, *with relatively low rank, will be a good approximation to the common factor part of* $\mathbf{R}$.

Suppose we actually knew each of the uniquenesses in $\mathbf{V}$. Subtracting $\mathbf{V}$ from both sides of the preceding equation gives us:

$$\mathbf{R}_c = \mathbf{R} - \mathbf{V}$$

which can be called the reduced form of $\mathbf{R}$. $\mathbf{R}_c$ *is the matrix that we shall actually factor.* Since $\mathbf{V}$ is a diagonal matrix of uniquenesses, $\mathbf{R}_c$ will differ from $\mathbf{R}$ only in terms of its diagonal elements. These are all unities in $\mathbf{R}$ but they now equal the communalities:

$$h_i^2 = 1 - v_i^2$$

in the case of $\mathbf{R}_c$.

If we applied eigenstructure decomposition to $\mathbf{R}_c$ we would get:

$$\mathbf{R}_c = \mathbf{U}_c\mathbf{D}_c\mathbf{U}'_c$$

where $\mathbf{D}_c$ is the diagonal matrix of eigenvalues of $\mathbf{R}_c$ and $\mathbf{U}_c$ is the matrix of associated eigenvectors of $\mathbf{R}_c$.

In short, this is what the common factor model is all about, namely finding the eigenstructure of $\mathbf{R}_c$ (rather than $\mathbf{R}$). The main operational problem is: how does one estimate the communalities h_i^2?

In practice one usually implements the common factor model by means of an iterative procedure that involves the following steps:

1. Estimate the initial communalities *and* the number of factors to extract.

2. Find the eigenstructure of $\mathbf{R}_c$ and record the new communalities. Insert these new communalities into the diagonal of $\mathbf{R}_c$ for the second iteration.

3. Repeat the iterative process until the communalities in iteration i are sufficiently close to those of iteration $i-1$ to terminate the process.

4. Examine the residuals $\mathbf{R}_c - \mathbf{R}_c^{(r)}$, where $\mathbf{R}_c^{(r)} = \mathbf{F}_{cr}\mathbf{F}'_{cr}$ for the r factors used. If these residuals are "too high," go through the *whole process* again with one additional factor, and so on, until the residuals are satisfactorily "small."

As can be surmised, application of the common factor model requires a number of decisions to be made regarding the starting communalities, the number of factors to extract and how long to iterate the communality convergence process within each specified number of factors.

In the BMD-08M factor program—the one used in Chapter 8's principal components analysis—the starting communalities can either be estimates that are read in by the user or the computed SMC (squared multiple correlation) of each variable with all of the other $n - 1$ variables.[3] Use of the SMC's as starting communalities is quite prevalent in the application of the common factor model, and has theoretical justification in the work by Guttman (1956). The SMC's provide a *lower bound* to the "true" communalities (the upper bound being 1.0) and, in this sense, are conservative

BMD-08M provides two criteria for the number of factors to extract. The user can specify the maximum number in advance as well as an eigenvalue cutoff (often set at 1.0) such that all factors with eigenvalues lower than the cutoff are dropped off. In any given run the program selects the smaller of the number of factors implied by the two criteria.

The user can specify (up to a maximum of 50) the number of communality iterations that he would like the program to run. The iterations will stop, either at the number prespecified or if an internal cutoff value is met. In this latter instance, the program stops if the maximum difference in any of the communalities of iteration i as compared to iteration i $- 1$ is no greater than 0.001.

Since BMD-08M can also do principal components analysis (with unities in the diagonal of the correlation matrix), it is relatively simple to run the program twice on the same data, should the researcher wish to compare the results of the components versus the common factor model.

Theoretically, the problem of separating common from specific variance is still unsolved. Common variance usually refers to a set of hypothetical constructs, not fully captured by the particular variables and the sample used in a specific study. From a pragmatic viewpoint, one still has to guess at the approximate rank of the matrix $\mathbf{R}_c$ when starting the iterative process. Fortunately, with problems of realistic size (e.g., 15 or more variables) the factor loadings are *not highly affected* by what entries, between the SMC's and unity, go into the main diagonal in the common factor model. Hence, the common factor model will often produce results that are similar to the principal components model.

From a philosophical viewpoint, however, the difference between principal components and the common factor model is appreciable. In principal components our interest is on common variance shared by all the variables. In the common factor model our interest is on reducing the (effective) rank of the correlation matrix from an initial rank of n to some much lower (approximate) rank of r. This is done by the insertion of h_i^2's in place of unities in the main diagonal. While $\mathbf{R}_c$ technically retains a rank of n, the incorporation of the h_i^2's leads to an *approximation* of $\mathbf{R}_c$ by a matrix that is much lower in rank than one by which we could approximate $\mathbf{R}$ if a diagonal of unities were used.

9.2.2 Applying the Common Factor Model

To provide contrast with the principal components analysis, the same demographic data in Chapter 8's analyses were factor analyzed by the common factor model. In this case BMD-08M was applied with SMC's (rather than unities) in the main diagonal. The iterative process for revising these initial communalities was set for a maximum of 50 iterations. Finally, so as to provide results comparable to Table 8.8 of Chapter 8, three factors were extracted and these were also rotated by Varimax.

Table 9.1 shows results of the computer run for comparison with Tables 8.7 and 8.8. We first note that the initial communalities, based on the SMC's all increase, so that after extracting three factors, the sum of the final communalities increases from an initial value of 2.56 to 3.87. The program required all 50 iterations to develop the final communalities.

Some idea of how well the iterative process has converged can be gained from examination of the mean absolute value of the residuals: $\mathbf{R}_c - \hat{\mathbf{R}}_c^{(3)}$. We can find these by first computing $\hat{\mathbf{R}}_c^{(3)} = \mathbf{F}_{c3}\mathbf{F}_{c3}'$ from the unrotated loadings of Table 9.1. As it turns out, the mean absolute value of the residuals is only 0.01. This suggests that a rank-3 approximation is an excellent description of the common variance in $\mathbf{R}_c$. In contrast, the mean absolute deviations in the case of the components model of Table 8.7 are 0.07. Here, however, we were not allowed the flexibility of altering main diagonal elements in order to provide a better rank-3 approximation to $\mathbf{R}$.

So much for the fit of the common factor model. We next inquire whether our *interpretation* of the Varimax-rotated loadings of Table 9.1 is enhanced over our earlier interpretation of Table 8.8, in the case of the components model.

When Table 9.1 is compared to Table 8.8, the rotated factor patterns are quite similar. However, in the present case the pattern is not quite as sharp as found in the Varimax-rotated components analysis. For example, the income and education loadings are not as prominent as their counterparts in Table 8.8. If anything, Table 8.8 seems to approximate a simple structure pattern more clearly than Table 9.1.

We conclude that in this example the common factor model produced results that were fairly close to those of the principal components analysis. Substantive interpretation seemed to be clearer in the components case but both rotated structures would probably lead to the same general kinds of conclusions. Based on the pattern of eigenvalues in both cases, it would seem that a rank-3 matrix represents an adequate approximation to the reduced correlation matrix $\mathbf{R}_c$.

As earlier discussed, the operational distinction between principal components and the common factor model concerns whether unities or some set of numbers (the cummunalities) between 0 and 1 are placed in the diagonal of the correlation matrix. From that point on, the procedures for finding the eigenstructure of $\mathbf{R}$ or $\mathbf{R}_c$ are the same.[4] Moreover, any of the transformation techniques, such as Varimax or Biquartimin, can be applied to the initial matrix in either model.

Table 9.1 Summary Output of Common Factor Analysis of Demographic Data

Variable	Initial SMC's	Final h_i^2	Factor Loadings Before Rotation			Factor Loadings After Rotation		
			I	II	III	I	II	III
1. Age	0.41	0.57	0.51	0.53	0.13	0.17	0.72*	0.12
2. Family Size	0.15	0.55	-0.08	0.02	-0.74	-0.08	-0.03	0.74*
3. Education	0.27	0.31	0.53	-0.16	-0.05	0.54*	0.13	0.06
4. Marital Status	0.42	0.70	0.49	0.68	-0.06	0.07	0.83*	0.07
5. White Collar	0.59	0.75	0.82	-0.24	0.07	0.84*	0.22	-0.06
6. Blue Collar	0.55	0.72	-0.72	0.45	0.01	-0.85*	0.01	-0.01
7. Income	0.17	0.26	0.29	0.02	-0.42	0.24	0.16	0.42*
	2.56	3.87	λ_j 2.09	1.03	0.75			

*Highest row loading after rotation.

Chapter 8 emphasized the use of the principal components model as being more directly related to information preservation under spatial reduction, as well as less esoteric in its application than most other factor models.[5] Clearly, however, the specific model chosen by the researcher should reflect the nature of the problem. The common factor model can be quite useful in various content areas where one wants to concentrate attention on the (assumed) common variance, net of unique factor and error contribution to the data matrix.

9.3 Other Factor Models and Methods

Still other factor models and methods are available. By *model* we mean the algebraic representation of the factoring approach while, by *method*, we mean the technique used to find the solution. For example, principal components and the common factor model are different models but, once formulated, the same method (viz., eigenstructure analysis) is used to find the loadings matrix.

9.3.1 Factor Models

A large number of alternative factor models are available for the more experienced user. Illustrative of these models are:

1. Canonical factor analysis.
2. Maximum likelihood factor analysis
3. Image analysis.

(No attempt will be made to describe these models in full detail.)

Canonical factor analysis (Rao, 1952) requires the researcher to be able to estimate a diagonal matrix $V^{-1/2}$ whose entries denote the reciprocals of the standard deviation of each (assumed) specific factor. Having done this, an eigenstructure decomposition of the following matrix product is undertaken:

$$[V^{-1/2}R^*V^{-1/2}]$$

where $R^* = R - V$ and R is the original correlation matrix.

One of the nice features of canonical factor analysis is that solutions based on different cross products matrices (e.g., the covariance matrix versus the correlation matrix) can be related to each other by simple transformations. This is *not* the case insofar as the principal components and common factor models are concerned.

One of the problems associated with canonical factor analysis is to estimate V, the diagonal matrix of uniquenesses. One possibility is to use the iterative model described earlier in the context of the common factor model. Since this procedure works with the communalities h_i^2, one could then find the uniquenesses v_i^2 as the complements: $1 - h_i^2$.

A more elegant approach is to use the *maximum likelihood* model (Lawley and Maxwell, 1963). Briefly put, in maximum likelihood analysis we assume that a sample has been drawn from some assumed known (such as a multivariate normal) distribution. What is not known are the parameters of the distribution, such as the centroid vector and covariance matrix. Having observed the sample distribution's statistics, we try to find universe parameters that would have the greatest likelihood of producing the statistics that we did get.

Maximum likelihood analysis has progressed rapidly from its early development by Lawley and Maxwell to the computationally efficient approaches of Jöreskog (1967). In particular, maximum likelihood analysis provides a way to solve for **V** that provides statistical goodness-of-fit tests, based on a chi square approximation, for the numbers of factors to retain.

Image analysis is a factor model proposed by Guttman (1953). The distinguishing characteristic of image analysis is that the common part of some variable j is represented by its multiple correlation $R^2_{j \cdot (n-1)}$ with all of the other $n - 1$ variables. The unique part of the variable is $1 - R^2_{j \cdot (n-1)}$. Guttman calls the common part the *image* and the unique part the *anti-image*.

Operationally, image analysis boils down to taking $R^2_{j \cdot (n-1)}$ as a variable's *exact* communality to be entered in the main diagonal of the correlation matrix. Unlike the various methods described earlier, these communalities are *not* modified by any subsequent iterative procedure. Other than this, one finds the eigenstructure of the reduced matrix in the same general way as was found in the common factor model.

The various models described thus far in the text:

1. Principal components
2. Common factor
3. Canonical factor
4. Maximum likelihood
5. Image

still fail to exhaust the possibilities. Other models, such as alpha factor analysis (Kaiser and Caffrey, 1965), the minimum residuals method (Harman, 1967), and weighted components (Mulaik, 1972), could be added. Even brief descriptions of these would carry us beyond the intended scope. The interested reader is urged to consult the specialized books by Harman (1967), Mulaik (1972) and Rummel (1970) on the subject.

However, with all of the alternative factor models, the fact still remains that the problems of estimating the number of factors to retain and the uniquenesses or communalities, in models other than principal components, are still not satisfactorily solved.

9.3.2 Factoring Methods

In contrast to the plethora of different factoring models listed in the preceding section, the methods used to find the factor loadings have all, with the exception of Jöreskog's maximum likelihood procedure, involved a single approach: eigenstructure analysis. The eigenstructure approach is computationally fast and efficient and well suited to computer programming.

Before the heyday of the computer, factor analysts often relied on the *square root method,* a factoring procedure that avoids the computation of matrix eigenstructures. The square root method illustrates a different type of matrix decomposition, namely one in which a symmetric matrix (e.g., a correlation matrix) is decomposed into the product of a lower triangular matrix and its transpose, an upper triangular matrix (Van de Geer, 1971). This form of decomposition has application to a variety of multivariate problems, such as the computation of stepdown F ratios in MANOVA (Stevens, 1973).

In the context of factor loadings the diagonal form implies that the first variable has zero correlation with all factors except the first, the second variable has zero correlation with all factors except the first two, and so on. As such, the square root factoring method requires specification of a prior ordering of the variables.

The central idea of the square root method is to choose the variables themselves as factors, subject to their linear association with earlier variables in the ordering being partialed out. In the square root method the first r factors are linear combinations of the first r variables only. The first variable becomes the first factor. The second factor is the second variable, after partialing out any association with the first. The third factor is the third variable with the association of the first two partialed out, and so on. (Another way of putting it is to say that square root factoring is as close as one can come to an uncorrelated set of factors that uses the adjusted variables themselves as factors.)

The *centroid method* represents an approximation to principal components factoring and was developed by Thurstone (1947) before the advent of computers. Geometrically, the centroid methods find the first factor by placing it through the centroid of the standardized variables. The first factor is used to provide a rank-1 approximation to **R** and the first residual matrix $\mathbf{R} - \hat{\mathbf{R}}^{(1)}$ is found. The centroid of the vectors in this reduced space is computed and the process is repeated until all desired factors are extracted.

While the eigenstructure procedure of principal components maximizes the variance at each step, the centroid method attempts to maximize the sum of the absolute values of the loadings (rather than the sum of their squares). Factor scores are mutually uncorrelated, just as in the case of principal components analysis. Since the centroid method has little to recommend it other than as a desk-calculator approximation to principal components factoring, it has received little attention of late.

The basic idea behind the *multiple group method* is to form *clusters* of variables, either through visual inspection of the correlation matrix or by cluster analytic methods (to be described later in the chapter). Each factor is then defined to be the equally weighted sum of the variables in its cluster. The factor loadings matrix is simply a factor structure matrix in which one finds the simple correlation of each variable in a specific cluster with the factor (equally weighted sum) representing that cluster. After finding the structure loadings matrix this matrix can be further transformed to either orthogonal or oblique orientation. The factor loadings tend to be approximately equal across factors.

In effect, the multiple group method is an extension of the centroid method in which several factors are extracted simultaneously, rather than one at a time, as in the centroid method. As such, the factors are passed through the centroids of several groups rather than passing one factor axis through the centroid of all of the variables, computing residuals, and then repeating the process, as is done in the centroid method. However, it should be borne in mind that the multiple group method does *not* approximate the results of a principal components analysis, either in terms of producing orthogonal factors or by extracting factors in the order of decreasing variance accounted-for.[6]

9.3.3 Principal Components Analysis Revisited

At this point a variety of factoring models and methods have been described. The topic began with the principal components model and a large portion of Chapter 8 was devoted to it. What are the merits of principal components analysis compared with alternative approaches?

As recalled, the principal components model does not consider the communalities question; unities appear along the diagonal of the correlation matrix. As also pointed out, in relatively large correlation matrices (say of order 15×15 or higher), the factor loadings matrix is not highly dependent on what appears in the main diagonal, within the practical limits of each variable's squared multiple correlation and unity. And, it is not unusual to have large matrices in behavioral and survey-type research.

The principal components model initially produces a unique decomposition (in the usual case of distinct eigenvalues) of the matrix, although further rotation can be made if one does not care to preserve the components orientation. Furthermore, the component scores can be computed exactly rather than requiring estimation via regression methods, as in the case of communalities-type models.

Further support for the principal components model concerns its behavior under cases where *equal* communalities can be assumed, even though they are all less than unity. In this special case, Harris (1975) shows that any communalities-type model (such as the common factor model) that utilizes an eigenstructure solution, will display factor loadings that bear a simple relationship to the factor loadings obtained from an ordinary principal components solution.

Specifically, let c denote the equal communality value across all variables. If so, the j-th column of factor loadings obtained from a communalities-type analysis is equal to the j-th column of loadings obtained from principal components if the latter's column is multiplied by an adjustment value:

$$w_j = \sqrt{1 - \frac{(1 - c)}{\lambda_j}}.$$

where c denotes the equal-communalities value for all variables and λ_j is the eigenvalue associated with the j-th principal component. Thus, to find the j-th column of the equal communalities structure one multiplies each entry in the j-th column of component loadings by w_j.

When one also bears in mind that the normalization procedure (described in Chapter 8) of Kaiser's Varimax method tends to make initially unequal communalities more equal, the rotated components are even less influenced by initial differences in communalties. Hence, unless one is dealing with a relatively small correlation matrix in which the communalities vary markedly across the variables, it would seem that the principal components model will produce results that are not in conflict with models that utilize communalities.[7]

In the author's view the principal components model has much to offer as a basic factor model; this is why Chapter 8 was devoted to it. Unless the researcher has very good reason to believe that communalities differ markedly across variables, it is probably a prudent approach to use the components model, followed by Varimax rotation or some other such orthogonal transformation procedure. Almost any computing package contains a principal components program and these programs are comparatively similar across various computing packages.

For the applications-oriented researcher, the principal components approach is probably the most easily applied procedure. Application of the more sophisticated models described in this chapter may wisely be deferred until more experience has been gained with the components model.

9.4 Related Topics in Factor Analysis

The subject of factor analysis is broad enough to justify a full book on the topic. Within the space limitations of this chapter we can only touch on a few of its ramifications. Again, references are made to more specialized writings on the topics described here.

9.4.1 R-Technique Versus Q-Technique

Up to this point, our discussion of factor analysis has emphasized what is often called R-technique. By this is meant that the correlation (or some other type of cross products) matrix formed from the *variables* of the data matrix represents the starting point of the analysis. Individuals are treated as replications.

There is, however, an obverse way of looking at the problem. In Q-technique it is the correlation matrix of *persons* that is factored; under this viewpoint variables become replications. Thus, if we have an $m \times n$ data matrix of m persons' scores on n variables, in R-technique the correlation matrix is $n \times n$. In Q-technique it is $m \times m$. As recalled from earlier chapters, the dimensionality of the derived matrix and, hence, the maximum number of factors that can be obtained will be $\min(m, n)$. In general, the retained number of factors r will be even smaller than the minimum of m and n.

In one sense—namely, computation of the singular value decomposition of a matrix (see section B.6 of Appendix B)—it does not matter whether one factors $\mathbf{XX'}$ or $\mathbf{X'X}$. If the data matrix $\mathbf{X}$ is written as:

$$\mathbf{X} = \mathbf{P\Delta Q'}$$

where $\mathbf{P'P} = \mathbf{I}$; $\mathbf{Q'Q} = \mathbf{I}$; and $\mathbf{\Delta}$ is an $r \times r$ diagonal matrix with ordered entries $\lambda_1 \geq \lambda_2 \geq \cdot \cdot \cdot \geq \lambda_r$, one could solve for the eigenstructure of $\mathbf{XX'}$, with $\mathbf{P}$ denoting its eigenvectors and $\mathbf{\Delta}$ denoting the square roots of its eigenvalues. The matrix $\mathbf{Q}$ is then found by:

$$\mathbf{Q} = \mathbf{X'P\Delta^{-1}}$$

Alternatively, we could first solve for the matrices $\mathbf{\Delta}$ and $\mathbf{Q}$ by finding the eigenstructure of $\mathbf{X'X}$ and then obtain $\mathbf{P}$ from:

$$\mathbf{P} = \mathbf{XQ\Delta^{-1}}$$

Indeed, one usually proceeds by finding the eigenstructure of the particular product moment matrix, $\mathbf{XX'}$ or $\mathbf{X'X}$, with the smaller order.

In factor analysis things are not as straightforward since one is usually factoring a correlation matrix in which certain standardizations of the original data are being carried out. Figure 9.1 illustrates the sort of situation that arises in factor analysis. In R-technique the data matrix is, in effect, standardized by columns through the computation of correlation coefficients; all columns have the same mean (zero) and the same standard deviation (unity).

In Q-technique it is the *rows* of the data matrix that are standardized to mean zero and unit standard deviation. However, Q-technique may be applied to a data matrix that has *previously* been:

1. Uncentered[8] or unstandardized by columns.
2. Column centered.
3. Column standardized.

Results will differ, depending upon which of the preceding options has been taken. Comrey (1973), among others, argues for a preliminary *standardization* of the data matrix (by columns) before computing Q-type correlations between rows.[9] Otherwise, the inter-row or Q-type correlations will reflect differences in the variables' means and standard deviations as well as differences in the shape of the person's profile. When the measurement scales differ markedly across variables, this preliminary standardization is a good idea.[10]

Q-type correlations remove differences in persons' averages and variabilities across the scores; only the profile "shapes" are preserved. Hence, if two individuals exhibit the same rank order of scores and about the same relative spacing,

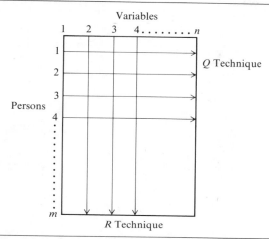

Figure 9.1 R-Type Versus Q-Type Factoring

their profiles will correlate highly, despite differences in mean levels or dispersion. Furthermore, the two individuals will load similarly on the same Q factors even though they differ in terms of mean level or variation in profile elements around the mean.

Another limitation of Q-type factoring is that each individual must be measured across a large number of variables in order to obtain stability in the between-person correlations. The more limited the sample of variables is, the less useful is the classification of individuals into "types," in accord with their factor loadings.[11]

Finally, one might raise the philosophical question of what a Q-type factor represents. Presumably it refers to a "pure" type of individual and the (rotated) loadings pattern of a real person purports to measure how much of each pure type the real person represents. It is not infrequently the case, however, that real individuals do not load highly on *any* of the pure types, despite attempts to rotate toward simple structure.[12]

Q-type factor analysis has typically been used as a type of clustering procedure in various research applications. This is carried out by simply assigning each person to the factor on which he loads most highly. As we shall point out in a later section of the chapter, other procedures are available for person or object clustering—procedures that are superior in many respects to Q-type factor analysis when used as a clustering device.[13]

Two other varieties of factor analysis—P-type and O-type (Cattell, 1966)—have been applied from time to time. P-type and O-type factoring derive from the idea of a data box (see Figure 9.2) of persons by variables by testing occasions.

In P-type analysis one factors the correlations between pairs of variables over occasions. P-type analysis can be useful in the study of the mood-like charac-

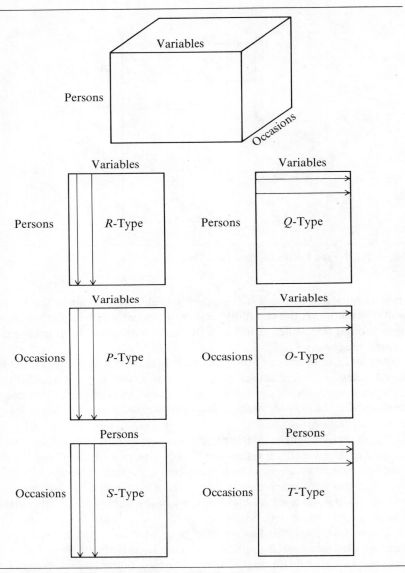

Figure 9.2 Six Correlation Matrices Obtained from a Three-Way Data Box

teristics of a person over time. If one reverses the idea and correlates pairs of occasions across variables, an O-type analysis can be undertaken. As performed on a single individual, O-type analysis could be used to examine a subject's learning behavior.

Of course, one could consider the other cases involving correlating pairs of people over occasions (S-type) or correlating pairs of occasions over people

(T-type), in each case over a single variable. These latter factoring procedures, as well as P-type and O-type, are used rather infrequently. Further information on the Q-type technique and related matters can be found in Comrey (1973) and Guertin and Bailey (1970).

9.4.2 Factor Analysis as an Adjunct to Other Multivariate Procedures

Factor analysis can serve a variety of purposes when used in conjunction with other multivariate techniques, such as multiple regression. For example, one might have a set of highly interdependent predictors and a single criterion variable. Factor analysis could be applied to the predictor set in order to find a set of uncorrelated factor scores on which the criterion variable can be regressed. Since the fit of a multiple regression is not affected by a linear transformation of the predictors, if all factors are extracted R^2 will not be affected by the preliminary factor analysis, whether or not the factor scores are subsequently rotated.

If the factor scores are not interpretable the researcher can still use factor analysis in a preliminary way to find some "best subset" of variables with reasonably low intercorrelations. This would be done by choosing the variable that loads most highly on each rotated factor; the actual variables are then used in the subsequent regression analysis. One must be careful that the multiple regression equation on the reduced set of predictors does not omit a variable that contributes highly to criterion-variable variance. This possibility can be checked by comparing R^2 before and after the predictor set has been reduced.

Similarly, factor analysis can be used as a preliminary transformation device in canonical correlation, MANOVA, MANCOVA or MDA if, for some reason, uncorrelated predictors or criteria are desired. In each case factor scores replace the original interval-scaled variables. With noisy or unreliable data the researcher may use a reduced set of factor scores (based on the highest-variance factors) as a data "clean-up" device.

In some types of applications one needs a single score on which persons can be ranked. If so, one could compute just the first principal component of a set of data and use the resulting component scores as values on an internally derived scale that has the property of maximally separating the individuals. Additional material on the use of factor analysis as an adjunct to other multivariate techniques can be found in Rummel (1970), Comrey (1973) and Green and Tull (1975).

9.4.3 Extensions of Factor Analysis

A number of extensions and generalizations have been made of traditional two-way (persons by variables) factoring procedures. These include:

1. Higher-order factor analysis.

2. Factoring of three-way (and higher-way) data matrices.

3. Hybrid models in which factor analysis is coupled with some other type of multivariate model.

Higher Order Factor Analysis   Higher order factor analysis involves successive rounds of factoring in which the factors at lower levels become variables for higher-level factoring. In order to be informative, this approach must use *oblique* factor solutions (such as the multiple group method) or else transform an initial orthogonal solution to an oblique orientation.

The matrix of oblique variable-by-factor pattern loadings obtained from the first round is then treated as a data matrix for a second round of factoring. In the second round the first-level factors are treated as variables in a loadings matrix whose columns represent second-level factors. The process can be continued to third or higher levels, although, in practice, one often stops with the second-level factors.

Applications of higher order factor analysis are still in their infancy. Furthermore, interpretation of higher order factors is often difficult and some researchers have felt that the method relies too heavily on the (possibly arbitrary) choice of some particular oblique transformation program. However, for the reader desirous of learning more about this approach, additional material can be found in Cattell (1966).

Three-Way Factor Analysis   Factoring three-way data matrices of the type represented by the data box of Figure 9.2 also represents a relatively recent development in factor analysis. Three-way (or three-mode) factor analysis was pioneered by Tucker (1966). The essence of three-way factor analysis is to analyze *simultaneously* all three facets of the data box, for example, persons, variables and occasions.

More recently, Carroll and Chang (1970) have introduced a model that is an important special case of Tucker's three-way factoring. To illustrate, consider the data box of Figure 9.2 in which we have several persons' scores on a set of variables over several different occasions. The general data entry in the three-way box can be denoted:

$$z_{ijk}$$

where $i = 1, 2, \ldots, I; j = 1, 2, \ldots, J; k = 1, 2, \ldots, K$

The Carroll and Chang model assumes that the general entry is modeled by:

$$z_{ijk} = \sum_{t=1}^{r} a_{it} b_{jt} c_{kt}$$

where $t = 1, 2, \ldots, r$ denotes the t-th underlying dimension or factor and a_{it}, b_{jt}, c_{kt} are parameters to be estimated by the approach.

Carroll and Chang call their model CANDECOMP, denoting *Canonical Decomp*osition. The parameters are solved iteratively by fixing the b's and c's and solving for the a's by regression techniques. One then finds better estimates of the b's while holding the a's and c's fixed and so on until no improvement in any of the parameters is noted.

An important property of CANDECOMP is its uniqueness of orientation. That is, the three r-dimensional configurations are not rotatable without nullifying the structure of the model.

Tucker's three-way analysis generalizes CANDECOMP as follows:

$$z_{ijk} = \sum_{t_1=1}^{r_1} \sum_{t_2=1}^{r_2} \sum_{t_3=1}^{r_3} a_{it1} b_{jt2} c_{kt3} g_{t_1 t_2 t_3}$$

where there are separate dimensionalities: r_1, r_2 and r_3 to contend with for each of the three modes. Moreover, there is an extra set of parameters:

$$g_{t_1 t_2 t_3}$$

that can be portrayed in a three-way matrix. Tucker calls this the *core matrix;* this matrix links the three separate configurations together.[14] Unlike CANDECOMP, Tucker's model does not produce a unique orientation of the three configurations; nor does their dimensionality have to be the same.

Neither model has received much application as yet, primarily because the models are relatively unknown outside of researchers who specialize in multivariate methodology. Moreover, the models are more complex mathematically and computer programs are still in the process of being disseminated. This situation should change over the next few years as the methodology receives greater publicity.[15]

Hybrid Models   Factor analysis can sometimes be used profitably in concert with other techniques. We have already commented on its use as a preliminary orthogonalization procedure, prior to employing other techniques, such as regression or MANOVA.

In a more direct way factor analysis can be employed as one part of a hybrid model. For example, Gollob (1968) has proposed an ANOVA-type model in which factor analysis is applied to the interaction numbers (in a two-way matrix) after the usual main effects have been removed. This approach provides a parsimonious modeling of the interaction numbers (considered as scalar products) that can be quite useful from a substantive viewpoint.

Green and Devita (1974) have adapted this model as a way to represent certain types of product complementarities in a consumer choice setting. Other possibilities for combinations of additive and factor analytic models or cluster and factor analytic models appear in the scaling of persons' perceptions and preferences, a topic covered later in the chapter.

9.4.4 Interpretative and Inferential problems

In a somewhat stereotyped way, two schools of thought on the role of factor analysis in research can be identified. One school would maintain that factor analysis is little more than a data reduction tool—a way to summarize associations across many variables in a few (often taken to be orthogonal) dimensions. This view would be strongly opposed by the second group. To them, factor analysis involves the systematic search for organizing constructs in some scientific domain of interest: personality, attitudes, demography, and so on.

Whatever one's predilections about the status of the technique it seems fair to say that most factor analyses are of an exploratory nature rather than confirmatory. As a matter of fact the methodological apparatus for carrying out confirmatory factor analysis—where one hypothesizes certain loadings to be high and others to be zero—is of quite recent origin (Jöreskog, 1969).

When one considers the difficulty of designing factor analytic studies, it is little wonder that the approach has exhibited a rather *ad hoc* flavor. First, one has to demarcate the content domain of interest. Assuming that the domain can be articulated, one has to consider the basic constructs that are reasonably independent and collectively exhaustive of the domain. Following this, one has the job of constructing variables (e.g., attitude statements, test items) that relate to each factor or construct.

Having sampled the universe of variables, one next has to sample the universe of persons. Is the same set of variables appropriate for everyone? Can one extend the results of one subset of variables to another subset? Can one generalize the results of one subpopulation of individuals to other subpopulations?

The problems of using factor analysis in theory testing are formidable. After a study has been completed one might find that factors thought to be present do not appear at all. Or new factors, not anticipated in the design of the variables, may have appeared. Are these new factors based on some artifact of the design? Will they continue to appear in other subpopulations of individuals?

Various rules of thumb have been proposed for the sampling of variables. Comrey (1973) recommends a minimum of three variables per hypothesized factor and would prefer at least five variables to pin down the factor reliably. He also suggests that one should include a sufficiently large number of factorially pure variables, each of which is anticipated to load on only a single factor.

When one considers that researchers often lack this kind of substantive knowledge to begin with, it is not surprising that factor analysis is used in a rather *ad hoc,* even casual, manner.

Factor analysts are frequently plagued by other problems as well. Some of these, drawn from Comrey (1973), are:

1. Using variables that display poor distributions:
   a. Badly skewed
   b. Bimodal in character
   c. Restricted range

2. Inclusion of linearly dependent variables, such as the sum of a subset of the included variables.

3. Using too few variables per hypothesized factor.

4. Including too few cases for the number of variables being analyzed.

5. Including too many factorially complex variables that load on more than one factor.

While other problems in applying factoring techniques could be mentioned, the preceding ones should be indicative of the difficulties incurred in carrying out factor analytic studies.

However, in spite of these difficulties there is little reason to expect factor analysis to disappear from the research scene. If anything, the behavioral and administrative sciences are placing increasing emphasis on multivariate analyses, where few or no alternatives to factor analysis exist. What *is* hoped, however, is that the future applications will take the viewpoint implied by confirmatory factor analysis in which one sets up specific hypotheses for test via the factor model. Otherwise, the outlook for the cumulative development of knowledge in content domains using factoring techniques appears dim.

9.5 Other Approaches to Dimensional Reduction

Up to this point our discussion of factoring methods has always assumed that a data matrix was the primary input data from which one computed some cross products matrix for subsequent factoring. For example, in principal components analysis we imagined that the rows of the data matrix, representing persons or objects, could be plotted as points in a space of correlated variables. The objective was then to find the axis of that space that resulted in the maximum variance of projected points. A second axis was then chosen, orthogonal to the first, that maximized residual variance, and so on. The original space of n variables was reduced to r $(r < n)$ components by neglecting the higher, smaller-variance dimensions.

However, suppose that what the researcher receives at the outset is a square matrix of correlations, covariances, interpoint distances, or some other such measure of pairwise proximity (or anti-proximity). In particular, suppose that these measures arise from respondents' *subjective judgments* of how dissimilar various pairs of objects are. One might be interested in taking the matrix of distance-like data and finding a space of relatively low dimensionality and a configuration of points in that space, one point for each object. The specific number of dimensions and configuration of points in the space would be chosen so as to achieve a "good match" between the computed interpoint distances and the original input data.

This is the task of multidimensional scaling (MDS), a set of techniques for transforming distance-like input data into coordinate representations of the points in a (typically Euclidean) space of low dimensionality. So-called *metric* MDS methods operate on the criterion of finding a best match between computed distances and the numerical values of the input data. So-called *nonmetric* MDS methods, a more recent development, attempt to find a configuration whose ranks of computed interpoint distances best match the ranks of the input data.

One class of metric MDS methods bears a close resemblance to the principal components factoring of cross-product matrices. We start our discussion with this class of procedures and then contrast the results with those obtained from the newer nonmetric approaches.

9.5.1 Metric Multidimensional Scaling

By way of illustration, consider the 15 points shown in Figure 9.3. For ease of identification the points have been chosen to trace out the letter *M*. By assigning a common, but arbitrary, unit to each of the two dimensions of the figure we find the coordinate values shown in Panel I of Table 9.2. Panel II of the table shows the computed Euclidean distances for the first six points (illustratively), obtained from the formula:

$$d_{ij} = \left[\sum_{t=1}^{r} (X_{it} - X_{jt})^2 \right]^{1/2}$$

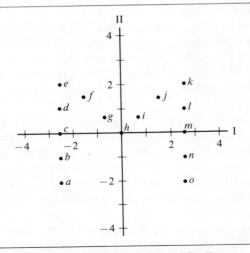

Figure 9.3 Sample Configuration for Multidimensional Scaling

**Table 9.2 Coordinate Representation of the Configuration of Figure 9.3
and Illustrative Euclidean Distances**

I Coordinates			II Interpoint Distances of First Six Points					
Observation	X_1	X_2	Observation	a	b	c	d	e
a	−2.5	−2	b	1				
b	−2.5	−1	c	2	1			
c	−2.5	0	d	3	2	1		
d	−2.5	1	e	4	3	2	1	
e	−2.5	2	f	3.64	2.69	1.80	1.12	1.12
f	−1.5	1.5						
g	−0.67	0.67						
h	0	0						
i	0.67	0.67						
j	1.5	1.15						
k	2.5	2						
l	2.5	1						
m	2.5	0						
n	2.5	−1						
o	2.5	−2						
$\overline{X}_j$	0	0.29						

For example, the distance between points e and f is:

$$d_{ef} = [(-2.5 + 1.5)^2 + (2 - 1.5)^2]^{1/2}$$
$$= 1.12$$

Note further that the centroid values of the configuration are $\overline{X}_1 = 0; \overline{X}_2 = 0.29$. This reference point is arbitrary in the sense that Euclidean distances are unaffected by a translation of origin.

In metric multidimensional scaling, one *starts* with the distances (or at least a set of distance-like numbers measured on either a ratio or interval scale) and the objective is to find a configuration of points in a small number of dimensions; the configuration is one whose computed distances best match the input data for that dimensionality. The key concept that connects this variety of metric MDS with factor analysis is the relationship between the scalar products of a pair of vectors and their squared Euclidean distance. If we let $\mathbf{x}_i$ and $\mathbf{x}_j$ denote two vectors, then their scalar product is:

$$\mathbf{x}_i'\mathbf{x}_j = \frac{\mathbf{x}_i'\mathbf{x}_i + \mathbf{x}_j'\mathbf{x}_j - d_{ij}^2}{2}$$

Now, if the d_{ij}^2's are assumed to be known, all that remains is to find the squared length of each vector from the origin of the space.

This, in turn, raises the problem of finding a suitable origin. One could make one of the actual points the origin but in real, errorful data one would expect the location of a single point to be rather unstable. An origin defined in terms of the centroid of all of the points would be expected to be more stable, as well as representing a handy reference point.

Torgerson (1958) and others, have shown that a simple formula relates d_{ij}^2 to the scalar product $\mathbf{x}_i'\mathbf{x}_j$ when all points are referred to a centroid-centered origin. The steps involved in finding the scalar products, in this case, are as follows:

1. Form the $m \times m$ matrix $\mathbf{E}$ with elements $e_{ij} = -\frac{1}{2}d_{ij}^2$

2. Form the $m \times m$ matrix $\mathbf{F}$ with elements $f_{ij} = e_{ij} - \bar{e}_{i.} - \bar{e}_{.j} + \bar{e}_{..}$. That is, doubly center the matrix $\mathbf{E}$ in order to obtain $\mathbf{F}$.

3. If the eigenstructure of $\mathbf{F}$ is found and the factor loadings are scaled (by columns) so that their sums of squares equal their respective eigenvalues, then the factor loadings provide the desired coordinate values for the m points.[16]

Accordingly, the full matrix of squared Euclidean distances was computed from the set of coordinate values shown in Table 9.2 and the entries were converted, via the procedure described above, into the matrix of scalar products, $\mathbf{F}$. The "factor loadings" of $\mathbf{F}$ are shown in Figure 9.4, just as they came out of the computer. Since the original configuration was already in principal components orientation, the solved-for configuration in Figure 9.4 is in a similar orientation. Aside from a constant of proportionality the original distances are matched perfectly.[17]

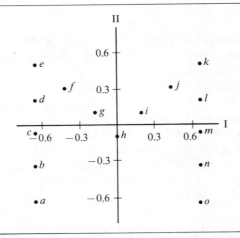

Figure 9.4 Configuration Computed from the Eigenstructure of the Scalar Products Matrix F

9.5.2 Distance Estimation and the Additive Constant

In the demonstration problem we made sure that the input data were bonafide squared Euclidean distances. In real applications the input data contain error and may not even be measured on a ratio scale. Since the concept of distance implies a unique zero (viz., the distance between a point and itself) the input data must be transformed in some way as to lead to a (possibly errorful) ratio scale. This is known as the *additive constant* problem in metric MDS.

If one adds some arbitrary constant c to each entry of the input data, the resulting solution will change, either in terms of configuration, dimensionality, or both. Indeed, some additive constants can even change the nature of the metric itself. Thus, before one can apply the method described in the preceding section to real data, one should find an additive constant that results in a set of estimated Euclidean distances, measured at the level of a *ratio* scale.

Estimation of the distance-like numbers used as input to metric MDS applications usually proceeds in stages. First, one attempts to obtain interval-scaled input data and then an additive constant c is estimated before the actual scaling is initiated.

To illustrate, in the domain of psychological scaling of perceptions and attitudes, respondents may be given all distinct pairs of a set of stimuli, such as toothpaste brands or production foremen, or political candidates, and so on. The respondent is then asked to rate each pair of stimuli on a magnitude estimation scale, ranging from 1—almost identical, to 9—completely dissimilar.

Alternatively, one pair may be singled out for reference and each test pair compared to it in terms of a subjective rating of relative similarity. However, even if one can obtain a reliable interval scale of similarity, one has the problem of converting this to a ratio scale prior to applying metric MDS.

Approaches to determining the constant c that should be added to each input entry to convert interval judgments to ratio judgments, typically proceed iteratively. By making the additive constant large enough, one can always find a set of distance-like numbers that can be related to a Euclidean space. The problem here is that with m stimuli the resulting dimensionality approaches $m - 1$ and, therefore, no parsimony is obtained.

Generally, one wants to work just the other way around; that is, to find the *smallest* constant that preserves Euclidean properties. Fortunately, the eigenstructure approach provides its own indicant. If at least one of the eigenvalues of the **F** matrix is less than zero, the space is non-Euclidean. A typical procedure for finding the additive constant proceeds as follows:

1. For convenience, set the smallest original input-data entry δ_{ij}^* to zero. Since distances must be non-negative, the to-be-solved-for additive constant must now be zero or positive.

2. Choose a trial value of c, the additive constant; for example, one could choose the mean value of all entries in the (transformed) input matrix.

3. Convert $[\delta_{ij} + c]$ to the matrix **E**, followed by transformation to **F** and find the eigenstructure of **F**.

4. If any λ_j of **F** is negative, repeat the process with a larger value of c.

5. If all λ_j's are non-negative, repeat the process with a smaller value of c.

6. Iterate until a solution is obtained that has the largest number of eigenvalues close to (and distributed about) zero.

In actuality the methods described by Messick and Abelson (1956) and Torgerson (1958) are more precise than this but are still based on the general principle of finding the smallest c that produces a matrix **F** with essentially non-negative eigenvalues.[18]

9.5.3 Ordinal Measures of Distance

In most behavioral science applications of MDS it is questionable whether respondents can supply reliable judgments of dissimilarity, even at the level of an interval scale, let alone a ratio scale. More recent developments in MDS relax the assumption of ratio-scaled distances as input. Rather, the input data are assumed to represent the δ_{ij}'s on only an ordinal or ranking scale. These computer-based methods are referred to as *nonmetric* and were pioneered by Shepard (1962).

The reader has already been introduced to a nonmetric procedure in the form of MONANOVA, a nonmetric algorithm for performing main effects ANOVA (as was discussed and illustrated in Chapter 5). Here, our interest focuses on a nonmetric analogue to the multidimensional scaling of (ordinally measured) proximities. Still, the similarity between this approach and MONANOVA will become apparent.

As motivation for applying a nonmetric scaling procedure, again consider the coordinates of the 15 points, as listed in Table 9.2. Let us compute the d_{ij}'s for each pair, a total of $15(14)/2 = 105$ interpoint distances. Now, however, suppose we compute the following monotonic increasing function of the d_{ij}'s:

$$\delta_{ij} = f_M(d_{ij}) = d_{ij}^4$$

That is, we raise each d_{ij} to the fourth power.

If we then tried to apply a metric MDS procedure to these input data (after transformation to the input matrix **F**), it is by no means the case that only two dimensions (two non-zero λ's) and perfect recovery of the configuration would be obtained.

This type of transformation was indeed carried out and the following eigenvalues were actually obtained:

$$
\begin{cases}
\lambda_1 = 10.13 \\
\lambda_2 = 6.32 \\
\lambda_3 = 0.31
\end{cases}
\qquad
\begin{cases}
\lambda_4 = 0.05 \\
\lambda_5 = 0.01 \\
\lambda_6 = 0.01
\end{cases}
$$

While the first two eigenvalues clearly overshadow the rest, we see that λ_3 is large enough that we might have been tempted to scale the data in three, rather than two, dimensions.

Assume, however, that we do choose two dimensions for plotting the scaled eigenvectors of the metric solution. Panel I of Figure 9.5 shows the results. The nonlinear (but monotonic) power function:

$$\delta_{ij} = d_{ij}^4$$

takes its toll with regard to solution recovery. As noted in Panel I of Figure 9.5, the metric solution tends to cave in the sides of the *"M"* rather severely. Hence, even with error-free data, subjecting the original distances to an ordinal transformation inhibits the recovery of the known configuration if one uses metric (linear) methods.

Alternatively, let us use a nonmetric algorithm, in this case, one called TORSCA (Young, 1968). This particular scaling algorithm is typical of a number of programs that are available for nonmetric MDS.

9.5.4 Nonmetric Multidimensional Scaling

The TORSCA nonmetric MDS program is an iterative program that can scale a set of rank-ordered dissimilarities in successive dimensionalities. For example, the researcher could request a scaling of the data from four dimensions down to one dimension. If so, the program will provide the "best" configuration for each dimensionality.

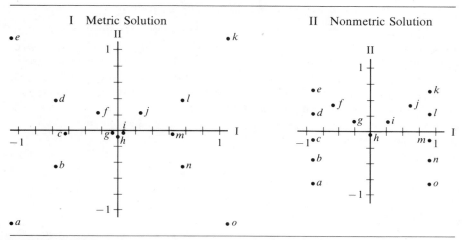

Figure 9.5 Metric Versus Nonmetric Solutions Involving an Increasing Monotonic Function of Distance

In general, higher dimensionalities provide better fits of the rank ordered distances to the input data, but less parsimony. The researcher is then faced with a trade-off between improved fit versus higher dimensionality. This problem is not unlike the problem in factor analysis concerning how many factors to retain.

A few details are in order concerning how TORSCA works. Assume that we have a set of *ranked* dissimilarities:

$$\delta_{ij};\, i,j = 1, 2, \ldots, m;\, i \neq j$$

which we can call psychological "distances." Similar to metric MDS our objective is to find a configuration, denoted by the $m \times r$ matrix $\mathbf{X}$, of m points in r dimensions. The coordinates of a given point $\mathbf{x}_i$ can be specified as:

$$\mathbf{x}_i' = (X_{i1}, X_{i2}, \ldots, X_{ir})$$

For each $\mathbf{x}_i$, $\mathbf{x}_j$ in $\mathbf{X}$ we may compute a distance d_{ij}. If $\mathbf{X}$ is a "good" configuration, in that the ranks of its distances d_{ij} reproduce the input ranks δ_{ij} well, then that configuration should be "final" or close thereto for representing the δ_{ij}.

The appropriate numbers (which may not be distances) that *are* perfectly monotonic with the δ_{ij} are denoted as $\hat{d}_{ij}$. The TORSCA algorithm considers relationships among the three sets of values:

1. The δ_{ij}—the input data ranks.

2. The d_{ij}—computed distances between all pars of points in the configuration $\mathbf{X}$ for a specific dimensionality.

3. The $\hat{d}_{ij}$—a set of numbers, chosen to be as close to their respective d_{ij} as possible, subject to being monotonic with the δ_{ij}. That is, $\hat{d}_{ij} \leq \hat{d}_{k\ell}$ whenever $\delta_{ij} < \delta_{k\ell}$.

All of these algorithms must consider two problems: (a) the development of an index of goodness of fit by which one can tell if the configuration $\mathbf{X}$ is an appropriate one for representing the input data δ_{ij} and (b) a procedure for moving the points $\mathbf{x}_i$, $\mathbf{x}_j$ to some "better" configuration if the current index of fit is poor.

Most of the nonmetric scaling algorithms provide an index of fit that represents some variant of Kruskal's stress measure:[19]

$$S = \sqrt{\dfrac{\displaystyle\sum_{\substack{i \neq j}}^{m} (d_{ij} - \hat{d}_{ij})^2}{\displaystyle\sum_{\substack{i \neq j}}^{m} d_{ij}^2}}$$

The numerator of this index consists of the sum of squares of the discrepancies between each computed d_{ij} for some configuration $\mathbf{X}$ and a set of numbers

$\hat{d}_{ij}$ chosen to be as close to their respective d_{ij} as possible, subject to being monotonic with the original δ_{ij}'s. If the d_{ij} equal their $\hat{d}_{ij}$ counterparts, the numerator of the expression becomes zero and, hence, stress is also zero, indicating a perfect fit.

The denominator of the expression is merely a normalizing value, computed to allow comparisons of the fit measure across different dimensionalities since, in general, the d_{ij} will increase with increasing dimensionality. More recent versions of the stress formula place the quantity:

$$\sum_{i \neq j}^{m} (d_{ij} - \bar{d})^2$$

(where $\bar{d}$ is the mean distance) in the denominator. While there are theoretical advantages ascribed to this second formulation, in practice both versions of the formula are used.[20]

Suppose, however, that the stress S of a particular configuration is high—that is, the monotonic fit is poor. The next feature of these algorithms consists of finding a new configuration whose ranks of interpoint distances are more closely monotonic to the original δ_{ij} than the previously ranked distances.

Improving the Configuration   Assume now that we wish to move the points around so that their distance ranks are closer to the input data ranks than those found in the previous configuration. In particular, consider a specific point i and its relationship to each of the points j in turn. We would like to move the point i so as to decrease its average discrepancy between the distances d_{ij} and the numbers $\hat{d}_{ij}$, the latter set of numbers being monotonic with the δ_{ij}.

If d_{ij} is larger than $\hat{d}_{ij}$ we could move point i toward point j by an amount proportional to the size of the discrepancy. Conversely, if $\hat{d}_{ij}$ is larger than d_{ij}, point i could be moved away from point j by an amount proportional to the discrepancy. Ordinarily we would not like to do this in just one step, since we may "overcorrect" in so doing. Suppose we let α represent the coefficient of proportionality ($0 \leq \alpha \leq 1$) or step size. (Often α is set at 0.2.)

To find a new coordinate $X_{it(j)}^*$ for point i on axis t, as related to point j, we can use the formula:

$$X_{it(j)}^* = X_{it} + \alpha \left(1 - \frac{\hat{d}_{ij}}{d_{ij}} \right) (X_{jt} - X_{it})$$

This formula will move point i in the appropriate direction with respect to point j, but we must consider all $m - 1$ points insofar as their effect on point i is concerned.

To do this we use the expression:

$$X_{it}^* = X_{it} + \frac{\alpha}{m-1} \sum_{\substack{j=1 \\ j \neq i}}^{m} \left(1 - \frac{\hat{d}_{ij}}{d_{ij}} \right) (X_{jt} - X_{it})$$

Note that we move point i along axis t in such a way as to take into account the discrepancies involving all other points. This is done for all points in all dimensions.

The procedure can be summarized as involving the following steps:

1. For a given dimensionality, select some initial configuration $\mathbf{X}_0$. TORSCA does this by means of the Torgerson metric procedure outlined earlier (without solving for an additive constant).

2. Compute d_{ij} between the vectors $\mathbf{x}_i$, $\mathbf{x}_j$ of the configuration $\mathbf{X}_0$ and also compute $\hat{d}_{ij}$, chosen to be as close as possible to the original d_{ij}, subject to being monotonic with the δ_{ij} (input data).

3. Evaluate the fit measure S, the stress of the configuration.

4. If $S > \epsilon$ (a "stopping" value), find a new configuration $\mathbf{X}_1$ whose ranks of the d_{ij} are closer to the δ_{ij}.

5. Repeat the process until successive configurations $\mathbf{X}_0, \mathbf{X}_1, \mathbf{X}_2, \ldots, \mathbf{X}_s$ converge such that S is satisfactorily "small."

6. Repeat the above process in the next lower dimensionality, and so on.

7. Choose the lowest dimensionality for which ϵ is satisfactorily "small."

Although specific fit measures and procedures for moving the points differ among various algorithms for performing nonmetric MDS, they all seem to be based on these concepts.

The TORSCA algorithm was applied to the monotonically transformed data:

$$\delta_{ij} = d_{ij}^4$$

with results appearing in Panel II of Figure 9.5. As observed, recovery is essentially perfect; the program's stress value was zero, to three decimal places. Only 16 iterations were required to reach the stress minimum.

As an added bonus the program also plots, in scatter diagram form, the δ_{ij}'s (original data) versus the d_{ij}'s and the $\hat{d}_{ij}$'s, the latter set of numbers being the best fitting monotonic transform of the original data. Although not shown here, in the present (error-free) case the exact function:

$$f_M(d_{ij}) = d_{ij}^4$$

appeared (in terms of discrete values) when the original data were plotted against the computed distances. Had the original d_{ij}'s been used as input data, the function would, of course, appear as linear.

9.5.5 Illustrative Applications of MDS

Although still of recent developnient, it has already become apparent that MDS provides an interesting and potentially useful methodology for portraying subjective judgments of diverse kinds. Various researchers have adapted different viewpoints about the methodology.

Perhaps the most pragmatic of these views is to consider MDS as basically concerned with *data reduction and display*. The primary motivation here is to search for pattern or structure in any kind of associative data whatsoever. One need not necessarily be concerned with dimensional representations or psychological interpretations of the space; MDS is strictly used as a way to heighten certain associations in the data. This view is akin to some factor analysts' use of factoring methods.

A more "psychological" way is to characterize MDS as a set of procedures for portraying perceptual or affective dimensions of substantive interest. That is, the intent here is to learn *psychological* things about the stimuli of interest, particularly in cases where the underlying objective dimensions are either quite diffuse (e.g., handwriting characteristics) or possibly interactive (e.g., psychological aspects of taste), even assuming that they might be measured at all. Here the emphasis is on the *discovery* of psychological dimensions and stimulus spacing on these dimensions. Hence, under this view the MDS models are considered as miniature psychological theories about certain aspects of perception and evaluation.

Still a third way to view MDS procedures is in terms of the *psychophysics* of the situation. Here, the researcher seeks to understand relationships between physically controllable dimensions and psychological responses—a kind of multidimensional psychophysics. Under this view, stimuli, such as products, are synthesized (or selected) in certain ways so that only a relatively few characteristics are systematically varied. The research task is then to develop psychological transformations of the "known" dimensions for predicting responses to new combinations of physical characteristics.

Most applications to date have explored the use of MDS in finding out the perceptual dimensions, and the spacing of stimuli along these dimensions, that people use in making judgments about the relative similarity of pairs of stimuli. Figure 9.6 shows an illustration (Green and Tull, 1975) of nonmetric MDS in the construction of perceptual maps of brands of beer sold in the Detroit, Michigan area. After consumers judged the relative dissimilarity of brands of beer sold in the area, they rated the beers that they were familiar with on various attributes, such as filling, expensive, light taste. Each set of ratings was then regressed on the coordinates of the nonmetric MDS configuration; the vector directions in the map are normalized beta weights from the separate regressions. As noted, the sponsor's brand is perceived as a strong tasting, filling beer.

MDS methods can be applied to a diversity of stimulus domains. For example, Carroll and Wish (1971) scaled 14 subjects' perceptions of the similarities of ten color chips that varied by hue (with saturation and brightness held constant). As is generally known, the perception of color tends to follow a circular arrangement, of the type illustrated in Panel I of Figure 9.7.

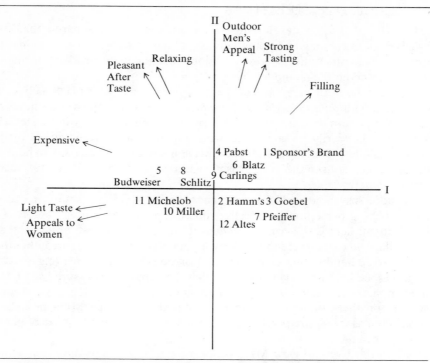

Figure 9.6 Nonmetric Multidimensional Scaling of Beer Brands (from Green and Tull, 1975)

The horizontal axis of this configuration is a yellow-blue dimension and the vertical axis is a green-red dimension.[21] Carroll and Wish used a special metric MDS program called INDSCAL (Carroll and Chang, 1970) that develops a uniquely oriented stimulus space (Panel I) and a subject space as well. In the latter space each subject is represented as a point.

The subjects of the experiment consisted of ten persons who were known to have normal color vision and four who were known to have color deficiency with respect to the green-red dimension. Interestingly enough, the INDSCAL program detected the differences among subjects, solely as they revealed themselves through individual differences in their perceived similarities of the color chips.

Panel II of Figure 9.7 shows the subject space. We first note that all four color deficient subjects have lower saliences for the green-red dimension than the normal subjects. (The relative salience of each subject is found by projecting each point onto the horizontal and vertical axes. The square roots of these saliences can be viewed as axis stretching or shrinking factors by which the group stimulus space in Panel I is modified to accommodate specific individuals' "private" spaces.)

It is satisfying to note that INDSCAL was able to capture physical differences in color vision from the various subjects' perceptual judgments. This particular

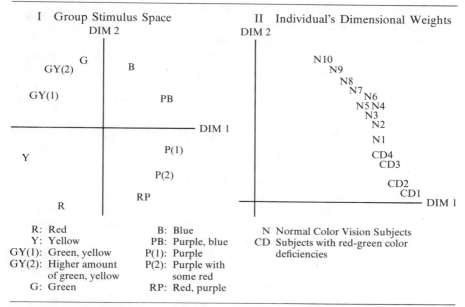

I   Group Stimulus Space

II   Individual's Dimensional Weights

R: Red
Y: Yellow
GY(1): Green, yellow
GY(2): Higher amount
    of green, yellow
G: Green

B: Blue
PB: Purple, blue
P(1): Purple
P(2): Purple with
    some red
RP: Red, purple

N   Normal Color Vision Subjects
CD Subjects with red-green color
    deficiencies

Figure 9.7 An INDSCAL Analysis of Judged Similarities of Color Chips
(from Carroll and Wish, 1971)

model provides a tool for the systematic exploration of perceptual domains across persons, occasions, data collection methods and other facets along which one could expect responses to vary.

Despite their recency of development, MDS methods have already enjoyed broad and varied application in psychology, sociology, political science, marketing, and other disciplines. Methods have been developed for analyzing preference data as well as similarities-type responses. Moreover, general interest in nonmetric approaches has led to advances in nonmetric variations of the more traditional multivariate techniques, such as multiple regression and canonical correlation. The reader interested in more detailed aspects of these methods should see the books by Shepard, Romney and Nerlove (1972), Green and Rao (1972), Green and Wind (1973), Lingoes (1973), and Roskam (1968).

9.6 Cluster Analysis

Up to this point our interest in interdependent data structures has emphasized factor analytic methods. Factor analysis—at least the more common R technique—focuses on the variables or columns of the $m \times n$ data matrix. The objective is to reduce the original set of correlated variables to some smaller set of orthogonal linear composites that preserve most of the original information. Representation of the data is spatial, with the factors (linear composites) serving as axes of the reduced space.

In cluster analysis we consider the obverse problem in which we wish to reduce the number of objects or rows of the data matrix to some smaller number by grouping the objects into clusters. We wish to perform this clustering so that objects in the same cluster are more like each other than they are to objects in other clusters. However, representation of the data is not spatial but classificatory.

The search for classifications or typologies pervades all of the sciences. In psychology one may wish to develop a classification of personality types. In political science one may want to classify legislators on the basis of their voting records over a varied sample of legislative actions. In marketing customers may be classified in terms of various demographic or purchase characteristics. In chemistry various compounds may be classified in terms of their performance properties. In production, alternative plant sites may be classified in terms of tax rates, labor availability, nearness to raw materials, and so on.

Classification is as old as science itself. However, *numerical* methods of classification—and their application to problems in zoology, botany, biology, and psychology, as well as business—are largely a development of the past two decades. The first book devoted exclusively to the topic appeared as late as the nineteen-sixties (Sokal and Sneath, 1963).

Cluster analysis is primarily concerned with three basic questions:

1. How can we measure interobject similarity or resemblance?

2. Assuming that one can measure the relative similarity of each object to every other object, how does one go about the process of placing similar objects in clusters?

3. After the grouping is completed how does one describe the clusters, and how does one know that the resulting clusters are real versus some kind of statistical artifact?

In the current state of the art, there are difficulties associated with answering each of these questions.

9.6.1 Similarity Measures

The choice of a similarity or resemblance measure is an interesting problem in cluster analysis. The concept of similarity always connotes the question: similarity with respect to what? Similarity measures are usually viewed in relative terms—two objects are similar if their profiles across variables are "close" or they exhibit "many" aspects in common, relative to those which other pairs of objects share.

Most clustering procedures use pairwise measures of resemblance. The choice of which objects and variables to use in the first place is largely a matter for the researcher's judgment. Even assuming that such choices have been made, the possible measures of pairwise resemblance are many. Generally speaking, these measures fall into two classes: (a) distance-type measures and (b) matching-type measures.

Distance-Type Measures   A surprisingly large number of resemblance measures can be viewed as closely related to interpoint distances in some type of metric space. As illustrated in section 9.5, the Euclidean distance between two points in a space of r dimensions is:

$$d_{ij} = \left[\sum_{t=1}^{r} (X_{it} - X_{jt})^2 \right]^{1/2}$$

where X_i, X_j are the projections of points i and j on variable t $(t = 1, 2, \ldots , r)$.

Inasmuch as the variables are often measured in different units, the above formula is frequently applied *after* each variable has been standardized to mean zero and unit standard deviation (although the initial step of centering to mean zero has no effect on the distances). Our subsequent discussion will assume that this preliminary step has been taken.

Some researchers feel that use of the Euclidean distance measure should be restricted to orthogonal, standardized variables. While this viewpoint seems unduly restrictive, it is useful to point out the implicit weighting of the principal components underlying the associated variables that occurs with the use of the traditional Euclidean measure:

1. Squared Euclidean distance in the original variables space has the effect of weighting each underlying principal component by that component's eigenvalue.

2. Squared Euclidean distance in the components space, if all components are first standardized to *unit* variance, has the effect of assiging equal weights to all components.

3. In terms of the geometry of the configuration, in the first case all points are rotated to orthogonal axes with no change in squared interpoint distance. The general effect is to portray the original configuration as a hyperellipsoid with principal components serving as the axes of that figure. Equating all axes to equal length has the effect of transforming the hyperellipsoid into a hypersphere where all axes become equal in length.[22]

As recalled from Chapter 4 we encountered the distinction between ordinary (squared) Euclidean distances and Mahalanobis' D^2 in the context of discriminant analysis. Mahalanobis' D^2 takes both differences in axis lengths and the correlatedness of axes into account. Mahalanobis' D^2 is the same as ordinary d^2 if the latter is computed in a space in which the configuration has been transformed into a hypersphere.

The preceding considerations can also be handled in terms of a modified squared distance model:

$$*d_{ij}^2 = \sum_{t=1}^{r} (Y_{it} - Y_{jt})^2$$

where Y_{it}, Y_{jt} denote unit-variance component scores of profiles i and j on component axis t ($t = 1, 2, \ldots, r$). If one weights the component scores according to the variances of the components, the expression is:

$$d_{ij}^2 = \sum_{t=1}^{n} \lambda_t (Y_{it} - Y_{jt})^2$$

where λ_t is the t-th component variance, or eigenvalue. This last expression is equivalent to d_{ij}^2 expressed in original variables space.

The above relationships assume that *all* principal components are extracted. If such is not the case, the squared interpoint distances will result in lost information due to the fact that they are computed in a components space of lower dimensionality than the original variables space.

Thus, both the squared distance d^2 in original variables space and the squared distance $*d^2$ in principal components space (assuming all components have been extracted) preserve all of the information in the original data matrix. Finally it should be pointed out that if in addition to being standardized to mean zero and unit variance, the original variables are uncorrelated, both d^2 and $*d^2$ will be equivalent.

Two other measures have often been proposed as resemblance measures. Both of these measures derive from historical clustering methods which used Q-type factor analysis to cluster objects. As recalled, in Q-type factor analysis the correlation (or covariance) matrix to be factored consists of interobject rather than intervariable associations.

The effect of a Q-type principal components analysis of either a covariance or a correlation matrix, as shown by Cronbach and Gleser (1953), is to reduce the dimensionality of the space underlying computation of the distance measures. Both procedures reduce the dimensionality of the original space to one less dimension by equating all profiles to a mean of zero. As such, profile differences in elevation are removed. In addition, a Q-type analysis applied to the inter-object correlation matrix will remove interprofile variation due to differences in dispersion.[23]

Matching-Type Measures   Quite often the analyst wishing to cluster profiles must contend with data that are only nominally scaled, in whole or in part. While dichotomous data can be expressed in terms of interpoint distances, the usual approach to nominally scaled data uses attribute matching coefficients. Intuitively speaking, two profiles are viewed as similar based on the extent to which they share common attributes.

As an illustration of this approach, consider the two profiles appearing below:

Object	1	2	3	Attribute 4	5	6
1	1	0	0	1	1	0
2	0	1	0	1	0	1

Each of the above objects is characterized by possession or non-possession of each of six attributes, where a 1 denotes possession and a 0 denotes non-possession. Suppose we just count up the total number of matches—either (1, 1) or (0, 0)—and divide by the total number of attributes; there are two such matches. A simple matching measure can then be stated as:

$$S_{12} = \frac{M}{N} = \frac{2}{6} = \frac{1}{3}$$

where M denotes the number of attributes held in common (matching 1's or 0's) and N denotes the total number of attributes.[24] We notice that this measure varies between zero and one.

If weak matches (non-possession of an attribute) are to be deemphasized the above measure can be modified to:

$$S_{ij}^* = \frac{\text{No. of attributes which are 1 for both objects } i \text{ and } j}{\text{No. of attributes which are 1 for either } i \text{ or } j \text{ or both}}$$

A variety of such matching-type coefficients are described in the book by Sokal and Sneath (1963).

Attributes need not be limited to dichotomies, however. In the case of un-ordered polytomies, matching coefficients are often developed by means similar to the above after recoding a k-state variable into $k - 1$ dummy (zero-one) variables. Such coefficients will be sensitive to variation in the number of states in each polytomy.

9.6.2 Clustering Algorithms

Once the analyst has settled on some pairwise measure of profile similarity, he must still use some type of computational routine for clustering the profiles. Hundreds of such computer programs already exist and more are being developed as interest in the field increases. Ball and Hall (1968) have made an extensive survey of clustering methods. The following categories are based, in part, on their classification:[25]

1. *Dimensionalizing the Proximity Matrix*—these approaches use principal components or some other factor analytic method to find a dimensional representation of points. Clusters are then developed visually or, sometimes, by a clustering method applied to a set of distances computed in components space.

2. *Non-Hierarchical Methods*—these methods start right from the similarity or distance matrix and can be characterized as:

    a. *Sequential Threshold*—in this case a cluster center is selected and all objects within a prespecified threshold value are grouped. Then a new cluster center is selected and the process is repeated for the unclustered points, and so on. (Once points enter a cluster they are removed from further processing.)

    b. *Parallel Threshold*—this method is similar except that several cluster centers are selected simultaneously and points within threshold level are assigned to the nearest center; threshold levels can then be adjusted to admit fewer or more points to clusters.

    c. *Optimizing Partitioning*—this method modifies (a) or (b) in that points can be later reassigned to clusters on the basis of optimizing some overall criterion measure, such as minimizing average within-cluster distance for a given number of clusters.[26]

3. *Hierarchial Methods*—these procedures are characterized by the construction of a hierarchy or tree-like structure. In some methods each point starts out as a unit (single-point) cluster. At the next level the two closest points are placed in a cluster. At the next level a third point joins the first two or else a second two-point cluster is formed, based on various criterion functions for assignment. Eventually all points are grouped into one large cluster. Variations on this procedure involve the development of a hierarchy from the top down. At the beginning the points are partitioned into two subsets based on some criterion measure related to average within-cluster distance. The subset with the highest average within-cluster distance is next partitioned into two subsets, and so on, until all points eventually become unit clusters.

While the above classes of programs are not exhaustive of the field,[27] most of the more widely used clustering routines can be typed as falling into one (or a combination) of the above categories. Criteria for grouping include such measures as average within-cluster distance and threshold cut-off values. The fact remains, however, that even the "optimizing" approaches achieve only conditional or local optima, since an unsettled question is *how many* clusters to form in the first place.

An illustration of one of the more popular clustering programs may be of interest. Consider the standardized (artificial) data of Figure 9.8. In this case we have 12 points portrayed in two dimensions. Visually, it would seem that four clusters are present:

    $\{a, b\}$
    $\{c, d, e, f, g\}$
    $\{h, i, j, k\}$
    $\{l\}$

However, in most practical problems of interest, we cannot fall back on visual clustering. We could have hundreds of points in several dimensions (where each dimension is a variable). Nonetheless, let us examine how various clustering rules would group the 12 objects represented as points in Figure 9.8.

BMD-P1M (Dixon, 1975) is one of the *P* series of Biomedical programs. This program can cluster up to 150 objects, assuming that other program constraints are met.[28] Various measures of resemblance, including correlation, covariance, or Euclidean distance can be accommodated.

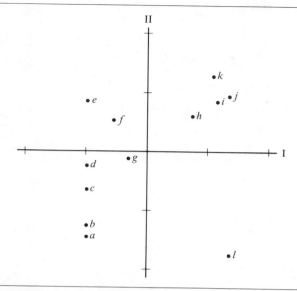

Figure 9.8 Initial Configuration of Points for Cluster Analysis

The program is a hierarchical algorithm that starts out with each point in its own unit cluster and eventually ends up with all points in one undifferentiated cluster. Three different amalgamation rules for building up the clusters are available.

Single Linkage   The single linkage or minimum distance rule starts out by finding the two points with the shortest distance. These are placed in the first cluster. At the next stage a third point joins the already-formed cluster of two if its shortest distance to the members of the cluster is smaller than the two closest unclustered points. Otherwise, the two closest unclustered points are placed in a cluster.

The process continues until all points end up in one cluster. The distance between two clusters is defined as the *shortest* distance from a point in the first cluster to a point in the second.

Complete Linkage   The complete linkage option starts out in just the same way by clustering the two closest points. However, the criterion for joining points to clusters or clusters to clusters involves maximum (rather than minimum) distance. In other words, the distance between two clusters is the *longest* distance from a point in the first cluster to a point in the second cluster.

Average Linkage   The average linkage option starts out in the same way as the other two. However, in this case the distance between two clusters is the *average* distance from points in the first cluster to points in the second cluster.

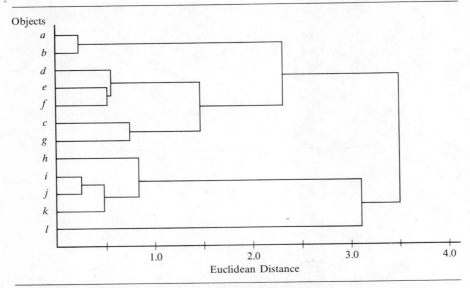

Figure 9.9 Dendrogram from Complete Linkage Clustering

Results The three amalgamation rules showed similar, but not identical, clusterings. For example, at the four-cluster level there was a difference in the placement of point c between the single and average linkage versus the complete linkage rules:

Single Linkage	Complete Linkage	Average Linkage
{a, b, c}	{a, b}	{a, b, c}
{d, e, f, g}	{c, d, e, f, g}	{d, e, f, g}
{h, i, j, k}	{h, i, j, k}	{h, i, j, k}
{*l*}	{*l*}	{*l*}

Since the program provides the full clustering sequence, it is easy to prepare a tree diagram (called a dendrogram). Illustratively, Figure 9.9 shows the dendrogram for the complete linkage rule.

We note that points a and b, the closest pair, first join at a distance of 0.23. The next pair to join are points i and j, and so on. The last two clusters to merge are {a, b, c, d, e, f, g} and {h, i, j, k, *l*} at a distance value of 3.5. We note that the dendrogram provides a succinct and convenient way to summarize the clustering sequence.

9.6.3 Cluster Description and Significance

Once clusters are developed, the analyst still faces the task of describing the clusters. One measure that is used frequently is the cluster's centroid or average value of the objects contained in the cluster on each of the variables making up the profiles. If the data are interval-scaled and clustering is performed in original variables space, this measure appears quite natural as a summary description. If the space consists of principal components dimensions, the axes usually are not capable of being described simply. Often in this case the analyst will want to go back to the original variables and compute average profile measures in these terms. If matching type coefficients are used, the analyst may describe a cluster in terms of the group's modal profile on each of the attributes.

In addition to central tendency, the researcher may compute some measure of the cluster's variability, such as the average interpoint distance of all members of the cluster from their centroid or the average interpoint distance between all pairs of points within the cluster.

Despite attempts made to construct various tests of statistical significance of clusters, in the author's view no defensible procedures are currently available. The lack of appropriate tests stems from the difficulty of specifying realistic null hypotheses. First, it is not clear just what the universe of content is. Quite often the researcher arbitrarily selects objects and variables and is often interested in confining his attention to only this sample. Second, the analyst is usually assuming that heterogeneity exists in the first place—otherwise why bother to cluster? Moreover, the clusters are formed from the data and not on the basis of outside criteria. Thus one would be placed in the uncomfortable statistical position of "testing" the significance between groups formed on the basis of the data themselves. Third, the distributions of the objects are largely unknown, and it could be dangerous to assume that they conformed to some tractable model like a multivariate normal distribution. (However, this assumption could provide a null hypothesis of some interest.)

In the present state of cluster analysis, this class of techniques might best be viewed as preclassification where the object is to formulate rather than test categorizations of data. After a classification has been developed and supported by theoretical research and subsequent reformulation of classes, other techniques like discriminant analysis might prove useful in the assignment of new members to groups identified on grounds that are not solely restricted to the original cluster analysis.[29]

One of the most useful aspects of cluster analysis is the simple manner in which the results can be portrayed. Figures 9.10 and 9.11 show two illustrations of the complete linkage clustering method, described earlier. Figure 9.10 shows a dendrogram of 19 words (Green and Tull, 1975). Eight of these words, designated as (S) in the figure, were used as stimulus words in a free association task. The objective of the task was to see how shampoo users reacted to various words used by advertisers to describe shampoos. In particular, interest centered on the word "body." As shown in the figure, "body" evokes two major classes of associations: one set dealing with hair appearance and texture (e.g., fullness) and

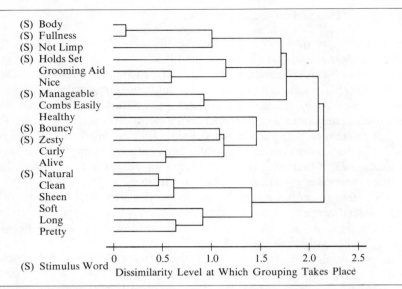

Figure 9.10 Hierarchical Clustering of Word Associations (from Green and Tull, 1975)

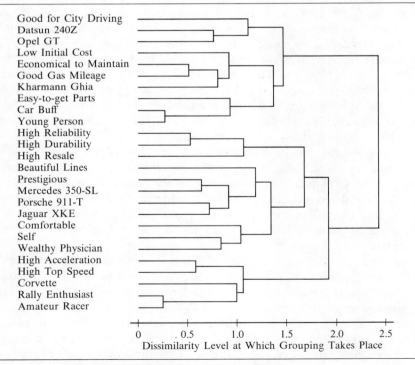

Figure 9.11 Hierarchical Clustering of Sports Car Attributes and Owner Stereotypes (from Green, 1974)

the other with hair control (e.g., holds set). This information was useful in the design of a new shampoo that emphasized the "control" aspects.

Figure 9.11 shows various associations among sports car brands, car features, and stereotyped car owners (Green, 1974). Input data to the clustering were obtained from respondents' ratings of the strength of relationship between various car features, brands, and stereotyped owners. The study illustrated how those various "images" were all interrelated.[30]

No attempt is made here to provide anything approaching an exhaustive description of applications of cluster analysis. The reader interested in pursuing the subject of cluster analysis more intensively is referred to specialized books on the subject by Sneath and Sokal (1973), Tryon and Bailey (1970), Cole (1969), Everitt (1974), and Hartigan (1975).

9.7 Summary

The purpose of this chapter has been to discuss a variety of topics in reduced space and cluster analysis. Naturally, given the limitations of a single chapter, all descriptions were more or less introductory; specialized references were provided in each section.

We first discussed factor analytic models, particularly the common factor model. This model differs both philosophically and operationally from the components model. Moreover, the common factor model represents a prototype of other models that require communalities estimation. We then described other types of factor models—canonical, maximum likelihood, image—and methods—square root, centroid, multiple group—in a much briefer way; references to more detailed discussions were included.

Other topics in factor analysis—Q-technique versus R-technique, higher order analysis, three-way factoring models, hybrid models using factor analysis—were described briefly in an effort to whet the reader's appetite for the wide variety of types and uses of factor analysis. We also summarized some of the virtues of the principal components model, whose features were emphasized in Chapter 8.

Discussion then turned to a related topic: metric and nonmetric MDS methods and their use in portraying distance-like input data in a reduced space of stimulus coordinate values. Each method was described mathematically and illustrative applications were presented.

The chapter concluded with an introductory discussion of cluster analysis. The major problems of cluster analysis were discussed and some illustrations of clustering were presented. Citations to specialized literature on the topic were also included for the reader desirous of going beyond the introductory comments of the chapter.

Review Questions

1. Using the data of Table 8.1 in Chapter 8, apply the common factor model, using SMC's in the main diagonal (no iterations) of the correlation matrix and extracting only one factor.

a. How do the loadings on the first factor compare to their counterparts in the principal components analysis of Table 8.7 in Chapter 8?

b. Find the approximation to $\hat{\mathbf{R}}_c^{(1)} = \mathbf{f}_{c1}\mathbf{f}'_{c1}$ by taking the matrix product of the first loadings vector $\mathbf{f}_{c1}$. How do the residuals $\mathbf{R}_c - \hat{\mathbf{R}}_c^{(1)}$ compare with those obtained from the rank-1 approximation to $\mathbf{R}$, based on the first vector $\mathbf{f}_1$ of loadings from the principal components analysis?

2. Consider the following matrix:

Observations	X_1	X_2	X_3	X_4
1	2	3	9	2
2	4	1	10	7
3	3	5	12	5
4	7	10	8	4
5	5	3	7	9
6	8	7	5	3
7	9	11	3	12
8	12	9	6	8
9	10	12	2	11
10	13	14	1	10

a. Standardize the data by columns and then regress X_1 on X_2, X_3 and X_4. Note the value of R^2 and the beta weights.

b. Find the principal components of the correlation matrix of predictors X_2, X_3 and X_4.

c. Regress (standardized) X_1 on the factor scores of X_2, X_3 and X_4. Compare this R^2 and the beta weights with those of part a.

3. Return to the data of question 2.

a. Standardize the columns of the 10×4 matrix and then compute squared Euclidean distances between all pairs of observations.

b. Convert the squared distances into scalar products via the method of section 9.5.

c. Find a metric multidimensional scaling of the 10 observations by computing the principal components of the matrix of scalar products.

d. Plot the coordinates of the first two dimensions.

e. If you have access to a nonmetric MDS program, scale the data this way after first transforming all squared distances to their integer ranks: 1, 2, . . . , 45.

f. Compare the two-dimensional nonmetric MDS solution to the two-dimensional metric MDS solution of part c.

4. Return to the data of question 2.

a. Choose the first six observations (only) and standardize the columns of this 6×4 matrix. Next, compute Euclidean distances between all 15 pairs of the six observations.

b. Cluster the distances via the complete linkage method.

c. Repeat the clustering with the average linkage method and compare the results.

d. Replace the distances by their integer ranks and repeat parts b and c. How do the results compare between the complete and average linkage methods?

5. Return to the data of question 2.

   a. Choose the first six observations (only) and standardize the columns of this 6×4 matrix.

   b. Next, find the (Q-type) correlations for the 6×6 matrix of observations.

   c. Convert these measures into Euclidean distances and cluster the distances via the complete linkage rule. Compare results with those of question 4.b.

   d. Repeat the clustering, this time using the single linkage rule and compare the results with those of part c, above.

Notes

1. We say "theoretical" for the reason that $\mathbf{T}$ is never really known. Rather, it represents a justification for reducing the main diagonal entries of $\mathbf{R}$ to numbers that are less than unity. Moreover, interest attaches to estimates of $\mathbf{F}_c$, the loadings for *common factors*, rather than estimates of $\mathbf{T}$.

2. Again, this is a theoretical assumption. In practice factor scores have to be *estimated* in the common factor model after the loadings matrix is computed. Unlike principal components analysis, there is no mathematical requirement that the estimated scores be uncorrelated. See Harman (1967).

3. BMD-08M also provides an option for using the largest (absolute-value) simple correlation of each variable with the $n - 1$ other variables as its initial communality.

4. While the relationships:

$$\mathbf{FF}' = \mathbf{R} \ (\text{or } \mathbf{F}_c\mathbf{F}'_c = \mathbf{R}_c)$$

$$\mathbf{F}'\mathbf{F} = \mathbf{D} \ (\text{or } \mathbf{F}'_c\mathbf{F}_c = \mathbf{D}_c)$$

hold for either model, the standardized factor scores for the common factor model *cannot* be computed in the simple and exact way:

$$\mathbf{Z}_s = \mathbf{X}_s\mathbf{U}\mathbf{D}^{-1/2}$$

described in Chapter 8's principal components model. In the case of the common factor model, regression procedures must be used to *estimate* the scores.

5. For example, the common factor model can result in reduced-form matrices $\mathbf{R}_c$ that are non-Gramian (i.e., have negative eigenvalues), a problem that does not arise in factoring $\mathbf{R}$, since $\mathbf{R}$ is of product moment form: $\mathbf{R} = \mathbf{X}'_s\mathbf{X}_s/(m - 1)$.

6. Computational details of both the centroid and multiple group methods can be found in Harman (1967).

7. The relatively recent development of *weighted* components analysis (Mulaik, 1972) should increase the flexibility of the principal components model even more. The weighted model allows the researcher to concentrate the error variance in the *lesser* components that are extracted later (and are usually dropped anyway).

8. By data matrix *centering* is meant subtraction of the mean of a given row (or column) from all entries in that row (or column). Doubly centered data, denoted X_{ij}^*, have both row *and* column means subtracted out:

$$X_{ij}^* = X_{ij} - \overline{X}_{.j} - \overline{X}_{i.} + \overline{X}_{..}$$

where we use the dot notation of Chapter 3 to indicate the index that is summed over in computing the mean of interest.

9. It should be noted that the initial standardization (by columns) will be lost when one *standardizes* by rows in the course of carrying out a Q-type factor analysis based on correlations. However, compared to no standardization at all, the effects of this upsetting of the column standardization are minuscule.

10. Without the initial columnwise standardization, interperson correlations will tend to be considerably higher, strictly as a consequence of differences in the means and standard deviations of the original variables' scales of measurement.

11. However, developing a sample of variables from a universe of variables is a more difficult problem than selecting a sample of persons from a person universe. Defining a "universe of content" can be quite a difficult conceptual and operational problem in Q-technique.

12. Another problem of Q-type factoring is that the correlations across variables are not invariant under admissible transformations of the variables. In R-type correlations reversing the scoring of a variable simply changes the sign of its correlations with other variables. In Q-type correlations a simple reversal of scoring in one or more dummy variables could change the *numerical* value of the correlation.

13. These cautionary comments are made about Q-type factor analysis largely because it continues to enjoy widespread acceptance in various applications, despite the limitations described above.

14. The "core" matrix in CANDECOMP is essentially a three-dimensional analogue of an identity matrix; hence there is no need to have it appear explicitly in the model.

15. A special variety of CANDECOMP, called INDSCAL (denoting *I*ndividual *Di*fferences *Scal*ing), has already received considerable application in multidimensional scaling. We comment briefly on this approach later in the chapter.

16. This formulation assumes that **F** is positive semi-definite; that is, all of its eigenvalues are non-negative.

17. It should be noted that any configuration obtained by (metric or nonmetric) multidimensional scaling can be freely: (a) translated; (b) rotated; (c) uniformly stretched or shrunk; or (d) reflected.

18. One particularly simple procedure (Torgerson, 1958) examines all triples of points and finds the smallest value of c satisfying:

$$c = \max_{[i,j,k]} \{\delta_{ik} - \delta_{ij} - \delta_{jk}\}$$

This constant (which is added to each δ_{ij}) guarantees satisfaction of the triangle inequality for all triples of points.

19. This form of stress is sometimes referred to as Stress-Form 1.

20. This form of stress is sometimes called Stress-Form 2. The reader will note similarities between this form of stress and the counterpart formula associated with the MONANOVA algorithm described in Chapter 5.

21. To call the axes *yellow-blue* and *green-red* is stretching the notion of "continuously varying dimensions" a bit. In many applications the labeling of dimensions is rather arbitrary. Furthermore, dimensional interpretations are frequently less appropriate than classificatory interpretations of stimulus configurations (as will be discussed later in the context of cluster analysis).

22. One could argue that distances in components space are only called for if the variables represent a *sample* from a larger domain. If all variables in the domain of interest are included, it could be argued that distance in the *original* space (of correlated axes) is more appropriate.

23. The squared Euclidean distance between *standardized* profiles i and j is related to their Q correlation by:

$$d^2_{ij(\text{cor})} = 2(1 - Q)$$

Thus, this particular type of squared distance measure varies between 0 and 4, since $-1 \le Q \le 1$.

24. Actually, however, this simple matching measure *is* a distance measure in disguise. For example, we can define it as:

$$S_{12} = \frac{1}{N}(N - d^2_{12})$$

where d^2_{12} is the squared Euclidean distance between objects 1 and 2 and $N = 6$, the number of "dimensions" (i.e., attributes). (This simple distance-like interpretation is *not* true for the S^*_{ij} measure, described next.)

25. Still other types of clustering algorithms can be found in books by Everitt (1974) and Hartigan (1975).

26. In some of the non-hierarchical methods, points are allowed to remain unclustered if they are outside of threshold distance from any cluster center.

27. In particular, clumping-type methods (Everitt, 1974) allow points to belong to more than a single cluster. This is also possible in the ADCLUS method of Shepard and Arabie (1975). Also, some methods (Hartigan, 1975) simultaneously cluster both objects and variables.

28. In actuality the program clusters *variables* rather than objects. In practice, however, one can easily use the program for either purpose.

29. Discriminant analysis, as an ad hoc device, may be used to find "optimal" weights for variables after performing the cluster analysis. In this case, however, its use would be limited strictly to description.

30. In the case of the complete linkage and single linkage rules, the clustering is *non-metric* in the sense that the same results are obtained over increasing monotonic functions of the input data. This is because complete linkage uses a maximum distance rule and single linkage uses a minimum distance rule. Neither the maximum nor the minimum of a set of numbers is affected by an order preserving transformation of those numbers.

The Alpha TV Commercial Study

A.1 Introduction

As first mentioned in Chapter 1, a data bank was assembled from an actual study involving the pretesting of four "semi-finished" television commercials for the Alpha Company. This data bank, made up from the responses of 252 male adults, has been used as a kind of "running case" throughout sections of the book in the course of demonstrating various multivariate procedures.

Here, we describe the nature of the study in somewhat more detail, provide supporting tables and other findings not covered in the chapters themselves. Finally, the substantive results of the study are recapitulated and the basic data are included to serve as a practice set, should the reader wish to replicate any of the analyses described in the book.

A.2 Study Overview

As mentioned in Chapter 1, the primary motivation for the study was to obtain consumer reactions to various semi-finished commercials regarding Alpha's brand of steel-belted radial tires. Evaluations were made in the context of commercials for three competing brands—Beta, Gamma, and Delta.

A.2.1 The Questionnaire

We first describe the questionnaire and then comment on various research questions that led to this particular survey instrument. The questionnaire is

439

shown in Figure A.1. This version represents a substantial abridgment of the one used in the real study. However, the essential characteristics of the study are retained and all data are those actually collected in the survey.

Part A (Pre-Exposure Data) Part A of the questionnaire consisted of a set of pre-exposure questions: interest in the product class of radial tires, brand of tire last purchased, and pre-exposure interest in each of the four test brands of steel-belted radials.

The first question was designed to obtain some measure of interest in the (generic) product class of radial tires, while the other two were designed to obtain information on brand predispositions for later comparison with post-exposure brand evaluations.

Part B (Conjoint Analysis Data) Part B of the questionnaire was designed to measure respondents' subjective trade-offs among various attributes of a tire purchase situation—tread mileage, brand name, tire price, and convenience of sales outlet. Each of these four factors was varied across five levels according to a graeco-latin square design. (These data were analyzed via MONANOVA, a type of nonmetric analysis of variance program that was described in Chapter 5.)

Part C (Post-Exposure, Single-Stimulus Rating Data) Part C provided two response ratings—believability of product claims and purchase interest—for each of the four advertised brands, as considered separately, after exposure to its particular test commercial.

Notice here that the evaluative ratings are made in terms of a *single-stimulus presentation.*

Part D (Post-Exposure, Stimulus-Comparison Data) In Part D the respondent was forced to choose among the test brands (including the choice of none of them), after having seen all four test commercials. This part of the study was of most interest to the Alpha Co. since *stimulus-comparison* data are involved. In particular, Alpha wanted to know what share of choices would be garnered by its brand and how the choice of its brand might be related to other responses in the questionnaire.

Part E (Demographics) Part E provided a short list of background variables—age, marital status, and so on—that were expected to be helpful in any detailed examination of respondents who were most attracted to Alpha's brand.

A.2.2 Limitations of Study

We have already commented on the fact that the questionnaire in Figure A.1 represents an abridgment of the actual questionnaire used in the study, one that nevertheless retains those aspects of most interest to the application of

Figure A.1 Questionnaire: Alpha TV Commercial Study

Respondent Name: _____

Address: _____ (Street)   Home Telephone: _____

_____ (City)   Date: _____

_____ (State)   Validated by: _____

Interviewer: _____   Validation Date: _____

Part A (Pre-Exposure Data)

1. If you were going to buy a replacement automobile tire today, what type of tire (not brand) would you purchase?

Steel belted radial ply _____ ⎫
Fiber glass radial ply _____ ⎬ (Coded 1)
Radial (not belted) _____ ⎭
Regular bias ply _____ ⎫
Fiber glass bias ply _____ ⎪
Other (please specify) _____ ⎬ (Coded 0)
Don't know _____ ⎭

2. Thinking about all the cars that you own in your household, what brand of replacement tire did you *last* buy?

Beta Co. _____   Other U.S. Producers _____
Gamma Co. _____   Foreign Producers _____
Delta Co. _____   Don't know _____
Alpha Co. _____

5. Imagine you were going to buy a steel belted radial tire today. (READ SHORT DESCRIPTION OF CHARACTERISTICS OF STEEL BELTED RADIAL TIRE.) Which number on this card best describes how interested you think you would be in buying a _____ (INSERT BRAND NAME ACCORDING TO RANDOMIZED ORDER) tire? (REPEAT FOR EACH BRAND LISTED BELOW.)

Brand	Not at all Interested ↓↓ 0	1	2	3	4	Rating 5	6	7	8	9	Extremely Interested ↓↓ 10
Beta Co.											
Gamma Co.											
Delta Co.											
Alpha Co.											

Part B (Conjoint Scaling Data)

Here is a group of 25 cards (SHOW RESPONDENT SET OF 25 BLUE CARDS). On each card is a description of various aspects involving the purchase of a steel belted radial tire: (a) advertised mileage; (b) brand name; (c) price per tire; and (d) average driving time from home to a store that sells the brand. You will notice that each card differs regarding the combination of characteristics that are offered.

For example there are 5 different levels of tread mileage—30,000; 40,000; 50,000; 60,000; and 70,000 miles. Five kinds of brands—Alpha; Beta; Delta; Gamma; and Epsilon*. Five prices—$40; $55; $70; $85 and $100 per tire. Five driving times from home to store—10; 20; 30; 40; and 50 minutes.

---

*Epsilon is a tire brand sold through station outlets by one of the major gasoline producers.   (continued)

Figure A.1 Continued

Please look through the cards and sort them into five piles (PLACE SORTER IN FRONT OF RESPONDENT) in terms of how likely you would be to buy each of these combinations:

Highly likely to buy	(Coded 5)
Somewhat likely to buy	(Coded 4)
May or may not buy	(Coded 3)
Somewhat unlikely to buy	(Coded 2)
Highly unlikely to buy	(Coded 1)

There is no need to place an equal number of cards in each pile but do try to use each category of the scale. (ALLOW TIME FOR SORT). Now that you have sorted all 25 cards, please look through each pile to satisfy yourself that the placement of the cards agrees with your judgments about the relative attractiveness of each combination. Otherwise, please feel free to shift cards across piles until you're satisfied with the final grouping.

Part C (Television Commercial Exposure)

Now we'd like to show you a commercial for _____ (INSERT BRAND NAME) steel-belted radial tires. This commercial is a test item and, hence, will seem different from what you're used to seeing on TV. Don't worry about this. Instead, just follow what the commercial is trying to convey. What we'd like you to do is form some impression of it, as you might imagine watching it on the air. (PLAY COMMER-CIAL _____ ACCORDING TO RANDOMIZED ORDER FOR THIS RESPONDENT.)

1. Now that you have seen the commercial, which number on this card best describes how *believable* the things they say about this brand are to you?

Not at all Believable ↓↓ ... Rating ... Extremely Believable ↓↓

0	1	2	3	4	5	6	7	8	9	10

2. Please look at this card and tell me which number best describes how *interested* you think you would be in buying the brand described in this commercial.

Not at all Interested ↓↓ ... Rating ... Extremely Interested ↓↓

0	1	2	3	4	5	6	7	8	9	10

(Repeat procedure for remaining three commercials according to specified presentation order for this respondent.)

Part D (Commercial Comparison)

Now that you have seen all four test commercials, try to think about the various brands advertised in them.

(continued)

Figure A.1 Continued

If you were going to buy steel-belted radial tires today, which, if any, of the four brands that were advertised in the commercials would you buy? Please check one.

Beta Co. _____
Gamma Co. _____
Delta Co. _____
Alpha Co. _____
None of these _____

Part E (Demographics)

Now we'd like to obtain a little background information about you.

1. What is your age? _____ Years

2. Including yourself, how many family members live at home?

_____ Number

3. Which of the categories listed on this card comes closest to the last grade that you completed?

_____ Some grammar school (06)   _____ Some college (14)
_____ Completed grammar school (08)   _____ Completed college (16)
_____ Some high school (10)   _____ Graduate studies or
_____ Completed high school (12)   advanced degree (18)

4. Marital Status? _____ Married (1)   _____ Single (0)

5. What is your occupation? _____

6. Please look at this card and tell me which one of these categories best describes your total household income, before taxes?

_____ Income (Code Letter)

multivariate techniques. Also, it was mentioned in Chapter 1 that the sample was not chosen by probability selection procedures. (This limitation appears to be a fact of life in most marketing research studies, despite the emphasis given by textbooks on the virtues of probability sampling procedures.)

From the point of view of design, perhaps the most serious flaw in the survey is that a single Alpha commercial is under evaluation, albeit in the context of three competing commercials. Strange as it may seem, the practice of testing only a single commercial is quite prevalent in advertising research. In this case, acceptance of the test commercial for national television airing would have to be based largely on the sponsor's judgment of how well it performed against competing commercials, relative to its share of pre-exposure interest.

Other limitations of the study have to do with more or less standard problems associated with the validity and reliability of survey-type studies. One can always question whether response to a single presentation of a commercial in an artificial test situation, even if shown in the context of other commercials, is indicative of reliable and valid attitude shifts in the context of real-world television viewing. Other problems, such as the cumulative impact of the commercial, the relationship of attitude shifts to purchase intentions, and the influence of point-of-purchase variables on brand choice, all impinge on the predictive accuracy of the study.

While these problems are not inconsiderable, from the sponsor's viewpoint, interest centered on the relative attractiveness of the sponsor's commercial, as viewed in a competitive framework. We can only conclude, then, that this type of performance is sufficient to decide whether to adopt the new commercial or not. (Given the specific situation described in Chapter 1, failure to adopt the new commercial would mean the preparation and testing of new alternatives.)

A.2.3 Research Questions

From Alpha's point of view several research questions were raised that were relevant to the firm's marketing strategy:

1. How is choice of Alpha's brand (in Part D of the questionnaire) associated with:
   a. Alpha's post-exposure ratings (Part C), on the believability of product claims and brand purchase interest?
   b. Alpha's pre-exposure data (Part A), on product class interest, brand last purchased, and interest in Alpha radials?
   c. Respondent demographic data (Part E)?

2. How are Alpha's ratings on purchase interest of the Alpha brand, after exposure to the commercial (C-2), related to:
   a. Alpha's pre-exposure brand rating, brand last purchased, and product class interest?
   b. Demographic data?

3. What is the brand representation (based on brand selected in Part D) in clusters of respondents who exhibit similar responses in the cost-benefit analysis of Part B? How are these respondent clusters described in terms of relative utilities for the various factors—tread mileage, brand, price and outlet convenience—used in making up the stimulus combinations?

While other questions of managerial interest were raised as well, the ones above provided the major rationale for the study.

A.3 Marginal Tabulations and Ancillary Analyses

In addition to the various multivariate analyses described throughout the book, various summary tabulations were made of the questionnaire responses, as well as some supplementary analyses of "brand-switching" responses.

A.3.1 Marginal Tabulations

In almost any market survey it is usually a routine matter to construct marginal tabulations of the data. Table A.1 shows marginal summaries for Parts A, B, and C of the questionnaire, dealing with pre- and post-exposure responses to the four test commercials.

Also shown in the table are the frequencies with which each brand is selected in the comparative choice situation of Part D. Finally, marginal tabulations of the demographic data are also shown. We note from Table A.1 that respondent interest in the product class of radials (A-1) is somewhat over 50 per cent. Insofar as the Alpha brand is concerned, about 19 per cent of the sample chose Alpha on their last replacement tire purchase. Moreover, Alpha ranks second (after Delta) in interest and believability scores in questions A-3, C-1, and C-2. Alpha's rating increase in pre- versus post-exposure purchase interest was $6.1 - 5.5 = 0.6$ of a rating point. This is the largest rating increase in the total sample.

Part D shows that Alpha's share of choices, after the respondents had seen all four radial tire commercials, is 31 per cent, a result that we comment on in more detail later. The demographics indicate that the sample contained a disproportionate percentage of young people (18–24 years in age) of higher socioeconomic class than the population at large. However, the sample demographics did come fairly close to those of the sponsor's customer group.

A.3.2 Brand-Switching Matrices

Of additional interest to the study was the question of how respondents might change their preferences for various brands of radial tires after being exposed to the test commercials. One can estimate each respondent's *pre-exposure* choice by assuming that he would select the radial tire brand that he rated the highest on purchase interest (Question A-3). The respondent's *post-exposure* choice can be obtained from Part D, where the respondent is asked to choose one brand after seeing all four commercials. Table A.2 shows the resulting brand switching matrix (excluding the 12 respondents who chose none of the four brands in Part D).

In addition to this summary switching matrix, individual rating-category switching matrices can be constructed—again on a pre- versus post-exposure basis—for each of the four test brands. This can be done by noting how each respondent changes his pre-exposure interest rating on each brand after exposure to that particular test commercial. Table A.3 shows these individual-brand matrices.

Brand Choice Behavior An examination of Table A.2 shows that before viewing the test commercials, 30 per cent of the respondent group choose Alpha as their first choice, as inferred from responses to A-3. Only Delta, with 40 per cent of the choices, ranks higher.

After having seen all four commercials, we find that 32.5 per cent of the respondents choose Alpha, a gain of 2.5 percentage points. Delta also registers a gain of 2.5 percentage points, while Beta and Gamma lose 3 and 2 percentage points, respectively. We tentatively conclude that the Alpha commercial appears to do quite well, relatively, in terms of post-exposure choice. Only the Delta commercial does as well in terms of share point change.

Table A.1 Marginal Tabulations of Parts A, C, D and E of the Questionnaire

A-1 Type of Replacement Tire
Respondent Would Plan to Purchase

	Frequency	Per Cent
Radial	146	(58)
Other	106	(42)
	252	(100)

A-2 Brand of Replacement Tire
Last Bought

	Frequency	Per Cent
Beta	26	(10)
Gamma	8	(3)
Delta	61	(24)
Alpha	48	(19)
Other U.S.	80	(32)
Foreign	10	(4)
Don't know	19	(8)
	252	(100)

A-3 Pre-Exposure Purchase Interest
in the Four Test Brands

	Mean Rating
Beta	5.2
Gamma	5.1
Delta	6.3
Alpha	5.5

C-1 Post-Exposure Belief in
Test Brand Claims

	Mean Rating
Beta	6.1
Gamma	6.0
Delta	7.2
Alpha	6.8

C-2 Post-Exposure Purchase Interest
in Test Brands

	Mean Rating
Beta	5.5
Gamma	4.5
Delta	6.6
Alpha	6.1

D First Choice Among the Brands

	Frequency	Per Cent
Beta	36	(14)
Gamma	24	(10)
Delta	102	(40)
Alpha	78	(31)
None	12	(5)
	252	(100)

Demographics

E-1 Age

Years	Frequency	Per Cent
18–24	88	(35)
25–31	48	(19)
32–38	33	(13)
39–45	33	(13)
46–52	18	(7)
53–64	32	(13)
	252	(100)

E-2 Number of Family Members
Living at Home

	Frequency	Per Cent
One	35	(14)
Two	48	(19)
Three	50	(20)
Four	61	(24)
Five	26	(10)
Six	16	(6)
Seven	9	(4)
Eight or more	7	(3)
	252	(100)

Table A.1 Continued

E-3 Education	Frequency	Per Cent
Some grammar school	3	(1)
Completed grammar school	3	(1)
Some high school	22	(9)
Completed high school	61	(24)
Some college	90	(36)
Completed college	41	(16)
Advanced	32	(13)
	252	(100)

E-4 Marital Status	Frequency	Per Cent
Married	143	(57)
Single	109	(43)
	252	(100)

E-5 Occupation	Frequency	Per Cent
Professional & White Collar	112	(44)
Blue Collar	98	(39)
Other	42	(17)
	252	(100)

E-6 Income	Frequency	Per Cent
Less than $5,000	15	(6)
$5,000–$ 9,999	40	(16)
$10,000–$13,999	57	(23)
$14,000–$17,999	42	(17)
$18,000–$19,999	21	(8)
$20,000 or over	77	(30)
	252	(100)

Table A.2 Switching Matrix Based on Part A-3 Versus Part D Responses

Derived Pre-Exposure Choice** from Question A-3	Stated First Choice* in Part D					Per Cent
	Beta	Gamma	Delta	Alpha	Total	
Beta	19	1	12	11	43	(18)
Gamma	3	8	11	7	29	(12)
Delta	4	9	66	17	96	(40)
Alpha	10	16	13	43	72	(30)
Total	36	24	102	78	240	
Per Cent	(15)	(10)	(42.5)	(32.5)		(100)

*Excludes all respondents (in A-3 and D) who responded "None" in Part D.

**First choice obtained from highest rating in A-3. In cases of tied highest ratings, ties were resolved on the basis of either: (a) last brand purchased (A-2) or (b) a random number draw in the cases where none of the above brands was purchased last.

Table A.3 Single-Brand Switching Matrices Based on A-3 Versus C-2 Responses

BETA

A-3 (Before)	C-2 (After)			Total	Per Cent
	0–2	3–7	8–10		
0–2	32	22	5	59	(23)
3–7	16	74	32	122	(49)
8–10	4	29	38	71	(28)
Total	52	125	75	252	
Per Cent	(21)	(50)	(29)		(100)

GAMMA

A-3 (Before)	C-2 (After)			Total	Per Cent
	0–2	3–7	8–10		
0–2	38	14	7	59	(23)
3–7	40	69	17	126	(50)
8–10	10	26	31	67	(27)
Total	88	109	55	252	
Per Cent	(35)	(43)	(22)		(100)

DELTA

A-3 (Before)	C-2 (After)			Total	Per Cent
	0–2	3–7	8–10		
0–2	19	15	5	39	(15)
3–7	8	52	31	91	(36)
8–10	5	29	88	122	(49)
Total	32	96	124	252	
Per Cent	(13)	(38)	(49)		(100)

ALPHA

A-3 (Before)	C-2 (After)			Total	Per Cent
	0–2	3–7	8–10		
0–2	24	30	6	60	(24)
3–7	7	60	39	106	(42)
8–10	6	24	56	86	(34)
Total	37	114	101	252	
Per Cent	(15)	(45)	(40)		(100)

Individual Switching Matrices   Alpha's performance also looks good in terms of the individual brand switching matrices of Table A.3. Each matrix was constructed by first tabulating the frequency of pre-exposure ratings (A-3) of a specific brand into one of three rating categories: 0–2, 3–7, or 8–10, poor, average, and excellent. Next, a tabulation was made, by specific brand, of the frequency with which respondents rated it 0–2, 3–7, or 8–10 on a post-exposure basis. Interest then centered on the switching that took place among categories, particularly switches that involved the "excellent" (8–10) category.

Table A.3 shows that Alpha fares well in this part of the analysis. The share of choices described as excellent (ratings of 8–10) on a pre-exposure basis is 34 per cent. This increases to 40 per cent on a post-exposure basis, a gain of 6 percentage points. On the other hand, Delta stays the same, Beta gains one point, while Gamma actually loses 5 points.

We conclude from all of this that the Alpha commercial appears to do well in terms of changes in brand attitudes—both on a single-stimulus rating basis and in terms of the Part D brand comparisons.

A.4 Multivariate Analysis Summary

Throughout the book the TV commercial testing data has served as a vehicle for describing various applications of multivariate tools. Here, a number of these findings are recapitulated from a substantive (more so than methodological) point of view. No exhaustive summary of the analyses shown earlier is attempted. Rather, we focus on those aspects that deal with content implications for the study.

A.4.1 Predicting Post-Exposure Interest in Alpha Radials

A series of analyses were made in which post-exposure interest in Alpha radials (response to C-2) served as a criterion variable. When C-2 responses were regressed on pre-exposure responses (A-1, A-2, and A-3) and the demographics in Chapter 2, we found the following results:

1. Pre-exposure interest in Alpha radials (A-3) is the predictor variable of most importance, followed by predictor variable A-1, interest in the product class of radial tires.

2. Somewhat surprisingly, predictor A-2 (whether Alpha was selected as last brand purchased) was not an important contributor to accounted-for variance in post-exposure interest ratings on Alpha. Apparently, past purchases of the Alpha brand (mostly consisting of non-radials, in view of the newness of the product) do not lead to the development of "carry-over" loyalty to the Alpha radial tire.

3. The demographics contribute virtually nothing to accounted-for variance in post-exposure ratings on purchase interest.

When a second criterion variable: belief in the product claims of Alpha (C-1), was introduced (in addition to C-2), the resulting canonical correlation in

Chapter 6 indicated pretty much the same thing. That is, the same predictors, A-3 and A-1, continued to remain important in accounting for variation in the linear composite of C-1 and C-2. Not surprisingly, C-1, belief in the commercial, was rather highly correlated with the major criterion variable, C-2, post-exposure interest in Alpha radials.

Next, a subset of the data was set up as a balanced research design, using data for only 68 out of the total 252 respondents. In this case C-2 represented the response variable and A-3 and A-2 served as "treatments." An ANOVA of these responses in Chapter 3 supported the effect of A-3 but not A-2 on post-exposure interest. And, again, addition of the demographics as covariates in an accompanying ANCOVA failed to account for any significant response variation in post-exposure interest in Alpha radials.

Finally, post-exposure belief in the product claims of Alpha radials was introduced as an additional criterion variable, with A-3 and A-2 continuing to serve as treatment variables. The resulting MANCOVA in Chapter 7 supported earlier findings: A-3 continued to show significant response differences, while A-2 did not. Inclusion of the demographics in a MANCOVA failed to account for any additional response variation.

A.4.2 Predicting Choice of Alpha Radials in a Forced-Choice Comparison

Respondents choosing the Alpha brand of radials versus those who did not were compared in a two-group discriminant analysis (Chapter 4) and also by means of AID (Chapter 5). The two-group discriminant analysis indicated that Alpha selectors in Part D were characterized by high pre-exposure interest in Alpha radials (A-3). The remaining predictors, A-2, C-1 and C-2, also contributed to the discrimination. The discriminant function correctly classified 73 per cent of the cases, a performance that is not particularly outstanding.

The AID analysis indicated that Alpha selectors in Part D displayed high ratings on the Alpha product claims (C-1), and also tended to choose Alpha (A-2) on their last replacement tire purchase. Non-selectors of Alpha in Part D displayed low C-1, C-2, and A-3 ratings on Alpha and tended not to choose Alpha (A-2) on their last replacement tire purchase.

A.4.3 Respondents' Views Regarding Cost-Benefit Profiles

In Part B of the questionnaire each respondent was shown 25 cards. On each card appeared various combinations of: (a) advertised mileage; (b) brand name; (c) price per tire; and (d) average driving time from home to retail store. The objective of this part of the study was to see if respondents with different first-choice brands in Part D also differed systematically in terms of the cost-benefit profiles being sought.

A four-group discriminant analysis in Chapter 7 indicated that inter-group differences were quite minor. While some small effect of brand name was evinced, in the main the average cost-benefit profiles among the four brand groups did not differ. The discriminant function managed to classify the brand chosen in Part D correctly in only 55 per cent of the cases, using the cost-benefit profiles in Part B as predictor variables.

A clustering routine was also used to segment respondents into more or less homogeneous groups, as obtained from the Part B data alone. However, the clusters that emerged from this analysis also displayed quite similar utility functions. In particular, no relationship was found with brand selected in Part D. The two most important factors (in all cases) were advertised tread mileage and price. Brand name and driving time played minor roles.

A.5 Summary and Implications

Not surprisingly, the major recommendation of this study was to implement the Alpha commercial. The brand-switching data indicated that Alpha's test commercial fared well within the competitive setting. Good performance was indicated by analyses at both the brand-comparison level and the single-stimulus rating level, involving pre-exposure versus post-exposure interest in Alpha radials. In each case Alpha fared as well or better than competing commercials.

The analysis of the cost-benefits data reinforced the sponsor's prior belief that tread mileage represented a highly important consumer "benefit" to promote. (The present Alpha test commercial played up this theme.) On the other hand, the finding that past purchases of Alpha tires exhibited little correlation with pre-exposure or post-exposure interest in Alpha radials was surprising to the sponsor. Based on an earlier analysis of customer complaint data, indicating relatively low complaints for Alpha purchasers compared to industry averages, the sponsor elected to believe that the introduction of the radial tire was a sufficiently major event to upset previously established brand loyalty patterns. Accordingly, a new program was initiated for building up, and monitoring, repeat purchasing of Alpha *radial* tires.

The finding that pre-exposure interest in Alpha radials represented a significant predictor variable in accounting for post-exposure interest and brand choice, was received with high interest by the sponsor. Apparently this type of data had not been collected routinely in past commercial effectiveness tests. Instead, the sponsor had relied almost exclusively on reports of brand last purchased. Accordingly, the sponsor decided to incorporate pre-exposure questions in future studies of commercial effectiveness.

Finally, the lack of association of demographics with the various criterion variables of interest was received as a mixed blessing. On the one hand, the lack of demographic segmentation obviated the need for tailoring promotional themes, or promotional media, to specific audiences. On the other hand, no opportunity existed for devising special products or promotions for selected demographic groups. It was decided, then, to direct future research to product benefits sought and consumer life styles as possible bases for segmentation.

The new Alpha commercial was, indeed, implemented and ran as the company's primary commercial for over two years. The new commercial was believed by the sponsor to be highly successful in helping the Alpha brand increase its share of market.

Card Columns

1–3	Respondent identification number (001, 002, . . . , 252)		
4	Blank		
5	A-1:	Radial ⇒ 1, otherwise 0	
6–11	A-2:	Beta	100000
		Gamma	010000
		Delta	001000
		Alpha	000100 (dummy codes)
		Other U.S. Producers	000010
		Foreign Producers	000001
		Don't Know	000000
12	Blank		

13–20	A-3:	13–14 Beta Rating
		15–16 Gamma Rating
		17–18 Delta Rating
		19–20 Alpha Rating
		(range from 0 to 10)

21–45   Part B: Integer Values Range from 1 (Least Likely to Buy) to 5 (Most Likely to Buy) for Stimulus Code Numbers 1, 2, . . . , 25.

46–61   C-1:   46–47 Beta Rating   C-2: 48–49 Beta Rating
                 50–51 Gamma Rating   52–53 Gamma Rating
                 54–55 Delta Rating   56–57 Delta Rating
                 58–59 Alpha Rating   60–61 Alpha Rating
                 (range from 0 to 10)   (range from 0 to 10)

62–65	D:	Beta	1000
		Gamma	0100
		Delta	0010 (dummy codes)
		Alpha	0001
		none	0000
66–67	E-1:	Age (in years)	
68	E-2:	Family Size (integer)	
69–70	E-3:	Education (in number of years)	
71	E-4:	Marital Status: Married ⇒ 1, otherwise 0	
72–73	E-5:	Occupation:	

                 Professional and White Collar   10 ⎱
                 Blue Collar   01 ⎬ (dummy codes)
                 Other   00 ⎰

74–76   E-6:   Income in $00's (according to nearest of nine categories)

Figure A.2 Coding Form: Alpha T.V. Commercial Test

A.6 Coding Form and Data Bank

Figures A.2 and A.3 present information for the reader desirous of using the data bank (appearing in Figure A.4) for various numerical exercises in multivariate analyses. All data have been coded so as to appear on a single card for each of the 252 respondents in the study.

Figure A.2 shows the coding format followed by the data of Figure A.4. Dummy-variable coding is used in various places (e.g., in card columns 6–11). If the reader reviews the questionnaire of Figure A.1, he should have no difficulty in following the coding format.

Insofar as Part B of the questionnaire is concerned, Figure A.3 shows the experimental design underlying the construction of the 25 cost-benefit cards. The description of each card, whose code number appears in the upper left-hand corner of each cell, is summarized for easy reference to the response data (card columns 21–45 in Figure A.2).

Coding of the demographic variables (in card columns 66–76) was straightforward. Education (E-3) was coded in numerical terms—years of education—according to the coding summarized in the questionnaire of Figure A.1. Income was also coded numerically in terms of nine categories (expressed in card columns 74–76, in hundred of dollars) as follows: $2,500; $6,000; $8,500; $11,000; $12,500; $15,000; $17,000; $19,000; and $22,000. All other codes are summarized in Figure A.2.

Tread Mileage	Brand Name				
	α	β	γ	δ	ϵ
30,000 miles	1 $40 10 minutes	4 $85 40 minutes	3 $70 30 minutes	2 $55 20 minutes	5 $100 50 minutes
40,000 miles	8 $85 30 minutes	6 $55 10 minutes	10 $40 50 minutes	9 $100 40 minutes	7 $70 20 minutes
50,000 miles	15 $55 50 minutes	13 $100 30 minutes	12 $85 20 minutes	11 $70 10 minutes	14 $40 40 minutes
60,000 miles	17 $100 20 minutes	20 $70 50 minutes	19 $55 40 minutes	18 $40 30 minutes	16 $85 10 minutes
70,000 miles	24 $70 40 minutes	22 $40 20 minutes	21 $100 10 minutes	25 $85 50 minutes	23 $55 30 minutes

Numbers in boxes denote card numbers used in Part B rating task:
1, 2, . . ., 25.

Figure A.3 Graeco-Latin Square Design Used in Part B of Questionnaire

```
 1 1000010    0  1  5 8423215231534155225544 5554 8 4  6  1 8 9 8 8100040412110190
 2 0000010    0  0  0 0551114321532155115531 5522 6 0  0  6 0 6 0000157212110220
 3 1000010 101010103533335553544355455543 5554 8 8  7  7 8 7 8 9000135116010220
 4 1000010    2  4  8 7321113132433243114311 4312 4 7  9  9 8 8 4 0010054314110220
 5 0001000    6  6  6 6441112111411124114211 4411 0 4  4  4 8 8 6 4001043514101110
 6 1001000 51010 7321113221142252335433 4344 2 2  8  8 8 8 8 7010019614000220
 7 0000010    7  0  7 9322112111211143115431 4221 8 5  5  0 9 710 8000038518110220
 8 0000010    6  9  8 5111114111411144215521 5521 5 5  5  8 7 5 5 5010042314110220
 9 0001000    6  6  6 6341125431452135214351 5523 0 5  8  2 8 9 8 8001021212010 25
10 1000100    8  9 71054221432235323422542 25432 5 4  8  8 6 5 7 7000019514000190
11 1000010    4  2  4 0432213221232132221432 15432 8 7  4  4 8 8 7 7001018414000110
12 1000001    5  0 0104111141114111431144115533 9 8  5  3 8 8 9 9000148318110190
13 1001000    4 410 4541115111111113311531 15422 0 5  1  0 7 7 5 5001030118010 25
14 0000000    0  0  5 5231111122111111111111 1111 8 5  5  0 8 5 5 5000028212110 85
15 1100001    9  9  9 5453425422343243325422 5532 8 8  9 610 9 8 9001020214000220
16 1000001    0  0  0 8442113232233222343225 4332 2 0  9  5 6 7 8 8000118514000220
17 1000100 71010 8334334333355434554545 5545 8 9  6  6 7 8101000012171400025
18 1000010    8  8  8 9421112111311131113311 4411 3 410 6 9 610 6100023216110170
19 1000100    3  5  3 8322224221422324322224 2422 8 3  7  6 8 7 7 8000128114010150
20 0001000    6  5  8 6321112111321122115431 5432 7 7  5  5 8 8 4 3001045318110110
21 1001000    6  4  8 8333324532452244435232 2424 9 6  6  4 8 8 8 6001048116001110
22 1000100    6  7 71023111111131311131521 15521 7 8  7  6 9 9 910000124212010110
23 0000010    0  6 410111111111423114111111 1114 4 2  8  7 8 6 8 8000133316110220
24 1000010    6  6  6 6443315441443145315541 5544 61010 610 6 8 9100047516110220
25 1000010    6  4  8 5221114111331143115511 5431 6 710 6 9 9 7 8001020414001220
26 1000010    0  6  6 0542114211511154115521 1111 8 1  8  7 8 6 8 1001057212101110
27 1000010    0  5 010111112222534225344 4211 1111 6 6  4  2 4 2 2 2100056218110190
28 0001000    0  2  6 6542134223421154325423 5443 4 2  8  0 8 8 8 6000137412110220
29 1001000    3  3  3 5422212111121221243112 3311 8 7  8  7 9 9 8 7001018514000220
30 1010000 010  2 0135114312521113215535 4541 2 010 9 8 6 8 4010021414110110
31 0001000    8  6  5 3322213221513153114411 5343 6 8  5  4 7 8 7 8100019712001150
32 1000100 610  8 6411114321222232413421 3422 8 8  8  4 8 8 8 8001027216100 25
33 1000001    8  8  9 8111112222233333444445 5555 8 8  7  8 8 9 9 9001023314001220
34 1000010    0  4  6 7441114311342152315511 5531 8 6  9  9 8 9 8 6001025416101125
35 1001000    8  6  8 8311113222322234224322 5422 6 6  0  8 9 7 9001023114010 60
36 1100000 10  6  6 6222215112431312123454 5123 9 9  4  4 8 8 8 7100030218110220
37 0001000    6  8  8 8524225442555355525442 3534 6 7  4  2 7 8 7 5001020512001150
38 0000100    4 610 2452235111434354311232 5431 8 7  5  5 8 8 7 4100018414000190
39 1001000 610  8 2224543533354435155553 1514 7 8  8 81010 7 7001018112001 60
40 1001000    4  2  8 6111113111221123115211 4312 4 6  2  2 7 7 7 7001022214101170
41 1000100    4  2  0 6451454431114144141141 1141 8 8  6  2 2 4 4100042512101150
42 1000010    6  6  8 8212111111121121112111 3111 9 7  9  7 8 7 8 7001019714001150
43 0000000    0  8  0 0111111111151111111551 1111 0 0  0  0 0 0 0 000001841000120
44 0000000    0  0  2 0541114111511154115411 5411 0 0  2  0 6 0 0 0001024411400 60
45 0000000    5  5  5 5411114111411144115511 5511 4 4  2  4 4 4 4 4000128114010125
46 1010000    4 810 1114521111544413411151 511210 8 7  9  7 8 8 6010021412001170
47 0001000 101010 8341114211541233135444 3421 1 3  9  9 3 3 9 9000133118110220
48 0000010    5  4  6 6541114111511151115111 5141 6 6  8  7 8 7 6001020714000220
49 0001000    6  6  8 5544114321522343325543 5422 3 4  8  8 7 8 7 8010018412000220
50 1000100    4  5  5 7322114531453255315433 5433 6 4  8  8 6 7 8 9010050418110190
51 1000010    8 310 3453314333443231322244 5145 9 9  3 110 9 9 9001023316000220
52 0000010    6  4  8 8453334343444334555545 4544 8 7  8  8 8 8 7 6010021912010125
53 0100010    8  4  6 2134225213533154115325 5531 8 6  6  2 8 8 7 7100023212101125
54 0000001    5  6  5 2211114321323242415433 5544 2 1  910 8 9 6 5010024814001190
55 1001000    4  3  9 2111111111124114311521 14111 7 7  4  3 910 6 3001024214001 85
56 0001000    6  6  8 8222224222222242225222 5222 6 6  6  6 8 8 8000132412110110
57 0100010    8  6  6 2242215321442153215252 5412 6 6  6  4 6 4 6100035514110220
58 1001000    0  0  8 8451112311341144415211 2535 7 310 0 9 8 9 8000125316101150
59 0000010    4  4 010512212411232142312334 4253 8 8  8  3 8 4 9900014241610125
60 0001000    6  2  8 2452525245252524245252 5245 6 6  2  2 6 6 6001025414101125
61 0010000 510  5 5434121232524123234414 1234 5 0  9  8 8 7 7 5010063216100150
62 1001000    4  4 410411115111411115511551 1554 8 8  9  6 8 6 910000144318110220
63 0000010    7  7  5 8222113221322143322321 5432 8 7  8  7 6 6 7 7000125112001190
64 1000010 91010 9432225222532354435543554310101010101010001036514110150
65 1000010    6  7  9 7111111111112141214511 5433 8 7  8  2 9 9 8 8001018714000220
66 1000001    2  2  2 0122112411113211411233 1411 2 1  0  0 2 0 0 0000023314000 60
67 1000100 10  9  8 8145353514245521511254 1111 8 6  8  6 8 7 8 8000164312101 25
68 0100010    6  6  6 6311324322433154435544 5542 2 4  4  4 6 6 2001029114001150
69 1000001    6 410 7344212221433144415213 4413 8 7  2  2 8 810 8000121314000190
70 1000100 101010 1431111111421142115311 5542 8 610 7 9 9 3 3001029318110220
```

Figure A.4 Basic Data Listing

```
 71 1100000   8  7  9105311133314332442144314432  6  5  8  6  8  7  9  9000142316110190
 72 1000100   0  7  9  94441334422432224242243333  8  8  6  5  8  8  8  7001036414110170
 73 0001000   6 610  34411142224411442252425422  5  6  7  3 810  5  6001036518110220
 74 0000100   0  0  0  82111111113111311133114211  1  0  1  0  4  4  0  1000030418110110
 75 1000010   6  6  6  52421151323443534332214425  5  6  8  7  6  7  7  7010021112001 25
 76 0000010  610  6  63333333333133313114115414  4 610  8  4  2  6  4010023114010 60
 77 0000010   6  1  8  01211121112211212152214211  6  6  1  1  8  8  2  0001041416110190
 78 1001000   4  4  4  43531122124321422154315421  5  3  3  3  8  8  4  3001040616110190
 79 1001000   3  6  3  75421121114223341154314421  5  4  4  2  8  7  8  6001045418110125
 80 0001000   6  8  9  31443243144542512155435532  5  9  6  9  8  9  7  4001024514101 60
 81 1010000   7  8  7  64111114414214221142154441  8  7  9  9  8  9  8100021414000 25
 82 1000010   0 010  23133333333333313454334333  8 610  91010  8  6001037412110220
 83 1000100   6  6  61054111111141411551144114511  6 610  6  6  61010000127312110 85
 84 1000010   7  5  6  64311122115231441254213533  3  5  9  7  6  6  8000144414110220
 85 0100000  10  6  2  65221252242222522242415422  8  8  8  6  8  6  8  8100045312110125
 86 0001000   6  9  4  91111151114511521154115551  8  8  8  8  81010000132214110220
 87 0000010   0  4  6  23111112214221552353111243  6  0  6  4  6  6  8  6001025114010 60
 88 1001000   6 610  64541323112513523254424513  8  8  8  6 810  8  8001026212010125
 89 0100000   2  2  2  24311121112112211221122114311  2  4  6  2  1  2  6  4000119612010190
 90 1000010   9  9101024544444445542445225522455  8 810 81010  9  9001026310101 60
 91 0100000   8  4  8  01534352132433125153415213  7  6  2  2  8  5  6001018414001220
 92 0001000   8  8  8  63111131123322432254225522  9  9  6  9  9  7  6001032314010125
 93 0000100   0  0  0  03433343343333333334333333  2 01010  2  0  6  00100421 8101170
 94 0001000   4  4  9  52211131114211442153215433  8  6  8  6  8  8  8  7001029316101190
 95 0000000   2  9  9  03333342332333433351345434  0  2  6  8  9  1  00102014140000110
 96 1000010   8  6  8  44433143313434343344533433  6  6  2  4  8  6  8001019312000 85
 97 1000010   8  8  0  23311142214331523115335421  9  9  8  9  6  7  810100002131400 0125
 98 1001000   4  4  8  01111242111333512142215414  4  6  8  4  6  4  2001034312101150
 99 1000010   6  6  6  65211111112112111421112111211  6  6  8  8  81010000151418110150
100 1000000   8  7  8  74211133111114411154125111  5  7  9  9  9  9  9  9010025316010220
101 0000000   6  6  6  42222222222222222222222222228  6  2  4  8  5  2  2001020314010 85
102 0000010   0  0  0  012111431114112131432154258  0  5  0  8  0  0001040918110220
103 1001000   6  6  8  41512351113423142553324121  6  7  8  8  8  8  8  8001018410000220
104 1001000   8 710  95443133335331153442335233  8  8  7 71010  8  9001028712101110
105 1000000   6  8  6  62333333333334344344334455  7  7  7  8  9  8  6001049312101220
106 0000010   6  6  6  65432154321543215412154321  5  2  1  9  8  5  0  00100464 8101110
107 1001000  010  0  01511111151511111151151115  6  0  8  0 610  6  0001027416101 60
108 0000000   6  6  6  63311131113331544254225533  8  8  8  4  7  6  8  8000134418110220
109 0001000   4 410  01111131113221411152214412  7  4  3  2 910  8  0001026214101150
110 1001000   4 210  65411133113511451152415431  9  7  2 010  9  8  6001019314000220
111 0000100   4  4  3  85532142215411431154115451  8  5  0  0  8  3  8  7000141516110190
112 1100000  10  5  5  74432152223222142144324232  3  4  6  6  5  5  5  6100053216110220
113 0000100   2  3  3  94111142214321532244315531  5  4  0  1  8  1  6  3000139318110220
114 1000010   6  9  9  84433344443224432543433443  8  4  7  8  9  8  7001019614001220
115 1001000   0 010  01511112314111111131111113  4  3  5  0 710  6  0001018112001 25
116 0000000   4  6  8  01221132113211321143215321  4  0  8  0  6  6  6  0001045316010170
117 1000010   6  6  8  01234133323343434333254314  8  8  5  1  9  9  8  7001019614010220
118 0100000   8  2  8  94414121115211531153115411  8 010  8  0  8  0000125312001 85
119 0000010   8  8  8  95311142115421552144425543  9  1  8  0  8  1  8  1000156416110190
120 0000010   7  1  0  74212141214332442154325432  7  7  5  1  5  5  5100037414110220
121 1000010   7  5  6  84211132124123443255225532  3  5  5  4  8  7  6  5001019510010220
122 1001000   2 110  53533331114211321142113311  3  4  2  1  8  9  9  9001061118010 85
123 1001000   4 410  33342143221431455225343445  5  6  6  5  7  6  5300010185100001170
124 0000001   2  5  5  53311143114124312111111111  3  5  3  4  6  5  5001032416110220
125 0000010   0  0  0  21111121115111441153214312  8  9  0  0  8 510  5000142416110220
126 1000100   9  9  8105211121533213252321325354  5  9  0  6  5101000001383140100125
127 1001000   8 610  81121144111411214152115511  8  8  8  6  6  6  6001058414110220
128 1000010   2  8  6  84233122315231454355414554  4  2  8  8  7  6  8  8010028314101125
129 0000010   4  4  4  45114554111444544154455511  4  6  4  6  6  4000134114001150
130 1001000   6 610  73111111111111113115313113  5  3  0  1  7  6  8  8000120312101 25
131 0000100   8  8  8  84111111114111221151115111 10  010101010101001001003 7612101125
132 1100000  10  6  0  84111111111111421132115111110101010  6 810  8000121314110220
133 1000010   8  0  8  62452252125222544152525552101010  8  8  81010000553161011 170
134 1100000   2  0  2  22115124531355442231151542310  4  4  2  6 610  6100057212100170
135 1000010   8  8  7  94211141141115424543155211  6  8  7  7  5  6  6  7000124218110 85
136 1000010   4  4  4  44311132122114321331143211  1  1  9  7  4  2  8  6010030416110220
137 0000100   6 110  95211141552151515152415151  8 810  9  9  6  9  9000140518110170
138 1000010   4  6  6  65511144114511444554555551  3  4  4  0  6  5  8  6000127216110220
139 0000010   6  6  6  61111141114111541145215521  8  8  2  0  6  6  4  8000129118010 60
140 1100000   4  2  4  84332343232233253354135413  8  8  7  7  8  710  83000130414101110
```

Figure A.4 Continued

```
141 1001000   0  6  9  42221232223241323254121543  2  1  9  6  5  8  6  7001026218110190
142 1000000   6  9  8  54211121112111521152115211  8  8  6  6  9  9  7  7001021714001220
143 1000010  10  5  8  24442153215511541154515542  6  8  4  6  6  6  8  8100019110001 25
144 1000000   6  8  8  63212133224311431142114311  8  7  8  4  8  7  8  5000126414000 25
145 1001000   0  0  0  81054111311131212311511153  8  8  6  5  8  8  6  8001024218110220
146 0100000   8  7  9  71524154411451324111211244  7  7  6  1  9  9  8  7001040916110125
147 1001000   8 910  65434154415424554555455555  9  9  9 91010  8  8001018312001190
148 1000010   8  9  9  84421144115411244122212243  8  9  8  8  8  8  8  8001052212110110
149 1001000   4  6  9  42222222222422424242222525  8  7  5  2  8  8  8  8001018514000 85
150 0100000  10  2  8  44221141214321141144435144  410  1  2  2  6  6  3100030214010 85
151 1001000   0  6  6  54421111325222552354325544  3  0  7  7  6  7  5  6001018412001170
152 0000010   8  8  8  82111132111143253143215422  6  6  6  6  7  6  6000121414000110
153 1000001   0  0  0  81111121113111511121113211  6  4  8  7  6  610  8000130514001125
154 0000010   6  6  8  63211131112111311311113211  8  7  9  8  9  9  6  6001022518010115
155 0001000   0  2  3  44321132113213213211112213321  8  7  8  4  8  5  8  2100022114001 85
156 0000010   0  01010152222222542255115541533  4  0  6  0  7  01010000127216110220
157 0000010   6  8  6  62111111111111111121112111  4  1  8  2  6  2  3  3000035514110220
158 1100000   6  4  8  02311141114221531154215421  8  8  7  7  8  9  2  2001028418000 85
159 0000010  21010  62411341115311521145114411  6  4  6  6  8  8  6  6001054312110220
160 1010000   8  9  8  85311131115211441144315322  9  8  6  7  9  8  8  8100020114001 60
161 0001000   6  810  01511141114111521154114411  6  6  2  2  8  6  8  5001020112001 60
162 1000100   0  0  01053333334333334343333353  0  0  4  0  2  0  610000126114010220
163 1000010   8  4  4  61334345521334445432534543  910  7  8  5  8  8  810004071210 1150
164 0000100   6  8  8  74242343325334543254225432  8  7  8  7  7  6  7  7100026312101 85
165 1000100   0  0  0  61221121125421233252114454  2  0  6  0  6  0  6  0000118410000 25
166 1000000   6  6  8  54121121111221222154115411  8  8  9  8  6  8  8001025112010 85
167 1001000   8  6  6  75422143115221432153215421  9  9  3  0  7  7  8  7100054518110220
168 1000100   9  610  94311132114322433243235432  4  6  3  2  7  8  8  9000120114000220
169 1000010   8  8  8  94221221313331433153115421  4  4  6  4  8  9  9  8000119414000110
170 0001000   6  7  8  01231121113111111132113211  6  6  6  2  8  6  7  6001060214100220
171 0000010   0  0  1  15432143215321542154315432  9  0  9  0  9  0  9  0000045618110125
172 0000010  10  1  8  31211141114111444144215522  5  610  8  9  8  2  5000063216110125
173 1000001   9  9  9  74332154314543435354325432  9  9  7  5  8  9  8001020616000220
174 1100000  10  8  8  84322242211331423144315321  8  7  7  6  8  810  7001063118010150
175 1000100   0  0  01051111151111115111111151  4  0  7  0  9  0  91000016421610 0110
176 1000100   7  4  9  94312142213211432254215532  8  8  2  2  8  8  7  7001020614000220
177 0000100   6  4  4  85321132213211352133214321  7  8  7  5  6  6  7  7000118210001 60
178 0000000  21010101111141113111531151115511  0101010  2  8  0  4010058210100110
179 0000010   4  5  4  52212123223431535144225542  4  0  4  1  6  0  8  5000118410000125
180 1001000   8  5  8  74111112114221523142115422  6  7  4  5  7  8  7  7001022914000150
181 0000000   2  2  2  22111121111111141114114211  2  2  8  4  6  3  6  4000024214001 85
182 1000010   8  810  81411142112214421433155310  5  8  8  8  2  0001038112001125
183 1010000   6  210  01323345144434212145434415  8  6  2  01010  8  6001021214110170
184 1100000   8  610  43534254223452315144335323  7  7  5  71010  6  3001024112001110
185 1000010  10  0  5  74111142213411422232434133  8  8  5  5  1  5  5  7000152216110220
186 0000010   6  8  6  32211133333454445554455555  8  9  8  8  8  8  9  9001020612001220
187 1000100  510101055421511151341553555345542  0  01010101010100015031010 1110
188 1000100   8  8  8  95314254143245231125345321  8  7  8  8  8  8  9000121512001220
189 1000010   6  610  44522242224222422254224222  810  2  41010  6  4001019310001110
190 1000000  10  9  910311112111422222112221511 1  6  6  2  3  6  81010000119310001170
191 1001000   4  2  8  01111121111421212125424331 4  6  6  2  1  8  7  8  3001021412101170
192 0000000   5  5  5  51111111111111141141115111  8  8  4  2  4  6  5100020414010 25
193 0000000   6  4  8  21111112112231443154415554  8  2  7  4  8  8  9  6000120114010 60
194 0100000   6  4  4  01432243125322512154325412  0  2  2  0  4  4  0  0001029314011125
195 0001000   2  4  8  44531112223521542255121442  5  0  5  3  7  8  7  4001023214101 85
196 0000010   6  4  8  5343214312334342424223552210101010101010101000020512010220
197 0000010   7  7  7  53222132214221442155252  9  910  8  9  8  9001024114001 25
198 0000001   0  6  6  01131113114321512144121513  3  2  8  5  5  4  7  2001023614001170
199 1001000   6  6  6  84111141114111121132135133  6  6  6  6  6  6  6  8000125214100110
200 0000010   4  4  6  43333333343334333533335333  6  0  8  0  7  0  6  6000121114001 85
201 1001000   7  5  9  42231122113411331154415541  7  7  7  5  7  7  7  7001020814001220
202 1000100   8  8  8  05532154215421552155315554  8  8  6  6  8  8  8  8000127418010220
203 0001000  10  6  7  22444525343555425233443231  010  4  21010  6  4100021610001 85
204 1100000   0  0  0  02221121113211332243235442  6  6  1  1  6  6  1  0100019610001220
205 1100000  10  6  8  63311145213434524154545545  8  9  7  6  8  9  8  9100031314110220
206 0000010   7  2  4  85211141211211511114115131  8  8  2  1  9  9  9000146412101125
207 0000010  10  7  0  01335111513135111133513313  8  6  2  4  3  5  2100039912101125
208 0000100  10  9  8  95543211121432153245424223101010101010100014061010 1 85
209 1000010   8  8  2  93244151415221452134415522  8  8  8  8  6  2  9  9000163418110220
210 1000010  10  810  01111111111111111111155555510  0  8  0101010100158216110220
```

Figure A.4 Continued

```
211 1000010    6   6   8  81231113311431123142334243  8  7  8  6  8  9  7  7001049414110220
212 1000000    7   7   7  61111114322233334444532543  3  5  4  2  710  2  2001036414101170
213 0100000    0   0   0  01111141113422223353545534  9  9  4  0  4  8  6  4100031314101190
214 1100000    9   8   7  83312122123222332243225322  810  1  6  8  8  6  8100052614101110
215 0001000    8  210  2 15333232132223212152324222  8  8  6  2  8  8  0  00010581 6001110
216 1000100    0   0   0  04332121413132141413212141  0  0  6  6  4  0  0  8000148316101190
217 0010000    4   0   8  81111111111111411151115111  0  0  4  8  4  2  0  0010033412101220
218 0000010    3   3   3  35432143215321552154315442  8  6  8  4  8  6  810000130310101125
219 1000100    0   0   0  02211121113111421141114111  1  2  6  2  4  2  2  2000030218100  25
220 1001000    5   3   7  02413321132411243454345324  8  6  1  0  5  5  8  7001034114010110
221 0000100    0   0   0  05111142135432454434355443  2  1  1  0  9  7  8  8000141416110110
222 1000010    010  01051111111151114111541155511  0  6  0  0  5  5  0  6000138416110110
223 1000000    9   9  91043211433234322531111254553  8  7  6  4  8  81010000130116010  85
224 0001000    7   5   8  52311131121121121154225133  7  7  4  0  7  7  7  7001031414110220
225 1100000    5   5 51054422411443255454541554 5  6  4  6  5  6  61010100064214100110
226 1010000   10   6   5  71123152113111323123225421 10  8  8  8  6  2  9  81000562 8100150
227 1000100    5   5   6  95421132314321332153215442  6  6  5  4  6  6  6  7001033112010  85
228 0000010    0   010  01522331425223131523 23315  5  2  0  0  9  9  0  0001060214110  60
229 0000010    0   0   0  01211121121211221111112111  8  5  5  5  5  2  6  5001064214110  25
230 1000010    0   5   5  6432114211331154425432543 3  4  4  8  5  8  6  9  900013941010122 0
231 1000100    2   0   8  95411143212331433153215321  6  6  2  4  6  6  8  8000131412101110
232 1000010    2   5   5  24211131114111431153114111  8  7  8  7  8  7  8  7000134412110  85
233 1000100    0   0  01051111115111111151555155555  0  0  0  0  4  0101000015321411022 0
234 1001000    9   2   2  84214455114343535153235421 1010  9  9  8  8  5  5001030314101125
235 0000000    4   4   4  45421132214211341143224131  3  2  1  0  3  4  8  5000144210101110
236 0000100    6   6   6  63445453235452112123445531  8  7  8  6  8  5  9  900014431210117 0
237 1000010    4   4   4  55211111141111511141115111  3  2  4  1  3  2  1  3000054212110  85
238 0000010    4   4   5  45211154413241554154315523  8  9  8  2  8  8  9  90001353 6101  85
239 1000100    0   1  11051111114113322424221223 51  1  1  5  1  3  1101000015421010111 0
240 1100000    5   5   5  55432141115221541154215331  5  4  4  0  3  5  8  6100044216110150
241 0000100    4   6  61011115323322223453523431155  2  6  0  01010  6  9001034318110220
242 1001000    8   6   6  54434442543545445422444554  8  8  4  11010 81000104531410112 5
243 0000100   91010105555535534153555355553553  8  4  2  0101010100001354121 01  85
244 1000010    8   8   8  81111141113141534154115541  4  4  8  6  6  6  4  6010046516110150
245 1000010    9   6   6  8333543534355535353333355551010  0  0101010100001416 6101150
246 0000100    4   4   4  54111142412144541152235411  7  7  4  5  6  6  2  4100032312001150
247 0001000    4   0   8  01111111111111111141115411  4  2  8  6  8  1  0  0001053712101220
248 0001000    9  810  73222242322223532254255432  2  7  6  4  8  5  600105441010122 0
249 1000100  101010102111214154412541554254344  7101010  810  81000014851411022 0
250 0000010    6   6   6  63211313133111222111111111  8  6  6  61010  0  6001062216110150
251 1000010    610   4  83333451112111551155111111  8  8  8  8  8  0  8010047512110170
252 1000010    6   8   4  12411122114234322125355543  7  7  5  6  6  7  4  4010030312101190
```

Figure A.4 Continued

Review Topics in Statistics and Matrix Algebra

B.1 Introduction

This technical appendix reviews a number of statistical and matrix algebra topics of general interest to multivariate analysis. By necessity our coverage is quite limited and the discussion brief. More extensive accounts of the statistical topics can be found in Morrison (1976). Further discussion of the matrix algebra topics can be found in either Graybill (1969) or, at a less technical level, Green (1976).[1]

The first part of the appendix describes four common univariate statistical distributions—Z, χ^2, t and F—whose tabled values for selected percentiles appear in Appendix C. This section also presents review material on statistical estimation and hypothesis testing.

The next part of the appendix reviews material on matrix inversion and eigenstructures. In particular, the main properties of eigenstructures of symmetric matrices and product moment matrices are outlined. Additional material on singular value decomposition is also included.

The appendix concludes with a review of the derivation of matrix equations that underlie the techniques of multiple regression, principal components analysis and multiple discriminant analysis.

Due to space restrictions and the fact that the discussion concerns review topics, the presentation style is considerably more concise than that utilized in the body of the book.

B.2 Statistical Distributions

Tables C.1 through C.4 in Appendix C show tabular values for selected percentiles of the:

1. Z (standard unit normal) distribution.
2. χ^2 distribution.
3. t distribution.
4. F distribution.

These are the principal distributions used in statistical estimation and hypothesis testing.

B.2.1 The Z Distribution

In Chapter 6's discussion of the multivariate normal distribution we had occassion to describe the special case of the univariate normal. As recalled, the normal density function of X, with parameters μ, the universe mean, and σ, the universe standard deviation, is:

$$f(X) = \frac{1}{\sqrt{2\pi}\sigma} \exp\left[\frac{-\frac{1}{2}(X - \mu)^2}{\sigma^2}\right]$$

where $-\infty < X, \mu < \infty; \sigma > 0$.

As also recalled from Chapter 6 the normal density function is bell-shaped with the highest ordinate at μ. Approximately 68 per cent of the area under the curve is contained in the interval $\mu \pm \sigma$; approximately 95 per cent of the area is contained in the interval $\mu \pm 2\sigma$.

The Z or standard unit normal distribution makes use of the transformation:

$$Z = \frac{(X - \mu)}{\sigma}$$

In this case $\mu(Z) = 0$ and $\text{var}(Z) = 1$. Accordingly, the density function of the standard unit normal is:

$$f(Z) = \frac{1}{\sqrt{2\pi}} \exp\left[\frac{-Z^2}{2}\right]$$

and is usually abbreviated as $N(0, 1)$. Panel I of Figure B.1 shows a plot of the standard unit normal density function.

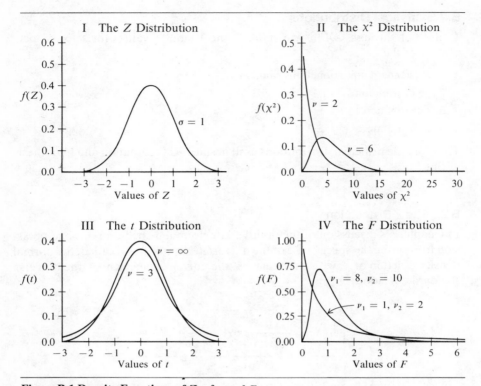

Figure B.1 Density Functions of Z, χ^2, t and F

Table C.1 in Appendix C shows probability values for the *cumulative* standard distribution over the range $0 \leq Z \leq 3.49$. For example, the probability that $Z \leq 2.42$ can be read off from the body of the table as 0.9922. The probability that $Z < -0.5$ is $1 - P(Z \geq 0.5) = 1 - 0.6915 = 0.3085$.

B.2.2 The χ^2 Distribution

The χ^2 distribution was employed in several places throughout the text, such as Bartlett's V statistic and the sphericity test. χ^2 has a wide variety of other uses that go far beyond its utilization in this book.

The χ^2 distribution deals with the sum of squared standard normal variates. Let $Z_1, Z_2, \ldots, Z_\nu$ denote a set of mutually independent variates where each is distributed $N(0, 1)$. Next, let us find the sum of their squares:

$$w = \sum_{i=1}^{\nu} Z_i^2$$

The sum w is distributed as χ^2 with ν degrees of freedom.[2]

Panel II shows the density of $w(= \chi^2)$ as a function of ν, the degrees of freedom, for two different cases. Notice that when ν is small the function is highly skewed to the right. However, as ν increases, the function becomes more symmetrical.

The mean and variance of the χ^2 distribution with ν degrees of freedom are:

$$\mu[\chi^2(\nu)] = \nu; \quad \text{var}[\chi^2(\nu)] = 2\nu$$

Selected percentiles of the χ^2 distribution with ν degrees of freedom appear in Table C.2 in Appendix C. For example, to find the 95 (probability) percentile for $\nu = 15$ in Table C.2, we go across the top of the table to $\chi^2_{[0.95]}$ and down the table to $\nu = 15$. The tabular value is 25.00. This means that 95 per cent of the samples based on $\nu = 15$ will exhibit χ^2 values less than 25.00 if the null hypothesis is true. If the sample value exceeds 25.00, then we reject the null hypothesis with an $\alpha = 0.05$ risk.

B.2.3 The t Distribution

Student's t distribution is useful in a variety of hypothesis testing cases where the variance σ^2 must be estimated from the sample. We can obtain the t distribution from the Z and χ^2 distributions. First, let us define a variate $Z = N(0, 1)$ and a variate $\chi^2(\nu)$.

If we further assume that Z and χ^2 are independent, then we can define a new variate t:

$$t = \frac{Z}{\sqrt{\dfrac{\chi^2}{\nu}}}$$

and call this variate Student's t with ν degrees of freedom.[3]

Panel III of Figure B.1 shows a plot of the density function of t for $\nu = 3$ and $\nu = \infty$. In the latter case the density of t is identical to that of Z. For small values of ν (≤ 30), the density shows higher ordinates in the tails of $f(t)$. As can also be noted, $f(t)$ is flatter than $f(Z)$. However, like $f(Z)$ it is symmetric and has a mean of zero.[4] Furthermore, as ν increases $f(t)$ comes closer and closer to $f(Z)$, so that they are identical when $\nu = \infty$.

Table C.3 in Appendix C shows selected upper percentiles of t. For example, assume that $\nu = 15$. The 95th (probability) percentile with 15 degrees of freedom is $t_{[0.95; 15]} = 1.753$. That is, 95 per cent of the samples with $\nu = 15$ will have t values less than 1.753 if the null hypothesis is true. If the sample value exceeds 1.753 for $\nu = 15$ we would reject the null hypothesis at the $\alpha = 0.05$ risk level.

Note that the t value for $\nu = 30$ and $1 - \alpha = 0.95$ is 1.697 in Table C.3. If we look up the Z value corresponding to $\alpha = 0.05$ in Table C.1, we note that its (interpolated) value is 1.645. Hence, even for $\nu = 30$, the Z variate represents a fairly close approximation to the t variate.

B.2.4 The F Distribution

We encountered the F distribution in a variety of places in the text, including discussions of regression, ANOVA and ANCOVA, Hotelling's T^2, and Rao's approximation to a function of Wilks' lambda.

If we define two separate χ^2 variates, $\chi_1^2(\nu_1)$ and $\chi_2^2(\nu_2)$ and further assume that they are independent, we can define F as:

$$F = \frac{\dfrac{\chi_1^2}{\nu_1}}{\dfrac{\chi_2^2}{\nu_2}}$$

and note that F has two parameters ν_1 and ν_2, usually referred to, respectively, as degrees of freedom for numerator and degrees of freedom for denominator.[5]

A plot of $f(F)$ appears in Panel IV of Figure B.1 for two sets of selected values of ν_1 and ν_2. For small values of ν_1 and ν_2, we see that the density function is highly skewed to the right. However, as ν_1 and ν_2 both increase, the density becomes more symmetrical.

It is of interest to note that:

$$[t(\nu)]^2 = F(1, \nu)$$

with percentiles $[t(1 - \alpha)/2; \nu] = F(1 - \alpha; 1, \nu)$. That is, the square of Student's t with ν degrees of freedom equals F with 1 degree of freedom in the numerator and ν degrees of freedom in the denominator.

Selected upper percentiles of F are shown in Table C.4 in Appendix C for various combinations of ν_1 and ν_2. For example, assume $\nu_1 = 4$ and $\nu_2 = 10$. The 95th (probability) percentile of F with 4 and 10 degrees of freedom is 3.48.[6] That is, 95 per cent of the samples (with 4 and 10 degrees of freedom) will exhibit values less than 3.48 if the null hypothesis is true. If the sample value of F exceeds 3.48 we reject the null hypothesis at the $\alpha = 0.05$ risk level.

B.3 Estimation and Hypothesis Testing

Two major aspects of statistical inference are the estimation of parameters and the testing of hypotheses. In the former case we are usually interested in two kinds of estimates: (a) a point estimate of the parameter and (b) an interval estimate with which we can make certain statements regarding the likelihood of its including the universe parameter.

B.3.1 Estimation Methods

Let us assume a random sample of $X_1, X_2, \ldots, X_m$ observations of some variable whose population parameters are given by $\theta_1, \theta_2, \ldots, \theta_n$. One method for estimating these parameters is called *maximum likelihood,* defined as the joint density of the sample observations:

$$L = \prod_{i=1}^{m} f(X_i; \theta_1, \theta_2, \ldots, \theta_n)$$

where the θ's are the parameters of the sampled universe distribution and f is a known function of the θ's. The maximum likelihood procedure finds estimates $\hat{\theta}_1, \hat{\theta}_2, \ldots, \hat{\theta}_n$ that maximize L with respect to $\theta_1, \theta_2, \ldots, \theta_n$. The values $\hat{\theta}_1, \hat{\theta}_2, \ldots, \hat{\theta}_n$ are called maximum likelihood estimates.

As is usually the case with problems of maximizing or minimizing functions of many variables, the solution procedure involves partial differentiation of L with respect to the θ's. Often one works with the natural logarithm of L rather that with L itself on the basis of mathematical convenience and the assurance that values which maximize L will maximize a strictly increasing function of L (such as its logarithm).

In any type of estimation technique we are interested in certain properties of the estimators:

1. *Unbiasedness*—the expectation of the estimates made by a particular estimating procedure equals the parameter being estimated.

2. *Efficiency*—an estimating procedure is more efficient than alternative procedures if the sampling variability of its estimates is less than the sampling variability associated with alternative procedures.

3. *Consistency*—an estimating procedure is said to be consistent if estimates produced by its application tend to approach the parameter value more and more closely as sample size approaches infinity.

4. *Sufficiency*—an estimating procedure is said to produce sufficient estimates if no other statistic that can be derived from the sample provides any additional information about the parameter being estimated.

Maximum likelihood estimates are usually consistent, efficient and sufficient. However, they are often biased as well, although this bias can sometimes be removed by simple algebraic operations, such as multiplication by a constant.

Least squares represents another type of estimation procedure. In this method we assume that the sample observations can be represented as:

$$X_i = f(\theta_1, \theta_2, \ldots, \theta_n) + \epsilon_i$$

where f is a known function of the parameters $\theta_1, \theta_2, \ldots, \theta_n$ and ϵ_i is a random variable.

In the least squares method, the objective is to minimize:

$$M = \sum_{i=1}^{m} [X_i - f_i(\theta_1, \theta_2, \ldots, \theta_n)]^2$$

Again, the calculus is used to find $\hat{\theta}$'s that minimize M. Least squares estimation was employed in a wide variety of places throughout the text.

A third method of estimation is called *minimum* χ^2. To use this approach we first assume that the sample outcome space consists of K exclusive and exhaustive categories. We assume a sample size of m_k for the k-th category and note that $m_1 + m_2 + \ldots + m_K = m$, the total sample size. We let $P_k(\theta_1, \theta_2, \ldots, \theta_n)$ be the probability of an observation being in the k-th category. Then we set up the function:

$$\chi^2 = \sum_{k=1}^{K} \frac{[m_k - mP_k(\theta_1, \theta_2, \ldots, \theta_n)]^2}{mP_k(\theta_1, \theta_2, \ldots, \theta_n)}$$

which is to be minimized, usually by computerized numerical methods.

Whatever the technique used to find some estimator $\hat{\theta}$, one may still wish to construct a confidence interval about it. This procedure—which was illustrated in Chapter 5—involves the selection of values $l(\hat{\theta})$ and $u(\hat{\theta})$ such that a probability statement can be made about the interval separating them containing θ:

$$P\{l(\hat{\theta}) \le \theta \le u(\hat{\theta})\} = 1 - \alpha$$

The interval $\{l(\hat{\theta}), u(\hat{\theta})\}$ is usually referred to as the $100(1 - \alpha)$ per cent confidence interval for θ. The values $l(\hat{\theta})$ and $u(\hat{\theta})$ are called, respectively, lower and upper confidence limits.

The following comments pertain to how one interprets confidence intervals:[7]

1. The confidence limits (end points of the interval) are computed from samples and, hence, are also subject to sampling fluctuations.

2. In repeated samples, therefore, the confidence interval will vary from sample to sample.

3. The parameter being estimated, whatever it is, is assumed to be fixed.

4. If the analyst were to repeat the process of taking random samples of a given size and computing, via a prescribed way, a large number of confidence intervals, any *single* interval will either include or not include the parameter.

5. Probability statements (i.e., levels of confidence) thus pertain to the *process* by which the intervals are computed.

6. The analyst can state: "If I were to take repeated random samples of a given size and compute (in a specified manner) confidence intervals, in the long run x percent of the intervals so computed will contain the population value. This x percent is the confidence coefficient associated with the process of computing intervals."

In general, confidence interval estimation goes hand in hand with point estimation. First, we seek some "best" single (point) estimate $\hat{\theta}$ and then we construct a confidence interval about it. In cases where we need to estimate a *vector* of parameters $\hat{\theta}$, advanced techniques are available for computing simultaneous confidence bounds for the multiple parameters.

B.3.2 Hypothesis Testing

Throughout the text occasions arose in which various hypotheses were tested. A statistical hypothesis is an assertion about some aspect of the population under study, typically some parameter(s) of it, such as its mean.

As recalled from basic statistics, any test involving two alternatives:

1. The null hypothesis H_0 is true
2. The alternative hypothesis H_1 is true

can be set up in terms of a 2×2 table in which we consider two types of potential errors:

Action	Actual Situation	
	H_0 is True	H_0 is False
Reject H_0	Type I error; α risk	No error; $1 - \beta$
Accept H_0	No error; $1 - \alpha$	Type II error; β risk

Type I error involves rejecting the null hypothesis when it is true while Type II error involves accepting H_0 when it is false. (Some authors prefer the phrase "not rejecting" H_0.)

Most statistical testing procedures operate on the basis of fixing the α risk at some suitably small level (say 0.05) and then finding a rule that minimizes the β risk, subject to the fixed level of α. The *power* of a test is defined as the probability of rejecting H_0 when it is false (i.e., when the alternative hypothesis H_1 is true). This probability is given by $1 - \beta$ in the preceding schema. The power $1 - \beta$ generally depends upon what the population parameter and specific alternative hypothesis are.

Statistical tests can be divided into two groups:

1. Tests of simple hypotheses where H_0 and H_1 each consists of a single parameter.
2. Tests of a composite hypothesis where either H_0 or H_1 contains more than a single parameter.

Suppose we consider tests of *simple* hypotheses first. A useful procedure for examining tests of simple hypotheses is based on the Neyman-Pearson likelihood ratio test (Morrison, 1976). In the general case of multiple parameters, one can characterize the two hypotheses by specific values assigned to the unknown θ's:

$$H_0: \theta_1 = y_1; \ \theta_2 = y_2; \ \ldots; \theta_n = y_n$$

$$H_1: \theta_1 = w_1; \ \theta_2 = w_2; \ \ldots; \theta_n = w_n$$

The following decision rule can then be set up:

$$\text{Accept } H_0 \text{ if } \lambda = \frac{f(X_1|\mathbf{y}) \cdot f(X_2|\mathbf{y}) \ldots \cdot f(X_m|\mathbf{y})}{f(X_1|\mathbf{w}) \cdot f(X_2|\mathbf{w}) \ldots \cdot f(X_m|\mathbf{w})} \geq c$$

$$\text{Accept } H_1 \text{ if } \lambda < c$$

The value c, denoting the critical level, is chosen so that:

$$P(\lambda < c \,|\, H_0 \text{ is true}) = \alpha$$

In the case of *composite* hypotheses, a generalization (also due to Neyman and Pearson) of the likelihood ratio criterion is needed. In the generalization one defines the ratio:

$$\lambda = \frac{L(\gamma)}{L(\Gamma)}$$

where $L(\gamma) = f(X_1|\mathbf{y}) \cdot f(X_2|\mathbf{y}) \ \ldots \cdot f(X_m|\mathbf{y})$. $L(\gamma)$ is a function that is maximized on the restricted basis that:

$$H_0: \{\theta_1, \theta_2, \ldots, \theta_n\} \epsilon \gamma$$

is true. On the other hand $L(\Gamma)$ is the maximized likelihood when $\{\theta_1, \theta_2, \ldots, \theta_n\}$ are free to vary over the entire parameter space of Γ. The alternative hypothesis H_1 can then be described by the difference:

$$H_1: \{\theta_1, \theta_2, \ldots, \theta_n\} \epsilon [\Gamma - \gamma]$$

The decision rule is:

$$\text{Accept } H_0 \text{ if } \lambda \geq c$$

$$\text{Accept } H_1 \text{ if } \lambda < c$$

where c is set so that $P(\lambda < c \,|\, H_0) = \alpha$.

It should be noted that λ is a random variable, since it depends on the observations $X_1, X_2, \ldots, X_m$. Moreover, γ represents a subset of Γ so that $0 \leq \lambda \leq 1$. As λ approaches 1 we are more likely to accept H_0.

In concluding this section, it should be mentioned that Bayesian estimation and hypothesis testing procedures are also available to the applied researcher. Some of the flavor of Bayesian decision theory was brought out in Chapter 4's discussion of two-group discriminant analysis. As recalled, in Bayesian inference one can introduce not only the conditional probabilities of making wrong decisions but the conditional costs or disutilities as well. Moreover, estimation and hypothesis testing can incorporate the researcher's prior probabilities over the parameter space. Readers who are interested in Bayesian methods should see books by Schlaifer (1959) and Savage (1954).

B.4 Matrix Inversion

One of the most common operations in multivariate analysis involves finding the inverse of a matrix. (Other vector and matrix operations are described in Green, 1976.) Only square matrices whose determinants are not zero possess (regular) inverses. If we have a square matrix $\mathbf{A}$, whose determinant does not equal zero, its inverse $\mathbf{A}^{-1}$ satisfies the conditions:

$$\boxed{\mathbf{A}\mathbf{A}^{-1} = \mathbf{A}^{-1}\mathbf{A} = \mathbf{I}}$$

For example, consider the matrix:

$$\mathbf{A} = \begin{bmatrix} 1 & 2 \\ 3 & 4 \end{bmatrix}$$

Its inverse is:

$$\mathbf{A}^{-1} = \begin{bmatrix} -2 & 1 \\ 3/2 & -1/2 \end{bmatrix}$$

and we see that:

$$\overset{\mathbf{I}}{\begin{bmatrix} 1 & 0 \\ 0 & 1 \end{bmatrix}} = \overset{\mathbf{A}}{\begin{bmatrix} 1 & 2 \\ 3 & 4 \end{bmatrix}} \overset{\mathbf{A}^{-1}}{\begin{bmatrix} -2 & 1 \\ 3/2 & -1/2 \end{bmatrix}} = \overset{\mathbf{A}^{-1}}{\begin{bmatrix} -2 & 1 \\ 3/2 & -1/2 \end{bmatrix}} \overset{\mathbf{A}}{\begin{bmatrix} 1 & 2 \\ 3 & 4 \end{bmatrix}}$$

We also note that the determinant of $\mathbf{A}$ is:

$$|\mathbf{A}| = (1 \times 4) - (3 \times 2)$$
$$= -2$$

which is non-zero.

Some of the most important properties of matrix inverses are:

1. If **A** has a (regular) inverse it is unique.

2. $(\mathbf{A}^{-1})^{-1} = \mathbf{A}$.

3. $(\mathbf{AB})^{-1} = \mathbf{B}^{-1}\mathbf{A}^{-1}$, assuming **A** and **B** have inverses themselves.

4. If k is a nonzero scalar:

$$(k\mathbf{A})^{-1} = \frac{1}{k}\mathbf{A}^{-1}$$

5. $(\mathbf{A}')^{-1} = (\mathbf{A}^{-1})'$.

If the square matrix **A** has a (regular) inverse, it is said to be nonsingular; otherwise it is said to be singular.

B.4.1 The 2 × 2 Case

With the advent of the computer, numerical techniques for finding matrix inverses have reached a high level of sophistication. Accordingly, desk calculator methods have decreased in importance. However, finding the inverse of a 2 × 2 matrix is particularly simple and involves the rule:

Transpose the diagonal entries and change the signs of the off-diagonal entries.

Then divide each entry by the determinant.[8]

To illustrate the rule, let us again consider the matrix:

$$\mathbf{A} = \begin{bmatrix} 1 & 2 \\ 3 & 4 \end{bmatrix}$$

As recalled, $|\mathbf{A}| = -2$. The inverse of **A** is then found by applying the preceding rule:

$$\mathbf{A}^{-1} = \frac{1}{-2}\begin{bmatrix} 4 & -2 \\ -3 & 1 \end{bmatrix} = \begin{bmatrix} -2 & 1 \\ 3/2 & -1/2 \end{bmatrix}$$

While this procedure is extremely simple in the case of 2 × 2 matrices, one needs a more complex technique for higher-order matrices. The pivotal method is useful for this purpose.

B.4.2 The Pivotal Method

The pivotal method of matrix inversion is a systematic procedure—readily adaptable for desk calculators—that can be used to compute all of the following:

1. The determinant of the matrix.

2. The solution to a set of simultaneous equations.

3. The inverse of the matrix.

In this illustration we shall find the determinant and the inverse. The easiest way to describe the pivotal method is by a numerical example. For illustrative purposes let us consider a fourth order matrix:

$$\mathbf{A} = \begin{bmatrix} 2 & 3 & 1 & 2 \\ 4 & 2 & 3 & 4 \\ 1 & 4 & 2 & 2 \\ 3 & 1 & 0 & 1 \end{bmatrix}$$

Evaluating the determinant and inverse of **A** proceeds in a step-by-step fashion, with the aid of a work sheet similar to that appearing in Table B.1.

The top, left-hand portion of Table B.1 shows the original matrix **A**, whose determinant and inverse we wish to evaluate. To the right of this matrix is shown an identity matrix of the same order (4 × 4) as the matrix **A**. The last column (column 9) is a check sum column, each entry of which represents the algebraic sum of the specific row of interest. (Other than for arithmetic checking purposes, column 9 plays no role in the computations.)

The objective behind the pivotal method is to reduce the columnar entries in **A** successively so that for each column of interest we have only one entry and this single entry is unity. Specifically, the boxed entry in row 01 (the number 2) serves as the first pivot. Row 10 is obtained from row 01 by dividing each entry in row 01 by 2, the pivot item. Note that *all* entries in row 01 are divided by the pivot, including the entries under the identity matrix and the check sum column. Dividing 2 by itself, of course, produces the desired entry of unity in the first column of row 10.

Row 11 is obtained from the results of two operations. First, we multiply each entry of row 10 by 4, the first entry in row 02. This particular step is not shown in the worksheet, but the nine products are:

$$4; \ 6; \ 2; \ 4; \ 2; \ 0; \ 0; \ 0; \ 18$$

These are then subtracted from their counterpart entries in row 02, so as to obtain row 11.

Note, particularly, that this subtraction has the desired effect of producing a zero (shown as a blank) in the first entry of row 11. Note further that the entries of row 11 add up to -4, the row check sum in the last column; the check sum column is provided for all rows and serves as an arithmetic check on the computations.

Row 12 is obtained in an analogous way; here, since the first element in row 03 is 1, we multiply row 10 by unity and then subtract the row 10 elements from their counterparts in row 03. Row 13 is also obtained in the same way. First the row 10 entries are multiplied by· 3, the first entry in row 04. Then these entries are subtracted from their counterparts in row 04. Finally, we see that in rows 10 through 13, all entries in column 1 are zero (and represented by blanks) except the first entry which is unity.

Table B.1 Finding the Determinant and Inverse By the Pivotal Method

Row No. 0	Original Matrix 1	2	3	4	Identity Matrix 5	6	7	8	Check Sum Column 9
01	[2]	3	1	2	1	0	0	0	9
02	4	2	3	4	0	1	0	0	14
03	1	4	2	2	0	0	1	0	10
04	3	1	0	1	0	0	0	1	6
10	1	1.5	0.5	1	0.5	0	0	0	4.5
11		[−4]	1	0	−2	1	0	0	−4
12		2.5	1.5	1	−0.5	0	1	0	5.5
13		−3.5	−1.5	−2	−1.5	0	0	1	−7.5
20		1	−0.25	0	0.5	−0.25	0	0	1
21			[2.125]	1	−1.75	0.625	1	0	3.0
22			−2.375	−2	0.25	−0.875	0	1	−4.0
30			1	0.471	−0.824	0.294	0.471	0	1.412
31				[−0.881]	−1.707	−0.177	1.119	1	−0.646
40				1	1.938	0.201	−1.270	−1.135	0.733
30*			1		−1.737	0.199	1.069	0.534	1.067
20*		1			0.066	−0.200	0.267	0.134	1.267
10*	1				−0.668	−0.001	0.334	0.667	1.332

$$|\mathbf{A}| = (2)(-4)(2.125)(-0.881)$$
$$= 15$$

At the next stage in the computations the first element in row 11 becomes the pivot. All entries in row 11 are divided by -4, the new pivot, and the results are shown in row 20. Row 20 now becomes the reference row. For example, row 21 is found in a way analogous to row 11. First we multiply all entries of row 20 by 2.5, the first entry of row 12. Although not shown in the worksheet, these are:

$$2.5;\ -0.625;\ 0;\ 1.25;\ -0.625;\ 0;\ 0;\ 2.5$$

These elements are then subtracted from their counterparts in row 12 and the results appear in row 21.

The procedure is then repeated by multiplying row 20 by -3.5, the leading element in row 13 and subtracting these new entries from their counterparts in row 13. Note that in rows 20 through 22, entries in columns 1 and 2 are all zero, except for the leading element of 1 in row 20.

The third pivot item is the entry 2.125 in row 21. All entries in row 21 are divided by 2.125 and the results listed in row 30. Finally, the entries of row 30 are multiplied by -2.375, the leading entry in row 22. Although not shown in the worksheet, these entries are:

$$-2.375;\ -1.119;\ 1.957;\ -0.698;\ -1.119;\ 0;\ -3.353$$

These entries are subtracted from their counterparts in row 22, providing row 31. The last pivot item is -0.881 and appears in row 31. Dividing each entry of row 31 by -0.881, we obtain row 40. Finally, the four pivots (boxed entries) are multiplied together, leading to the determinant:

$$|\mathbf{A}| = (2)(-4)(2.125)(-0.881)$$
$$= 15$$

At this point the reader may well wonder what is the role played by the various changes being made in the identity matrix as the pivot procedure is applied. As will be shown, in the course of modifying the original matrix $\mathbf{A}$ to find its determinant, the companion identity matrix in columns 5 through 8 is being modified to find the inverse of $\mathbf{A}$.

To find $\mathbf{A}^{-1}$, let us first set up a matrix by taking the first four columns of the rows numbered 10, 20, 30 and 40 in Table B.1:

$$\mathbf{H} = \begin{bmatrix} 1 & 1.5 & 0.5 & 1 \\ 0 & 1 & -0.25 & 0 \\ 0 & 0 & 1 & 0.471 \\ 0. & 0 & 0 & 1 \end{bmatrix}$$

Next, if we treat column 5 as a vector of constants, we can obtain the following values of x_4, x_3, x_2, x_1 via back substitution:

$$x_4 = 1.938 \qquad \text{(from row 40 and column 5)}$$

$$x_3 + 0.471(1.938) = -0.824 \qquad \text{(from row 30 and column 5)}$$
$$x_3 = -1.737$$

$$x_2 - 0.25(-1.737) + 0(1.938) = 0.5 \qquad \text{(from row 20 and column 5)}$$
$$x_2 = 0.066$$

$$x_1 + 1.5(0.066) + 0.5(-1.737) + 1(1.938) = 0.5 \qquad \text{(from row 10 and column 5)}$$
$$x_1 = -0.668$$

These entries are shown as a column vector in rows 40, 30*, 20*, and 10* and column 5.

What are we doing here is building up the inverse of **A** one column at a time.[9] Next, we repeat the whole procedure, using the entries of column 6, followed by column 7 and finally by column 8. The result is the inverse $\mathbf{A}^{-1}$, shown in rows, 40, 30*, 20*, and 10*, and columns 5, 6, 7 and 8.

However, it is important to note that the inverse $\mathbf{A}^{-1}$ is shown in *reverse row order*. That is, the *first* row of $\mathbf{A}^{-1}$ is given by:

$$(-0.668 \quad -0.001 \quad 0.334 \quad 0.667)$$

as designated by the row 10*: Hence, if one follows the order 10*, 20*, 30*, and 40, the inverse $\mathbf{A}^{-1}$ will be in correct row order.

Furthermore, had an explicit set of constants been present, in the usual form of a set of simultaneous equations, the same procedure could have been applied to this column as well. As indicated, the pivotal method can be used for:

1. Finding the determinant of **A**.
2. Solving an explicit set of simultaneous equations.
3. Finding the inverse of **A** (if it exists).

In applying the pivotal method one should be on the lookout for cases in which a zero appears in the leading diagonal position. If such does occur, we cannot, of course, divide that row by the pivot. If the matrix is nonsingular—which will usually be the case in data-based applications—the presence of a leading zero suggests that we want to move to the next row and select it as the pivot (i.e., permute rows) before proceeding. This has no effect on the inverse but will reverse the sign of the determinant (if an odd number of such transpositions appear).[10]

While a variety of more specialized procedures are available, the pivotal method exhibits the virtues of simplicity and directness and is well suited for desk calculator application.[11] In general, however, the computation of matrix inverses has reached a high state of the art insofar as computer routines are concerned.[12]

B.5 Matrix Eigenstructures

In various chapters throughout the book we had occasion to discuss matrix eigenstructures. By way of more formal introduction to the topic, let us now consider a square matrix $\mathbf{A}$ of order $n \times n$, with all real entries. This matrix is assumed to represent the transformation of some set of vectors in Euclidean n-space. We seek a particular direction in that space, denoted by the vector $\mathbf{x}_j$, such that $\mathbf{x}_j$ is transformed (by $\mathbf{A}$) into a multiple of itself:

$$\boxed{\mathbf{A}\mathbf{x}_j = \lambda_j \mathbf{x}_j}$$

where λ_j is a scalar denoting the multiple.

If $\mathbf{x}_j$ and λ_j exist, by algebraic transposition we can write:

$$\boxed{(\mathbf{A} - \lambda_j \mathbf{I})\mathbf{x}_j = \mathbf{0}}$$

While $\mathbf{x}_j = \mathbf{0}$ satisfies this equation trivially, what we really want is some *nonzero* vector (and scalar) that also satisfies the equation. However, if the determinantal equation:

$$\boxed{|\mathbf{A} - \lambda_j \mathbf{I}| = 0}$$

with unknown λ_j can be satisfied, then we shall be able to find the nonzero $\mathbf{x}_j$ that we seek.[13] The immediately preceding equation is called the *characteristic equation* of $\mathbf{A}$. If $\mathbf{A}$ is $n \times n$, then there will be exactly n roots (eigenvalues, denoted by λ_j) and n associated eigenvectors, denoted by $\mathbf{x}_j$.[14] Each $\mathbf{x}_j$ that we find can be subjected to multiplication by an arbitrary scalar k without affecting the relation. For various purposes we shall sometimes find it useful to scale the $\mathbf{x}_j$'s to unit sum of squares; that is, $\mathbf{x}_j' \mathbf{x}_j = 1$.

As a numerical illustration of finding the eigenstructure of a (square) matrix, consider the matrix:

$$\mathbf{A} = \begin{bmatrix} -3 & 5 \\ 4 & -2 \end{bmatrix}$$

To find the eigenstructure of $\mathbf{A}$ we first write:

$$(\mathbf{A} - \lambda_j \mathbf{I})\mathbf{x}_j = \mathbf{0}$$

$$\left\{ \begin{bmatrix} -3 & 5 \\ 4 & -2 \end{bmatrix} - \lambda_j \begin{bmatrix} 1 & 0 \\ 0 & 1 \end{bmatrix} \right\} \begin{bmatrix} X_{1j} \\ X_{2j} \end{bmatrix} = \begin{bmatrix} 0 \\ 0 \end{bmatrix}$$

Next, we set up the characteristic equation:

$$\begin{vmatrix} -3-\lambda_j & 5 \\ 4 & -2-\lambda_j \end{vmatrix} = 0$$

and expand the determinant to get:

$$\lambda_j^2 + 5\lambda_j - 14 = 0$$

We then find the roots of this quadratic equation by simple factoring:

$$(\lambda_j + 7)(\lambda_j - 2) = 0$$
$$\lambda_1 = -7$$
$$\lambda_2 = 2$$

Next, substitute $\lambda_1 = -7$ in the equation $(\mathbf{A} - \lambda_j\mathbf{I})\mathbf{x}_j = \mathbf{0}$:

$$\left\{ \begin{bmatrix} -3 & 5 \\ 4 & -2 \end{bmatrix} - \begin{bmatrix} -7 & 0 \\ 0 & -7 \end{bmatrix} \right\} \begin{bmatrix} X_{11} \\ X_{21} \end{bmatrix} = \begin{bmatrix} 0 \\ 0 \end{bmatrix}$$

$$\begin{bmatrix} 4 & 5 \\ 4 & 5 \end{bmatrix} \begin{bmatrix} X_{11} \\ X_{21} \end{bmatrix} = \begin{bmatrix} 0 \\ 0 \end{bmatrix}$$

The obvious solution to the two equations, each of which is:

$$4X_{11} + 5X_{21} = 0$$

is the vector:

$$\mathbf{x}_1 = \begin{bmatrix} 5 \\ -4 \end{bmatrix}$$

or, more generally:

$$\mathbf{x}_1 = \begin{bmatrix} 5k \\ -4k \end{bmatrix}$$

where k is an arbitrary scalar. In many statistical applications, eigenvectors are normalized, *by convention,* to unit length. Even so, however, their sign is indeterminate; for example, we can multiply $\mathbf{x}_1$ by -1 without disturbing the preceding equation. Next, let us substitute $\lambda_2 = 2$ in the same way, so as to find:

$$\left\{ \begin{bmatrix} -3 & 5 \\ 4 & -2 \end{bmatrix} - \begin{bmatrix} 2 & 0 \\ 0 & 2 \end{bmatrix} \right\} \begin{bmatrix} X_{12} \\ X_{22} \end{bmatrix} = \begin{bmatrix} 0 \\ 0 \end{bmatrix}$$

$$\begin{bmatrix} -5 & 5 \\ 4 & -4 \end{bmatrix} \begin{bmatrix} X_{12} \\ X_{22} \end{bmatrix} = \begin{bmatrix} 0 \\ 0 \end{bmatrix}$$

A solution to these two equations:

$$-5X_{12} + 5X_{22} = 0$$

$$4X_{12} - 4X_{22} = 0$$

is evidently the vector:

$$\mathbf{x}_2 = \begin{bmatrix} 1 \\ 1 \end{bmatrix}$$

or, again more generally,

$$\mathbf{x}_2 = \begin{bmatrix} 1k \\ 1k \end{bmatrix}$$

Hence, insofar as $\mathbf{x}_1$ and $\mathbf{x}_2$ are concerned, any vector whose components are in the ratio of either:

$$5 : -4 \quad \text{or} \quad 1 : 1$$

represents an eigenvector of the transformation given by the matrix $\mathbf{A}$.

Suppose we now represent the two eigenvectors of $\mathbf{A}$ as columns in a new matrix:

$$\mathbf{U} = \begin{bmatrix} 5 & 1 \\ -4 & 1 \end{bmatrix}$$

Next, let us find the inverse of $\mathbf{U}$ via the adjoint method described in section B.4:

$$\mathbf{U}^{-1} = \frac{1}{|\mathbf{U}|} \, \text{adj}(\mathbf{U})$$

$$= \frac{1}{9} \begin{bmatrix} 1 & -1 \\ 4 & 5 \end{bmatrix} = \begin{bmatrix} 1/9 & -1/9 \\ 4/9 & 5/9 \end{bmatrix}$$

Then, let us form the triple product:

$$\mathbf{A} = \overset{\mathbf{U}}{\begin{bmatrix} 5 & 1 \\ -4 & 1 \end{bmatrix}} \overset{\mathbf{D}}{\begin{bmatrix} -7 & 0 \\ 0 & 2 \end{bmatrix}} \overset{\mathbf{U}^{-1}}{\begin{bmatrix} 1/9 & -1/9 \\ 4/9 & 5/9 \end{bmatrix}}$$

$$= \begin{bmatrix} -3 & 5 \\ 4 & -2 \end{bmatrix}$$

We note that **A** has been decomposed into the product of:

1. A nonsingular matrix **U** whose columns are the eigenvectors of **A**.
2. A diagonal matrix **D** whose diagonal entries are the eigenvalues of **A**.
3. A matrix $\mathbf{U}^{-1}$, the inverse of **U**.

By algebraic transposition we can also examine the relation between the diagonal matrix **D** and the triple product:

$$\mathbf{D} = \mathbf{U}^{-1}\mathbf{A}\mathbf{U}$$

This relation is known in matrix algebra as a similarity relation. In more general terms, two matrices **D** and **A** are *similar* if there exists a nonsingular matrix **U** such that the above equation holds.[15] If all eigenvalues of **A** are real, **D** will be a real diagonal matrix and all entries in **U** will be real-valued as well.[16]

If **A** is not only square but *symmetric* as well, then the preceding diagonalization becomes particularly simple:

$$\mathbf{D} = \mathbf{T}'\mathbf{A}\mathbf{T}$$

where **T** is orthogonal.[17] As recalled, if **T** is orthogonal, then:

$$\mathbf{T}^{-1} = \mathbf{T}'$$

and we see that this represents a special case of $\mathbf{D} = \mathbf{U}^{-1}\mathbf{A}\mathbf{U}$. To illustrate the symmetric diagonalization case, let:

$$\mathbf{A} = \begin{bmatrix} 4 & 1 \\ 1 & 4 \end{bmatrix}$$

Its eigenvalues and eigenvectors are given, respectively, by:

$$\overset{\mathbf{D}}{\begin{bmatrix} 5 & 0 \\ 0 & 3 \end{bmatrix}} \quad \text{and} \quad \overset{\mathbf{T}}{\begin{bmatrix} 0.707 & 0.707 \\ 0.707 & -0.707 \end{bmatrix}}$$

We then have the diagonalization:

$$\overset{\mathbf{D}}{\begin{bmatrix} 5 & 0 \\ 0 & 3 \end{bmatrix}} = \overset{\mathbf{T}'}{\begin{bmatrix} 0.707 & 0.707 \\ 0.707 & -0.707 \end{bmatrix}} \overset{\mathbf{A}}{\begin{bmatrix} 4 & 1 \\ 1 & 4 \end{bmatrix}} \overset{\mathbf{T}}{\begin{bmatrix} 0.707 & 0.707 \\ 0.707 & -0.707 \end{bmatrix}}$$

It is easy to see that **T** is orthogonal ($\mathbf{TT}' = \mathbf{T}'\mathbf{T} = \mathbf{I}$) and represents a 45° rotation.

The two forms of diagonalization noted above are quite useful in multivariate analysis. The first case, involving the nonsymmetric matrix $\mathbf{A}$ in which:

$$\mathbf{D} = \mathbf{U}^{-1}\mathbf{A}\mathbf{U}$$

is relevant to certain kinds of problems encountered in multiple discriminant analysis, MANOVA and MANCOVA. The second case, where $\mathbf{A}$ is considered to be symmetric:

$$\mathbf{D} = \mathbf{T}'\mathbf{A}\mathbf{T}$$

is of the type encountered in principal components analysis.

B.5.1 Some Properties of Eigenstructures

Let us now consider some of the properties of eigenstructures, particularly those that are appropriate in multivariate statistical applications. Two quite general properties of eigenstructures that apply to either the nonsymmetric or symmetric cases are:

1. The sum of the eigenvalues of the eigenstructure of a matrix equals the sum of the main diagonal elements (called the *trace*) of the matrix. That is, for some matrix $\mathbf{W}_{n \times n}$:

$$\sum_{j=1}^{n} \lambda_j = \sum_{j=1}^{n} w_{jj}$$

2. The product of the eigenvalues of $\mathbf{W}$ equals the determinant of $\mathbf{W}$:

$$\prod_{j=1}^{n} \lambda_j = |\mathbf{W}|$$

Notice here that if $\mathbf{W}$ is singular, at least one of its eigenvalues must be zero in order that $|\mathbf{W}| = 0$, the condition that must be met in order for $\mathbf{W}$ to be singular.[18] However, even though $\mathbf{W}$ may be singular, $\mathbf{T}$ in the expression:

$$\mathbf{D} = \mathbf{T}'\mathbf{W}\mathbf{T}$$

can still be orthogonal (and, hence, nonsingular). However, if there is more than one zero eigenvalue in $\mathbf{W}$, $\mathbf{T}$ is not unique.

In addition to the above, a number of other properties related to sums, products, powers and roots should also be mentioned. (The last two properties that are listed below pertain only to symmetric matrices with non-negative eigenvalues.)

3. If we have the matrix $\mathbf{B} = \mathbf{A} + k\mathbf{I}$, where k is a scalar, then the eigenvectors of $\mathbf{B}$ are the same as those of $\mathbf{A}$ and the j-th eigenvalue of $\mathbf{B}$ is:

$$\boxed{\lambda_j + k}$$

where λ_j is the j-th eigenvalue of $\mathbf{A}$.

4. If we have the matrix $\mathbf{C} = k\mathbf{A}$, where k is a scalar, then $\mathbf{C}$ has the same eigenvectors as $\mathbf{A}$ and:

$$\boxed{k\lambda_j}$$

is the j-th eigenvalue of $\mathbf{C}$, where λ_j is the j-th eigenvalue of $\mathbf{A}$.

5. If we have the matrix $\mathbf{A}^p$ (where p is a positive integer), then $\mathbf{A}^p$ has the same eigenvectors as $\mathbf{A}$ and:

$$\boxed{\lambda_j^p}$$

is the j-th eigenvalue of $\mathbf{A}^p$, where λ_j is the j-th eigenvalue of $\mathbf{A}$.

6. If $\mathbf{A}^{-1}$ exists, then $\mathbf{A}^{-p}$ has the same eigenvectors as $\mathbf{A}$ and:

$$\boxed{\lambda_j^{-p}}$$

is the eigenvalue of $\mathbf{A}^{-p}$ corresponding to the j-th eigenvalue of $\mathbf{A}$. In particular, $1/\lambda_j$ is the eigenvalue of $\mathbf{A}^{-1}$ corresponding to λ_j, the j-th eigenvalue of $\mathbf{A}$.

7. If a *symmetric* matrix $\mathbf{A}$ can be written as the product:

$$\boxed{\mathbf{A} = \mathbf{TDT'}}$$

where $\mathbf{D}$ is diagonal with all entries non-negative and $\mathbf{T}$ is an orthogonal matrix of eigenvectors, then:

$$\boxed{\mathbf{A}^{1/2} = \mathbf{TD}^{1/2}\mathbf{T'}}$$

and it is the case that $\mathbf{A}^{1/2}\mathbf{A}^{1/2} = \mathbf{A}$.

8. If a *symmetric* matrix $\mathbf{A}^{-1}$ can be written as the product:

$$\boxed{\mathbf{A}^{-1} = \mathbf{TD}^{-1}\mathbf{T'}}$$

where $\mathbf{D}$ is diagonal with all entries non-negative and $\mathbf{T}$ is an orthogonal matrix of eigenvectors, then:

$$\mathbf{A}^{-1/2} = \mathbf{T}\mathbf{D}^{-1/2}\mathbf{T}'$$

and it is the case that $\mathbf{A}^{-1/2}\mathbf{A}^{-1/2} = \mathbf{A}^{-1}$.

We can illustrate these properties of eigenstructures by means of a simple example:

$$\mathbf{A} = \begin{bmatrix} 2 & 1 \\ 1 & 2 \end{bmatrix}$$

The eigenvalues of $\mathbf{A}$ are $\lambda_1 = 3$ and $\lambda_1 = 1$. The associated (and normalized to unit sums of squares) eigenvectors are:

$$\mathbf{t}_1 = \begin{bmatrix} 0.707 \\ 0.707 \end{bmatrix}; \mathbf{t}_2 = \begin{bmatrix} 0.707 \\ -0.707 \end{bmatrix}$$

Since $\mathbf{A}$ is symmetric, we can write the diagonalization as:

$$\mathbf{D} = \mathbf{T}'\mathbf{A}\mathbf{T}$$
$$= \begin{bmatrix} 0.707 & 0.707 \\ 0.707 & -0.707 \end{bmatrix}\begin{bmatrix} 2 & 1 \\ 1 & 2 \end{bmatrix}\begin{bmatrix} 0.707 & 0.707 \\ 0.707 & -0.707 \end{bmatrix}$$
$$= \begin{bmatrix} 3 & 0 \\ 0 & 1 \end{bmatrix}$$

The various properties, discussed above, are illustrated in Table B.2.

B.5.2 An Alternative Approach to Finding the Eigenstructure of a Nonsymmetric Matrix

In section 6.5 of Chapter 6 we utilized direct polynomial expansion to find the eigenstructure of a nonsymmetric matrix. Alternatively, it is possible to apply a preliminary transformation to "symmetrize" the original matrix $\mathbf{F}$, as computed in Table 6.2.

This alternative is sketched out in Table B.3. Note that we make use of the transformations:

$$\mathbf{Q} = \mathbf{R}_{xx}^{1/2}\mathbf{V}$$
$$\mathbf{V} = \mathbf{R}_{xx}^{-1/2}\mathbf{Q}$$

Table B.2 Illustrations of Selected Eigenstructure Properties

Trace of A	Determinant of A
$tr(\mathbf{A}) = a_{11} + a_{22} = \lambda_1 + \lambda_2$	$\lvert \mathbf{A} \rvert = \lambda_1 \lambda_2$
$= 3 + 1 = 4$	$= 3(1) = 3$

Eigenvalues of $\mathbf{A} + 2\mathbf{I}$

$$\mathbf{A} + 2\mathbf{I} = \begin{bmatrix} 4 & 1 \\ 1 & 4 \end{bmatrix}$$

$$\lambda^2 - 8\lambda + 15 = 0$$
$$\lambda_1 = 5$$
$$\lambda_2 = 3$$

Eigenvalues of $2\mathbf{A}$

$$2\mathbf{A} = \begin{bmatrix} 4 & 2 \\ 2 & 4 \end{bmatrix}$$

$$\lambda^2 - 8\lambda + 12 = 0$$
$$\lambda_1 = 6$$
$$\lambda_2 = 2$$

Eigenvalues of $\mathbf{A}^2$

$$\mathbf{A}^2 = \begin{bmatrix} 5 & 4 \\ 4 & 5 \end{bmatrix}$$

$$\lambda^2 - 10\lambda + 9 = 0$$
$$\lambda_1 = 9$$
$$\lambda_2 = 1$$

Eigenvalues of $\mathbf{A}^{-2}$

$$\mathbf{A}^{-1} = \begin{bmatrix} 2/3 & -1/3 \\ -1/3 & 2/3 \end{bmatrix} \quad ; \quad \mathbf{A}^{-2} = \begin{bmatrix} 5/9 & -4/9 \\ -4/9 & 5/9 \end{bmatrix}$$

$$\lambda^2 - 4/3\lambda + 1/3 = 0 \qquad \lambda^2 - 10/9\lambda + 1/9 = 0$$
$$\lambda_2 = 1/3 \qquad \lambda_1 = 1/9$$
$$\lambda_2 = 1 \qquad \lambda_2 = 1$$

Table B.2 (*continued*)

The Square Root of $\mathbf{A}$	The Square Root of $\mathbf{A}^{-1}$

$\mathbf{A}^{1/2} = \mathbf{TD}^{1/2}\mathbf{T}'$

$$= \begin{bmatrix} 0.707 & 0.707 \\ 0.707 & -0.707 \end{bmatrix}\begin{bmatrix} \sqrt{3} & 0 \\ 0 & \sqrt{1} \end{bmatrix}\begin{bmatrix} 0.707 & 0.707 \\ 0.707 & -0.707 \end{bmatrix}$$

$$= \begin{bmatrix} 1.366 & 0.366 \\ 0.366 & 1.366 \end{bmatrix}; \quad \mathbf{A}^{1/2}\mathbf{A}^{1/2} = \begin{bmatrix} 2 & 1 \\ 1 & 2 \end{bmatrix}$$

$\mathbf{A}^{-1/2} = \mathbf{TD}^{-1/2}\mathbf{T}'$

$$= \begin{bmatrix} 0.707 & 0.707 \\ 0.707 & -0.707 \end{bmatrix}\begin{bmatrix} 1/\sqrt{3} & 0 \\ 0 & 1/\sqrt{1} \end{bmatrix}\begin{bmatrix} 0.707 & 0.707 \\ 0.707 & -0.707 \end{bmatrix}$$

$$= \begin{bmatrix} 0.788 & -0.211 \\ -0.211 & 0.788 \end{bmatrix}; \quad \mathbf{A}^{-1/2}\mathbf{A}^{-1/2} = \begin{bmatrix} 2/3 & -1/3 \\ -1/3 & 2/3 \end{bmatrix}$$

$$\begin{array}{ccc} \mathbf{A}^{1/2} & \mathbf{A}^{-1/2} & \mathbf{I} \end{array}$$

$$\begin{bmatrix} 1.366 & 0.366 \\ 0.366 & 1.366 \end{bmatrix}\begin{bmatrix} 0.788 & -0.211 \\ -0.211 & 0.788 \end{bmatrix} = \begin{bmatrix} 1 & 0 \\ 0 & 1 \end{bmatrix}$$

Table B.3 An Alternative Formulation of the Eigenstructure Problem in Canonical Correlation

Basic Expression

$$\mathbf{F} = \mathbf{R}_{xx}^{-1}\mathbf{R}_{xy}\mathbf{R}_{yy}^{-1}\mathbf{R}_{yx} \qquad \text{(This matrix is nonsymmetric.)}$$

However, define:

$$\mathbf{Q} = \mathbf{R}_{xx}^{1/2}\mathbf{V}; \quad \mathbf{V} = \mathbf{R}_{xx}^{-1/2}\mathbf{Q}$$

where, if $\mathbf{R}_{xx}$ is symmetric and positive definite (which it is), we have:

$$\mathbf{R}_{xx}^{1/2} = \mathbf{K}\mathbf{D}^{1/2}\mathbf{K}'$$

$$\mathbf{R}_{xx}^{-1/2} = \mathbf{K}\mathbf{D}^{-1/2}\mathbf{K}'$$

where $\mathbf{D}^{1/2}$, $\mathbf{D}^{-1/2}$ are each diagonal and $\mathbf{K}$ is orthogonal.

Original Equation:

$$[\mathbf{R}_{xx}^{-1}\mathbf{R}_{xy}\mathbf{R}_{yy}^{-1}\mathbf{R}_{yx}]\mathbf{V} = \mathbf{V}\mathbf{D}$$

Substitute $\mathbf{R}_{xx}^{-1/2}\mathbf{Q}$ for $\mathbf{V}$:

$$[\mathbf{R}_{xx}^{-1}\mathbf{R}_{xy}\mathbf{R}_{yy}^{-1}\mathbf{R}_{yx}]\mathbf{R}_{xx}^{-1/2}\mathbf{Q} = \mathbf{R}_{xx}^{-1/2}\mathbf{Q}\mathbf{D}$$

Premultiply both sides by $\mathbf{R}_{xx}^{1/2}$:

$$[\mathbf{R}_{xx}^{-1/2}\mathbf{R}_{xy}\mathbf{R}_{yy}^{-1}\mathbf{R}_{yx}\mathbf{R}_{xx}^{-1/2}]\mathbf{Q} = \mathbf{Q}\mathbf{D}$$

Matrix on the left is now symmetric. Solve for $\mathbf{Q}$, the matrix of eigenvectors.

Finally, substitute:

$$\mathbf{V} = \mathbf{R}_{xx}^{-1/2}\mathbf{Q}$$

to obtain $\mathbf{V}$, the matrix of right-hand eigenvectors of $\mathbf{F}$.

and, then, decompose $\mathbf{R}_{xx}^{-1/2}$ into the triple product $\mathbf{K}\mathbf{D}^{-1/2}\mathbf{K}'$, as illustrated earlier in Table B.2.

This procedure is particularly useful for those situations in which computer programs for finding eigenstructures are limited to symmetric matrices.

B.5.3 Special Characteristics of
Product Moment Matrices

Product moment matrices, such as the SSCP, covariance and correlation matrices, play an important role in multivariate analysis. As an example, let us take the covariance matrix:

$$C(X) = \frac{1}{m-1}[X_d'X_d]$$

which has been discussed at various points in the book.

Not only is $C(X)$ symmetric but, in addition, it is constructed as a (minor) product moment of X_d (which is then divided by a constant $m - 1$).

Product moment matrices exhibit all of the eigenstructure properties of symmetric matrices and, in addition, display some special properties as well:

1. Like any symmetric matrix, $C(X)$ can be written as the triple product:

$$C(X) = TDT'$$

where D is a (real) diagonal matrix of eigenvalues of $C(X)$ and T is an orthogonal matrix of associated eigenvectors.

2. Since $C(X)$ is of product moment form all of its eigenvalues are *non-negative* and we can therefore order the eigenvalues of $C(X)$ from large to small.

3. Whether nonsingular or not, the rank of $C(X)$ equals the number of *positive* eigenvalues (since all eigenvalues in this case must be non-negative).

4. It is generally the case that if *any* square matrix A is symmetric with non-negative eigenvalues, then $A = B'B$.[19] One way of writing the relationship involves defining B' as:

$$B' = TD^{1/2}$$

where $D^{1/2}$ is a diagonal matrix of the square roots of the eigenvalues of A and T is the orthogonal matrix whose columns are the associated eigenvectors of A.

5. Since all of the eigenvalues of $C(X)$ are either positive or zero, in this special case there exists a matrix B such that:

$$C(X) = B'B$$

6. The earlier definition of $B' = TD^{1/2}$ is, however, not unique. If $B_1 = VB$ is the product of an orthogonal matrix V and B, then A can also be written, equally appropriately, as:

$$A = B_1'B_1 = (VB)'(VB) = B'V'VB$$

This latter property:

$$\boxed{C(X) = B'B = B'V'VB}$$

where V *is orthogonal* $(V'V = VV' = I)$ *is of particular relevance to factor analysis and has to do with the so-called "rotation" (or indeterminacy) problem.*

In summary, product moment matrices such as $C(X)$ have eigenvalues that are all non-negative. The number of such non-negative eigenvalues equals the rank of $C(X)$. (The indeterminacy associated with the $B'B$ form of $C(X)$ was discussed in the context of principal components analysis in Chapter 8.)

B.6 Singular Value Decomposition

Up to this point the decompositions of interest:

1. $A = UDU^{-1}$ where A is square but nonsymmetric
2. $A = TDT'$ where A is symmetric
3. $A = B'B$ where A is symmetric with nonnegative eigenvalues

have all involved square matrices.

As we saw in Chapter 8, however, an extremely useful type of decomposition, known as singular value decomposition, is applicable to *both* rectangular and square matrices. Here, we elaborate on some of the material covered earlier in Chapter 8.

We first examine some of the algebra of this type of decomposition. Then we apply the decomposition to an illustrative problem. Since the material that follows can become a bit complex, let us set down our objective at the outset. And that is: given an arbitrary rectangular transformation matrix A, we wish to find a way to express A in terms of the product of three, relatively simple, matrices:

1. An $m \times k$ matrix P which is orthonormal by columns and, hence, satisfies the relation $P'P = I$.
2. A $k \times k$ matrix Δ which is diagonal and consists of k positive diagonal entries ordered from large to small.[20]
3. An $n \times k$ matrix Q which is orthonormal by columns and, hence, satisfies the relation $Q'Q = I$.

B.6.1 The Algebra of Singular Value Decomposition

Given an arbitrary "vertical" matrix $A_{m \times n}$, with $m > n$ and rank $r(A) = k \leq n < m$, its basic structure is defined by the triple product:

$$\mathbf{A}_{m \times n} = \mathbf{P}_{m \times k} \mathbf{\Delta}_{k \times k} \mathbf{Q}'_{k \times n}$$

where $\mathbf{P}$ and $\mathbf{Q}$ are each orthonormal by columns ($\mathbf{P}'\mathbf{P} = \mathbf{I}_{k \times k}$; $\mathbf{Q}'\mathbf{Q} = \mathbf{I}_{k \times k}$) and $\mathbf{\Delta}_{k \times k}$ is diagonal with ordered positive entries.

Note in particular that $\mathbf{Q}$ cannot be orthogonal (where $\mathbf{Q}\mathbf{Q}' = \mathbf{I}_{n \times n}$) unless $k = n$. Moreover, $\mathbf{P}$ cannot be orthogonal (where $\mathbf{P}\mathbf{P}' = \mathbf{I}_{m \times m}$) unless $k = n = m$. As such, $\mathbf{P}_{m \times k}$ and $\mathbf{Q}_{n \times k}$ are called orthonormal sections in which all *columns* are of unit length and mutually orthogonal.

First, let us comment on $\mathbf{\Delta}_{k \times k}$, which is generally called the *basic diagonal*. First, all elements of $\mathbf{\Delta}_{k \times k}$ can be:

1. Taken to be positive.
2. Ordered from large to small.

Moreover, it will turn out that there is one and only one basic diagonal for any given matrix; that is, the basic diagonal part of the decomposition is always unique and this will be true regardless of whether $\mathbf{A}$ is of full or less than full rank, or square or rectangular.

Next, let us examine the relationship of $\mathbf{A} = \mathbf{P}\mathbf{\Delta}\mathbf{Q}'$ to its major and minor product moments, $\mathbf{A}\mathbf{A}'$ and $\mathbf{A}'\mathbf{A}$, respectively:

$\mathbf{A}\mathbf{A}' = (\mathbf{P}\mathbf{\Delta}\mathbf{Q}')(\mathbf{P}\mathbf{\Delta}\mathbf{Q}')'$

$= \mathbf{P}\mathbf{\Delta}\mathbf{Q}'\mathbf{Q}\mathbf{\Delta}\mathbf{P}'$, but since $\mathbf{Q}'\mathbf{Q} = \mathbf{I}$ and letting $\mathbf{\Delta}^2 = \mathbf{D}$, we see that $\mathbf{D}$ is still diagonal. Thus, we get:

$$\mathbf{A}\mathbf{A}' = \mathbf{P}\mathbf{D}\mathbf{P}'$$

Furthermore:

$\mathbf{A}'\mathbf{A} = (\mathbf{P}\mathbf{\Delta}\mathbf{Q}')'(\mathbf{P}\mathbf{\Delta}\mathbf{Q}')$

$= \mathbf{Q}\mathbf{\Delta}\mathbf{P}'\mathbf{P}\mathbf{\Delta}\mathbf{Q}'$, but since $\mathbf{P}'\mathbf{P} = \mathbf{I}$ and $\mathbf{\Delta}^2 = \mathbf{D}$, we get:

$$\mathbf{A}'\mathbf{A} = \mathbf{Q}\mathbf{D}\mathbf{Q}'$$

Note that in both cases we have the eigenstructure formulation shown earlier for the case of symmetric matrices, namely, the triple product of an orthogonal, diagonal, and transposed orthogonal matrix. This is not surprising since both $\mathbf{A}\mathbf{A}'$ and $\mathbf{A}'\mathbf{A}$ are symmetric. What is important, however, is that the diagonal matrix $\mathbf{D}$ of each triple product *is the same* for both product moments, $\mathbf{A}\mathbf{A}'$ and $\mathbf{A}'\mathbf{A}$. And equally interesting are the facts that:

1. All entries of $\mathbf{D}$ are real.
2. All entries of $\mathbf{D}$ are non-negative.
3. All positive entries of $\mathbf{D}$ can be ordered from large to small.

We take advantage of these facts in describing Δ, the $k \times k$ portion of D that has positive (as opposed to zero) entries, in terms of the following definition:

$$\boxed{\Delta_{k \times k} = D_{k \times k}^{1/2}}$$

At this point, then, we are starting to get some hints about how to find $P_{m \times k}$, $\Delta_{k \times k}$ and $Q'_{k \times n}$. As observed above, $P_{m \times k}$ represents the first k columns of the orthogonal matrix $P_{m \times m}$ obtained from the eigenstructure of AA' while $Q_{n \times k}$ represents the first k columns of the orthogonal matrix $Q_{n \times n}$ obtained from the eigenstructure of $A'A$. Their common diagonal matrix D has all non-negative entries. Furthermore, we can order the positive entries of D from large to small, until we get k of them. The remaining entries on the main diagonal will be zero. The columns of P and Q can be made to correspond to the ordered diagonal elements in D and, hence, to their square roots in Δ.

Note further that if A is *symmetric* to begin with, we have the special case:

$$\boxed{A = TDT'}$$

since $AA' = A'A$, and therefore $P' = Q' = T'$. Hence, this same approach to matrix decomposition can be applied even to the more familiar case of a symmetric matrix. However, the diagonal D is to be interpreted as Δ in the current context since we refer to A rather than to its product moments, AA' or $A'A$.

B.6.2 Illustrating the Procedure

To illustrate singular value decomposition, let us take a particularly simple case involving a 3×2 matrix:

$$A = \begin{bmatrix} 1 & 2 \\ 0 & 2 \\ -1 & 1 \end{bmatrix}$$

Since $m > n$, we first find the (smaller) minor product moment matrix:

$$A'A = \begin{bmatrix} 2 & 1 \\ 1 & 9 \end{bmatrix} = \begin{bmatrix} 1 & 0 & -1 \\ 2 & 2 & 1 \end{bmatrix} \begin{bmatrix} 1 & 2 \\ 0 & 2 \\ -1 & 1 \end{bmatrix}$$

Table B.4 shows a summary of the computations involved in finding the eigenstructure of $A'A$. Note that this part of the problem is a standard one in finding the eigenstructure of a symmetric matrix.

After finding $D = \Delta^2$ and Q by solving the characteristic equation, we find Δ and then Δ^{-1}. The last step is to solve for P in the equation:[21]

Table B.4 Finding the Eigenstructure of A'A

Minor Product Moment Matrix	Characteristic Equation

$$A'A = \begin{bmatrix} 2 & 1 \\ 1 & 9 \end{bmatrix} \qquad |(A'A) - \lambda I| = \begin{vmatrix} 2 - \lambda & 1 \\ 1 & 9 - \lambda \end{vmatrix} = 0$$

$$\lambda^2 - 11\lambda + 17 = 0$$

Quadratic Formula   Substitution in General Quadratic

$$y = ax^2 + bx + c \qquad \frac{-b \pm \sqrt{b^2 - 4ac}}{2a} = \frac{11 \pm \sqrt{(-11)^2 - 4(17)}}{2}$$

Eigenvalues of A'A   Matrix of Eigenvectors of A'A

$$\lambda_1 = 9.14$$
$$\lambda_2 = 1.86$$

$$Q = \begin{bmatrix} -0.14 & 0.99 \\ -0.99 & -0.14 \end{bmatrix}$$

Basic Diagonal

$$\Delta = \begin{bmatrix} (9.14)^{1/2} & 0 \\ 0 & (1.86)^{1/2} \end{bmatrix} = \begin{bmatrix} 3.02 & 0 \\ 0 & 1.36 \end{bmatrix}$$

Solve for the Matrix P

$$P = \overset{A}{\begin{bmatrix} 1 & 2 \\ 0 & 2 \\ -1 & 1 \end{bmatrix}} \overset{Q}{\begin{bmatrix} -0.14 & 0.99 \\ -0.99 & -0.14 \end{bmatrix}} \overset{\Delta^{-1}}{\begin{bmatrix} 1/3.02 & 0 \\ 0 & 1/1.36 \end{bmatrix}} = \begin{bmatrix} -0.70 & 0.52 \\ -0.65 & -0.20 \\ -0.28 & -0.83 \end{bmatrix}$$

$$\boxed{P = AQ\Delta^{-1}}$$

These results also appear in Table B.4.
Finally, we assemble the triple product:

$$A = \begin{bmatrix} 1 & 2 \\ 0 & 2 \\ -1 & 1 \end{bmatrix} = \overset{P}{\begin{bmatrix} -0.70 & 0.52 \\ -0.65 & -0.20 \\ -0.28 & -0.83 \end{bmatrix}} \overset{\Delta}{\begin{bmatrix} 3.02 & 0 \\ 0 & 1.36 \end{bmatrix}} \overset{Q'}{\begin{bmatrix} -0.14 & -0.99 \\ 0.99 & -0.14 \end{bmatrix}}$$

As can be noted, after taking the transpose of Q, the matrix A has been decomposed into the product of an orthonormal (by columns) matrix P times a diagonal matrix Δ times a (square) orthogonal matrix Q'. In general, however, Q' will be orthonormal by rows if A is not of full column rank.

B.7 Derivation of Statistical Equations

Throughout the text, various references were made to the use of the calculus (and Lagrange techniques) in deriving various matrix equations utilized in multivariate analysis. In this section of the appendix we discuss the derivation of three sets of equations related to:

1. Multiple regression
2. Principal components analysis
3. Multiple discriminant analysis

as illustrative of the general procedure of function optimization, with or without side constraints.

B.7.1 Multiple Regression

The so-called normal equations of multiple regression theory represent a straightforward application of function minimization that utilizes the least squares criterion. In multiple regression we have the case in which the matrix equation:

$$\boxed{\mathbf{y} \cong \mathbf{Xb}}$$

has more equations (one for each case) than unknowns. As recalled from Chapter 2, $\mathbf{X}$ is the data matrix of predictors (augmented by a column vector of unities), $\mathbf{y}$ is the data vector representing the criterion variable, $\mathbf{b}$ is the to-be-solved-for vector of regression coefficients, including the intercept term. The symbol $\cong$ denotes least squares approximation.

The vector of prediction errors can be written as:

$$\mathbf{e} = \mathbf{y} - \hat{\mathbf{y}}$$

where $\hat{\mathbf{y}}$ denotes the set of predicted values for $\mathbf{y}$. As we know, the least squares criterion seeks a vector $\mathbf{b}$ that minimizes:

$$f = (\mathbf{y} - \hat{\mathbf{y}})'(\mathbf{y} - \hat{\mathbf{y}})$$

Since $\hat{\mathbf{y}} = \mathbf{Xb}$, we have:

$$f = (\mathbf{y} - \mathbf{Xb})'(\mathbf{y} - \mathbf{Xb})$$
$$= \mathbf{y}'\mathbf{y} - \mathbf{b}'\mathbf{X}'\mathbf{y} - \mathbf{y}'\mathbf{Xb} + \mathbf{b}'\mathbf{X}'\mathbf{Xb}$$
$$= \mathbf{b}'\mathbf{X}'\mathbf{Xb} - 2\mathbf{y}'\mathbf{Xb} + \mathbf{y}'\mathbf{y}$$

where $\mathbf{y}'\mathbf{Xb} = \mathbf{b}'\mathbf{X}'\mathbf{y}$ since each term is a scalar. Our objective is to find a vector of parameters $\mathbf{b}$ that minimizes f. This suggests finding the symbolic derivative (Green, 1976) and setting it equal to the $\mathbf{0}$ column vector:

$$\frac{\partial f}{\partial \mathbf{b}} = 2\mathbf{X}'\mathbf{X}\mathbf{b} - 2\mathbf{X}'\mathbf{y} = \mathbf{0}$$

We note that $\mathbf{X}'\mathbf{X}$ in $\mathbf{b}'\mathbf{X}'\mathbf{X}\mathbf{b}$ is symmetric, with derivative $2\mathbf{X}'\mathbf{X}\mathbf{b}$. Furthermore, we observe that the partial derivative with respect to the row vector $\mathbf{b}'$ is being found; hence, we take the transpose of $2\mathbf{y}'\mathbf{X}$ to obtain $2\mathbf{X}'\mathbf{y}$, the second term in the preceding equation. Dividing both sides by 2 and transposing leads to:

$$\mathbf{X}'\mathbf{X}\mathbf{b} = \mathbf{X}'\mathbf{y}$$

and solving for $\mathbf{b}$, we get:

$$\boxed{\mathbf{b} = (\mathbf{X}'\mathbf{X})^{-1}\mathbf{X}'\mathbf{y}}$$

We now recognize the matrix equation as that appearing in Chapter 2's discussion of multiple regression. Although no check of sufficiency conditions has been made here, it turns out that $\mathbf{b}$ is the vector of parameters that does, indeed, minimize the function f.

B.7.2 Principal Components Analysis

In principal components analysis, we recall that interest centers on rotation of a deviation-from-mean data matrix $\mathbf{X}_d$ so as to maximize the expression.[22]

$$\boxed{f = \mathbf{t}'(\mathbf{X}_d'\mathbf{X}_d)\mathbf{t}}$$

where we denote the SSCP matrix by $\mathbf{X}_d'\mathbf{X}_d$, the minor product moment of $\mathbf{X}_d$. (Alternatively, we could use the raw cross products, covariance or correlation matrix.) Furthermore, we want the vector $\mathbf{t}$ to be a set of direction cosines that define the vector of linear composites:

$$\mathbf{y} = \mathbf{X}_d\mathbf{t}, \text{ where } \mathbf{t}'\mathbf{t} = 1$$

If we let $\mathbf{A} = \mathbf{X}_d'\mathbf{X}_d$, the principal components problem is to maximize:

$$f = \mathbf{t}'\mathbf{A}\mathbf{t}$$

subject to the constraint that $\mathbf{t}'\mathbf{t} = 1$.[23]

We can formalize the task by writing the Lagrangian equation:

$$u = \mathbf{t}'\mathbf{A}\mathbf{t} - \lambda(\mathbf{t}'\mathbf{t} - 1)$$

where $\mathbf{t}'\mathbf{t} - 1 = 0$ represents the constraint equation and λ is a Lagrange multiplier. The problem is to find the symbolic partial derivative of u with respect to **t** and set this equal to the **0** vector. Remembering that **A** is symmetric, we obtain:

$$\frac{\partial u}{\partial \mathbf{t}} = 2\mathbf{A}\mathbf{t} - 2\lambda\mathbf{t} = \mathbf{0}$$

Next, dividing through by 2 and factoring out **t**, we get:

$$(\mathbf{A} - \lambda\mathbf{I})\mathbf{t} = \mathbf{0}$$

This represents the necessary condition to be satisfied by a stationary point **t** in which the constraint equation $\mathbf{t'}\mathbf{t} = 1$ is also satisfied. Again, we do not delve into the more complex topic of checking on sufficiency conditions, other than to say that the eigenvector $\mathbf{t}_1$, associated with the largest eigenvalue λ_1 of **A**, is the vector of direction cosines that maximizes the function u.[24]

B.7.3 Multiple Discriminant Analysis

As recalled from Chapter 7, in multiple discriminant analysis we seek a vector **v** with the property of maximizing the ratio:

$$\boxed{\lambda = \frac{\mathbf{v'}\mathbf{A}\mathbf{v}}{\mathbf{v'}\mathbf{W}\mathbf{v}}}$$

where **A** is the among-groups SSCP matrix and **W** is the pooled within-groups SSCP matrix. Again, we place some restriction on the vector **v**, such as $\mathbf{v'}\mathbf{v} = 1$. Note, however, that λ, the discriminant ratio in the present context, is simply the quotient of two functions. We can then find the symbolic derivative of λ with respect to **v**, by means of the quotient rule, and set it equal to the **0** vector:

$$\frac{\partial \lambda}{\partial \mathbf{v}} = \frac{2[(\mathbf{A}\mathbf{v})(\mathbf{v'}\mathbf{W}\mathbf{v}) - (\mathbf{v'}\mathbf{A}\mathbf{v})(\mathbf{W}\mathbf{v})]}{(\mathbf{v'}\mathbf{W}\mathbf{v})^2} = \mathbf{0}$$

This can be simplified by dividing numerator and denominator by $(\mathbf{v'}\mathbf{W}\mathbf{v})$ and making the substitution:

$$\lambda = \frac{\mathbf{v'}\mathbf{A}\mathbf{v}}{\mathbf{v'}\mathbf{W}\mathbf{v}}$$

to obtain:

$$\frac{2[\mathbf{A}\mathbf{v} - \lambda\mathbf{W}\mathbf{v}]}{\mathbf{v'}\mathbf{W}\mathbf{v}} = \mathbf{0}$$

Next, we multiply both sides by the scalar $v'Wv/2$ and further simplify to:

$$(A - \lambda W)v = 0$$

Next, assuming that W is nonsingular, we can premultiply both sides of the equation by W^{-1} to obtain:

$$\boxed{(W^{-1}A - \lambda I)v = 0}$$

where, as we know, $W^{-1}A$ is nonsymmetric. Again, we omit discussion of the sufficiency conditions, indicating that a maximum has been found. Suffice it to say that all three procedures:

1. Multiple regression
2. Principal components analysis
3. Multiple discriminant analysis

involve aspects of the calculus that deal with the optimization of functions of multivariable arguments—either unconstrained or constrained optimization, as the case may be.

B.8 Summary

This appendix has touched on a number of statistical and matrix algebra topics of interest to multivariate data analysis. The beginning part of the appendix described the commonly used distributions: Z, χ^2, t and F, whose tabular values for selected percentiles appear in Appendix C. Also discussed were some basic concepts related to estimation and hypothesis testing.

Discussion then turned to the central matrix algebra topics of matrix inversion, eigenstructures and singular value decomposition. The pivotal method was described as a systematic way to find determinants and inverses (if they exist) as well as to solve simultaneous equations. Some of the properties of eigenstructures of nonsymmetric, symmetric and product moment matrices were described and illustrated numerically. Material on singular value decomposition was also reviewed and a numerical example was presented.

The appendix concluded with derivations of the matrix equations used in regression, principal components analysis and multiple discriminant analysis. As noted, these all involve optimization-type applications of the calculus (with or without side constraints).

Notes

1. Sections in this appendix dealing with various matrix algebra topics are drawn from Green (1976).

2. The density function, $f(w)$, can be written as:

$$f(w) = \frac{w^{(\nu/2)-1}e^{-w/2}}{2^{\nu/2}\Gamma\left[\dfrac{\nu}{2}\right]}$$

where Γ denotes the gamma function. Tables of the gamma function can be found in Beyer (1966).

The gamma function itself is defined as the definite integral:

$$\Gamma(n) = \int_0^\infty e^{-x}x^{n-1}\,dx$$

and has the property that $\Gamma(n + 1) = n\Gamma(n)$. However, if n is a positive integer, then:

$$\Gamma(n) = (n - 1)!$$

If $n > 0$ but not an integer, then:

$$\Gamma(n) = (n - 1)(n - 2)\ldots\delta\Gamma(\delta)$$

where $0 < \delta < 1$.

3. The density function, $f(t)$, can be written as:

$$f(t) = \frac{\Gamma\left[\dfrac{\nu + 1}{2}\right]}{\sqrt{\nu\pi}\,\Gamma\left[\dfrac{\nu}{2}\right]\left[1 + \dfrac{t^2}{\nu}\right]^{(\nu+1)/2}}$$

by recalling that $\Gamma(n) = (n - 1)!$ if n is a positive integer.

4. The variance of t is $\nu/(\nu - 2)$; $\nu > 2$.

5. The density function, $f(F)$, can be written as:

$$f(F) = \frac{\Gamma\left[\dfrac{\nu_1 + \nu_2}{2}\right]}{\Gamma\left[\dfrac{\nu_1}{2}\right]\Gamma\left[\dfrac{\nu_2}{2}\right]}\cdot\left[\dfrac{\nu_1}{\nu_2}\right]^{1/2(\nu_1)}\cdot\frac{F^{(\nu_1-2)/2}}{\left[1 + \dfrac{\nu_1}{\nu_2}F\right]^{(\nu_1+\nu_2)/2}}$$

The mean and variance of F are:

$$\mu(F) = \frac{\nu_2}{\nu_2 - 2}; \nu_2 > 2$$

$$\text{var}(F) = \frac{2\nu_2^2(\nu_1 + \nu_2 - 2)}{\nu_1(\nu_2 - 2)^2(\nu_2 - 4)}; \nu_2 > 4$$

6. The 5th percentile can be obtained from the 95th percentile as follows:

$$F_{[0.05;4,10]} = \frac{1}{F_{[0.95;10,4]}} = \frac{1}{5.96} = 0.168$$

In general, to find the α lower percentile, we use the equation:

$$F_{[\alpha;\nu_1,\nu_2]} = \frac{1}{F_{[1-\alpha;\nu_2,\nu_1]}}$$

For the most part, however, it will be the *upper* percentiles that are of most relevance to applied work.

7. This interpretation is in accord with so-called traditional (Neyman-Pearson) inference. Bayesian statisticians adopt a different viewpoint (Schlaifer, 1959).

8. This rule is based on the more general adjoint method in which $\mathbf{A}^{-1}$ is found from:

$$\mathbf{A}^{-1} = \frac{\text{adj}(\mathbf{A})}{|\mathbf{A}|}$$

where $\text{adj}(\mathbf{A})$ is the adjoint matrix. The adjoint of a matrix is defined as the transpose of a matrix of cofactors of the original matrix. A cofactor, in turn, is a signed minor with sign given by $(-1)^{i+j}$ for the ij-th element. The minor of the ij-th element of $\mathbf{A}$ is defined as the determinant of the submatrix formed by deleting the i-th row and the j-th column of $\mathbf{A}$.

In the case where $\mathbf{A}$ is of order 2×2 the cofactors are:

$$\begin{bmatrix} a_{22} & -a_{21} \\ -a_{12} & a_{11} \end{bmatrix}$$

and the adjoint of $\mathbf{A}$ is simply the transpose of this.

9. What we have also done is solve a set of equations in which the appropriate entries of column 5 serve as the vector of constants $\mathbf{b}$ in the usual expression:

$$\mathbf{Ax} = \mathbf{b}$$

in which we solve for the vector $\mathbf{x}$.

10. If the matrix is singular, this fact will be revealed by the pivotal method through the appearance of *all zeroes* in some row to be pivoted. If this happens the procedure is terminated.

11. In particular, the square root method (Graybill, 1969) is quite useful and efficient for finding the inverse of a symmetric matrix.

12. More recently, various procedures have been developed for finding *generalized* inverses in cases where the original matrix is singular (Green, 1976).

13. In addition, we assume that all eigenvalues of $\mathbf{A}$ are real, rather than complex. (We shall also assume that all eigenvalues of $\mathbf{A}$ are distinct.)

14. In the general case, not all eigenvalues (and their associated eigenvectors) need be real-valued. However, in the applications of interest to us, all eigenvalues and eigenvectors *will* be real-valued.

15. Similar matrices ($\mathbf{D}$ and $\mathbf{A}$) have the same eigenvalues; these are on the main diagonal of $\mathbf{D}$.

16. In all cases of interest here, $\mathbf{D}$ and $\mathbf{U}$ will be real valued.

17. In addition, all eigenvalues and eigenvectors of $\mathbf{A}$ are *necessarily* real. (We shall assume that all eigenvalues of $\mathbf{A}$ are distinct and all eigenvectors have been normalized to unit length.)

18. As might be surmised, one way to determine the rank of a square matrix is to find out how many nonzero eigenvalues it has.

19. A symmetric matrix with non-negative eigenvalues is often referred to as Gramian.

20. To simplify the exposition we shall assume that all nonzero eigenvalues of **A** are distinct, a realistic assumption in the case of data-based matrices. While singular value decomposition is equally applicable to the case where some of the eigenvalues are tied, various uniqueness properties regarding eigenvectors no longer hold within tied blocks of eigenvalues.

21. We use the equation:

$$P = AQ\Delta^{-1}$$

to solve for **P** so as to avoid the need to solve another eigenstructure problem (based on the major product moment **AA'**). This also resolves the problem of sign indeterminacy in **P**. In general, **P** and **Q** are uniquely determined up to a reflection of each vector (assuming that permutation indeterminacy in the columns of **P** and **Q** is resolved by ordering the entries in Δ from large to small and correspondingly ordering the associated columns of **P** and **Q**). Defining **P** this way also resolves the problem of finding a set of compatible vectors in the case where some entries in Δ may be tied.

22. As noted in Chapter 6, this type of expression is often called a quadratic form.

23. Because of the side constraint, a Lagrange multiplier procedure is used in solving the calculus problem.

24. Finding successive orthogonal axes that maximize residual variance utilizes the same general approach, although additional Lagrangian functions are needed.

Statistical
Tables

Table C.1 Cumulative Standard Unit Normal Distribution

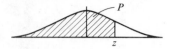

Values of P corresponding to Z for the normal curve. Z is the standard normal variable. The value of P for $-Z$ equals one minus the value of P for $+Z$, e.g., the P for -1.62 equals $1 - .9474 = .0526$.

Z	.00	.01	.02	.03	.04	.05	.06	.07	.08	.09
.0	.5000	.5040	.5080	.5120	.5160	.5199	.5239	.5279	.5319	.5359
.1	.5398	.5438	.5478	.5517	.5557	.5596	.5636	.5675	.5714	.5753
.2	.5793	.5832	.5871	.5910	.5948	.5987	.6026	.6064	.6103	.6141
.3	.6179	.6217	.6255	.6293	.6331	.6368	.6406	.6443	.6480	.6517
.4	.6554	.6591	.6628	.6664	.6700	.6736	.6772	.6808	.6844	.6879
.5	.6915	.6950	.6985	.7019	.7054	.7088	.7123	.7157	.7190	.7224
.6	.7257	.7291	.7324	.7357	.7389	.7422	.7454	.7486	.7517	.7549
.7	.7580	.7611	.7642	.7673	.7704	.7734	.7764	.7794	.7823	.7852
.8	.7881	.7910	.7939	.7967	.7995	.8023	.8051	.8078	.8106	.8133
.9	.8159	.8186	.8212	.8238	.8264	.8289	.8315	.8340	.8365	.8389
1.0	.8413	.8438	.8461	.8485	.8508	.8531	.8554	.8577	.8599	.8621
1.1	.8643	.8665	.8686	.8708	.8729	.8749	.8770	.8790	.8810	.8830
1.2	.8849	.8869	.8888	.8907	.8925	.8944	.8962	.8980	.8997	.9015
1.3	.9032	.9049	.9066	.9082	.9099	.9115	.9131	.9147	.9162	.9177
1.4	.9192	.9207	.9222	.9236	.9251	.9265	.9279	.9292	.9306	.9319
1.5	.9332	.9345	.9357	.9370	.9382	.9394	.9406	.9418	.9429	.9441
1.6	.9452	.9463	.9474	.9484	.9495	.9505	.9515	.9525	.9535	.9545
1.7	.9554	.9564	.9573	.9582	.9591	.9599	.9608	.9616	.9625	.9633
1.8	.9641	.9649	.9656	.9664	.9671	.9678	.9686	.9693	.9699	.9706
1.9	.9713	.9719	.9726	.9732	.9738	.9744	.9750	.9756	.9761	.9767
2.0	.9772	.9778	.9783	.9788	.9793	.9798	.9803	.9808	.9812	.9817
2.1	.9821	.9826	.9830	.9834	.9838	.9842	.9846	.9850	.9854	.9857
2.2	.9861	.9864	.9868	.9871	.9875	.9878	.9881	.9884	.9887	.9890
2.3	.9893	.9896	.9898	.9901	.9904	.9906	.9909	.9911	.9913	.9916
2.4	.9918	.9920	.9922	.9925	.9927	.9929	.9931	.9932	.9934	.9936
2.5	.9938	.9940	.9941	.9943	.9945	.9946	.9948	.9949	.9951	.9952
2.6	.9953	.9955	.9956	.9957	.9959	.9960	.9961	.9962	.9963	.9964
2.7	.9965	.9966	.9967	.9968	.9969	.9970	.9971	.9972	.9973	.9974
2.8	.9974	.9975	.9976	.9977	.9977	.9978	.9979	.9979	.9980	.9981
2.9	.9981	.9982	.9982	.9983	.9984	.9984	.9985	.9985	.9986	.9986
3.0	.9987	.9987	.9987	.9988	.9988	.9989	.9989	.9989	.9990	.9990
3.1	.9990	.9991	.9991	.9991	.9992	.9992	.9992	.9992	.9993	.9993
3.2	.9993	.9993	.9994	.9994	.9994	.9994	.9994	.9995	.9995	.9995
3.3	.9995	.9995	.9995	.9996	.9996	.9996	.9996	.9996	.9996	.9997
3.4	.9997	.9997	.9997	.9997	.9997	.9997	.9997	.9997	.9997	.9998

Table C.2 Selected Percentiles of the χ^2 Distribution

Values of χ^2 corresponding to P

ν	$\chi^2_{.005}$	$\chi^2_{.01}$	$\chi^2_{.025}$	$\chi^2_{.05}$	$\chi^2_{.10}$	$\chi^2_{.90}$	$\chi^2_{.05}$	$\chi^2_{.975}$	$\chi^2_{.99}$	$\chi^2_{.995}$
1	.000039	.00016	.00098	.0039	.0158	2.71	3.84	5.02	6.63	7.88
2	.0100	.0201	.0506	.1026	.2107	4.61	5.99	7.38	9.21	10.60
3	.0717	.115	.216	.352	.584	6.25	7.81	9.35	11.34	12.84
4	.207	.297	.484	.711	1.064	7.78	9.49	11.14	13.28	14.86
5	.412	.554	.831	1.15	1.61	9.24	11.07	12.83	15.09	16.75
6	.676	.872	1.24	1.64	2.20	10.64	12.59	14.45	16.81	18.55
7	.989	1.24	1.69	2.17	2.83	12.02	14.07	16.01	18.48	20.28
8	1.34	1.65	2.18	2.73	3.49	13.36	15.51	17.53	20.09	21.96
9	1.73	2.09	2.70	3.33	4.17	14.68	16.92	19.02	21.67	23.59
10	2.16	2.56	3.25	3.94	4.87	15.99	18.31	20.48	23.21	25.19
11	2.60	3.05	3.82	4.57	5.58	17.28	19.68	21.92	24.73	26.76
12	3.07	3.57	4.40	5.23	6.30	18.55	21.03	23.34	26.22	28.30
13	3.57	4.11	5.01	5.89	7.04	19.81	22.36	24.74	27.69	29.82
14	4.07	4.66	5.63	6.57	7.79	21.06	23.68	26.12	29.14	31.32
15	4.60	5.23	6.26	7.26	8.55	22.31	25.00	27.49	30.58	32.80
16	5.14	5.81	6.91	7.96	9.31	23.54	26.30	28.85	32.00	34.27
18	6.26	7.01	8.23	9.39	10.86	25.99	28.87	31.53	34.81	37.16
20	7.43	8.26	9.59	10.85	12.44	28.41	31.41	34.17	37.57	40.00
24	9.89	10.86	12.40	13.85	15.66	33.20	36.42	39.36	42.98	45.56
30	13.79	14.95	16.79	18.49	20.60	40.26	43.77	46.98	50.89	53.67
40	20.71	22.16	24.43	26.51	29.05	51.81	55.76	59.34	63.69	66.77
60	35.53	37.48	40.48	43.19	46.46	74.40	79.08	83.30	88.38	91.95
120	83.85	86.92	91.58	95.70	100.62	140.23	146.57	152.21	158.95	163.64

ν = degrees of freedom

Adapted with permission from *Introduction to Statistical Analysis* (2d ed.) by W. J. Dixon and F. J. Massey, Jr., McGraw-Hill Book Company, Inc., 1957.

Table C.3 Upper Percentiles of the *t* Distribution

ν	.75	.90	.95	.975	.99	.995	.9995
1	1.000	3.078	6.314	12.706	31.821	63.657	636.619
2	.816	1.886	2.920	4.303	6.965	9.925	31.598
3	.765	1.638	2.353	3.182	4.541	5.841	12.941
4	.741	1.533	2.132	2.776	3.747	4.604	8.610
5	.727	1.476	2.015	2.571	3.365	4.032	6.859
6	.718	1.440	1.943	2.447	3.143	3.707	5.959
7	.711	1.415	1.895	2.365	2.998	3.499	5.405
8	.706	1.397	1.860	2.306	2.896	3.355	5.041
9	.703	1.383	1.833	2.262	2.821	3.250	4.781
10	.700	1.372	1.812	2.228	2.764	3.169	4.587
11	.697	1.363	1.796	2.201	2.718	3.106	4.437
12	.695	1.356	1.782	2.179	2.681	3.055	4.318
13	.694	1.350	1.771	2.160	2.650	3.012	4.221
14	.692	1.345	1.761	2.145	2.624	2.977	4.140
15	.691	1.341	1.753	2.131	2.602	2.947	4.073
16	.690	1.337	1.746	2.120	2.583	2.921	4.015
17	.689	1.333	1.740	2.110	2.567	2.898	3.965
18	.688	1.330	1.734	2.101	2.552	2.878	3.922
19	.688	1.328	1.729	2.093	2.339	2.861	3.883
20	.687	1.325	1.725	2.086	2.528	2.845	3.850
21	.686	1.323	1.721	2.080	2.518	2.831	3.819
22	.686	1.321	1.717	2.074	2.508	2.819	3.792
23	.685	1.319	1.714	2.069	2.500	2.807	3.767
24	.685	1.318	1.711	2.064	2.492	2.797	3.745
25	.684	1.316	1.708	2.060	2.485	2.787	3.725
26	.684	1.315	1.706	2.056	2.479	2.779	3.707
27	.684	1.314	1.703	2.052	2.473	2.771	3.690
28	.683	1.313	1.701	2.048	2.467	2.763	3.674
29	.683	1.311	1.699	2.045	2.462	2.756	3.659
30	.683	1.310	1.697	2.042	2.457	2.750	3.646
40	.681	1.303	1.684	2.021	2.423	2.704	3.551
60	.679	1.296	1.671	2.000	2.390	2.660	3.460
120	.677	1.289	1.658	1.980	2.358	2.617	3.373
∞	.674	1.282	1.645	1.960	2.326	2.576	3.291

ν = degrees of freedom

Taken with permission from Table III of R. A. Fisher and F. Yates: *Statistical Tables for Biological, Agricultural, and Medical Research*, published by Oliver & Boyd Ltd., Edinburgh, 1963.

Table C.4 Selected Percentiles of the *F* Distribution

$\nu_1 =$ degrees of freedom for numerator

$F_{.90(\nu_1, \nu_2)} \qquad \alpha = 0.1$

ν_2 \ ν_1	1	2	3	4	5	6	7	8	9	10	12	15	20	24	30	40	60	120	∞
1	39.86	49.50	53.59	55.83	57.24	58.20	58.91	59.44	59.86	60.19	60.71	61.22	61.74	62.00	62.26	62.53	62.79	63.06	63.33
2	8.53	9.00	9.16	9.24	9.29	9.33	9.35	9.37	9.38	9.39	9.41	9.42	9.44	9.45	9.46	9.47	9.47	9.48	9.49
3	5.54	5.46	5.39	5.34	5.31	5.28	5.27	5.25	5.24	5.23	5.22	5.20	5.18	5.18	5.17	5.16	5.15	5.14	5.13
4	4.54	4.32	4.19	4.11	4.05	4.01	3.98	3.95	3.94	3.92	3.90	3.87	3.84	3.83	3.82	3.80	3.79	3.78	3.76
5	4.06	3.78	3.62	3.52	3.45	3.40	3.37	3.34	3.32	3.30	3.27	3.24	3.21	3.19	3.17	3.16	3.14	3.12	3.10
6	3.78	3.46	3.29	3.18	3.11	3.05	3.01	2.98	2.96	2.94	2.90	2.87	2.84	2.82	2.80	2.78	2.76	2.74	2.72
7	3.59	3.26	3.07	2.96	2.88	2.83	2.78	2.75	2.72	2.70	2.67	2.63	2.59	2.58	2.56	2.54	2.51	2.49	2.47
8	3.46	3.11	2.92	2.81	2.73	2.67	2.62	2.59	2.56	2.54	2.50	2.46	2.42	2.40	2.38	2.36	2.34	2.32	2.29
9	3.36	3.01	2.81	2.69	2.61	2.55	2.51	2.47	2.44	2.42	2.38	2.34	2.30	2.28	2.25	2.23	2.21	2.18	2.16
10	3.29	2.92	2.73	2.61	2.52	2.46	2.41	2.38	2.35	2.32	2.28	2.24	2.20	2.18	2.16	2.13	2.11	2.08	2.06
11	3.23	2.86	2.66	2.54	2.45	2.39	2.34	2.30	2.27	2.25	2.21	2.17	2.12	2.10	2.08	2.05	2.03	2.00	1.97
12	3.18	2.81	2.61	2.48	2.39	2.33	2.28	2.24	2.21	2.19	2.15	2.10	2.06	2.04	2.01	1.99	1.96	1.93	1.90
13	3.14	2.76	2.56	2.43	2.35	2.28	2.23	2.20	2.16	2.14	2.10	2.05	2.01	1.98	1.96	1.93	1.90	1.88	1.85
14	3.10	2.73	2.52	2.39	2.31	2.24	2.19	2.15	2.12	2.10	2.05	2.01	1.96	1.94	1.91	1.89	1.86	1.83	1.80
15	3.07	2.70	2.49	2.36	2.27	2.21	2.16	2.12	2.09	2.06	2.02	1.97	1.92	1.90	1.87	1.85	1.82	1.79	1.76
16	3.05	2.67	2.46	2.33	2.24	2.18	2.13	2.09	2.06	2.03	1.99	1.94	1.89	1.87	1.84	1.81	1.78	1.75	1.72
17	3.03	2.64	2.44	2.31	2.22	2.15	2.10	2.06	2.03	2.00	1.96	1.91	1.86	1.84	1.81	1.78	1.75	1.72	1.69
18	3.01	2.62	2.42	2.29	2.20	2.13	2.08	2.04	2.00	1.98	1.93	1.89	1.84	1.81	1.78	1.75	1.72	1.69	1.66
19	2.99	2.61	2.40	2.27	2.18	2.11	2.06	2.02	1.98	1.96	1.91	1.86	1.81	1.79	1.76	1.73	1.70	1.67	1.63
20	2.97	2.59	2.38	2.25	2.16	2.09	2.04	2.00	1.96	1.94	1.89	1.84	1.79	1.77	1.74	1.71	1.68	1.64	1.61
21	2.96	2.57	2.36	2.23	2.14	2.08	2.02	1.98	1.95	1.92	1.87	1.83	1.78	1.75	1.72	1.69	1.66	1.62	1.59
22	2.95	2.56	2.35	2.22	2.13	2.06	2.01	1.97	1.93	1.90	1.86	1.81	1.76	1.73	1.70	1.67	1.64	1.60	1.57
23	2.94	2.55	2.34	2.21	2.11	2.05	1.99	1.95	1.92	1.89	1.84	1.80	1.74	1.72	1.69	1.66	1.62	1.59	1.55
24	2.93	2.54	2.33	2.19	2.10	2.04	1.98	1.94	1.91	1.88	1.83	1.78	1.73	1.70	1.67	1.64	1.61	1.57	1.53
25	2.92	2.53	2.32	2.18	2.09	2.02	1.97	1.93	1.89	1.87	1.82	1.77	1.72	1.69	1.66	1.63	1.59	1.56	1.52
26	2.91	2.52	2.31	2.17	2.08	2.01	1.96	1.92	1.88	1.86	1.81	1.76	1.71	1.68	1.65	1.61	1.58	1.54	1.50
27	2.90	2.51	2.30	2.17	2.07	2.00	1.95	1.91	1.87	1.85	1.80	1.75	1.70	1.67	1.64	1.60	1.57	1.53	1.49
28	2.89	2.50	2.29	2.16	2.06	2.00	1.94	1.90	1.87	1.84	1.79	1.74	1.69	1.66	1.63	1.59	1.56	1.52	1.48
29	2.89	2.50	2.28	2.15	2.06	1.99	1.93	1.89	1.86	1.83	1.78	1.73	1.68	1.65	1.62	1.58	1.55	1.51	1.47
30	2.88	2.49	2.28	2.14	2.05	1.98	1.93	1.88	1.85	1.82	1.77	1.72	1.67	1.64	1.61	1.57	1.54	1.50	1.46
40	2.84	2.44	2.23	2.09	2.00	1.93	1.87	1.83	1.79	1.76	1.71	1.66	1.61	1.57	1.54	1.51	1.47	1.42	1.38
60	2.79	2.39	2.18	2.04	1.95	1.87	1.82	1.77	1.74	1.71	1.66	1.60	1.54	1.51	1.48	1.44	1.40	1.35	1.29
120	2.75	2.35	2.13	1.99	1.90	1.82	1.77	1.72	1.68	1.65	1.60	1.55	1.48	1.45	1.41	1.37	1.32	1.26	1.19
∞	2.71	2.30	2.08	1.94	1.85	1.77	1.72	1.67	1.63	1.60	1.55	1.49	1.42	1.38	1.34	1.30	1.24	1.17	1.00

$\nu_2 =$ degrees of freedom for denominator

Adapted with permission from *Biometrika Tables for Statisticians*, Vol. I (2nd ed.), edited by E. S. Pearson and H. O. Hartley, Cambridge University Press, 1958.

Table C.4—Continued
Selected Percentiles of the F Distribution

$$F_{.95}(\nu_1, \nu_2) \qquad \alpha = 0.05$$

ν_1 = degrees of freedom for numerator

ν_2 \ ν_1	1	2	3	4	5	6	7	8	9	10	12	15	20	24	30	40	60	120	∞
1	161.4	199.5	215.7	224.6	230.2	234.0	236.8	238.9	240.5	241.9	243.9	245.9	248.0	249.1	250.1	251.1	252.2	253.3	254.3
2	18.51	19.00	19.16	19.25	19.30	19.33	19.35	19.37	19.38	19.40	19.41	19.43	19.45	19.45	19.46	19.47	19.48	19.49	19.50
3	10.13	9.55	9.28	9.12	9.01	8.94	8.89	8.85	8.81	8.79	8.74	8.70	8.66	8.64	8.62	8.59	8.57	8.55	8.53
4	7.71	6.94	6.59	6.39	6.26	6.16	6.09	6.04	6.00	5.96	5.91	5.86	5.80	5.77	5.75	5.72	5.69	5.66	5.63
5	6.61	5.79	5.41	5.19	5.05	4.95	4.88	4.82	4.77	4.74	4.68	4.62	4.56	4.53	4.50	4.46	4.43	4.40	4.36
6	5.99	5.14	4.76	4.53	4.39	4.28	4.21	4.15	4.10	4.06	4.00	3.94	3.87	3.84	3.81	3.77	3.74	3.70	3.67
7	5.59	4.74	4.35	4.12	3.97	3.87	3.79	3.73	3.68	3.64	3.57	3.51	3.44	3.41	3.38	3.34	3.30	3.27	3.23
8	5.32	4.46	4.07	3.84	3.69	3.58	3.50	3.44	3.39	3.35	3.28	3.22	3.15	3.12	3.08	3.04	3.01	2.97	2.93
9	5.12	4.26	3.86	3.63	3.48	3.37	3.29	3.23	3.18	3.14	3.07	3.01	2.94	2.90	2.86	2.83	2.79	2.75	2.71
10	4.96	4.10	3.71	3.48	3.33	3.22	3.14	3.07	3.02	2.98	2.91	2.85	2.77	2.74	2.70	2.66	2.62	2.58	2.54
11	4.84	3.98	3.59	3.36	3.20	3.09	3.01	2.95	2.90	2.85	2.79	2.72	2.65	2.61	2.57	2.53	2.49	2.45	2.40
12	4.75	3.89	3.49	3.26	3.11	3.00	2.91	2.85	2.80	2.75	2.69	2.62	2.54	2.51	2.47	2.43	2.38	2.34	2.30
13	4.67	3.81	3.41	3.18	3.03	2.92	2.83	2.77	2.71	2.67	2.60	2.53	2.46	2.42	2.38	2.34	2.30	2.25	2.21
14	4.60	3.74	3.34	3.11	2.96	2.85	2.76	2.70	2.65	2.60	2.53	2.46	2.39	2.35	2.31	2.27	2.22	2.18	2.13
15	4.54	3.68	3.29	3.06	2.90	2.79	2.71	2.64	2.59	2.54	2.48	2.40	2.33	2.29	2.25	2.20	2.16	2.11	2.07
16	4.49	3.63	3.24	3.01	2.85	2.74	2.66	2.59	2.54	2.49	2.42	2.35	2.28	2.24	2.19	2.15	2.11	2.06	2.01
17	4.45	3.59	3.20	2.96	2.81	2.70	2.61	2.55	2.49	2.45	2.38	2.31	2.23	2.19	2.15	2.10	2.06	2.01	1.96
18	4.41	3.55	3.16	2.93	2.77	2.66	2.58	2.51	2.46	2.41	2.34	2.27	2.19	2.15	2.11	2.06	2.02	1.97	1.92
19	4.38	3.52	3.13	2.90	2.74	2.63	2.54	2.48	2.42	2.38	2.31	2.23	2.16	2.11	2.07	2.03	1.98	1.93	1.88
20	4.35	3.49	3.10	2.87	2.71	2.60	2.51	2.45	2.39	2.35	2.28	2.20	2.12	2.08	2.04	1.99	1.95	1.90	1.84
21	4.32	3.47	3.07	2.84	2.68	2.57	2.49	2.42	2.37	2.32	2.25	2.18	2.10	2.05	2.01	1.96	1.92	1.87	1.81
22	4.30	3.44	3.05	2.82	2.66	2.55	2.46	2.40	2.34	2.30	2.23	2.15	2.07	2.03	1.98	1.94	1.89	1.84	1.78
23	4.28	3.42	3.03	2.80	2.64	2.53	2.44	2.37	2.32	2.27	2.20	2.13	2.05	2.01	1.96	1.91	1.86	1.81	1.76
24	4.26	3.40	3.01	2.78	2.62	2.51	2.42	2.36	2.30	2.25	2.18	2.11	2.03	1.98	1.94	1.89	1.84	1.79	1.73
25	4.24	3.39	2.99	2.76	2.60	2.49	2.40	2.34	2.28	2.24	2.16	2.09	2.01	1.96	1.92	1.87	1.82	1.77	1.71
26	4.23	3.37	2.98	2.74	2.59	2.47	2.39	2.32	2.27	2.22	2.15	2.07	1.99	1.95	1.90	1.85	1.80	1.75	1.69
27	4.21	3.35	2.96	2.73	2.57	2.46	2.37	2.31	2.25	2.20	2.13	2.06	1.97	1.93	1.88	1.84	1.79	1.73	1.67
28	4.20	3.34	2.95	2.71	2.56	2.45	2.36	2.29	2.24	2.19	2.12	2.04	1.96	1.91	1.87	1.82	1.77	1.71	1.65
29	4.18	3.33	2.93	2.70	2.55	2.43	2.35	2.28	2.22	2.18	2.10	2.03	1.94	1.90	1.85	1.81	1.75	1.70	1.64
30	4.17	3.32	2.92	2.69	2.53	2.42	2.33	2.27	2.21	2.16	2.09	2.01	1.93	1.89	1.84	1.79	1.74	1.68	1.62
40	4.08	3.23	2.84	2.61	2.45	2.34	2.25	2.18	2.12	2.08	2.00	1.92	1.84	1.79	1.74	1.69	1.64	1.58	1.51
60	4.00	3.15	2.76	2.53	2.37	2.25	2.17	2.10	2.04	1.99	1.92	1.84	1.75	1.70	1.65	1.59	1.53	1.47	1.39
120	3.92	3.07	2.68	2.45	2.29	2.17	2.09	2.02	1.96	1.91	1.83	1.75	1.66	1.61	1.55	1.50	1.43	1.35	1.25
∞	3.84	3.00	2.60	2.37	2.21	2.10	2.01	1.94	1.88	1.83	1.75	1.67	1.57	1.52	1.46	1.39	1.32	1.22	1.00

ν_2 = degrees of freedom for denominator

Table C.4—Continued
Selected Percentiles of the F Distribution

$F_{.975}(\nu_1, \nu_2)$

$F_{.975}(\nu_1, \nu_2)$ $\alpha = 0.025$

ν_1 = degrees of freedom for numerator

ν_2 \ ν_1	1	2	3	4	5	6	7	8	9	10	12	15	20	24	30	40	60	120	∞
1	647.8	799.5	864.2	899.6	921.8	937.1	948.2	956.7	963.3	968.6	976.7	984.9	993.1	997.2	1001	1006	1010	1014	1018
2	38.51	39.00	39.17	39.25	39.30	39.33	39.36	39.37	39.39	39.40	39.41	39.43	39.45	39.46	39.46	39.47	39.48	39.49	39.50
3	17.44	16.04	15.44	15.10	14.88	14.73	14.62	14.54	14.47	14.42	14.34	14.25	14.17	14.12	14.08	14.04	13.99	13.95	13.90
4	12.22	10.65	9.98	9.60	9.36	9.20	9.07	8.98	8.90	8.84	8.75	8.66	8.56	8.51	8.46	8.41	8.36	8.31	8.26
5	10.01	8.43	7.76	7.39	7.15	6.98	6.85	6.76	6.68	6.62	6.52	6.43	6.33	6.28	6.23	6.18	6.12	6.07	6.02
6	8.81	7.26	6.60	6.23	5.99	5.82	5.70	5.60	5.52	5.46	5.37	5.27	5.17	5.12	5.07	5.01	4.96	4.90	4.85
7	8.07	6.54	5.89	5.52	5.29	5.12	4.99	4.90	4.82	4.76	4.67	4.57	4.47	4.42	4.36	4.31	4.25	4.20	4.14
8	7.57	6.06	5.42	5.05	4.82	4.65	4.53	4.43	4.36	4.30	4.20	4.10	4.00	3.95	3.89	3.84	3.78	3.73	3.67
9	7.21	5.71	5.08	4.72	4.48	4.32	4.20	4.10	4.03	3.96	3.87	3.77	3.67	3.61	3.56	3.51	3.45	3.39	3.33
10	6.94	5.46	4.83	4.47	4.24	4.07	3.95	3.85	3.78	3.72	3.62	3.52	3.42	3.37	3.31	3.26	3.20	3.14	3.08
11	6.72	5.26	4.63	4.28	4.04	3.88	3.76	3.66	3.59	3.53	3.43	3.33	3.23	3.17	3.12	3.06	3.00	2.94	2.88
12	6.55	5.10	4.47	4.12	3.89	3.73	3.61	3.51	3.44	3.37	3.28	3.18	3.07	3.02	2.96	2.91	2.85	2.79	2.72
13	6.41	4.97	4.35	4.00	3.77	3.60	3.48	3.39	3.31	3.25	3.15	3.05	2.95	2.89	2.84	2.78	2.72	2.66	2.60
14	6.30	4.86	4.24	3.89	3.66	3.50	3.38	3.29	3.21	3.15	3.05	2.95	2.84	2.79	2.73	2.67	2.61	2.55	2.49
15	6.20	4.77	4.15	3.80	3.58	3.41	3.29	3.20	3.12	3.06	2.96	2.86	2.76	2.70	2.64	2.59	2.52	2.46	2.40
16	6.12	4.69	4.08	3.73	3.50	3.34	3.22	3.12	3.05	2.99	2.89	2.79	2.68	2.63	2.57	2.51	2.45	2.38	2.32
17	6.04	4.62	4.01	3.66	3.44	3.28	3.16	3.06	2.98	2.92	2.82	2.72	2.62	2.56	2.50	2.44	2.38	2.32	2.25
18	5.98	4.56	3.95	3.61	3.38	3.22	3.10	3.01	2.93	2.87	2.77	2.67	2.56	2.50	2.44	2.38	2.32	2.26	2.19
19	5.92	4.51	3.90	3.56	3.33	3.17	3.05	2.96	2.88	2.82	2.72	2.62	2.51	2.45	2.39	2.33	2.27	2.20	2.13
20	5.87	4.46	3.86	3.51	3.29	3.13	3.01	2.91	2.84	2.77	2.68	2.57	2.46	2.41	2.35	2.29	2.22	2.16	2.09
21	5.83	4.42	3.82	3.48	3.25	3.09	2.97	2.87	2.80	2.73	2.64	2.53	2.42	2.37	2.31	2.25	2.18	2.11	2.04
22	5.79	4.38	3.78	3.44	3.22	3.05	2.93	2.84	2.76	2.70	2.60	2.50	2.39	2.33	2.27	2.21	2.14	2.08	2.00
23	5.75	4.35	3.75	3.41	3.18	3.02	2.90	2.81	2.73	2.67	2.57	2.47	2.36	2.30	2.24	2.18	2.11	2.04	1.97
24	5.72	4.32	3.72	3.38	3.15	2.99	2.87	2.78	2.70	2.64	2.54	2.44	2.33	2.27	2.21	2.15	2.08	2.01	1.94
25	5.69	4.29	3.69	3.35	3.13	2.97	2.85	2.75	2.68	2.61	2.51	2.41	2.30	2.24	2.18	2.12	2.05	1.98	1.91
26	5.66	4.27	3.67	3.33	3.10	2.94	2.82	2.73	2.65	2.59	2.49	2.39	2.28	2.22	2.16	2.09	2.03	1.95	1.88
27	5.63	4.24	3.65	3.31	3.08	2.92	2.80	2.71	2.63	2.57	2.47	2.36	2.25	2.19	2.13	2.07	2.00	1.93	1.85
28	5.61	4.22	3.63	3.29	3.06	2.90	2.78	2.69	2.61	2.55	2.45	2.34	2.23	2.17	2.11	2.05	1.98	1.91	1.83
29	5.59	4.20	3.61	3.27	3.04	2.88	2.76	2.67	2.59	2.53	2.43	2.32	2.21	2.15	2.09	2.03	1.96	1.89	1.81
30	5.57	4.18	3.59	3.25	3.03	2.87	2.75	2.65	2.57	2.51	2.41	2.31	2.20	2.14	2.07	2.01	1.94	1.87	1.79
40	5.42	4.05	3.46	3.13	2.90	2.74	2.62	2.53	2.45	2.39	2.29	2.18	2.07	2.01	1.94	1.88	1.80	1.72	1.64
60	5.29	3.93	3.34	3.01	2.79	2.63	2.51	2.41	2.33	2.27	2.17	2.06	1.94	1.88	1.82	1.74	1.67	1.58	1.48
120	5.15	3.80	3.23	2.89	2.67	2.52	2.39	2.30	2.22	2.16	2.05	1.94	1.82	1.76	1.69	1.61	1.53	1.43	1.31
∞	5.02	3.69	3.12	2.79	2.57	2.41	2.29	2.19	2.11	2.05	1.94	1.83	1.71	1.64	1.57	1.48	1.39	1.27	1.00

ν_2 = degrees of freedom for denominator

Table C.4—Continued
Selected Percentiles of the F Distribution

$$F_{.99}(v_1, v_2) \qquad \alpha = 0.01$$

v_1 = degrees of freedom for numerator

v_2 = degrees of freedom for denominator

v_2 \ v_1	1	2	3	4	5	6	7	8	9	10	12	15	20	24	30	40	60	120	∞
1	4052	4999.5	5403	5625	5764	5859	5928	5982	6022	6056	6106	6157	6209	6235	6261	6287	6313	6339	6366
2	98.50	99.00	99.17	99.25	99.30	99.33	99.36	99.37	99.39	99.40	99.42	99.43	99.45	99.46	99.47	99.47	99.48	99.49	99.50
3	34.12	30.82	29.46	28.71	28.24	27.91	27.67	27.49	27.35	27.23	27.05	26.87	26.69	26.60	26.50	26.41	26.32	26.22	26.13
4	21.20	18.00	16.69	15.98	15.52	15.21	14.98	14.80	14.66	14.55	14.37	14.20	14.02	13.93	13.84	13.75	13.65	13.56	13.46
5	16.26	13.27	12.06	11.39	10.97	10.67	10.46	10.29	10.16	10.05	9.89	9.72	9.55	9.47	9.38	9.29	9.20	9.11	9.02
6	13.75	10.92	9.78	9.15	8.75	8.47	8.26	8.10	7.98	7.87	7.72	7.56	7.40	7.31	7.23	7.14	7.06	6.97	6.88
7	12.25	9.55	8.45	7.85	7.46	7.19	6.99	6.84	6.72	6.62	6.47	6.31	6.16	6.07	5.99	5.91	5.82	5.74	5.65
8	11.26	8.65	7.59	7.01	6.63	6.37	6.18	6.03	5.91	5.81	5.67	5.52	5.36	5.28	5.20	5.12	5.03	4.95	4.86
9	10.56	8.02	6.99	6.42	6.06	5.80	5.61	5.47	5.35	5.26	5.11	4.96	4.81	4.73	4.65	4.57	4.48	4.40	4.31
10	10.04	7.56	6.55	5.99	5.64	5.39	5.20	5.06	4.94	4.85	4.71	4.56	4.41	4.33	4.25	4.17	4.08	4.00	3.91
11	9.65	7.21	6.22	5.67	5.32	5.07	4.89	4.74	4.63	4.54	4.40	4.25	4.10	4.02	3.94	3.86	3.78	3.69	3.60
12	9.33	6.93	5.95	5.41	5.06	4.82	4.64	4.50	4.39	4.30	4.16	4.01	3.86	3.78	3.70	3.62	3.54	3.45	3.36
13	9.07	6.70	5.74	5.21	4.86	4.62	4.44	4.30	4.19	4.10	3.96	3.82	3.66	3.59	3.51	3.43	3.34	3.25	3.17
14	8.86	6.51	5.56	5.04	4.69	4.46	4.28	4.14	4.03	3.94	3.80	3.66	3.51	3.43	3.35	3.27	3.18	3.09	3.00
15	8.68	6.36	5.42	4.89	4.56	4.32	4.14	4.00	3.89	3.80	3.67	3.52	3.37	3.29	3.21	3.13	3.05	2.96	2.87
16	8.53	6.23	5.29	4.77	4.44	4.20	4.03	3.89	3.78	3.69	3.55	3.41	3.26	3.18	3.10	3.02	2.93	2.84	2.75
17	8.40	6.11	5.18	4.67	4.34	4.10	3.93	3.79	3.68	3.59	3.46	3.31	3.16	3.08	3.00	2.92	2.83	2.75	2.65
18	8.29	6.01	5.09	4.58	4.25	4.01	3.84	3.71	3.60	3.51	3.37	3.23	3.08	3.00	2.92	2.84	2.75	2.66	2.57
19	8.18	5.93	5.01	4.50	4.17	3.94	3.77	3.63	3.52	3.43	3.30	3.15	3.00	2.92	2.84	2.76	2.67	2.58	2.49
20	8.10	5.85	4.94	4.43	4.10	3.87	3.70	3.56	3.46	3.37	3.23	3.09	2.94	2.86	2.78	2.69	2.61	2.52	2.42
21	8.02	5.78	4.87	4.37	4.04	3.81	3.64	3.51	3.40	3.31	3.17	3.03	2.88	2.80	2.72	2.64	2.55	2.46	2.36
22	7.95	5.72	4.82	4.31	3.99	3.76	3.59	3.45	3.35	3.26	3.12	2.98	2.83	2.75	2.67	2.58	2.50	2.40	2.31
23	7.88	5.66	4.76	4.26	3.94	3.71	3.54	3.41	3.30	3.21	3.07	2.93	2.78	2.70	2.62	2.54	2.45	2.35	2.26
24	7.82	5.61	4.72	4.22	3.90	3.67	3.50	3.36	3.26	3.17	3.03	2.89	2.74	2.66	2.58	2.49	2.40	2.31	2.21
25	7.77	5.57	4.68	4.18	3.85	3.63	3.46	3.32	3.22	3.13	2.99	2.85	2.70	2.62	2.54	2.45	2.36	2.27	2.17
26	7.72	5.53	4.64	4.14	3.82	3.59	3.42	3.29	3.18	3.09	2.96	2.81	2.66	2.58	2.50	2.42	2.33	2.23	2.13
27	7.68	5.49	4.60	4.11	3.78	3.56	3.39	3.26	3.15	3.06	2.93	2.78	2.63	2.55	2.47	2.38	2.29	2.20	2.10
28	7.64	5.45	4.57	4.07	3.75	3.53	3.36	3.23	3.12	3.03	2.90	2.75	2.60	2.52	2.44	2.35	2.26	2.17	2.06
29	7.60	5.42	4.54	4.04	3.73	3.50	3.33	3.20	3.09	3.00	2.87	2.73	2.57	2.49	2.41	2.33	2.23	2.14	2.03
30	7.56	5.39	4.51	4.02	3.70	3.47	3.30	3.17	3.07	2.98	2.84	2.70	2.55	2.47	2.39	2.30	2.21	2.11	2.01
40	7.31	5.18	4.31	3.83	3.51	3.29	3.12	2.99	2.89	2.80	2.66	2.52	2.37	2.29	2.20	2.11	2.02	1.92	1.80
60	7.08	4.98	4.13	3.65	3.34	3.12	2.95	2.82	2.72	2.63	2.50	2.35	2.20	2.12	2.03	1.94	1.84	1.73	1.60
120	6.85	4.79	3.95	3.48	3.17	2.96	2.79	2.66	2.56	2.47	2.34	2.19	2.03	1.95	1.86	1.76	1.66	1.53	1.38
∞	6.63	4.61	3.78	3.32	3.02	2.80	2.64	2.51	2.41	2.32	2.18	2.04	1.88	1.79	1.70	1.59	1.47	1.32	1.00

References

Aaker, D. A. (ed.) *Multivariate Analysis in Marketing: Theory & Application.* Belmont, CA: Wadsworth, 1971.

Afifi, A. A. and Azen, S. P. *Statistical Analysis: A Computer Oriented Approach.* New York: Academic Press, 1972.

Alpert, M. I. and Peterson, R. A. "On the Interpretation of Canonical Analysis." *Journal of Marketing Research,* 9, (1972), 187–192.

Anderberg, M. R. *Cluster Analysis for Applications.* New York: Academic Press, 1973.

Anderson, N. H. "Functional Measurement and Psychophysical Judgment." *Psychological Review,* 77 (1970), 153–170.

Anderson, T. W. *Introduction to Multivariate Statistical Analysis.* New York: Wiley, 1958.

Andrews, F. M., Morgan, J. N., and Sonquist, J. A. *Multiple Classification Analysis: A Report on a Computer Program for Multiple Regression Using Categorical Predictors.* Institute for Social Research, The University of Michigan, 1967.

Appelbaum, M. I. and Cramer, E. M. "Some Problems on the Nonorthogonal Analysis of Variance." *Psychological Bulletin,* 81 (1974), 335–343.

Ball, G. H. and Hall, D. J. "Background Information on Clustering Techniques." Menlo Park, CA: Stanford Research Institute, July 1968.

Bartlett, M. S. "Multivariate Analysis." *Journal of the Royal Statistical Society,* Series B, 9 (1947), 176–197.

———"Tests of Significance of Factor Analysis." *British Journal of Psychology* (Statistical section), 3 (1950), 77–85.

Bennett, S. and Bowers, D. *An Introduction to Multivariate Techniques for Social and Behavioral Sciences.* New York: Halsted, 1976.

Beyer, W. H. *Handbook of Tables for Probability and Statistics.* Cleveland: Chemical Rubber Co., 1966.

Birnbaum, M. H. "The Devil Rides Again: Correlation as an Index of Fit." *Psychological Bulletin,* 79 (1973), 239–242.

Blalock, H. M. *Social Statistics.* New York: McGraw-Hill, 1960.

Bock, R. D. and Haggard, E. A. "The Use of Multivariate Analysis of Variance in Behavioral Research," in D. K. Whitla (ed.), *Handbook of Measurement and Assessment in Behavioral Sciences.* Reading, MA: Addison-Wesley, 1968, 100–142.

————*Multivariate Statistical Methods in Behavioral Research.* New York: McGraw-Hill, 1975.

Bolch, B. W. and Huang, C. J. *Multivariate Statistical Methods for Business and Economics.* Englewood Cliffs, NJ: Prentice-Hall, 1974.

Box, G. E. P. "A General Distribution Theory for a Class of Likelihood Criteria." *Biometrika,* 36 (1949), 317–346.

Box, G. E. P. and Draper, N. R. *Evolutionary Operation.* New York: Wiley, 1969.

Browne, M. W. "On Oblique Procrustes Rotation." *Psychometrika,* 32 (1967), 125–132.

Bryant, E. H. and Atchley, W. R. (eds.). *Multivariate Statistical Methods: Within-Groups Covariation.* Stroudsburg Pa.: Dowden, Hutchinson and Ross, 1975.

————, *Multivariate Statistical Methods: Among-Groups Covariation.* Stroudsburg, Pa.: Dowden, Hutchinson and Ross, 1975.

Carroll, J. B. "An Analytical Solution for Approximating Simple Structure in Factor Analysis." *Psychometrika,* 18 (1953), 23–28.

————"Biquartimin Criterion for Rotation to Oblique Simple Structure in Factor Analysis." *Science,* 126 (1957), 1114–1115.

Carroll, J. D. "A Generalization of Canonical Correlation to Three or More Sets of Variables." *Proceedings of the 76th Annual Convention of the American Psychological Association* (1968), 227–228.

————"Models and Algorithms for Multidimensional Scaling, Conjoint Measurement and Related Techniques," in *Multiattribute Decisions in Marketing: A Measurement Approach,* P. E. Green and Y. Wind. Hinsdale, IL: Dryden, 1973.

Carroll, J. D. and Chang, J. J. "Analysis of Individual Differences in Multidimensional Scaling Via an N-Way Generalization of 'Eckart-Young' Decomposition." *Psychometrika,* 35 (1970), 283–320.

Carroll, J. D. and Wish, M. "Measuring Preference and Perception with Multidimensional Models," *Bell Laboratories Record,* 49 (1971), 147–154.

Cattell, R. B. "The Meaning and Strategic Use of Factor Analysis," in R. B. Cattell (ed.), *Handbook of Multivariate Experimental Psychology.* Chicago: Rand McNally, 1966.

Cliff, N. "Orthogonal Rotation to Congruence." *Psychometrika,* 31 (1966), 33–42.

Cliff, N. and Krus, D. J. "Interpretation of Canonical Analysis: Rotated Vs. Unrotated Solutions." *Psychometrika,* 41 (1976), 35–42.

Clyde, D. J., Cramer, E. M., and Sherin, R. J. *Multivariate Statistical Programs.* Coral Gables, Florida: University of Miami, Biometric Laboratory, 1966.

Cochran, W. G. "Analysis of Covariance: Its Nature and Uses." *Biometrics,* 13 (1957), 261–281.

Cohen, J. "Multiple Regression as a General Data Analytic System." *Psychological Bulletin,* 70 (1968), 426–443.

Cohen, J. and Cohen, P. *Applied Multiple Regression/Correlation Analysis for the Behavioral Sciences.* New York: Wiley, 1975.

Cohen, P. and Cohen, J. "Inefficient Redundancy." *Multivariate Behavioral Research,* 12 (1977), 167–170.

Cole, A. S. (ed.) *Numerical Taxonomy.* New York: Academic Press, 1969.

Comrey, A. L. *A First Course in Factor Analysis.* New York: Academic Press, 1973.

Cooley, W. W. and Lohnes, P. R. *Multivariate Data Analysis.* New York: Wiley, 1971.

Coombs, C. H. *A Theory of Data.* New York: Wiley, 1964.

Cox, D. R. "The Use of a Concomitant Variable in Selecting an Experimental Design." *Biometrika,* 44 (1957), 150–158.

————*The Analysis of Binary Data.* London: Methuen, 1970.

Cramer, E. M. "Significance Tests and Tests of Models in Multiple Regression." *The American Statistician,* 26 (1972), 26–29.

————, "The Relation Between Rao's Paradox in Discriminant Analysis and Regression Analysis." *Multivariate Behavioral Research,* 10 (1975), 99–107.

Crawford, C. B. and Koopman, P. "A Note on Horn's Test for the Number of Factors in Factor Analysis." *Multivariate Behavioral Research,* 8 (1973), 117–125.

Cronbach, L. J. and Gleser, G. C. "Assessing Similarity Between Profiles." *Psychological Bulletin,* 50 (1953), 456–473.

Daniel, C. and Wood, F. S. with the assitance of Gorman, J. W. *Fitting Equations to Data.* New York: Wiley, 1971.

Darlington, R. B. "Multiple Regression in Psychological Research and Practice." *Psychological Bulletin,* 69 (1968), 161–182.

Dempster, A. P. *Elements of Continuous Multivariate Analysis.* Reading, MA: Addison-Wesley, 1969.

Dixon, W. J. (ed.) *Biomedical Computer Programs.* Berkeley, CA: University of California Press, 1973.

————*BMDP Biomedical Computer Programs.* Los Angeles, CA: University of California Press, 1975.

Draper, N. R. and Smith, H. *Applied Regression Analysis.* New York: Wiley, 1966.

Dunn, O. J. and Clark, V. A. *Applied Statistics: Analysis of Variance and Regression.* New York: Wiley, 1974.

Dwyer, P. S. and MacPhail, M. S. "Symbolic Matrix Derivatives." *Annals of Mathematical Statistics,* 19 (1948), 517–534.

Eckart, C. and Young, G. "The Approximation of One Matrix by Another of Lower Rank." *Psychometrika,* 1 (1936), 211–218.

Eisenbeis, R. A. and Avery, R. B. *Discriminant Analysis and Classification Procedures: Theory and Applications.* Lexington, MA: Heath, 1972.

Elashoff, J. D. "Analysis of Variance: A Delicate Instrument." *American Educational Research Journal,* 6 (1969), 383–401.

Evans, S. H. and Anastasio, E. J. "Misuse of Analysis of Covariance When Treatment Effect and Covariate are Confounded." *Psychological Bulletin*, 69 (1968), 225–234.

Everitt, B. *Cluster Analysis.* New York: Wiley, 1974.

Finn, J. D. *A General Model for Multivariate Analysis.* New York: Holt, Rinehart and Winston, 1974.

Finney, D. J. *Probit Analysis.* Second edition. Cambridge: Cambridge University Press, 1952.

Fisher, R. A. "The Use of Multiple Measurements in Taxonomic Problems." *Annals of Eugenics*, 7 (1936), 179–188.

Frank, R. E., Massy, W. F., and Morrison, D. G. "Bias in Multiple Discriminant Analysis." *Journal of Marketing Research*, 2 (1965), 250–258.

Freeman, L. C. *Elementary Applied Statistics.* New York: Wiley, 1965.

Gilbert, E. S. "On Discrimination Using Qualitative Variables." *Journal of the American Statistical Association*, 63 (1968), 1399–1418.

————"The Effect of Unequal Variance-Covariance Matrices on Fisher's Linear Discriminant Function." *Biometrics*, 25 (1969), 505–516.

Glass, G. V. and Hakstian, A. F. "Measures of Association in Comparative Experiments: Their Development and Interpretation." *American Educational Research Journal*, 6 (1969), 403–414.

Gleason, T. C. "On Redundancy in Canonical Analysis." *Psychological Bulletin*, 83 (1976) 1004–1006.

Gnadadesikan, R. *Methods for Statistical Data Analysis of Multivariate Observations.* New York: John Wiley & Sons, 1976.

Goldberger, A. S. *Topics in Regression Analysis.* New York: Macmillan, 1968.

Gollob, H. T. "A Statistical Model Which Combines Features of Factor Analytic and Analysis of Variance Techniques." *Psychometrika*, 33 (1968), 73–116.

Good, I. J. "Some Applications of the Singular Decomposition of a Matrix." *Technometrics*, 11 (1969), 823–831.

Gorman, J. W. and Toman, R. J. "Selection of Variables for Fitting Equations to Data." *Technometrics*, 8 (1966), 27–51.

Gorsuch, R. L. "Data Analysis of Correlated Independent Variables." *Multivariate Behavioral Research*, 8 (1973), 89–107.

Gower, J. C. "Some Distance Properties of Latent Root and Vector Methods Used in Multivariate Analysis," *Biometrika*, 53 (1966), 325–338.

————"Generalized Procrustes Analysis." *Psychometrika*, 40 (1975), 33–51.

Graybill, F. A. *An Introduction to Linear Statistical Models,* Vol. I. New York: McGraw-Hill, 1961.

————*Introduction to Matrices with Applications in Statistics.* Belmont, CA: Wadsworth, 1969.

Green, Jr., B. F. "Best Linear Composites with a Specified Structure." *Psychometrika*, 34 (1969), 301–318.

Green, P. E. "On the Analysis of Interactions in Marketing Research Data." *Journal of Marketing Research*, 10 (1973), 410–420.

————"A Multidimensional Model of Product Features Association." *Journal of Business Research,* 2 (1974), 107–118.

————with contributions by Carroll, J. D. *Mathematical Tools for Applied Multivariate Analysis.* New York: Academic Press, 1976.

Green, P. E. and Devita, M. T. "A Complementarity Model of Consumer Utility for Item Collections." *Journal of Consumer Research,* 1 (1974), 56–67.

Green, P. E. and Rao, V. R. *Applied Multidimensional Scaling: A Comparison of Approaches and Algorithms.* New York: Holt, Rinehart and Winston, 1972.

Green, P. E. and Tull, D. S. *Research for Marketing Decisions,* third edition. Englewood Cliffs, N.J.: Prentice-Hall, 1975.

Green, P. E. and Wind, Y. *Multiattribute Decisions in Marketing: A Measurement Approach.* Hinsdale, IL: Dryden, 1973.

Guertin, W. H. and Bailey, Jr., J. P. *Introduction to Modern Factor Analysis.* Ann Arbor, Mich.: Edwards Brothers, 1970.

Guttman, L. "Image Theory for the Structure of Quantitative Variates." *Psychometrika,* 18 (1953), 277–296.

————"Best Possible Systematic Estimates of Communalities." *Psychometrika,* 21 (1956), 273–285.

Hanushek, E. A. and Jackson, J. E. *Statistical Methods for Social Scientists.* New York: Academic Press, 1977.

Harman, H. H. *Modern Factor Analysis,* second edition. Chicago: University of Chicago Press, 1967.

Harris, C. W. "Some Rao-Guttman Relationships." *Psychometrika,* 27 (1962), 247–263.

Harris, R. J. *A Primer of Multivariate Statistics.* New York: Academic Press, 1975.

————"The Invalidity of Partitioned-U Tests in Canonical Correlation and Multivariate Analysis of Variance." *Multivariate Behavioral Research,* II (1976), 353–365.

Hartigan, J. A. *Clustering Algorithms.* New York: Wiley, 1975.

Hays, W. L. and Winkler, R. L. *Statistics: Probability, Inference and Decision.* New York: Holt, Rinehart and Winston, 1971.

Heck, D. L. "Charts of Some Upper Percentage Points of the Distribution of the Largest Characteristic Root." *Annals of Mathematical Statistics,* 31 (1960), 625–642.

Hoerl, A. E. "Fitting Curves to Data," in J. H. Perry (ed.) *Chemical Business Handbook.* New York: McGraw-Hill, 1954, 55–77.

Horn, J. L. "A Rationale and Test for the Number of Factors in Factor Analysis." *Psychometrika,* 30 (1965), 179–186.

Horst, P. "Relations Among m Sets of Measures." *Psychometrika,* 26 (1961), 129–150.

————*Matrix Algebra for Social Scientists.* New York: Holt, Rinehart and Winston, 1963.

————"An Overview of the Essentials of Multivariate Analysis Methods," in R. B. Cattell (ed.), *Handbook of Multivariate Experimental Psychology.* Chicago: Rand McNally, 1966, 129–152.

Hotelling, H. "The Generalization of Student's Ratio." *Annals of Mathematical Statistics,* 2 (1931), 360–378.

————"Analysis of a Complex of Statistical Variables into Principal Components." *Journal of Educational Psychology,* 24 (1933), 417–441; 498–520.

————"A Generalized T-Test and Measure of Multivariate Dispersion." *Proceedings of the Second Berkeley Symposium of Mathematical Statistics and Probability,* 2 (1951), 23–41.

Hsu, P. L. "On the Generalized Analysis of Variance." *Biometrika,* 31 (1940), 221–237.

Huang, D. S. *Regression and Econometric Methods.* New York: Wiley, 1970.

Huberty, C. J. "Multivariate Indices of Strength of Association." *Multivariate Behavioral Research,* 7 (1972), 523–526.

Hurley, J. L. and Cattell, R. B. "The Procrustes Program: Producing Direct Rotation to Test a Hypothesized Factor Structure." *Behavioral Science,* 7 (1962), 258–262.

Jardine, N. and Sibson, R. *Mathematical Taxonomy.* New York: Wiley, 1971.

Johnston, J. *Econometric Methods.* Second edition. New York: McGraw-Hill, 1972.

Jöreskog, K. G. "Some Contributions to Maximum Likelihood Factor Analysis." *Psychometrika,* 32 (1967), 443–382.

————"A General Approach to Maximum Likelihood Factor Analysis." *Psychometrika,* 34 (1969), 183–202.

Kaiser, H. F. "The Varimax Criterion for Analytic Rotation in Factor Analysis." *Psychometrika,* 23 (1958), 187–200.

————"The Application of Electronic Computers to Factor Analysis." Symposium on the application of computers to psychological problems, American Psychological Association, 1959.

Kaiser, H. F. and Caffrey, J. "Alpha Factor Analysis." *Psychometrika,* 30 (1965), 1–14.

Kelley, T. L. "An Unbiased Correlation Ratio Measure." *Proceedings of the National Academy of Sciences,* 21 (1935), 554–559.

Kendall, M. G. *A Course in Multivariate Analysis.* London: Charles Griffen, 1957.

————"Discrimination and Classification," in P. R. Krishnaiah (ed.), *Multivariate Analysis.* New York: Academic Press, 1966, 165–185.

Kerlinger, F. N. and Pedhazur, E. J. *Multiple Regression in Behavioral Research.* New York: Holt, Rinehart and Winston, 1973.

Kettenring, J. R. "Canonical Analysis of Several Sets of Variables." *Biometrika,* 58 (1972), 433–451.

Kirk, R. E. Experimental Design: *Procedures for the Behavioral Sciences.* Belmont, CA: Wadsworth, 1969.

Krantz, D. H. and Tversky, A. "Conjoint Measurement Analysis of Composition Rules in Psychology." *Psychological Review,* 78 (1971), 151–169.

Kristof, W. and Wingersky, B. "Generalization of the Orthogonal Procrustes Rotation Procedure to Two or More Matrices." *Proceedings of the 79th Annual Convention of the American Psychological Association* (1971), 81–90.

Kruskal, J. B. "Multidimensional Scaling by Optimizing Goodness of Fit to a Nonmetric Hypothesis." *Psychometrika,* 29 (1964), 1–27.

————"Analysis of Factorial Experiments by Estimating Monotone Transformations of the Data." *Journal of the Royal Statistical Society,* Series B, 27 (1965), 251–263.

Kruskal, J. B. and Carmone, F. J. "Use and Theory of MONANOVA, a Program to Analyze Factorial Experiments by Estimating Monotone Transformations of the Data." Unpublished paper, Bell Laboratories, 1968.

Kshirsagar, A. M. and Arseven, E. "A Note on the Equivalency of Two Discrimination Procedures," *The American Statistician,* 29 (1975), 38–39.

Lachenbruch, P. A. and Mickey, M. R. "Estimation of Error Rates in Discriminant Analysis." *Technometrics,* 10 (1968), 1–11.

Lawley, D. N. and Maxwell, A. E. *Factor Analysis as a Statistical Method.* London: Butterworths, 1963.

Lingoes, J. C. *The Guttman-Lingoes Nonmetric Program Series.* Ann Arbor, Michigan: Mathesis Press, 1973.

Linn, R. L. and Werts, C. E. "Assumptions in Making Causal Inferences from Part Correlations, Partial Correlations and Partial Regression Coefficients," *Psychological Bulletin,* 72 (1969), 307–310.

McDonald, R. P. "A Unified Treatment of the Weighting Problem." *Psychometrika,* 33 (1968), 351–381.

Mahalanobis, P. C. "On the Generalized Distance in Statistics." *Proceedings of the National Institute of Science, Calcutta,* 12 (1936), 49–55.

Malinvaud, E. *Statistical Methods of Econometrics.* Chicago: Rand McNally, 1966.

Maxwell, A. E. *Multivariate Analysis in Behavioral Research.* New York: Halsted, 1977.

Mendenhall, W. *Introduction to Linear Models and the Design and Analysis of Experiments.* Belmont, CA: Wadsworth, 1968.

Messick, S. J. and Abelson, R. P. "The Additive Constant Problem in Multidimensional Scaling." *Psychometrika,* 21 (1956), 1–17.

Moore, D. H. "Evaluation of Five Discrimination Procedures for Binary Variables." *Journal of the American Statistical Association,* 68 (1973), 399–404.

Morgan, J. N. and Messenger, R. C. *THAID, A Sequential Analysis Program for the Analysis of Nominal Scale Dependent Variables.* Ann Arbor, Mich.: Survey Research Center, University of Michigan, 1973.

Morrison, D. F. *Multivariate Statistical Methods.* Second edition. New York: McGraw-Hill, 1976.

Morrison, D. G. "On the Interpretation of Discriminant Analysis." *Journal of Marketing Research,* 6 (1969), 156–163.

Mosteller, F. and Tukey, J. W. *Data Analysis and Regression.* Reading, Mass.: Addison-Wesley, 1977.

Mulaik, S. A. *The Foundations of Factor Analysis.* New York: McGraw-Hill, 1972.

Myers, J. L. *Fundamentals of Experimental Design.* Boston: Allyn and Bacon, 1972.

Namboodiri, N. K., Carter, L. F. and Blalock, Jr., H. M. *Applied Multivariate Analysis and Experimental Designs.* New York: McGraw-Hill, 1975.

Neter, J. and Wasserman, W. *Applied Linear Statistical Models.* Homewood, Ill.: Irwin, 1974.

Nicewander, W. A. and Wood, D. "Comments on a General Canonical Correlation Index." *Psychological Bulletin,* 81 (1974), 92–94.

Nunnally, J. C. *Psychometric Theory*. New York: McGraw-Hill, 1967.

Overall, J. E. and Klett, C. J. *Applied Multivariate Analysis*. New York: McGraw-Hill, 1972.

Overall, J. E. and Woodward, J. R. "Common Misperceptions Concerning the Analysis of Variance." *Multivariate Behavioral Research*, 12 (1977), 171–186.

Peng, K. C. *The Design and Analysis of Scientific Experiments*. Reading, Mass.: Addison-Wesley, 1967.

Pillai, K. C. S. *Statistical Tables for Tests of Multivariate Hypotheses*. Manila: Statistical Service Center, University of the Philippines, 1960.

Plackett, R. L. "A Note on Interactions in Contingency Tables." *Journal of the Royal Statistical Society*, Series B-24 (1962), 162–166.

Press, S. J. *Applied Multivariate Analysis*. New York: Holt, Rinehart and Winston, 1972.

Pruzek, R. M. "Methods and Problems in the Analysis of Multivariate Data." *Review of Educational Research*, 41 (1971), 163–190.

Raghavarao, D. *Constructions and Combinatorial Problems in Design of Experiments*. New York: Wiley, 1971.

Rao, C. R. *Advanced Statistical Methods in Biometric Research*. New York: Wiley, 1952.

————*Linear Statistical Inference and Its Applications*. New York: Wiley, 1965.

Rao, C. R. and Mitra, S. K. *Generalized Inverse of Matrices and Its Applications*. New York: Wiley, 1971.

Rao, C. R. and Slater, P. "Multivariate Analysis Applied to the Differences Between Neurotic Groups." *British Journal of Psychology* (Statistical section), 2 (1949), 17–29.

Rao, V. R., McCann, J. M. and Craig, C. S. "Identifying Market Segments with AID III," *1976 Educators' Proceedings of the American Marketing Association*, K. L. Bernhardt, (ed.), Chicago: American Marketing Association, 1976, 393–397.

Roskam, E. E. C. I. *Metric Analysis of Ordinal Data in Psychology*. The Hague: Voorschoten, 1968.

Roy, J. "Step-Down Procedures in Multivariate Analysis." *Annals of Mathematical Statistics*, 29, (1958), 1177–1187.

Roy, S. N. *Some Aspects of Multivariate Analysis*. New York: Wiley, 1957.

Rulon, P. J., Tiedeman, D. V., Tatsuoka, M. M., and Langmuir, C. R. *Multivariate Statistics for Personnel Classification*. New York: Wiley, 1957.

Rummel, R. J. *Applied Factor Analysis*. Evanston, Ill.: Northwestern University Press, 1970.

Savage, L. J. *The Foundations of Statistics*. New York: Wiley, 1954.

Schatzoff, M. "Exact Distributions of Wilks' Likelihood Ratio Criteria." *Biometrika*, 53, (1966), 347–358.

Scheffe, H. *The Analysis of Variance*, Appendices I and II. New York: Wiley, 1969.

Schlaifer, R. *Probability and Statistics for Business Decisions*. New York: McGraw-Hill, 1959.

Schönemann, P. H. "A Generalized Solution of the Orthogonal Procrustes Problem." *Psychometrika*, 31 (1966), 1–10.

Schönemann, P. H. and Carroll, R. M. "Fitting One Matrix to Another Under Choice of a Central Dilation and a Rigid Motion." *Psychometrika*, 35 (1970), 245–256.

Searle, S. R. *Linear Models.* New York: Wiley, 1971.

Sebestyen, G. S. *Decision-Making Processes in Pattern Recognition.* New York: Macmillan, 1962.

Shepard, R. N. "The Analysis of Proximities: Multidimensional Scaling with an Unknown Distance Function," Part One. *Psychometrika*, 27 (1962), 125–139.

Shepard, R. N. and Arabie, P. "Additive Cluster Analysis of Similarity Data," *Theory, Methods and Applications of Multidimensional Scaling and Related Techniques*, National Science Foundation and Japan Society for Promotion of Science Proceedings, San Diego, Calif., August 1975.

Shepard, R. N., Romney, A. K., and Nerlove, S. (eds.). *Multidimensional Scaling: Theory and Application in the Behavioral Sciences*, Vols. I and II. New York: Seminar Press, 1972.

Skinner, H. A. "Exploring Relationships Among Multiple Data Sets." *Multivariate Behavioral Research*, 12 (1977), 199–200.

Sneath, P. H. A. and Sokal, R. R. *Numerical Taxonomy.* San Francisco: Freeman, 1973.

Sokal, R. R. and Sneath, P. H. A. *Principles of Numerical Taxonomy.* San Francisco: Freeman, 1963.

Sonquist, J. A., Baker, E. L., and Morgan, J. N. *Searching for Structure (Alias, AID-III).* Ann Arbor, Mich.: Institute for Social Research, University of Michigan, 1971.

Sonquist, J. A. and Morgan, J. N. *The Detection of Interaction Effects.* Ann Arbor, Mich.: Institute for Social Research, University of Michigan, 1964.

Stevens, J. P. "Four Methods of Analyzing Between Variation for the K-Group MANANOVA Problem." *Multivariate Behavioral Research*, 7 (1972), 499–522.

———"Step-Down Analysis and Simultaneous Confidence Intervals in MANOVA." *Multivariate Behavioral Research*, 8 (June 1973), 391–402.

Stewart, D. K. and Love, W. A. "A General Canonical Correlation Index." *Psychological Bulletin*, 70 (1968), 160–163.

Stewart, G. W. *Introduction to Matrix Computations.* New York: Academic Press, 1973.

Tatsuoka, M. M. *Discriminant Analysis.* Champaign, Ill.: Institute for Personality and Ability Testing, 1970.

———*Multivariate Analysis: Techniques for Educational and Psychological Research.* New York: Wiley, 1971.

Tatsuoka, M. M. and Tiedeman, D. V. "Statistics as an Aspect of Scientific Method in Research on Teaching," in N. L. Gage (ed.), *Handbook of Research on Teaching.* Skokie, Ill.: Rand McNally (1963), 142–170.

Thurstone, L. L. *Multiple Factor Analysis.* Chicago: University of Chicago Press, 1947.

Timm, N. H. *Multivariate Analysis with Applications in Education and Psychology.* Belmont, CA: Wadsworth, 1975.

Torgerson, W. S. *Theory and Methods of Scaling.* New York: Wiley, 1958.

Tryon, R. C. and Bailey, D. E. *Cluster Analysis.* New York: McGraw-Hill, 1970.

512 References

Tucker, L. R. "Some Mathematical Notes on Three-Mode Factor Analysis." *Psychometrika,* 31 (1966), 279–311.

Tucker, L. R. and Messick, S. "An Individual Difference Model for Multidimensional Scaling." *Psychometrika,* 28 (1963), 333–367.

Tufte, E. R. *Data Analysis for Politics and Policy.* Englewood Cliffs, NJ: Prentice-Hall, 1974.

Van de Geer, J. P. *Introduction to Multivariate Analysis for the Social Sciences.* San Francisco: Freeman, 1971.

Ward, Jr., J. H. and Jennings, E. *Introduction to Linear Models.* Englewood Cliffs, NJ: Prentice-Hall, 1973.

Wilkinson, J. H. *The Algebraic Eigenvalue Problem.* Oxford: Clarendon Press, 1965.

Wilks, S. S. "Certain Generalizations in the Analysis of Variance." *Biometrika,* 24 (1932), 471–494.

Winer, B. J. *Statistical Principles in Experimental Design.* New York: McGraw-Hill, 1971.

Wish, M. and Carroll, J. D., "Applications of INDSCAL to Studies of Human Perception and Judgment," in E. C. Carterette and M. P. Friedman (eds.), *Handbook of Perception.* New York: Academic Press, 1973.

Wonnacott, R. J. and Wonnacott, T. H. *Econometrics.* New York: Wiley, 1970.

Young, F. W. "TORSCA, An IBM Program for Nonmetric Multidimensional Scaling." *Journal of Marketing Research,* 5 (1968), 319–321.

Index

SOCIAL SCIENCE LIBRARY

Manor Road Building
Manor Road
Oxford OX1 3UQ
Tel: (2)71093 (enquiries and renewals)
http://www.ssl.ox.ac.uk

WITHDRAWN

This is a NORMAL LOAN item.

We will email you a reminder before this item is due.

Please see http://www.ssl.ox.ac.uk/lending.html
for details on:

- loan policies; these are also displayed on the notice boards and in our library guide.

- how to check when your books are due back.

WITHDRAWN

- how to renew your books, including information on the maximum number of renewals. Items may be renewed if not reserved by another reader. Items must be renewed before the library closes on the due date.

- level of fines; fines are charged on overdue books.

Please note that this item may be recalled during Term.